Lecture Notes in Computer Science   11448

Commenced Publication in 1973
Founding and Former Series Editors:
Gerhard Goos, Juris Hartmanis, and Jan van Leeuwen

More information about this series at http://www.springer.com/series/7409

Guoliang Li · Jun Yang ·
Joao Gama · Juggapong Natwichai ·
Yongxin Tong (Eds.)

Database Systems
for Advanced Applications

DASFAA 2019 International Workshops:
BDMS, BDQM, and GDMA
Chiang Mai, Thailand, April 22–25, 2019
Proceedings

 Springer

Editors
Guoliang Li
Tsinghua University
Beijing, China

Joao Gama 🆔
University of Porto
Porto, Portugal

Yongxin Tong
Beihang University
Beijing, China

Jun Yang
Duke University
Durham, NC, USA

Juggapong Natwichai
Chiang Mai University
Chiang Mai, Thailand

ISSN 0302-9743   ISSN 1611-3349 (electronic)
Lecture Notes in Computer Science
ISBN 978-3-030-18589-3   ISBN 978-3-030-18590-9 (eBook)
https://doi.org/10.1007/978-3-030-18590-9

LNCS Sublibrary: SL3 – Information Systems and Applications, incl. Internet/Web, and HCI

This Springer imprint is published by the registered company Springer Nature Switzerland AG
The registered company address is: Gewerbestrasse 11, 6330 Cham, Switzerland

Preface of Workshops

The International Conference on Database Systems for Advanced Applications DASFAA is an annual international database conference that showcases state-of-the-art R&D activities in database systems and their applications. It provides a forum for technical presentations and discussions among database researchers, developers, and users from academia, business, and industry. As the 24th event in the increasingly popular series, DASFAA 2019 was held in Chiang Mai, Thailand, during April 22–25, and it attracted more than 300 participants from all over the world.

Along with the main conference, DASFAA workshops intend to provide an international forum for researchers to discuss and share research results. This DASFAA 2019 workshop volume contains the papers accepted for the following three workshops that were held in conjunction with DASFAA 2019. These three workshops were selected after a public call for proposals process, each of which focuses on a specific area that contributes to the main themes of the DASFAA conference. The three workshops were as follows:

- The 6th International Workshop on Big Data Management and Service BDMS 2019
- The 4th International Workshop on Big Data Quality Management BDQM 2019
- The Third International Workshop on Graph Data Management and Analysis GDMA 2019

All the organizers of the previous DASFAA conferences and workshops have made DASFAA a valuable trademark, and we are proud to continue their work. We would like express our thanks to all the workshop organizers and Program Committee members for their great effort in making the DASFAA 2019 workshops a success. In total, 14 papers were accepted for into the workshops. In particular, we are grateful to the main conference organizers for their generous support and help.

March 2019

Qun Chen
Jun Miyazaki

BDMS Workshop Organization

Program Co-chairs

Xiaoling Wang	East China Normal University, China
Kai Zheng	University of Electronic Science and Technology of China, China
An Liu	Soochow University, China

Program Committee

Muhammad Aamir Cheema	Monash University, Australia
Cheqing Jin	East China Normal University, China
Qizhi Liu	Nanjing University, China
Bin Mu	Tongji University, China
Yaqian Zhou	Fudan University, China
Xuanjing Huang	Fudan University, China
Yan Wang	Macquarie University, Australia
Lizhen Xu	Southeast University, China
Xiaochun Yang	Northeastern University, China
Kun Yue	Yunnan University, China
Dell Zhang	University of London, UK
Xiao Zhang	Renmin University of China, China
Bolong Zheng	Huazhong University of Science and Technology, China

BDQM Workshop Organization

Program Co-chairs

Xin Wang	Tianjin University, China
Jianxin Li	Deakin University, Australia

Program Committee

Zhifeng Bao	RMIT, Australia
Laure Berti-Equille	Institut de Recherche pour le Dèveloppement (IRD), France
Yingyi Bu	Couchbase, USA
Gao Cong	Nanyang Technological University, Singapore
Yunpeng Chai	Renmin University of China, China
Qun Chen	Northwestern Polytechnical University, China
Yueguo Chen	Renmin University of China, China
Yongfeng Dong	Hebei University of Technology, China
Rihan Hai	Lehrstuhl Informatik 5, Germany
Cheqing Jin	East China Normal University, China
Guoliang Li	Tsinghua University, China
Lingli Li	Heilongjiang University, China
Hailong Liu	Northwestern Polytechnical University, China
Xianmin Liu	Harbin Institute of Technology, China
Xueli Liu	Tianjin University, China
Shuai Ma	Beihang University, China
Zhijing Qin	Pinterest, USA
Chuitian Rong	Tianjin Polytechnic University, China
Nan Tang	Qatar Computing Research Institute, Qatar
Hongzhi Wang	Harbin Institute of Technology, China
Jiannan Wang	Simon Fraser University, Canada
Xiaochun Yang	Northeast University, China
Yajun Yang	Tianjin University, China
Rui Zhang	The University of Melbourne, Australia
Wenjie Zhang	University of New South Wales, Australia

GDMA Workshop Organization

Program Co-chairs

Xiaowang Zhang   Tianjin University, China
Peng Peng   Hunan University, China

Program Committee

Robert Brijder   Hasselt University, Belgium
George H. L. Fletcher   Eindhoven University of Technology, The Netherlands
Liang Hong   Wuhan University, China
Egor V. Kostylev   University of Oxford, UK
Zechao Shang   The University of Chicago, USA
Hongzhi Wang   Harbin Institute of Technology, China
Kewen Wang   Griffith University, Australia
Xin Wang   Tianjin University, China
Guohui Xiao   Free University of Bozen-Bolzano, Italy
Zhiwei Zhang   Hong Kong Baptist University, SAR China

Contents

**The Third International Workshop on Graph Data Management
and Analysis (GDMA 2019)**

Posters

Tutorials

The 6th International Workshop on Big Data Management and Service (BDMS 2019)

A Probabilistic Approach for Inferring Latent Entity Associations in Textual Web Contents

Lei Li[1], Kun Yue[1(✉)], Binbin Zhang[1], and Zhengbao Sun[2]

[1] School of Information Science and Engineering, Yunnan University,
Kunming, China
kyue@ynu.edu.cn
[2] School of Resource Environment and Earth Science, Yunnan University,
Kunming, China

Abstract. Latent entity associations (EA) represent that two entities associate with each other indirectly through multiple intermediate entities in different textual Web contents (TWCs) including e-mails, Web news, social network pages, etc. In this paper, by adopting Bayesian Network as the framework to represent and infer latent EAs as well as the probabilities of associations, we propose the concept of entity association Bayesian Network (EABN). To construct EABN efficiently, we employ self-organizing map for TWC dataset division to make the co-occurrence-based dependence of each pair of entities concern just a small set of documents. Using probabilistic inferences of EABN, we evaluate and rank EAs in all possible entity pairs, by which novel latent EAs could be found. Experimental results show the effectiveness and efficiency of our approach.

Keywords: Entity Association · Bayesian Network · Self-organizing map · Knowledge base · Probabilistic inference

1 Introduction

Including e-mails, web news, social network pages, etc., textual Web content (TWC) is composed of words, some of which refer to entities. Figure 1 shows an example of entity associations (EAs) in TWC documents. Rounded rectangles represent entities and edges indicate that two entities co-occur in the same TWC document. Intuitively, EAs among entities could be divided into two categories: direct EAs and latent EAs. The former represents that two entities co-occur in the same TWC document, while the latter represents that two entities associate with each other indirectly through other intermediate entities in different TWC documents. As shown in Fig. 1, a direct EA exists between "GitHub" and "Developer", while a latent EA exist between "GitHub" and "Microsoft". Both a direct EA and a latent EA exist (through "Visual Studio") between "Microsoft" and "Developer". Actually in realistic TWC documents, more latent EAs are embodied than direct ones. Expressing latent EAs appropriately is the basis of data acquisition [1], relationship strength estimation [2] and social network analysis [3]. In this paper, we are to discover latent EAs from textual Web contents while focusing on the implied dependencies upon the co-occurrence of entity pairs.

© Springer Nature Switzerland AG 2019
G. Li et al. (Eds.): DASFAA 2019, LNCS 11448, pp. 3–18, 2019.
https://doi.org/10.1007/978-3-030-18590-9_1

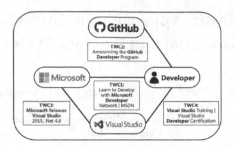

Fig. 1. EAs in TWC documents

From the characteristics of latent EAs in practical applications, several critical challenges are faced taking as input TWC documents.

(1) Latent EAs are uncertain in general cases. As shown in Fig. 1, whether "Microsoft" is associative with "GitHub" is uncertain unless all co-occurrences are discovered already.
(2) Latent EAs are likely to exist between two arbitrary entities. It is necessary to construct a global model that contains most entities in TWC documents while efficiently fulfilling the evaluation of co-occurrences on exponential numbered entities.
(3) Ranking of latent EAs is significant, since not all latent EAs are useful for subsequent tasks w.r.t. the situations with massive TWC and massive entities.

It is known that Bayesian Network (BN) is a well-adopted framework for representing and inferring uncertain knowledge, as well as prediction, diagnosis, and decision [4–6]. By using BN as the knowledge framework, we propose entity association Bayesian Network (EABN), in which entities are regarded as variables, edges describe the associations between entities from the co-occurrence point of view, and conditional probability tables (CPTs) quantify the dependencies. By EABN, the dependence relationships among entities could be described from the co-occurrences of pairs of entities, and latent EAs could be inferred by BN's probabilistic inferences.

Self-organizing map (SOM) is a type of artificial neural network that is trained using unsupervised learning to produce a low-dimensional (typically two-dimensional) and discretized representation of the input space of training samples [7–9]. To construct the EABN from TWC efficiently, we employ SOM to divide a TWC dataset into several subsets. Most co-occurrences for each entity pair could be found by traversing a TWC subset which includes much fewer documents. Thus, the execution time of EABN construction will be not increased sharply w.r.t. the exponential increase of costly traversal on TWC documents caused by the linear increase of entities.

To rank the entities based on latent EAs for the given specific entities, we use probabilistic inferences [10] with an EABN to evaluate the association quantitatively. To this end, we translate the task of latent EA discovery to conditional probability computations. For example, we regard X as "evidence variables" and Y as "query variables" to discover the latent EA when given X, denoted as $X \rightarrow Y$. The result of probabilistic inferences contains the set Y of entities and corresponding probabilities as the association degree, by which we rank the latent EAs of all possible entity pairs in descending order.

The results on real datasets show that our approach can be used to express both direct and latent EAs. It is consistent with practical situations that most EAs are latent EAs, which cannot be found by the classic association rule (AR) mining algorithm [17–19]. Experimental results show that our proposed approach for inferring EAs are effective by comparing the results with those by AR mining. Moreover, novel latent EAs could be discovered by our method. The execution time of EABN has been reduced greatly by using the SOM-based TWC division, and the efficiency of our approach is verified. The capability of inferring latent EAs with probabilistic associations and expressing latent EAs distinguish our approach compared with the classic algorithms for entity association discovery based on Latent Dirichlet Allocation (LDA) [11–13], relation extraction (RE) [14–16], and AR.

The rest of this paper is organized as follows. In Sect. 2, we introduce related work. In Sect. 3, we give definitions and problem formalization. In Sect. 4, we introduce the details of generating samples of EABN nodes. In Sect. 5, we give the algorithm for EABN learning and EA ranking. In Sect. 6, we show experimental results. In Sect. 7, we conclude and discuss future work.

2   Related Work

Latent Dirichlet Allocation of TWC Analysis. Zhang et al. [11] obtained the different topic distributions on different categories in each microblog, and calculated the similarities between users. Wood et al. [12] translated knowledge sources into a topic distribution and ensured that the topic inference process is consistent with existing knowledge. Poria et al. [13] proposed a LDA method leveraged on the semantics associated with words and multi-word expressions to improve LDA. These methods learn BNs to express the joint probability distribution of entities in TWC documents while our approach learns an EABN and find EAs based on probabilistic inferences of the EABN.

Relation Extraction from TWC. Mintz et al. [14] used Freebase, a sizable semantic database of several thousand relations, to provide distant supervision for a relation extraction task. Surdeanu et al. [15] proposed a novel approach to multi-instance and multi-label learning for relation extraction. Ren et al. [16] proposed a novel domain-independent framework which jointly embeds entity mentions, relation mentions into two low-dimensional spaces, and then estimate the similarity between a new relation and relations in low-dimensional spaces. These methods aim to find instances for a predefined relation type between two entities, but our approach can express latent EAs.

Association Rule Mining from TWC. Idoudi et al. [17] proposed to enrich existing ontology base by using the discovered association rules. Ahmed et al. [18] introduced an approach for enhanced association rule derivation, where two main categories of knowledge are employed in the mining process. Erlandsson et al. [19] proposed association learning to detect relationships between users in online social networks. These methods are effective to discover direct EAs. In contrast, our approach is designed to express both direct EAs and latent EAs.

3 Definitions and Problem Formalization

In this section, we give the definition of entity Bayesian Network based on the idea of a general Bayesian network [10], and then formalize our problem to be solved.

Definition 1. An entity association Bayesian Network (EABN) is a pair $B_e = (G_e, P_e)$. $G_e = (V, E)$ is a DAG. V is the set of nodes and E is the set of edges in EABN. A node in V represents an entity as a variable in EABN, and V's possible values express frequencies of entities qualitatively. Edges correspond to the dependency of one node (an entity) on another in TWC documents. P_e is specified as a set of conditional probability tables (CPTs) associated with the nodes in G_e.

The problem to be solved is formalized as follows:

(1) Wikidata [20–22] and supplementary entities are integrated into a knowledge base K to recognize entities in a TWC dataset T. Then, frequent entities are selected to serve as nodes of an EABN, denoted as B_e. Samples of EABN nodes are generated by normalizing frequencies of frequent entities in each TWC document and adopted as the training data for learning B_e.

(2) Learning B_e contains structure selection and parameter estimation. In structure selection, SOM is employed to divide T into multiple subsets, which we choose to provide samples thus the time consumption could be reduced. In parameter estimation, CPTs are generated for each EABN node by maximum likelihood estimation.

(3) For each pair of nodes $\langle X, Y \rangle$ in B_e, we set $X = x$ as the "evidence" and set Y as the "query variables". By probabilistic inferences with an EABN, $P(Y|X = x)$ could be obtained, as the conditional probability over the values y of Y. The values of y express frequencies of Y qualitatively. We then provide $W_{Y,x} = \sum y * P(Y = y)$ as the quantitative evaluation for the latent EA of y given x. Finally, we produce $W_{Y,x}$ for all entity pairs $\langle X, Y \rangle$ and rank all $W_{Y,x}$ in descending order.

4 Generating Samples of EABN Nodes

We integrate Wikidata and supplementary entities into a knowledge base K. Entity frequencies could be sampled but vary greatly in different TWC documents. To reduce the complexity of EABN construction, entity frequencies will be normalized into bounded values. We check whether each word in TWC documents is an entity (whether the word exists in K) and add it into an entity set. Then some most frequent entities serve as the EABN nodes. Finally, we record and normalize the frequencies of most frequent entities in each TWC document as integers. Figure 2 shows an example that we generate 2 samples from 2 TWC documents for 3 most frequent entities.

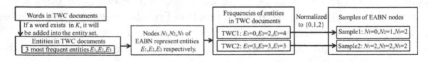

Fig. 2. Generating 2 samples from 2 TWC documents for 3 most frequent entities

The above ideas are summarized in Algorithm 1.

Algorithm 1. Generating Samples of EABN Nodes

Input:
$B_e=(G_e, P_e)$, $G_e=(V, E)$, an EABN
$T=\{T_1,..., T_M\}$, a TWC dataset containing M documents
K, a knowledge base consisting of Wikidata and supplementary entities
n, EAs among n entities to be evaluated and ranked
Variables:
$T_i.E$, entity array of T_i; $T_i.E$.append(Ent), the function to append entity Ent into $T_i.E$
H, a hash table storing frequencies of entities (default values are set to 0)
Output:
$T_i.S$, samples of V in T_i

---

1. FOR $i \leftarrow 1$ TO M DO // find frequent entities
2.   FOR $j \leftarrow 1$ TO $|T_i|$ DO // $T_{i,j}$ is the j-th word in T_i;
3.   IF $T_{i,j} \in K$ THEN // words in K serve as nodes of EABN
4. $T_i.E$.append($T_{i,j}$)
5.   IF $H[T_{i,j}]$ is undefined THEN $H[T_{i,j}] \leftarrow 0$
6.   ELSE $H[T_{i,j}] \leftarrow H[T_{i,j}]+1$ // frequency of $T_{i,j}$ is increased
7. Sort H by entity frequencies
8. $V \leftarrow \{$top n frequent entities in $H\}$
9. FOR $i \leftarrow 1$ TO M DO // generating samples for frequent entities
10.   FOR $j \leftarrow 1$ TO $|T_i.E|$ DO
11.   FOR $k \leftarrow 1$ TO $|V|$ DO
12.   IF $T_i.E[j]=V[k]$ THEN
13.   IF $T_i.S[k]$ is undefined THEN $T_i.S[k] \leftarrow 0$
14.   ELSE $T_i.S[k] \leftarrow T_i.S[k]+1$
15. $X_M \leftarrow$ Max$\{T_i.S[k]\}$
16.   FOR $k \leftarrow 1$ TO $|V|$ DO
17. $T_i.S[k] \leftarrow$ Round$[(T_i.S[k] / X_M) * (Z-1)]$ // normalization, each node of EABN
       can be set to Z possible values
18. RETURN $T_i.S$

Suppose the average of $|T_i|$ is C_1, so the number of iterations for steps 1–6 is MC_1. The number of iterations of step 7 is $|H|\log|H|$. The number of iterations for steps 10–13 is $|T_i.E|n$. The number of iterations of step 16 will be executed for n times. The number of iterations in Algorithm 1 is $MC_1 + |H|\log|H| + M * (|T_i.E|n + n+n)$, and thus the complexity of Algorithm 1 is $O(M|T_i.E|n)$.

5   Learning an EABN and Ranking EAs

5.1   BIC Metric and Division of TWC Dataset

As a classic method for learning a BN from data, scoring & search based structure learning defines a scoring function that measures how well the structure fits samples, and then finds the highest-scoring structure from a hypothesis space of potential models [10].

To exhibit the trade-off between the degree of fitting and structure complexity, the Bayesian Information Criterion (BIC) [10] is employed as the scoring function in our method. The stronger the dependence of an EABN node on its parents, the higher the score will be. The BIC scoring function is given as Formula (1).

$$\begin{cases} \text{score}_{\text{BIC}}(G_e : T) = M \sum_{i=1}^{n} I_{\hat{P}}(X_i; Pa_{X_i}) - \frac{\log M}{2} |V| \\ I_{\hat{P}}(X_i; Pa_{X_i}) = \sum_{X_i, Pa_{X_i}} \hat{P}(X_i, Pa_{X_i}) \log \frac{\hat{P}(Pa_{X_i}|X_i)}{\hat{P}(Pa_{X_i})} \end{cases} \tag{1}$$

where G_e denotes an EABN structure and $|V|$ is the number of nodes in G_e. T is a TWC dataset and $\hat{P}$ is the empirical distribution observed in samples. X_i is a node of G_e and Pa_{X_i} are parents of X_i.

In EABN construction, the execution time of BIC evaluation is increased exponentially with the linear increase of entities. To this end, we propose a strategy called subset BIC (SBIC) to reduce the execution time of BIC evaluation so that massive TWC documents from various domains could be leveraged. SBIC consists of two phases. In phase 1, SOM is employed to divide the TWC dataset into several subsets. A unique index is assigned to each entity in the TWC dataset, and the indexes of entities are used to build a feature vector for each TWC document. SOM will assign a TWC document into multiple two-dimensional output vectors. Thus we treat an output vectors as a TWC subset containing some documents. The numbers of entities in each subset are also recorded. The above ideas are given in Algorithm 2.

Algorithm 2. TWC Division by SOM

Input:
T, a TWC dataset
N_s, the number of output vectors in SOM
Variables:
$T_i.E$, the entity array of T_i
$F_i=[0,\ldots,0]$, the feature vector of T_i, $|F_i|=|V|$
H_I, a hash table storing indexes of entities
I_m, the iteration limit of SOM
Output:
S, the array containing output vectors of SOM; $S[k]$, the k-th output vector in S
$S[k].D$, the array contains TWC documents assigned to $S[k]$
$S[k].H$, the hash table contains the number of entities in $S[k]$
Function:
Euc(X_1, X_2), calculating the Euclidean distance between X_1 and X_2
$S[k].D$.append(doc), appending the TWC document doc into $S[k].D$

---

1. $I \leftarrow 0$ // I records the number of SOM iterations
2. $C_E \leftarrow 0$ // an auto-incrementing index starting
3. FOR $i \leftarrow 1$ TO M DO
4.   FOR $j \leftarrow 1$ TO $|T_i.E|$ DO
5.   IF $H_I[T_i.E[j]]$ is undefined THEN
6. $I \leftarrow 0$ // I records the number of SOM iterations

7. $I \leftarrow 0$ // I records the number of SOM iterations
8. $C_E \leftarrow 0$ // an auto-incrementing index starting
9. FOR $i \leftarrow 1$ TO M DO
10.   FOR $j \leftarrow 1$ TO $|T_i.E|$ DO
11.   IF $H_I[T_i.E[j]]$ is undefined THEN
12. $H_I[T_i.E[j]] \leftarrow C_T \leftarrow C_E$ // C_T is a temporary variable of C_E
13. $C_E \leftarrow C_E + 1$
14.   ELSE $C_T \leftarrow H_I[T_i.E[j]]$
15. $F_i[C_T] \leftarrow 1$
16.   WHILE $I < I_m$ DO // assign T_i to an output vector of SOM
17. $I \leftarrow I + 1$
18.   FOR $k \leftarrow 1$ TO N_s DO
19.   IF D_T is undefined OR $\mathrm{Euc}(F_i, S[k]) < D_T$ THEN // D_T stores the minimal Euclidean distance between F_i and the nearest output vector, if D_T is undefined (in the first iteration) or we find a nearer output vector, we assign $\mathrm{Euc}(F_i, S[k])$ to D_T
20. $D_T \leftarrow \mathrm{Euc}(F_i, S[k])$
21. $k_B \leftarrow k$ // k_B records the index of best matching output vector
22. $S[k_B].D.\mathrm{append}(T_i)$
23.   FOR $k \leftarrow 1$ TO N_s DO // updating the output vectors
24. $a \leftarrow 1 - I/I_m$ // learning rate of SOM
25. $\theta \leftarrow [(N_s{}^{\wedge}0.5)/2]*a$ // vectors within the range θ of $S[k_B]$ approach F_i
26.   IF $\mathrm{Euc}(S[k_B], S[k]) < \theta$ THEN
27. $S[k] \leftarrow S[k] + \theta a[F_i - S[k]]$ // neighbors of $S[k_B]$ approaches F_i
28. FOR $i \leftarrow 1$ TO N_s DO // recording the numbers of entities in each subset
29.   FOR $j \leftarrow 1$ TO $|S[i].D|$ DO
30.   FOR $k \leftarrow 1$ TO $|S[i].D[j].E|$ DO
31. $Ent \leftarrow S[i].D[j].E[k]$
32. $S[i].H[Ent] \leftarrow S[i].H[Ent] + 1$
33. RETURN S

Suppose the average of $|T_i.E|$ is C_2, so the number of iterations of steps 4–9 is C_2. The number of iterations for WHILE loop is I_m, and the number of iterations of steps 10–21 is $I_m * 2N_s$. Suppose the average of $|S[i].D|$ is C_3 and the average of $|S[i].D[j].E|$ is C_4, and the number of iterations of steps 22–26 is $N_s C_3 C_4$. The number of iterations in Algorithm 2 is $M*(C_2 + I_m * 2N_s) + N_s C_3 C_4$, and thus the complexity of Algorithm 2 is $O(MI_m N_s)$.

We now give an example in Fig. 3 to show the evaluation of SBIC. For demonstration purposes, the learning rate of SOM is assigned as 1 and output vectors are initialized to [0.5, 0.5, 0.5, 0.5].

In phase 1 of SBIC: (1) We assign 0, 1, 2 and 3 to entities "Microsoft", "Visual Studio", "Developer" and "Github" respectively. (2) Each TWC document produce a feature vector based on corresponding entities. (3) Each feature vector of TWC document is assigned to an output vector with the minimal Euclidean distance. TWC1 and TWC3 are assigned to output Vector1. TWC2 and TWC4 are assigned to output Vector2. (4) We also record the number of entities in each output vector.

In phase 2 of SBIC, we consider a candidate structure "Microsoft→Developer→ GitHub", $I_{\widehat{P1}}$(Developer(i); Microsoft$_{\text{Developer}(i)}$) and $I_{\widehat{P2}}$ (Github(i); Developer$_{\text{Github}(i)}$) need to be computed where Developer(i) represents the i-th value of "Developer". $I_{\widehat{P1}}$ can be computed by the samples in Vector1 since "Developer" and "Microsoft" do not co-occur in Vector2. 50% time consumption is reduced since only two TWC documents (TWC1 and TWC3) in Vector1 are engaged in computing $I_{\widehat{P1}}$. In practical situations, Vector1 is chosen to compute $I_{\widehat{P1}}$ since it has the maximal sum of frequency of "Developer" and "Microsoft". By SBIC, a subset is chosen to compute each $I_{\widehat{P1}}(X_i; Pa_{X_i})$ and provides an approximation of BIC by less execution time.

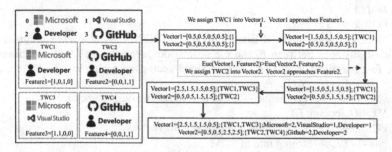

Fig. 3. Dividing a TWC dataset into two subsets (output vectors)

5.2 Obtaining Structure and Parameters with the Highest Score

We use a heuristic search technique to find the structure of EABN with the highest SBIC score starting from an empty structure G. By applying edge addition, deletion or edge reversal, we could generate G's neighbors and evaluate their SBIC score respectively. We then apply the change that leads to the best improvement in the score. This process will be continued until no modification improves the score. The above idea is given in Algorithm 3.

Algorithm 3. Obtaining an EABN structure with the highest SBIC score

Input:
$B_e=(G_e, P_e)$, $G_e=(V, E)$, an EABN
G_c, the empty structure of an EABN B_e
Output:
G_c, the highest-scoring structure of B_e

---

1. $X_S \leftarrow 0$ // X_S record the maximal SBIC score
2. $X_B \leftarrow$ TRUE // X_B indicates whether we find a better structure
3. WHILE X_B = TRUE DO
4. $X_B \leftarrow$ FALSE
5.   FOR EACH $G_c{'}$ DO // $G_c{'}$ is generated by edge addition, deletion or edge reversal
     IF SBIC($G_c{'}$) > X_S THEN $X_S \leftarrow$ SBIC($G_c{'}$), $X_B \leftarrow$ TRUE, $G_c \leftarrow G_c{'}$
6. RETURN G_c

EABN includes $|V|$ nodes and $|E|$ edges. The number of edge addition, deletion and reversal are $|V|(|V| - 1) - |E|$, $|E|$ and $|E|$ respectively. The total number of edge operations is $V|(|V| - 1) + |E|$. Thus, the complexity of Algorithm 3 is $O(M|V|^2)$.

Based on the structure G_c generated by Algorithm 3, we use Maximum Likelihood Estimation (MLE) to implement parameter estimation, which satisfies the global decomposition property of the likelihood function [10]. Furthermore, we can decompose the local likelihood function for a CPT into a product of simple likelihood function $P(X_i|Pa_{X_i})$, calculated by Formula (2). For a node X_i and the set of parent nodes Pa_{X_i}, N_{pr} is the number of instances in the TWC dataset T that $X_i = p$ and $Pa_{X_i} = r$ while N_r is the number of instances that $Pa_{X_i} = r$.

$$P(X_i|Pa_{X_i}) = \frac{N_{pr}}{N_r} \tag{2}$$

In Fig. 4, we give an example to show the process of EABN construction. On the generated samples, Formula (2) is employed to generate CPTs for each edge in the DAG of EABN. For instance, we employ $Entity(i)$ to represent that an $Entity$ in EABN sets to its i-th value. The numbers of instances for $Developer(0)$ and $Developer(1)$ are 4 and 6 respectively given $Microsoft(0)$. The numbers of instances for $Developer(0)$ and $Developer(1)$ are 3 and 7 respectively given $Microsoft(1)$. Thus we can compute four conditional probabilities

$P(Developer(0)|Microsoft(0)) = 4/(4 + 6) = 0.4$, $P(Developer(1)|Microsoft(0)) = 0.6$, $P(Developer(0)|Microsoft(1)) = 0.3$, $P(Developer(1)|Microsoft(1)) = 0.7$.

These four conditional probabilities constitute the CPT for the directed edge from $Microsoft$ to $Developer$, shown as the table in Fig. 4.

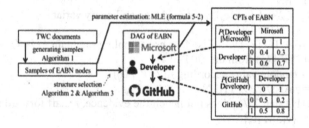

Fig. 4. EABN construction

5.3 Ranking EAs by Probabilistic Inferences of EABN

To produce quantitative evaluations of EAs, we set $X = x$ as the "evidence" and set Y as the "query variables" for each pair of nodes $<X, Y>$ in B_e. By probability inferences of EABN, we obtain $P(Y|X = x)$ as the conditional probability over the values y of Y. The values of y can express frequencies of Y qualitatively. We provide $W_{Y,x} = \sum y * P(Y = y)$ as the quantitative evaluation for the EA $A_{x \rightarrow Y}$ of Y depending x. Finally, we produce $W_{Y,x}$ for all entity pairs $<X, Y>$ and rank all $W_{Y,x}$ in descending order. Top $W_{Y,x}$ indicates significant EAs in TWC documents.

To rank EAs, we employ forward sampling [10] to implement efficient approximate inference of EABN, since an exact inference algorithm will be executed in exponential time and will not work with respect to a large number of entities in TWC documents. M samplings of EABN nodes were generated by assigning values to non-evidence nodes based on CPTs of EABN in the topological order which could be obtained by employing the depth-first search (DFS). DFS find parent nodes of an input node recursively and the node visited in this recursion will be added to the tail of a list. The final list is the topological order of EABN nodes [23]. A random value can be generated according to the CPT of the node V_j since the nodes are processed in the topological ordering and the values of V_j's parent nodes U_j are known.

For instance, $P(V_j = 0|U_j) = 0.2$ and the range 0 to 0.2 is assigned to $V_j = 0$. $P(V_j = 1|U_j) = 0.8$ and the range 0.2 to 1.0 is assigned to $V_j = 1$. A random value between 0.0 to 1.0 is generated as 0.48 which is in the range of 0.2 to 1.0. Thus V_j set to 1. We count M_y by representing the number of instances for $Y = y$ and we calculate $P(Y = y) = M_y/M$. We summarize above ideas in Algorithm 4.

Algorithm 4. Ranking EAs by probabilistic inference of EABN

Input:
$B_e=(G_e, P_e)$, $G_e=(V, E)$, an EABN; M, the number of forward samplings
Variables:
$F=\{F[1],\ldots, F[M]\}$, the set of forward samplings; $Y[k]$, the value of Y in $F[k]$
Output:
R, A ranked list for $W_{Y,x}$

---

1. $V_1, \ldots , V_{|V|}$←a topological ordering of V
2. FOR $p \leftarrow 1$ TO $|V|$ DO
3.   FOR $q \leftarrow 1$ TO $|V|$ DO
4.   IF $p{\neq}q$ THEN
5. $Y\leftarrow\{ V_p \}, X\leftarrow\{ V_q = v_q\}$ // set evidence and query variables
6.   FOR $i\leftarrow1$ TO M DO
7.   FOR $j\leftarrow1$ TO $|V|$ DO // Use forward sampling to obtain $P(Y|X)$
8.   Find $P(V_j|U_j)$ from the CPT of v_j
9. v_j←a random value according to $P(V_j|U_j)$
10.   IF $j{=}q$ AND $v_j{\neq}v_q$ THEN
11.   GOTO step 7 // v_j does not match the evidence, restart forward sampling
12. $F[i] \leftarrow\{v_1, \ldots ,v_n\}$
13. $P(Y{=}y | X{=}x) \leftarrow \frac{1}{M}\sum_{k=1}^{M}\{Y[k] = y\}$
14. $W_{Y,x}\leftarrow\sum y*P(Y{=}y)$
15. $R\leftarrow R\cup \{(A_{x\to Y},W_{Y,x})\}$ // append the pair $(A_{x\to Y},W_{Y,x})$ into R
16. Sort R by $W_{Y,x}$
17. RETURN R

The complexity of producing the topological order of EABN nodes is $O(|V| + |E|)$. The complexity of steps 5–10 is $O(M|V|)$. We can produce $L_C = |V| * (|V| - 1)/2$ entity pairs and the number of iterations of step 15 is $L_C \log(L_C)$. Thus, the complexity of Algorithm 4 is $O(M|V|^3)$.

On the EABN in Fig. 4, an example is provided for ranking EAs including EA1 $(A_{Microsoft \rightarrow Developer})$, EA2 $(A_{Developer \rightarrow Github})$ and EA3 $(A_{Microsoft \rightarrow Github})$. EA1 and EA2 are direct EAs while EA3 is a latent EA. In Table 1, 100 forward samplings were generated for EA3 when Microsoft set to 1. Among 100 forward samplings, we can obtain 32 instances that Github = 0 and 68 instances that Github = 1. Thus, we can compute the conditional probabilities $P(Github = 0|Microsoft = 1) = 32/100 = 0.32$ and $P(Github = 1| Microsoft = 1) = 0.68$. After all $W_{Y,x}$ were computed, we can rank EAs in descending order as [EA2(0.8), EA3(0.68), EA1(0.6)].

Table 1. Conditional probability distributions and EABN evaluations

Generating 100 forward samplings for each EA (val = value, ins = instance, prob = probability)								
EA1 (Microsoft = 1)			EA2 (Developer = 1)			EA3 (Microsoft = 1)		
Developer	val	0	1	GitHub	val	0	1	GitHub	val	0	1
	ins	40	60		ins	20	80		ins	32	68
	prob	0.4	0.6		prob	0.2	0.8		prob	0.32	0.68
$W_{Developer, Microsoft = 1}$ $= 0 * 0.4 + 1 * 0.6 = 0.6$			$W_{Github, Developer = 1}$ $= 0 * 0.2 + 1 * 0.8 = 0.8$			$W_{Developer, Microsoft = 1}$ $= 0 * 0.32 + 1 * 0.68 = 0.68$		

6 Experimental Results

6.1 Experiment Setup

Configuration. Python 2.7 is employed to implement all the algorithms mentioned in this paper. All algorithms are executed by a machine with Intel Xeon CPU (4×3.6 GHz) and 128 GB RAM.

Datasets. "Groceries dataset" [24] serves as the standard data, which contains 1 month (30 days) of real-world point-of-sale transaction data from a typical local grocery outlet, with 9835 transactions and the items are aggregated to 169 categories. The realistic "News Popularity in Multiple Social Media Platforms Data Set" [25] which comprises 4 topics of news headlines was adopted as our test dataset. The topic "Microsoft" with 21858 documents is abbreviated as the "Microsoft dataset".

Metrics. The effectiveness of our approach is tested in two aspects. First, two rankings produced by EABN and Apriori respectively for the same EA is compared. Second, we divide a TWC dataset equally into several subsets and propose a metric named consistency of ratio (COR), shown as follows

$$COR(A_{X \rightarrow Y}) = STDEV(N_Y/N_X) \tag{3}$$

where N_Y is an array of Y's number in each TWC subset and $N_{Y,i}$ represent Y's number in the i-th subset. STDEV is the function to compute a standard deviation.

Section 6.3 shows the time consumption of EABN including the total time of learning (using BIC metric and SBIC metric), the total time and average time of probabilistic inference with respect to each EA.

6.2   Effectiveness

AR (Apriori algorithm with *minSupport* = 0.04 and *minConfidence* = 0.2) and EABN (top 32 frequent entities in Groceries dataset as nodes) are employed to find EAs in Groceries dataset. We summarized the results of AR and EABN in Table 2(a). AR (the Apriori algorithm with *minSupport* = 0.013 and *minConfidence* = 0.2) and EABN (top 20 frequent entities in Microsoft dataset as nodes) are employed to find EAs in Microsoft dataset. The results are summarized in Table 2(b).

From Table 2, we note that the order of EAs obtained by the EABN ranking is consistent with that by the Apriori ranking. EABN could be used to infer latent EAs, and thus more EAs could be found upon EABN than those found by Apriori (just direct EAs). The index of a direct EA in EABN ranking is greater than that in Apriori ranking, since latent EAs do not exist in the latter.

Table 2. EAs found by EABN and AR in Groceries dataset and Microsoft dataset

<table>
<tr><td colspan="4">(a) Groceries dataset</td><td colspan="4">(b) Microsoft dataset</td></tr>
<tr><td colspan="2">$A_{X \to Y}$</td><td>EABN</td><td>Apriori</td><td colspan="2">$A_{X \to Y}$</td><td>EABN</td><td>Apriori</td></tr>
<tr><td>X</td><td>Y</td><td>Ranking</td><td>ranking</td><td>X</td><td>Y</td><td>Ranking</td><td>ranking</td></tr>
<tr><td>Root vegetables</td><td>Whole milk</td><td>5</td><td>1</td><td>Google</td><td>Microsoft</td><td>5</td><td>1</td></tr>
<tr><td>Root vegetables</td><td>Other vegetables</td><td>7</td><td>2</td><td>Apple</td><td>Microsoft</td><td>6</td><td>2</td></tr>
<tr><td>Tropical fruit</td><td>Whole milk</td><td>9</td><td>3</td><td>Windows</td><td>Microsoft</td><td>13</td><td>3</td></tr>
<tr><td>Other vegetables</td><td>Whole milk</td><td>14</td><td>5</td><td>Company</td><td>Microsoft</td><td>14</td><td>6</td></tr>
<tr><td>Yogurt</td><td>Whole milk</td><td>15</td><td>4</td><td>Software</td><td>Microsoft</td><td>16</td><td>5</td></tr>
<tr><td>Whole milk</td><td>Other vegetables</td><td>30</td><td>8</td><td>Users</td><td>Microsoft</td><td>17</td><td>4</td></tr>
<tr><td>Yogurt</td><td>Other vegetables</td><td>33</td><td>6</td><td></td><td></td><td></td><td></td></tr>
<tr><td>Rolls/buns</td><td>Whole milk</td><td>36</td><td>7</td><td></td><td></td><td></td><td></td></tr>
</table>

COR is used to express whether the ratio $N_{Y,i}/N_{X,i}$ remains consistent in different TWC subsets. If the ratio $N_{Y,i}/N_{X,i}$ remains consistent while $N_{Y,i}$ and $N_{X,i}$ change randomly in different TWC subsets, then the tendencies of N_Y and N_X are similar and COR metric would be relatively small. Thus, we conclude that smaller COR metrics are more valuable for subsequent EA-based tasks.

Experiments are conducted to whether EABN could find EAs with small COR metric. In the "Groceries dataset", $A_{e3(root\ vegetables) \to e4(whole\ milk)}$ ranks first in the Apriori ranking and $A_{e1(chicken) \to e2(other\ vegetables)}$ ranks first in the EABN ranking. The statistics of some entities in "Groceries dataset" are listed in Table 3. N_{e1}/N_{e2}, N_{e3}/N_{e4} and corresponding averages are shown in Fig. 5(a). We can intuitively know that the ratio N_{e1}/N_{e2} is more consistent than the ratio N_{e3}/N_{e4}. In fact, COR($A_{e1(chicken) \to e2(other\ vegetables)}$) = 0.029 is less than COR($A_{e3(root\ vegetables) \to e4(whole\ milk)}$) = 0.041. The EA between N_{e1} and N_{e2} has not been found by Apriori, since N_{e1} and N_{e2} do not co-occur frequently. However, N_{e1}/N_{e2} is more consistent than N_{e3}/N_{e4} when N_{e1} and N_{e3} change in different TWC subsets. Thus $A_{e1(chicken) \to e2(other\ vegetables)}$ could be found successfully by EABN with smaller COR values for subsequent EA-based tasks than $A_{e3(root\ vegetables) \to e4(whole\ milk)}$.

Table 3. Statistics of some entities in Groceries dataset

	Subsets of Groceries dataset									
	1	2	3	4	5	6	7	8	9	10
N_{e1}	32	40	37	50	42	52	38	37	48	46
N_{e2}	186	197	185	197	201	226	179	156	176	200
N_{e1}/N_{e2}	0.17	0.20	0.20	0.25	0.21	0.23	0.21	0.24	0.27	0.23
N_{e3}	110	117	99	109	140	114	89	104	92	98
N_{e4}	269	248	252	227	284	274	231	262	235	231
N_{e3}/N_{e4}	0.41	0.47	0.39	0.48	0.49	0.42	0.39	0.40	0.39	0.42

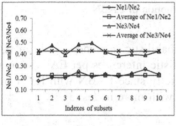

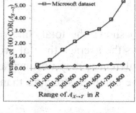

(a) N_{e1}/N_{e2} and N_{e3}/N_{e4} (b) Average of 100 COR($A_{X \to Y}$)

Fig. 5. COR analysis

We computed COR($A_{X \to Y}$) for each EA $A_{X \to Y}$ in the ranked list R produced by Algorithm 4. Figure 5(b) shows the average of 100 COR($A_{X \to Y}$) values for ranges in the Groceries dataset and Microsoft dataset, since the number of entity Y is usually influenced by many entities rather than only X and COR($A_{X \to Y}$) is not strictly ordered. It could be concluded from Fig. 5(b) that the COR value is increased generally in the list R ranked by $W_{Y,x}$ in descending order and the Groceries dataset has a smaller average of COR than the Microsoft dataset in the same range of the list R.

6.3 Efficiency

The time consumption of EABN is shown in Fig. 6. In Fig. 6(a), the total time of learning an EABN (BIC metric) is mainly determined by the number of EABN nodes and the number of samples. The total time of learning increases exponentially while the nodes of EABN increase linearly. In Fig. 6(b), we show the total time of learning an EABN (SBIC metric). The time consumption of different combinations of nodes and samples is about 30% compared with those by using the classic BIC metric.

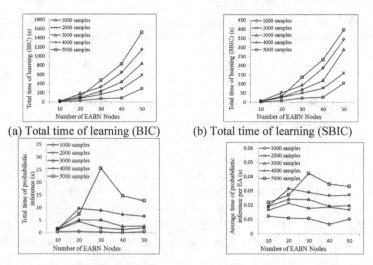

(a) Total time of learning (BIC)   (b) Total time of learning (SBIC)

(c) Total time of probabilistic inference   (d) Average time of probabilistic inference per EA

Fig. 6. Time consumption of EABN

Figure 6(c) shows the total time of probabilistic inferences with an EABN and Fig. 6(d) shows the average time of probabilistic inferences per EA. The total time of probabilistic inferences may not increase strictly since EABN with fewer nodes may produce a more complex structure than that with more nodes.

7   Conclusions and Future Work

EABN, SBIC and COR are incorporated in this paper to discover and rank both direct and latent EAs implied in TWC documents. A similar order in the EABN ranking and the Apriori ranking of direct EAs could be achieved. SBIC makes the time consumption of EABN construction reduced so that EABN can incorporate more entities and EA with smaller COR values valuable for subsequent EA-based tasks. The COR value is increased generally in the ranked list of EAs. As an initial exploration, serial implementation is provided in algorithm implementation and experiments. In future work, parallel and distributed implementation for our proposed algorithms will be given to find and rank massive EAs in TWC datasets from various domains.

Acknowledgement. This paper was supported by the National Natural Science Foundation of China (U1802271), Program for the second Batch of Yunling Scholar of Yunnan Province (C6153001), Donglu Scholar Cultivation Project of Yunnan University, and Research Foundation of Educational Department of Yunnan Province (2016ZZX006).

References

1. Yin, Z., Yue, K., Wu, H., Su, Y.: Adaptive and parallel data acquisition from online big graphs. In: Pei, J., Manolopoulos, Y., Sadiq, S., Li, J. (eds.) DASFAA 2018, Part I. LNCS, vol. 10827, pp. 323–331. Springer, Cham (2018). https://doi.org/10.1007/978-3-319-91452-7_21

2. Liao, C., Xiong, Y., Kong, X., Zhu, Y., Zhao, S., Li, S.: Functional-oriented relationship strength estimation: from online events to offline interactions. In: Pei, J., Manolopoulos, Y., Sadiq, S., Li, J. (eds.) DASFAA 2018, Part I. LNCS, vol. 10827, pp. 442–459. Springer, Cham (2018). https://doi.org/10.1007/978-3-319-91452-7_29

3. Zhang, J., Tan, L., Tao, X., Zheng, X., Luo, Y., Lin, J.C.-W.: SLIND: identifying stable links in online social networks. In: Pei, J., Manolopoulos, Y., Sadiq, S., Li, J. (eds.) DASFAA 2018, Part II. LNCS, vol. 10828, pp. 813–816. Springer, Cham (2018). https://doi.org/10.1007/978-3-319-91458-9_54

4. Liu, W., Yue, K., Yue, M., et al.: A Bayesian network-based approach for incremental learning of uncertain knowledge. Int. J. Uncertain. Fuzziness Knowl.-Based Syst. 26(1), 87–108 (2018)

5. Teye, M., Azizpour, H., Smith, K.: Bayesian uncertainty estimation for batch normalized deep networks. In: Proceedings of the 35th International Conference on Machine Learning, pp. 4914–4923. ACM, New York (2018)

6. Ishak, R., Messaouda, F., Hafida, B.: Toward a general formalism of fuzzy multi-entity Bayesian networks for representing and reasoning with uncertain knowledge. In: Proceedings of the 19th International Conference on Enterprise Information Systems, ICEIS 2017, vol. 1, pp. 520–528. SciTePress, Setúbal (2017)

7. Kuo, R.J., Rizki, M., Zulvia, F.E., et al.: Integration of growing self-organizing map and bee colony optimization algorithm for part clustering. Comput. Ind. Eng. 120, 251–265 (2018)

8. Li, Z., Fang, H., Huang, M., et al.: Data-driven bearing fault identification using improved hidden markov model and self-organizing map. Comput. Ind. Eng. 116, 37–46 (2018)

9. Saraswati, A., Nguyen, V.T., Hagenbuchner, M., et al.: High-resolution self-organizing maps for advanced visualization and dimension reduction. Neural Netw. 105, 166–184 (2018)

10. Daphne, K., Nir, F.: Probabilistic Graphical Models: Principles and Techniques, 1st edn. The MIT Press, Cambridge (2009)

11. Zhang, W., Pan, T., Wang, Y., et al.: UT-LDA based similarity computing in microblog. In: 2015 IEEE International Conference on Software Quality, Reliability and Security, pp. 197–201. IEEE, Piscataway (2015)

12. Wood, J., Tan, P., Wang, W., et al.: Source-LDA: enhancing probabilistic topic models using prior knowledge sources. In: 33rd IEEE International Conference on Data Engineering, pp. 411–422. IEEE, Piscataway (2017)

13. Poria, S., Chaturvedi, I., Bisio, F., et al.: Sentic LDA: improving on LDA with semantic similarity for aspect-based sentiment analysis. In: 2016 International Joint Conference on Neural Networks, pp. 4465–4473. IEEE, Piscataway (2016)

14. Mintz, M., Bills, S., Snow, R., et al.: Distant supervision for relation extraction without labeled data. In: ACL 2009, Proceedings of the 47th Annual Meeting of the Association for Computational Linguistics and the 4th International Joint Conference on Natural Language Processing of the AFNLP, pp. 1003–1011. ACL, Pennsylvania (2009)

15. Surdeanu, M., Tibshirani, J., Nallapati, R., et al.: Multi-instance multi-label learning for relation extraction. In: Proceedings of the 2012 Joint Conference on Empirical Methods in Natural Language Processing and Computational Natural Language Learning, pp. 455–465. ACL, Pennsylvania (2012)
16. Ren, X., Wu, Z., He, W., et al.: CoType: joint extraction of typed entities and relations with knowledge bases. In: Proceedings of the 26th International Conference on World Wide Web, pp. 1015–1024. ACM, New York (2017)
17. Idoudi, R., Ettabaâ, K.S., Solaiman, B.: Association rules-based ontology enrichment. Int. J. Web Appl. **8**(1), 16–25 (2016)
18. Ahmed, E.B., Gargouri, F.: Enhanced association rules over ontology resources. Int. J. Web Appl. **7**(1), 10–22 (2015)
19. Erlandsson, F., Bródka, P., Borg, A., et al.: Enhanced association rules over ontology resources. Entropy **18**(5), 164–178 (2016)
20. Wikidata. https://www.wikidata.org/wiki/Wikidata:Main_Page. Accessed 01 Oct 2018
21. Färber, M., Bartscherer, F., Menne, C., et al.: Linked data quality of DBpedia, Freebase, OpenCyc, Wikidata, and YAGO. Seman. Web **9**(1), 77–129 (2018)
22. Ismayilov, A., Kontokostas, D., Auer, S., et al.: Wikidata through the Eyes of DBpedia. Seman. Web **9**(4), 493–503 (2018)
23. Cormen, T.H., Leiserson, C.E., Rivest, R., et al.: Introduction to Algorithms, 2nd edn. The MIT Press, Cambridge (2001)
24. Machine-Learning-with-R-datasets/groceries.csv. https://github.com/stedy/MachineLearning-with-R-datasets/blob/master/groceries.csv. Accessed 15 Nov 2018
25. News Popularity in Multiple Social Media Platforms Data Set. http://archive.ics.uci.edu/ml/datasets/News+Popularity+in+Multiple+Social+Media+Platforms. Accessed 01 Nov 2018

UHRP: Uncertainty-Based Pruning Method for Anonymized Data Linear Regression

Kun Liu[1], Wenyan Liu[1], Junhong Cheng[1], and Xingjian Lu[2(✉)]

[1] School of Computer Science and Software Engineering,
East China Normal University, Shanghai, China
{kunliu,wyliu,51184501101}@stu.ecnu.edu.cn
[2] School of Information Science and Engineering,
East China University of Science and Technology, Shanghai, China
luxj@ecust.edu.cn

Abstract. Anonymization method, as a kind of privacy protection technology for data publishing, has been heavily researched during the past twenty years. However, fewer researches have been conducted on making better use of the anonymized data for data mining. In this paper, we focus on training regression model using anonymized data and predicting on original samples using the trained model. Anonymized training instances are generally considered as hyper-rectangles, which is different from most machine learning tasks. We propose several hyper-rectangle vectorization methods that are compatible with both anonymized data and original data for model training. Anonymization brings additional uncertainty. To address this issue, we propose an Uncertainty-based Hyper-Rectangle Pruning method (UHRP) to reduce the disturbance introduced by anonymized data. In this method, we prune hyper-rectangle by its global uncertainty which is calculated from all uncertain attributes. Experiments show that a linear regressor trained on anonymized data could be expected to do as well as the model trained with original data under specific conditions. Experimental results also prove that our pruning method could further improve the model's performance.

Keywords: Machine learning · Anonymization · Interval value

1 Introduction

Data is the cornerstone of data mining technology. The demand for data sharing between different departments and systems has been upsurging. There is a trend to publish data on open platforms, such as Kaggle [1] and Tianchi [2] whereas raising the risk of privacy leakage. In order to publish data with higher utility under the premise of protecting users' privacy, the main methods adopted at present are anonymization algorithms, such as k-anonymity [3,4], l-diversity [5], t-closeness [6] and *Anotomy* [12]. Generalization-based anonymization has been

© Springer Nature Switzerland AG 2019
G. Li et al. (Eds.): DASFAA 2019, LNCS 11448, pp. 19–33, 2019.
https://doi.org/10.1007/978-3-030-18590-9_2

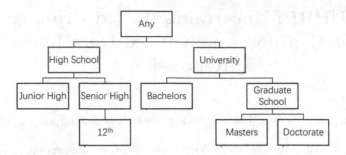

Fig. 1. Value generalization hierarchies for education

introduced to hospitals and enterprises for privacy concern [7,8]. In addition, we think anonymization is better than differential privacy [9] for publishing relational data.

There is a concern that generalization-based anonymization using Value Generalization Hierarchy (as shown in Fig. 1) will heavily reduce data utility in high dimensional spaces [12]. Non-homogeneous generalization, such as [16], takes this issue into account and significantly improve the utility of anonymized data while better keeping truth compared with *Anotomy*. Worries also exist that the above anonymize methods may fail to guarantee perfect privacy against all kinds of attacks when facing a sufficiently strong adversary. However, U.S. Healthcare Insurance Portability and Accountability Act (HIPAA) thinks limited re-identification risk is bearable and allows data release so long as to extract valuable knowledge [13].

For the time being, there are much fewer researches concentrating on how to better utilize the anonymized data, compared with privacy-preserving data publishing. In this paper, we focus on how to train the regression model on anonymized training data and use the model to predict on original test data. A similar scenario exists in real life. For example, several hospitals may collaborate to create one large anonymized dataset which is then shared among all collaborating hospitals. For a researcher in a participating hospital who is interested in using machine learning for auxiliary diagnosis that may help a lot. She/he

	Table 1. Original dataset				**Table 2.** 3-anonymous dataset	
R	Degree	Age		R'	Degree	Age
r1	Bechelors	20		r1'	Bechelors	[19, 21]
r2	Bechelors	21		r2'	Bechelors	[19, 21]
r3	Bechelors	19		r3'	Bechelors	[19, 21]
r4	Masters	23		r4'	Grad School	[22, 25]
r5	Masters	22		r5'	Grad School	[22, 25]
r6	Doctorate	25		r6'	Grad School	[22, 25]

is able to build a model using the large anonymized dataset and diagnose new patients [21] with the model.

Compared with the original data, anonymized data is less specific. Generalization of a numerical value yields a range while a set for a category value like Table 1. Two main challenges for model training on anonymized data are as follow:

(1) Representation of uncertain data. The lack of a unified and reliable feature vector representing method adds the difficulty for further training and prediction.
(2) The noise brought by anonymization method. The uncertainty caused by noise may mislead model training.

In this paper, we explore how to use anonymized data for regression. The trained model can be used to predict on both original and anonymized data.

This paper has the following key contributions:

- We study the regression task on anonymized data. The problem is under a practical scenario where model is trained on anonymized data while doing prediction on original data with the trained model.
- We propose several methods to represent anonymized data and compare their effectiveness with experiments.
- We innovate an Uncertainty-based Hyper-Rectangle Pruning (UHRP) method, which improve regression model performance under certain conditions by filtering out samples with excessive noise.

2   Related Work

The problem of Privacy-Preserving Data Publishing (PPDP) has been heavily researched since it was first introduced in [14,15]. The main purpose of PPDP is to improve the utility of published data while ensuring privacy security.

Methods for anonymizing sensitive data is first proposed by Sweeney [4]. After that, l-diversity and t-closeness model are presented successively. To meet the requirements of privacy protection, $DataFly$, μ-argus [4], $Incognito$ [10], $Mondrian$ [11], non-uniform generalization [16] have been proposed in succession. Most existing anonymize methods rely on the Value Generalization Hierarchy (VGH) in advance. However, it increases workload of data publisher and lacks flexibility. Some people have implemented an anonymize algorithm [17] based on $Mondrian$ which does not rely on VGH to solve this problem. An original dataset like Table 1 can be transformed into Table 2 in order to meet the requirements of k-anonymity ($k = 3$).

Generally, the machine learning algorithms take one input sample as a point in an m-dimensional space where m is the number of features. However, the anonymized data with q quasi-identifier attributes can be regarded as q-dimensional hyper-rectangles. This kind of hyper-rectangle is called as generalized exemplars by Salzberg [22] in his Nested Generalized Exemplar (NGE) theory. In NGE theory, Salzberg proposes a definition D_{EH} to calculate the distance between an example and generalized exemplar. LeFevre et al. [10] map the m-dimensional hyper-rectangle to a $2m$-dimensional space in order to enable machine learning algorithms to support classification learning from m-dimension hyper-rectangle dataset.

Anonymized data has some similarities with interval-valued fuzzy sets (IVFS) in the fuzzy system. In recent years, many studies about IVFS appeared like [23–25]. However, due to the limitation of privacy model, we cannot obtain anonymized data's membership function which is vital for IVFS prediction. It is currently hard to employ IVFS methods to solve the regression problem on anonymized data. At the same time, the generalized samples can be regarded as uncertain data. There are researches paying attention to the anonymized data classification problem [18–21, 26]. We refer to two basic hyper-rectangle representation methods proposed by Inan [21] and use them for regression tasks. Mancuhan et al. train linear classification model upon anatomized data [26], with a pruning method reducing the noise samples. Mancuhan et al. prune training data based on location, which is different from our uncertainty-based pruning method.

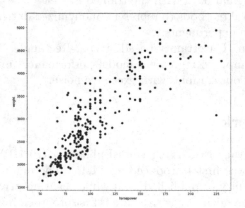

Fig. 2. Example of original training data with two attributes Horsepower (horizontal axis) and Weight (vertical axis). Points represent the position, not size.

3   Preliminaries

We restate several concepts:

(1) **Original dataset.** A dataset D is called a person-specific dataset for population P if each instance $x_i \in D$ belongs to a unique individual $p \in P$. We call a D without anonymization as original dataset.

(2) **Quasi-identifying identifier.** A set of attributes are called quasi-identifying identifier (QID) if there is background knowledge available to the adversary that associates the quasi-identifying attributes with a unique individual $p \in P$.

(3) **Equivalence class.** Equivalence class refer to a group of tuples. The tuples in a equivalence class share the same quasi-identifier values.
(4) **Original test dataset.** An original test dataset O is a subset of D but has no intersection with the original training dataset.
(5) **Anonymized training data.** A training dataset D that satisfies the following conditions is said to be anonymized training data D_k:
    (1) The training data D_k does not contain any unique identifying attributes.
    (2) Every instance $x_i \in D_k$ is indistinguishable from at least other $k - 1$ instances in D_k with respect to its QID.
(6) **Feature Representation method.** Feature representation method fr is a function to vectorize an anonymized instance or original instance.

Anonymity has a significant impact on the form of data. For the convenience of understanding, a two-dimensional(horsepower and weight) dataset is selected to illustrate. We selected 2 attributes of *Auto MPG* database from UCI dataset [29]. We then randomly selected 300 records. With *horsepower* as the horizontal axis and *weight* as the vertical axis, original data were drawn as shown in Fig. 2. After that, original data were anonymized to different degrees ($k = 2, 3, 4, 5$). Hence we obtained the four graphs in Fig. 3. Each rectangle in Fig. 3 represents an anonymized equivalence class. Each rectangle contains at least k instances but the exact location of these instances in a rectangle is unknown. As shown in Fig. 3, the average size of all rectangles is increasing with k growing up.

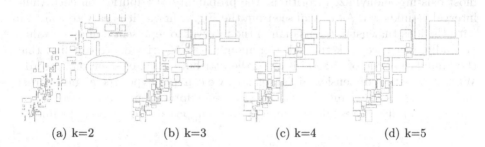

      (a) k=2   (b) k=3   (c) k=4   (d) k=5

Fig. 3. Example of anonymized training data with two attributes: horsepower (horizontal axis) and weight (vertical axis). Rectangle represents the range of possible values for a sample point. Attribute values are processed with MinMaxScaler in Scikit-Learn [30] for illustration.

4 Problem Definition and Solution

Most learning algorithms take the input data as points in an m-dimensional space, where m is the number of attributes. For anonymized data instances, attribute values are generalized into numerical value intervals or set of category values like Table 1. An instance can be seen as a hyper-rectangle. The main problem we address is how to utilize anonymized dataset for data-mining purposes, mainly for regression here. Clearly, we can represent each uncertain hyper-rectangle with an explicit point and implement regression tasks based on those points. That raises the problem of how to select the most appropriate point(s) as the representative(s) of hyper-rectangles.

In this paper, we study how to use anonymized data to train a regression model with better prediction performance on original data. We also study how to improve the training performance by introducing Uncertainty-based Hyper-Rectangle Pruning method (UHRP).

4.1 Feature Representation for Anonymized Data

The first problem is how to represent hyper-rectangles with numerical vectors. At present, there is no recognized and effective method to represent anonymized data, so we propose several heuristic methods to represent the generalized data after anonymization. The regression model is more concerned with numerical attributes. Three different numerical data representation methods are proposed. For categorical data, we choose a more effective representation method according to past experience [21]. The attributes in test dataset may be specific or generalized. Details are as follow:

(1) Numerical Attributes: Generalization makes a specific numeric value a range. Diverse from the classification problem, we should try to maintain the accuracy of numerical data in the regression problem, so different numerical intervals shall not be considered as various categories. For the released data of most existing anonymize algorithms, the probability distribution on each value interval is unknown. As we can see from the Fig. 4, firstly, it is easy to come up with the idea that using the median of the interval to represent a numeric value. Secondly, when we look through the anonymized algorithm, we can find that the boundary value of interval is often the attribute value of some real records. When the feature dimension of data is low, we can add hyper-rectangles' corner into training set which may provide more useful information. Thirdly, an interval can be naturally expressed by a tuple with its upper and lower bound values.

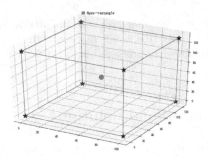

Fig. 4. 3D hyper-rectangle with a central point and corner points. Black **Stars** represent corner points while green **Circle** represents a center point. (Color figure online)

(2) Categorical Attributes: We treat each value that appears in the original data category attribute as a class and represent it with one-hot in the feature vector. Similarly, we use multi-hot to represent the generalized value. At the same time, considering the uncertain world hypothesis, according to probability values to deal with it. For example, for education attributes, "Masters" can be expressed as vectors <0, 0, 0, 1, 0> and "Doctorate" can be expressed as <0, 0, 0, 0, 1>. The generalized education attribute Graduate School can be expressed as <0, 0, 0, 1, 1>. If the probability distribution of different genders is unknown, it can be expressed as <0.5, 0.5> according to *the principle of maximum entropy* [27].

From the above, we propose three Feature Representation methods as follows:

- **FR1.** Each data interval is represented by its average. If the numerical probability distribution on a given interval is unknown, we can choose to represent the interval data with the midpoint of the interval. For example, for the interval value of weight [2250, 2750), we can use the mid-point 2500 to express it. For the data without generalization, such as 2250, we can directly use the original accurate value 2250 to express it.
- **FR2.** A hyper-rectangle is represented by its center point in *FR1*. In *FR2*, we add corners (the **stars** as shown in the Fig. 4) on the basis of *FR1*. This method is employed for anonymized training set.
- **FR3.** Each interval data is represented by the value of the upper and lower bounds of the interval. For each numeric data type, this method can be used regardless of generalization. For example, the above [2250,2750) generalization data can be expressed as <2250, 2750> and occupy two positions in the eigenvector; while 2250 can also be expressed as <2250, 2250> and occupy two positions in feature vector.

4.2   Uncertainty-Based Hyper-Rectangle Pruning Method

The philosophy behind regression is that some data generating function, $f(x)$, exists, combined with additive noise, to produce observable target values y,

$$y = f(x) + \epsilon \tag{1}$$

The ϵ is termed *irreducible noise* or *data noise*. Generally the goal of regression is to produce an estimate $\hat{f}(x)$, which allows prediction of point estimates (ϵ is assumed to have zero mean) [28]. However, when estimating the uncertainty of y, additional terms must be estimated. Given that both terms of Eq. (1) have associated sources of uncertainty, and assuming they are independent, the total variance of observations is given by,

$$\sigma_y^2 = \sigma_{model}^2 + \sigma_{noise}^2 \tag{2}$$

with σ_{model}^2 termed *model uncertainty* or *epistemic uncertainty* and σ_{noise}^2 termed *data uncertainty* variance. Intuitively, hyper-rectangles with bigger "size" have more "uncertainty" (aka measurement errors in linear regression). Given the anonymize algorithm and anonymity privacy model into account, we realize that different equivalence class contains approximately the same number of samples. Thus hyper-rectangles with larger uncertainty tend to introduce more disturbance for model fitting. Our UHRP method filters out samples that bring excessive noise and thus reduces the *data noise* and improves model performance.

How to quantify the uncertainty of hyper-rectangle still needs discussion. For numerical attribute value, we can directly use the upper bound of the interval minus the lower bound of the interval, and then normalmize it by dividing value range just like Min-Max Normalization; For category attribute value, Shannon entropy can be used to calculate uncertainty of the value according to its probability distribution. The hyper-rectangle's uncertainty is obtained by sum the uncertainty values of different attributes. We come up with two methods to calculate total uncertainty of hyper-rectangle according to different attributes:

(1) Referring to the method of calculating cube volume, multiply all attributes' uncertainty to get hyper-rectangle's total uncertainty. For an instance x_i from anonymized dataset D_k with $|QID| = q$ and u_{ij}, $j = 1...q$ represents the uncertainty of j_{th} anonymized attribute of x_i. Global uncertainty of x_i can be calculated as follows:

$$U(x_i) = \prod_{j=1}^{q} u_{ij} \tag{3}$$

There are problems with this approach. On the one hand, when the sums of all attributes' uncertainty of two hyper-rectangles are the same, this approach tends to choose "slender" hyper-rectangle. On the other hand, when the upper and lower bounds of an interval are very close or even the same, the uncertainty result of hyper-rectangle is near 0, which seriously reduce the comparability.

(2) Adding up the uncertainties of different attributes. The uncertainty of x_i can be calculated as follows:

$$U(x_i) = \sum_{j=1}^{q} u_{ij} \tag{4}$$

This approach still has a problem. Of the same volume, it favors more symmetrical hyper-rectangle when attributes' uncertainties are nearly the same.

Considering the characteristics of anonymized data itself, problems approach (1) is more likely to happen. Therefore, we choose the option (2). To be more adaptable, the total uncertainty of a hyper-rectangle can be adjusted by adding weight w_i for each attribute as below:

$$U(x_i) = \sum_{j=1}^{q} w_i u_{ij} \tag{5}$$

In UHRP, we will select and remove hyper-rectangles with too much noise as circled in red in Fig. 3. The selection of hyper-rectangles is mainly by calculating values' uncertainty. Algorithm with *O(n)* complexity is as follows:

Algorithm 1. Hyper-rectangle pruning algorithm

Input: Hyper-rectangles array $[R_i] = R_1, R_2, ..., R_n$ and
   Remainder-Ratio β. Each record has d uncertain attributes.
Output: Pruned hyper-rectangles
1   initialization global_range and the global record uncertainty $[U_i]$ of $[R_i]$,
  $[U_i] = 0$, $i=1...n$;
2   **for** $k = 1; k \leq d; k+ = 1$ **do**
3    $global\_max = \max\{x[k]\}$;
4   // $x[k]$ is k_{th} uncertain attribute values of all records;
5    $global\_min = \min\{x[k]\}$;
6    $global\_range_k = global\_max - global\_min$;

7   **for** $j = 1; j \leq n; j+ = 1$ **do**
8   **for** $k = 1; k \leq d; k+ = 1$ **do**
9     $sup = Upper\{x_j[k]\}$;
10     $inf = Lower\{x_j[k]\}$;
11   // $x_j[k]$ is k_{th} uncertain attribute values of R_j;
12     $U_j = U_j + \frac{sup - inf}{global\_range_k}$;

13   Sort hyper-rectangles array $[R_i]$ by their uncertainty $[U_i]$ ascendingly;
14   return $R[1...\lfloor \beta * n \rfloor]$;

4.3 Illustration of Regression Models

Given training data $(x_i, y_i), i = 1...n$, a regression model h with parameter w, we try to solve the following optimization problem:

$$w, b = \underset{w,b}{\arg\min} \sum_{i=1}^{n} \Theta(x_i, y_i, fr, h_{w,b}) \tag{6}$$

Linear Regression Model. The linear regression model is a simple and extensible regression model, which is helpful for understanding our problems. With the help of higher order features, nonlinear regression tasks can be achieved. We select 1-order features and 3-order features respectively to train the model and test on the same original test dataset. Given training data $(x_i, y_i), i = 1...n$, where $x_i \in R^d$ is a feature vector and $y \in R$ indicates the predicted value of x_i, 1-order (7) and 3-order (8) linear regression model can be defined as follows:

$$h_{w,b}^1(x) = wx_i + b \tag{7}$$

$$h_{w_{1,2,3},b}^3(x) = w_1 x_i + w_2 x_i^2 + w_3 x_i^3 + b \tag{8}$$

For a regression model h, we train it with anonymized data D_k and test it on original test dataset O. For each instance x_i with label y_i, we represent it with

fr and give it to the model for training or prediction. Given a loss function L, the cost Θ is as follows:

$$\Theta(x_i, y_i, fr, H) = \sum_{i=1}^{n} L(h(fr(x_i)), y_i) \tag{9}$$

5   Experiments

5.1   Datasets and Anonymity Algorithm

We conducted experiments on *Auto MPG* dataset[1] and *Air Quality* dataset[2] from UCI Repository [29]. The *Auto MPG* dataset contains 350 records, including the "model year of the car", "horsepower", "displacement", "weight", "number of cylinders", and "fuel consumption of 100 km". It has been used to explore the relationship between vehicle fuel consumption and other parameters. We randomly selected 300 records as training set and another 50 records as the test set. *Air Quality* dataset contains more than 9000 pieces of data. Each piece of data includes content information of air's composition. Code is available on Github[3].

Establishing VGH for each numerical attribute is unconvincing and significantly increase the workload of data publisher so we chose a reformed *Mondrian* algorithm [17]. It satisfies the requirement of k-anonymity without the need for manual VGH. Reformed Mondrian makes data publishers free from designing VGH for all attributes before releasing dataset. Reformed *Mondrian* makes the generalized values difficult to know in advance. We took this into consideration when representing category attributes.

5.2   Model Setup

Evaluation Method. In order to evaluate the prediction ability on original test dataset, we selected the well-known Mean Absolute Error (MAE) as the measure. Given predicting $\hat{y}_i$ and target y_i, $i = 1...n$, MAE is calculated as follows:

$$MAE = \frac{\sum_{i=1}^{n} |\hat{y}_i - y_i|}{n} \tag{10}$$

Linear Regression Model. We chose the linear regression model provided in *sklearn* [30] tool and used the default parameters. The 3-order linear regression models would be marked with the suffix 'H' following feature representation method, just like *"FR1-H"*.

5.3   Analysis of Results

Evaluation of Feature Representation. In this part, we compared the performance of linear regression models trained with anonymized data and original data. At the same time, we evaluated the actual effects of three feature representation methods (FR1, FR2, FR3).

---

[1] http://archive.ics.uci.edu/ml/datasets/Auto+MPG.
[2] http://archive.ics.uci.edu/ml/datasets/Air+Quality.
[3] https://github.com/build2last/UHRP.

(a) The degree of anonymity and feature representation. We studied the relationship between model performance and the parameter k of k-anonymity. For
the smaller *Auto MPG* dataset, we set the value range of k as [1(original), 2, 3, 4,
5]. For *Air Quality* dataset, we set it [1(original), 8, 32, 64, 128]. In experiment,
we used three attributes of the dataset as features and a linear regression model
for prediction. Pruning mechanism was not employed here.

Table 3. Linear regression model performance (MAE) using different feature representation methods with various privacy parameter k.

Data set	k	FR1	FR1-H	FR2	FR2-H	FR3	FR3-H
Auto MPG (size 300+)	1(origin)	6.54	103.31	4.22	3.57	3.82	3.25
	2	4.82	98.77	4.13	3.63	4.06	3.55
	3	7.31	121.32	4.14	3.71	3.99	3.71
	4	8.4	130.36	4.12	3.81	4.07	3.42
	5	4.13	85.65	4.08	3.91	4.4	5.48
AirQuality (size 9000+)	1(origin)	461.05	297.22	97.22	99.95	86.93	86.9
	8	362.81	383.17	94	95.31	79.09	85.13
	32	191.17	1044.96	92.73	94.72	79.64	117.03
	64	221.54	5220.48	101.04	121.03	118.2	184.03
	128	194.62	2740.58	130.17	141.9	146.41	150.76

Experimental results in Table 3 show that *FR1* is most sensitive to the change
of k. As the privacy protection parameter k growing, the performance of *FR1* is
nearly always the worst. For small dataset (*Auto MPG*), the performance of the
models trained by anonymized data is of the same magnitude with the original
model. For large dataset (*Air Quality*), the anonymized model performance is
close to the original model when k is less than 32. When k is greater than 32,
model train with anonymized data significantly deteriorate. That is where our
UHRP methods will play a role.

(b) Data dimension and feature representation. We have carried out experiments
on three different feature representation methods. Firstly, we conducted experiments on original training dataset without anonymization. We train models on
the original training dataset and verified them on the original test dataset. Then
we conducted experiments on the anonymized dataset: for the smaller *Auto MPG*
dataset, we set the privacy parameter k to 2; For the larger *Air Quality* dataset,
we set k as 16 and 40. Then we conducted experiments respectively. In each set
of experiments, we adjusted the number of attributes involved in feature vectors
and implemented experiments on the linear regression models in Sect. 4.3. The
experimental results are shown in Tables 4 and 5.

FR1 is still the worst. On small datasets, *FR3* performs better. *FR2* performs better on a larger dataset than others when anonymity is high. It can
be concluded that *FR1* is a relatively poor method to represent hyper-rectangle

Table 4. Performance of different feature representation methods on original training data

Data set	K	M	FR1	FR2	FR3	FR1-H	FR2-H	FR3-H
Auto MPG	Original	1	16.43	5.91	**3.89**	171.52	4.26	**3.53**
		2	12.77	5.66	**3.81**	128.28	4.48	**3.34**
		3	6.54	4.22	**3.82**	103.31	3.57	**3.25**
		4	6.53	4.04	**3.58**	101.58	3.63	**3.14**
		5	119.71	4.11	**3.59**	3283.54	3.45	3
Air Quality	Original	1	97.68	94.88	**94.54**	271.01	94.9	**94.16**
		2	447.91	97.54	**86.99**	327.61	99.05	**87.04**
		3	461.05	97.22	**86.93**	297.22	99.95	**86.9**
		4	693.34	100.06	**85.51**	433.78	99.77	**85.46**
		5	513.51	91.68	**85.01**	187.68	83.75	84.75

Table 5. Performance of different feature representation methods on anonymized training data

Data set	K	M	FR1	FR2	FR3	FR1-H	FR2-H	FR3-H
Auto MPG	2	1	16.78	5.92	**3.88**	173.04	4.33	**3.62**
		2	12.96	5.91	**3.96**	112.67	4.78	**3.56**
		3	4.82	4.13	**4.06**	98.77	3.63	**3.55**
		4	11.75	4.06	**3.66**	89.78	3.71	3.19
		5	7.76	4.02	**3.66**	96.7	**3.52**	10.17
Air Quality	16	1	91.01	91.28	**89.26**	452.33	**89.18**	95.79
		2	236.28	92.03	**80.87**	631.23	90.96	**87.33**
		3	239.3	92.95	**78.26**	943.79	**90.69**	90.97
		4	209.93	95.64	**79.53**	736.45	**91.73**	94.49
		5	171.89	89.86	**76.87**	549.57	**84.18**	100.17
Air Quality	40	1	86.6	90.51	**89.84**	623.38	**92.43**	95.71
		2	164.01	**91.38**	140.83	556.53	**95.75**	175.87
		3	162.88	**91.56**	117.32	169.08	**91.46**	175.59
		4	163.35	**93.65**	133.99	565.79	**93.76**	216.69
		5	155.65	88.83	123.8	3058.54	**89.75**	335.65

which only uses the geometric center. *FR2* method added hyper-rectangle corner data on the basis of *FR1* which significantly improve the model performance. Experiments of *FR3* show that the method of using the lower and upper bound to represent the interval value can achieve much better results than *FR1*. At the same time, the experimental results show that *FR2* and *FR3* are feasible methods to represent anonymized data.

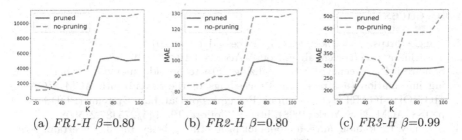

(a) *FR1-H* $\beta=0.80$ (b) *FR2-H* $\beta=0.80$ (c) *FR3-H* $\beta=0.99$

Fig. 5. Model's performance before and after pruning for *Air Quality* dataset with 5 features. The horizontal axis represents the **k** from 20 to 100. The vertical axis is the test results **MAE** on original test dataset.

Evaluation of UHRP Method. In this part, we compared model's performance changes before and after pruning.

It should be emphasized here that *remainder-ratio* (β) is still determined according to experience. β is not fixed for different representation methods. We choose *remainder-ratio* (β) from 0.5 to 1 (interval = 0.1) with the average best performance by fixing the anonymity degree ($k = 40$ and $k = 80$). For *FR1-H* and *FR2-H* we set $\beta = 0.8$. For *FR3-H* we set $\beta = 0.99$ with more precise parameter tuning between 0.9 to 1 with interval 0.01 because of its curve changes gently.

As shown in Fig. 5, our UHRP method can effectively reduce more than 20% MAE on average in most cases (except for *FR1-H* with k less than 30), especially when k over 70. It means that our UHRP method can effectively improve the performance of linear regression model on anonymized dataset when data uncertainty is large.

6 Conclusions

In this paper, we do researches on how to better utilize anonymized dataset, for regression model training and predicting original data. We propose several feature representation methods that are compatible with anonymized data and original data. Experiments show the regression model trained with anonymized data can be expected to do as well as the model trained on original dataset under certain conditions. When training set's anonymity degree is high, the pruning method UHRP improve linear regression model's performance effectively. At the same time, UHRP can reduce the number of training samples, thus improve the efficiency of model training.

For future work, methods to filter anonymized data for model training will be further studied. Although pruning method leads to performance improvement, the adjustment of *remainder-ratio* needs more research to improve UHRP's applicability and interpretability in different scenarios.

Acknowledgments. This work is supported by National Key R&D Program of China (No. 2017YFC0803700), NSFC grants (No. 61532021), Shanghai Knowledge Service Platform Project (No. ZF1213) and SHEITC.

References

1. Kaggle. https://www.kaggle.com/. Accessed 1 Dec 2018
2. Tianchi. https://tianchi.aliyun.com/. Accessed 1 Dec 2019
3. Samarati, P., Sweeney, L.: Generalizing data to provide anonymity when disclosing information (abstract). In: Proceedings of the Seventeenth ACM SIGACT-SIGMOD-SIGART Symposium on Principles of Database Systems (PODS 1998), p. 188. ACM, New York (1998). https://doi.org/10.1145/275487.275508
4. Sweeney, L.: k-anonymity: a model for protecting privacy. Int. J. Uncertain. Fuzziness Knowl.-Based Syst. **10**(05), 557–570 (2002)
5. Machanavajjhala, A., Gehrke, J., Kifer, D., Venkitasubramaniam, M.: L-diversity: privacy beyond k-anonymity. In: International Conference on Data Engineering (2006)
6. Li, N., Li, T., Venkatasubramanian, S.: t-closeness: privacy beyond k-anonymity and l-diversity. In: IEEE International Conference on Data Engineering (2007)
7. Google Privacy Terms. https://policies.google.com/technologies/anonymization. Accessed 14 Jan 2019
8. Gal, T., Chen, Z., Gangopadhyay, A.: A privacy protection model for patient data with multiple sensitive attributes. Int. J. Inf. Secur. Priv. **2**(3), 28–44 (2008)
9. Dwork, C.: Differential privacy: a survey of results. In: Agrawal, M., Du, D., Duan, Z., Li, A. (eds.) TAMC 2008. LNCS, vol. 4978, pp. 1–19. Springer, Heidelberg (2008). https://doi.org/10.1007/978-3-540-79228-4_1
10. LeFevre, K., DeWitt, D.J., Ramakrishnan, R.: Incognito: efficient full-domain k-anonymity. In: SIGMOD 2005, Baltimore, MD, USA, pp. 49–60 (2005)
11. Lefevre, K., DeWitt, D.J., Ramakrishnan, R.: Mondrian multidimensional k-anonymity. In: ICDE 2006, Atlanta, GA, USA, pp. 25–36 (2006)
12. Xiao, X., Tao, Y.: Anatomy: simple and effective privacy preservation. In: Proceedings of the 32nd International Conference on Very Large Data Bases, Seoul, Korea, 12–15 September 2006. VLDB Endowment (2006)
13. Standard for privacy of individually identifiable health information (HIPAA). Fed. Reg. **67**(157), 53181–53273 (2002)
14. Samarati, P.: Protecting respondents' identities in microdata release. IEEE Trans. Knowl. Data Eng. **13**(6), 1010–1027 (2001)
15. Sweeney, L.: k-anonymity: a model for protecting privacy. Int. J. Uncertain. Fuzziness Knowl.-Based Syst. **10**(5), 557–570 (2002)
16. Wong, W.K., Mamoulis, N., Cheung, D.W.L.: Non-homogeneous generalization in privacy preserving data publishing. In: Proceedings of the 2010 ACM SIGMOD International Conference on Management of Data, pp. 747–758. ACM, June 2010
17. Mondrian. https://github.com/qiyuangong/Mondrian. Accessed 1 Dec 2018
18. Buratović, I., Miličević, M., Žubrinić, K.: Effects of data anonymization on the data mining results. In: 2012 Proceedings of the 35th International Convention MIPRO, pp. 1619–1623. IEEE, May 2012
19. Prasser, F., Eicher, J., Bild, R., Spengler, H., Kuhn, K.A.: A tool for optimizing de-identified health data for use in statistical classification. In: 2017 IEEE 30th International Symposium on Computer-Based Medical Systems (CBMS), pp. 169–174. IEEE, June 2017
20. Lin, B.R., Kifer, D.: Information measures in statistical privacy and data processing applications. ACM Trans. Knowl. Discov. Data (TKDD) **9**(4), 28 (2015)
21. Inan, A., Kantarcioglu, M., Bertino, E.: Using anonymized data for classification. In: IEEE 25th International Conference on Data Engineering, ICDE 2009, pp. 429–440. IEEE, March 2009

22. Salzberg, S.: A nearest hyperrectangle learning method. Mach. Learn. **6**(3), 251–276 (1991)
23. Akbari, M.G., Hesamian, G.: Linear model with exact inputs and interval-valued fuzzy outputs. IEEE Trans. Fuzzy Syst. **26**(2), 518–530 (2018)
24. Akbari, M.G., Hesamian, G.: Signed-distance measures oriented to rank interval-valued fuzzy numbers. IEEE Trans. Fuzzy Syst. **26**(6), 3506–3513 (2018)
25. Huang, Y., Li, T., Luo, C., Fujita, H., Horng, S.J.: Dynamic fusion of multi-source interval-valued data by fuzzy granulation. IEEE Trans. Fuzzy Syst. **26**(6), 3403–3417 (2018)
26. Mancuhan, K., Clifton, C.: Statistical learning theory approach for data classification with l-diversity. In: Proceedings of the 2017 SIAM International Conference on Data Mining, pp. 651–659. Society for Industrial and Applied Mathematics, June 2017
27. Jaynes, E.T.: Information theory and statistical mechanics. Phys. Rev. **106**(4), 620 (1957)
28. Pearce, T., Zaki, M., Brintrup, A., Neely, A.: High-quality prediction intervals for deep learning: a distribution-free, ensembled approach. arXiv preprint arXiv:1802.07167 (2018)
29. Dua, D., Karra Taniskidou, E.: UCI Machine Learning Repository (2017). School of Information and Computer Science, University of California, Irvine, CA. http://archive.ics.uci.edu/ml
30. Pedregosa, F., Gramfort, A., Michel, V., Thirion, B., Grisel, O., Blondel, M., et al.: Scikit-learn: machine learning in python. J. Mach. Learn. Res. **12**(10), 2825–2830 (2013)

Meta-path Based MiRNA-Disease Association Prediction

Hao Lv[1], Jin Li[1(⊠)], Sai Zhang[1], Kun Yue[2], and Shaoyu Wei[1]

[1] School of Software, Yunnan University, Kunming, China
lijin@ynu.edu.cn
[2] School of Information, Yunnan University, Kunming, China

Abstract. Predicting the association of miRNA with disease is an important research topic of bioinformatics. In this paper, a novel meta-path based approach MPSMDA is proposed to predict the association of miRNA-disease. MPSMDA uses experimentally validated data to build a miRNA-disease heterogeneous information network (MDHIN). Thus, miRNA-disease association prediction is transformed into a link prediction problem on a MDHIN. Meta-path based similarity is used to measure the miRNA-disease associations. Since different meta-paths between a miRNA and a disease express different latent semantic association, MPSMDA make full use of all possible meta-paths to predict the associations of miRNAs with diseases. Extensive experiments are conducted on real datasets for performance comparison with existing approaches. Two case studies on lung neoplasms and breast neoplasms are also provided to demonstrate the effectiveness of MPSMDA.

Keywords: Heterogeneous Information Network · Meta-paths · MiRNA · Disease · Association prediction · Link prediction

1 Introduction

With the rapid development of molecular biology and biotechnology, mounting biological observations and studies have indicated that microRNAs (miRNAs) plays a very important role in many biological processes [1–5]. In addition, emerging evidences imply the strong links between miRNAs and diseases [6–8]. Therefore, predicting the association of potential miRNAs with disease is important for understanding the etiology and pathogenesis of the disease. Early studies on the association of miRNA with disease were based on biological experiments, but the experimental cycle was long and costly. Therefore, computational biology methods to predict the association of miRNA with disease have become an important research topic of bioinformatics.

MiRNA-disease association is in fact a link prediction problem on a miRNA-disease network [9]. At present, there are a few of methods for solving this problem. These methods are mainly summarized as: (1) **Neighbor information based prediction**. Xuan et al. [10] proposed HDMP which integrates the similarity between the disease terms and diseases to infer functional similarity between miRNAs. Luo et al. [11] proposed a prediction method based on transduction learning using local neighbors of different node types. (2) **Random walks based prediction.** Chen et al. [12].

© Springer Nature Switzerland AG 2019
G. Li et al. (Eds.): DASFAA 2019, LNCS 11448, pp. 34–48, 2019.
https://doi.org/10.1007/978-3-030-18590-9_3

presented the RWRMDA model to predict potential miRNA-disease association by adopting random walks on the miRNA functional similarity network. Liu et al. [13] and Liao [14] proposed a method for predicting miRNA-disease association by random walks in a network composed of multiple data sources. (3) **Machine learning based predictions**. A semi-supervised method (RLSMDA) proposed by Chen et al. [15] which uses a regularized least squares to construct a continuous classification function that represents the probability that each miRNA is associated with a given disease. Chen et al. [16] also proposed an association prediction method (RBMMMDA) based on restricted Boltzman model. (4) **Path-based prediction**. You [17] et al. proposed an efficient path-based prediction in 2017, which uses a depth-first search algorithm to integrate disease semantic similarity, miRNA functional similarity, and known human miRNAs-association and Gaussian kernel similarity to predict potential miRNA-disease associations. Zhang et al. [18] proposed a path-based approach based on the Katz method to predict associations. In addition to the main types of methods, Chen et al. proposed methods such as HGIMDA [19] and WBSMDA [20] to predict associations based on miRNA-disease heterogeneous networks.

There have been many research results for the prediction of miRNAs with disease association. However, existing methods still have following shortcomings. (1) The neighbor-based predictions only consider the k neighbor nodes of a certain node when making predictions, thus ignoring some other useful information in heterogeneous networks. (2) The existing random walks based predictions do not adequately consider different biological data sets. For example, the relationship between miRNA and disease, and the like. Thus, the association of new diseases not associated with any other known miRNAs cannot be predicted. (3) Machine learning-based predictions are generally limited by the representation of miRNA and disease features and how to deal with model designs with positive sample data. (4) The existing path-based methods have also certain limitations. First, in order to improve the efficiency of the method, PBMDA [17] set the length of the path to be 3, and the edge with a weight less than 0.5 in the network is deleted. Although it saves a lot of time, it does not make full use of the valuable path information in the constructed miRNA-disease network. In the Katz-based prediction proposed by Zhang et al., [18] the number of paths between miRNA nodes and disease nodes at different lengths is counted and then combined as the final predicted score. The higher the predicted score, the more likely the association between miRNA and disease. However, similar to PBMDA, Zhang's method also does not fully utilize the path information between nodes, and the prediction effect is not very satisfactory.

To overcome the mentioned shortcomings of the existing methods, this paper proposes a novel meta-path similarity based miRNA-disease associations prediction approach (MPSMDA). MPSMDA uses a variety of the latest biological datasets to construct a heterogeneous miRNA-disease network (HMDN) based on the verified miRNA-miRNA, disease-disease, and miRNA-disease relationships. Then, a novel meta-path based similarity proposed to measure the associations between a miRNA and a disease. Since different meta-paths between miRNAs and diseases express different latent semantic associations, MPSMDA makes full use of all possible meta-paths to predict the associations of miRNAs with diseases. Finally, extensive experiments are conducted on real datasets for performance comparison with existing approaches. Two case studies on lung neoplasms and breast neoplasms also provide to verify the effectiveness of our methods.

2   Construction of MiRNAs-Diseases Heterogeneous Network

In this section, we present how to construct a miRNA-disease heterogeneous network (HMDN). A HMDN consists of three relationships: miRNAs similarity, diseases similarity, and miRNA-disease associations. In the follows, we describe how to establish these relationships based on datasets.

(1) The experimentally verified miRNA-disease association datasets was downloaded from two popular databases, HMDD v2.0 [21] and miR2Disease [22]. A total of 6,313 associations between 577 miRNAs and 336 diseases are available after removing duplications. The adjacency matrix $\mathbf{Y}$ is constructed to describe the confirmed association of miRNA with disease. Namely, if miRNA m is recorded to be associated with disease d, the entry of $\mathbf{Y}(m, d)$ is equal to 1, otherwise 0.

(2) The experimentally valid miRNA similarity datasets have small scale and data sparse problems, According to the method proposed by Zhang et al. [18]. miRNA-gene data, verified miRNA-disease associations information, miRNA family data, miRNA cluster information are combined to establish the miRNA similarity network.

$$\begin{aligned} \text{RS}(m_1, m_2) = \alpha * \text{RSG}(m_1, m_2) + \beta * \text{RSC}(m_1, m_2) + \gamma \\ * \text{RSF}(m_1, m_2) + (1 - \alpha - \beta - \gamma) * \text{RSD}(m_1, m_2) \end{aligned} \tag{1}$$

We consider that miRNAs sharing the same gene are more likely to be similar, so the similarity between miRNAs can be calculated by association data between miRNA and target genes and the similarity value is recorded as RSG. MiRNAs can influence and regulate the occurrence and development of diseases. We believe that miRNAs affecting the same or similar diseases are more likely to have a potential association. So the miRNA similarity based on disease can be calculated as RSD. In order to make the calculation of similarity between miRNAs more accurate, we consider to propose the similarity between miRNAs from the perspective of miRNA itself. We selected family information of miRNA and cluster information of miRNA. miRNAs belonging to the same family and same cluster are more likely to be similar to miRNAs which belonging to different family and clusters. We can calculate the similarity score RSF and RSC based the family and cluster information of miRNAs. Then the final miRNA similarity is composed of four weighted parts. We finally obtain a miRNA similarity network which contains 71,053 associations among 577 miRNAs.

(3) The similarity between two diseases $\text{DS}(d_1, d_2)$ is established by integrating the semantic similarity (SS) and functional similarity (FS) of two diseases

$$\text{DS}(d_1, d_2) = \alpha * \text{FS}(d_1, d_2) + (1 - \alpha) * \text{SS}(d_1, d_2) \tag{2}$$

Wang et al. [23] used a the disease structure of a directed acyclic graph(DAG) to calculate the semantic similarity (SS) of diseases. In a DAG, first, we choose two diseases t and d, Suppose t is the ancestor node of d or t is the same as d, we define the semantic contribution of t to d:

$$C_d(t) = f(x) = \begin{cases} 1, t = d \\ \max\{0.5 \times C_d(t') | t' \in children\ of\ t\}, & t \neq d \end{cases} \quad (3)$$

According to $C_d(t)$ (the semantic contribution of diseases), we can calculates the semantic contribution of any disease to other diseases. By comparing disease $C_d(t)$ of two disease, we can calculate the semantic similarity score between each disease as follow

$$SS(d_1, d_2) = \frac{\sum_{t \in T(d_1) \cap T(d_2)} (C_{d_1}(t) + C_{d_2}(t))}{\sum_{t \in T(d_1)} C_{d_1}(t) + \sum_{t \in T(d_2)} C_{d_2}(t)} \quad (4)$$

Where $T(d)$ is the set of diseases, which include all disease nodes in DAG.

However, if we only use the semantic information of the disease to construct a semantically similar network, there will be data sparsity problems. Therefore, according to the method of Zhang et al., disease function information can be considered. The disease functional similarity is measured by disease-target interactions. In the DisGeNet database [24], which contains the probability value and log likelihood scores (LLS) of each interaction between two genes. We calculate the similarity of diseases, based on the similarity of genes, according to the theory that functionally similar genes have a greater probability of regulating similar diseases, The function similarity score between each disease as follow

$$FS(d_1, d_2) = \frac{\sum_{g \in \mathbb{S}(d_1)} LLS_N(g, \mathbb{S}(d_2)) + \sum_{g \in \mathbb{S}(d_2)} LLS_N(g, \mathbb{S}(d_1))}{|\mathbb{S}(d_1)| + |\mathbb{S}(d_2)|} \quad (5)$$

where $\mathbb{S}(d_1)$ and $\mathbb{S}(d_2)$ represent the gene sets related to disease d_1 and d_2 respectively. $|\mathbb{S}(d_1)|$ and $|\mathbb{S}(d_2)|$ are the numbers of genes in disease sets $\mathbb{S}(d_1)$ and $\mathbb{S}(d_2)$, respectively. LLS_N is a max-min normalization of the LLS value. $LLS_N(g, \mathbb{S}(d))$ is the maximum LLS_N value between g and genes from the set $\mathbb{S}(d)$.

Then, by integrating disease function similarity (FS) and disease semantic similarity (SS) of diseases, we obtain a disease similarity network which contains 112896 similar associations between 336 diseases (DSN).

Finally, we combine miRNA similarity network, disease similarity network and experimentally valid miRNA-disease interactions to get the whole miRNAs-diseases heterogeneous information network.

3 MiRNA-Disease Association Prediction

In this section, we first introduce a novel meta-path based similarity of miRNA-disease. Then, miRNA-disease association prediction is carried on based on this similarity measure.

3.1   Meta-path Similarity of MiRNA-Disease

For the convenience of description, we first introduce heterogeneous information network and meta-path.

A heterogeneous information network (HIN) is defined as a multi-typed directed network $G(V, E, \varphi, A, R)$, where V is a set of $|V|$ vertices and E is a set of $|E|$ edges between vertices. There is a type mapping function $\varphi : V \to A$ with A as the set of vertex types, i.e., each vertex $v \in V$ belongs to a particular vertex type in A. Analogously, there is also an edge type mapping function $\psi : E \to R$ with R as the set of edge types, i.e., each edge $e \in E$ belongs to a particular edge type in R.

The constructed miRNA-disease network is a typical HIN with two types of vertices: miRNA vertex and disease vertex. Accordingly, edges in this network can be defined between different types of vertices. For example, an association from a miRNA to a disease can be represented by edge$(m, d) \in E$, where $m, d \in V$, $\varphi(m) = $ miRNA, $\varphi(d) = $ disease, and $\psi(m, d) = $ miRNA $\to$ disease; symmetrically, the edge(m, d) represents that the disease d is associated with the miRNA m. An illustration of miRNA-disease HIN is shown in Fig. 1.

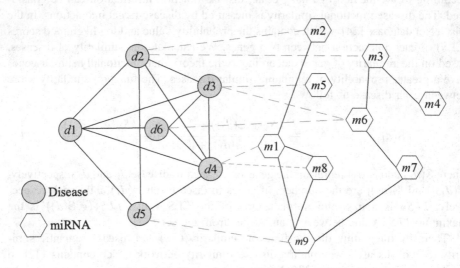

Fig. 1. An illustration of miRNA-disease HIN

In a MDHIN, the paths connecting two vertices present different semantics. In many cases, it is necessary to know the semantics of the paths, which may help understand the reasons why two vertices are closely related to each other. Because of the multiple types for vertices and edges, the paths from one vertex to another can also be associated with multiple types. We use the concept meta-path to represent the type/semantics of a path.

A L-length meta-path in a MDHIN is an ordered sequence of L edge types connecting two vertices with type A_1 and type A_{T+1}, denoted by $P_M = A_1 \to^{R_1}$ $A_2 \to^{R_2} \cdots \to^{R_T} A_{T+1}$, where $A_2 \in A$ and $R_i \in R$. An instantiation of P_M is a path in G, denoted by $p = (v_1, v_2, \ldots v_{T+1})$, satisfying $\varphi(v_i) = A_i, \forall i = 1, 2, \ldots T+1$ and $\psi(v_i, v_{i+1}) = R_i, \forall i = 1, 2, \ldots T$.

For example, in our miRNA-disease HIN, P_M = (miRNA-disease-miRNA-disease) represents a meta-path, where *miRNA, disease* $\in A$ and *disease* $\to$ miRNA, miRNA $\to$ disease $\in R$. An instantiation of this meta-path connects a miRNA and a disease by their directly connected miRNA and disease. The two simple examples of meta-path is shown in Fig. 2(a, left) and (b, right). Different meta-paths usually convey different semantics. For example, "M-M-D" path shows that if a miRNA is associated with a disease, then other miRNA similar to the miRNA will be potential associated with the disease; "M-D-D" path shows that if a disease associated with a miRNA, then other disease similar to the disease will be potential associated with the miRNA.

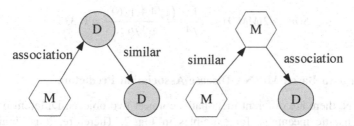

Fig. 2. The meta-path examples: (a) miRNA-disease-disease (M-D-D) (b) miRNA-miRNA-disease (M-M-D)

In order to measure the association of a miRNA with a disease in a MDHIN, a novel meta-path based similarity is proposed and defined as follows.

Definition 1. Let P_M be a meta-path in a MDHIN. The similarity based P_M between a miRNA m and a disease d is denoted by $S(m, d|P_M)$ and defined as

$$S(m, d|P_M) = \frac{\text{pathcount}(m, d|P_M) \times \left(\frac{1}{\circ(m)} + \frac{1}{\circ(d)} \right)}{\frac{1}{\circ(m)} \times \sum_j \text{pathcount}(m, j|P_M) + \frac{1}{\circ(d)} \times \sum_i \text{pathcount}(i, d|P_M)} \quad (6)$$

In Eq. (6), pathcount$(m, d|P_M)$ means the number of paths connecting a miRNA m and a disease d following the specified meta-path P_M. deg(m) and deg(d) are degrees of m and d. $\sum_j \text{pathcount}(m, j|P_M)$ is the total number of paths that satisfy the meta-path P_M with m as the source node. $\sum_i \text{pathcount}(i, d|P_M)$ is the total number of paths specified by the meta path P_M with d as the target node.

The idea of similarity measure defined in Eq. (6) is initiative. In real-world, two objects are said to be highly related to each other when the strength of their relationship is high and the number of relationships with other objects is less. i.e., two objects are said to be highly related to when the exclusivity and the strength of their relationship are high. To mathematically represent the idea, we take the sum of the inverse of vertex degree of the source and target objects and multiply it by all the number of the paths connecting the source and target object. Then for normalization, we divide it by sum of the average the number of the paths between source object and target object.

We use an example to demonstrate our definition. Considering a meta-path P_M = "miRNA-miRNA-disease (M-M-D)" in Fig. 1, given a source miRNA $m1$ and a target disease $d2$, we have pathcount$(m1, d2|P_M) = 1$, $°(m1) = 4$ and $°(d2) = 6$, $\sum_j$ pathcount$(m1, j|MMD) = 3$. i.e. there are a total 3 of M-M-D meta-paths with $m1$ as the source. We also have $\sum_i$ pathcount$(i, d2|MMD) = 3$. Thus the similarity scores of $m1$ and $d2$ with the meta-path M-M-D can be calculated as 0.333 by following equations.

$$S(m1, d2|MMD) = \frac{1 \times (1/4 + 1/6)}{1/4 \times 3 + 1/6 \times 3} = 0.333$$

3.2   Meta-path Based MiRNA-Disease Association Prediction

In a MDHIN, there are different meta-paths connect two objects. Different paths have different semantic meanings, for examples in Fig. 2. Therefore, it is significant to consider different paths in the procedure of similarity calculation. We then introduce a systematic approach to measure the similarity between objects in a MDHIN.

The MPSMDA uses the meta-path based similarity defined in Eq. 6 to measure the association between objects in MDHIN. The meta-path similarity scores of different paths are combined with a constant that dampens contributions from longer paths. Because of the proposed similarity is based on the meta-path-based relevance framework, it can capture effectively the subtle semantics of search meta-paths. Consequently, we combined the similarity score of different meta-paths with a non-negative coefficient α to adjust the contributions of the different meta-paths with distinct path length. The similarity score between a miRNA m and disease d is then defined as

$$score(m, d) = \sum_{L=1}^{k} \alpha^L \times \sum_{P_i \in \varphi_L} S(m, d|P_i)) \tag{7}$$

where L is the length of meta-path. φ_L is the set of all types of meta-paths from miRNA m to disease d with path length L. P_i is a meta-path instance of φ_L.

All possible meta-paths with path length L = 3 are shown in Fig. 3. There are four types of meta-paths from a miRNA node m to a disease node d, such as miRNA-diseasemiRNA-disease (M-D-M-D), miRNA-disease-disease-disease (M-D-D-D) and so on. Usually, a short path may contribute more than a long path. Different choices for α lead to diverse similarity values between miRNAs and disease. Given a disease d and a miRNA m (m isn't associated with d), when k = 4, the similarity score is calculated as follows

$$
\begin{aligned}
score(m,d) = {} & \alpha * (S(m,d|MD)) \\
& + \alpha^2 (S(m,d|MMD) + S(m,d|MDD)) \\
& + \alpha^3 (S(m,d|MDDD) + S(m,d|MMDD) \\
& + S(m,d|MMMD) + S(m,d|MDMD)) \\
& + \alpha^4 (S(m,d|MMMDD) \\
& + S(m,d|MMMMD) + \ldots)
\end{aligned}
$$

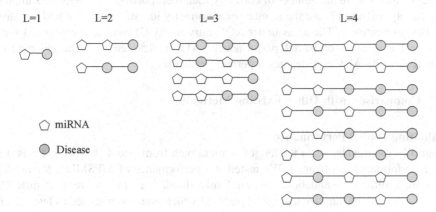

Fig. 3. The different paths with distinct lengths

After calculating all possible miRNA-disease pairs in a MDHIN, each pair of miRNA-disease could be obtained a final score representing the association confidence between this miRNA and disease, i.e., the higher score they obtain, the more closely related they should be.

4   Experiments and Results

4.1   Evaluation of Prediction Performance

We evaluate the result of our method to predict potential disease-related miRNAs by 5-fold-cross validation. For a given disease d, the associated relationships are randomly divided into five subsets, four of which are used as known information to predict candidate miRNAs; the last subset is used for testing. All the unknown miRNAs are contained in the disease-related miRNA candidate pool. The similarity scores of the disease for each miRNA are computed using MPSMDA. We rank the candidate miRNAs according to their relevance scores. The higher the ranking of the candidate miRNAs, the better the prediction performance will be. For a certain threshold, a miRNA with a relevance score higher than the threshold indicates that our method predicts that the miRNA is associated with a disease d. Otherwise, the miRNA is identified to be unconnected with a disease, d. The true positive rate (TPR) and the false positive rate (FPR) are calculated by varying the threshold to obtain the receiver operating characteristic (ROC) curves.

$$TPR = \frac{TP}{TP + FN} \tag{8}$$

$$FPR = \frac{FP}{FP + TN} \tag{9}$$

where TP and TN are the number of correctly identified positive and negative samples, respectively. FP and FN are the number of incorrectly identified positive and negative samples, respectively. The areas under ROC curves (AUC) were also calculated for a numerical evaluation of model performance. AUC = 0.5 denotes a purely random prediction while AUC = 1 denotes a perfect prediction.

4.2   Comparison with Other Existing Methods

Evaluating Global Performance

In our all experiments, when the length of meta path from 1 to 4 ($k = 4$) and α is 0.1, better performance is obtained. We tested the performance of MPSMDA for all diseases in 5-fold-cross-validation. We randomly divided all known verified miRNA-disease associations into five uncrossed parts, of which one was regarded as test set and the other four were used for training set in turns.

We selected the following four representative method with high performance to compare with the proposed study: WBSMDA [20], RLSMDA [15], HDMP [10], PBMDA [17]. RLSMDA is representative of machine learning methods. WBSMDA is a method to uncover the potential miRNA-disease associations by integrating several heterogeneous biological datasets. HDMP is a neighborhood-based method, it considers similarities of k nearest neighbors. PBMDA is also a novel path-based method. As a result, the reliable AUC = **0.929** were obtained by MPSMDA, other four methods, PBMDA, RLSDAM, HDMP, WBSMDA achieved AUCs of 0.917, 0.854, 0.835 and 0.814. Respectively, which was observed that MPSMDA obtained the best performance based on 5-fold-CV. The ROC curves of MPSMDA and other methods are shown in Fig. 4.

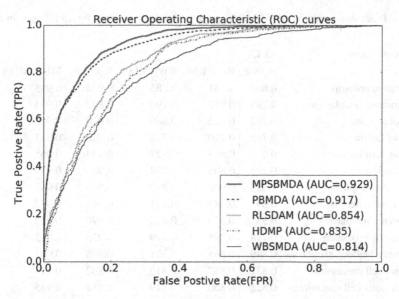

Fig. 4. The comparison results between MPSBMDA and other four computational models in terms of global 5-fold-CV

Evaluating Specific Diseases Performance

To further evaluate the ability of our method, we selected HDMP [10], RWRMDA [12], RLSMDA [15], Katz-ML [18] with our method MPSMDA to compare their performance in the same fifteen specific diseases by 5-fold-cross-validation. RWRMDA is one of classical prediction methods based random walk; Katz-ML is also a path-based method proposed by Zhang et al. Because our method MPSMDA is also belong to a path-based type method. We can compare prediction performance with different other path-based methods to prove the advantage of our method. For each in these fifteen diseases, the number of associations is above 75 and maximum and minimum are 250 (in hepatocellular carcinoma) and 76 (in acute myeloid leukemia) respectively. The number of known associations of those 15 diseases is 243, 171, 118, 137, 157, 165, 129, 117, 134, 205, 106, 133, 93, 76, 250 in order.

We obtained superior performance with average AUC values of for MPSMDA, respectively. The predicted results of our methods and the other methods (HDMP, RLSMDA, RWRMDA, Katz-ML) are shown in Table 1. The average AUC values of HDMP, RLSMDA, RWRMDA, Katz-ML are 0.815, 0.825, 0.799 and 0.919 respectively. As indicated in bold font, the average AUC values of MPSMDA, is better than the average value of these methods by 11.9%, 10.9%, 13.5%, 13.5% and 1.5%.

Table 1. Prediction results of MPSBMDA and the other four methods for 15 diseases in terms of AUC

Disease name	AUC				
	HDMP	RLSMDA	RWRMDA	Katz-ML	MPSBMDA
Breast neoplasms	0.803	0.831	0.785	0.944	0.983
Colorectal neoplasms	0.796	0.832	0.793	0.892	0.934
Glioblastoma	0.702	0.712	0.680	0.910	0.938
Heart failure	0.765	0.740	0.722	0.920	0.937
Lung neoplasms	0.835	0.853	0.827	0.949	0.965
Melanoma	0.795	0.805	0.784	0.906	0.959
Ovarian neoplasms	0.883	0.905	0.882	0.890	0.944
Pancreatic neoplasms	0.893	0.885	0.871	0.945	0.943
Prostatic neoplasms	0.855	0.836	0.823	0.890	0.914
Stomach neoplasms	0.778	0.795	0.779	0.925	0.912
Urinary bladder neoplasms	0.849	0.848	0.821	0.898	0.903
Renal cell carcinoma	0.832	0.842	0.815	0.882	0.875
Squamous cell carcinoma	0.822	0.850	0.819	0.912	0.935
Acute myeloid leukemia	0.854	0.852	0.839	0.925	0.929
Hepatocellular carcinoma	0.762	0.795	0.749	0.944	0.945
The Average AUC	0.815	0.825	0.799	0.919	**0.934**

4.3   Evaluating Performance on a New Disease Prediction

For a miRNA-disease association prediction method, it is important to be able to predict unknown diseases and their association relationship. We evaluated the performance of MPSMDA to predict new diseases. We deleted all known relations of the concerned disease with miRNAs. Thus, we treat this disease as a new disease, which has no information about related miRNAs. Then, we performed predictions for each new disease. Then we validated the prediction result with dbDEMC [25], miRCancer [26] and miRdSNP [27]. Then we use the result on Colorectal neoplasms for demonstration as shown in following Table 2. As a result, 48 out of top-50 predicted miRNAs have been verified to be associated with colorectal neoplasms by three database. Based on prediction situation, we can conclude that MPSMDA can achieve the reliable prediction performance for a new disease. Most importantly, it also demonstrates that our model is indeed applicable to predict association relationship of a new disease.

Table 2. The top 50 Colorectal neoplasms - related candidates.

Rank	miRNAs	Evidence	Rank	miRNAs	Evidence
1	has-mir-425	dbDEMC, miRCancer	26	has-mir-452	dbDEMC, miRCancer
2	has-mir-519e	dbDEMC	27	has-mir-513a	dbDEMC
3	has-mir-296	dbDEMC, miRCancer	28	has-mir-506	miRdSNP
4	has-mir-520b	dbDEMC, miRCancer	29	has-mir-488	dbDEMC, miRdSNP
5	has-mir-493	dbDEMC	30	has-mir-520c	dbDEMC
6	has-mir-365b	dbDEMC	31	has-mir-361	dbDEMC, miRdSNP
7	has-mir-1290	dbDEMC	32	has-mir-1249	dbDEMC
8	has-mir-450b	miRCance	33	has-mir-204	dbDEMC, miRCancer
9	has-mir-450a	dbDEMC	34	has-mir-302d	dbDEMC
10	has-mir-320e	dbDEMC, miRCancer	35	has-mir-502	dbDEMC
11	has-mir-508	dbDEMC	36	has-mir-153	dbDEMC
12	has-mir-526b	dbDEMC	37	has-mir-519d	dbDEMC
13	has-mir-661	dbDEMC	38	has-mir-30d	dbDEMC
14	has-mir-650	dbDEMC	39	has-mir-516a	dbDEMC
15	has-mir-191	dbDEMC, miRCancer	40	has-mir-147a	**unconfirmed**
16	has-mir-519c	dbDEMC	41	has-mir-320d	dbDEMC, miRdSNP
17	has-mir-517a	dbDEMC, miRCancer	42	has-mir-302a	dbDEMC
18	has-mir-744	dbDEMC	43	has-mir-515	dbDEMC
19	has-mir-320b	dbDEMC, miRCancer	44	has-mir-365a	dbDEMC
20	has-mir-202	dbDEMC, miRCancer	45	has-mir-383	dbDEMC
21	has-mir-302c	dbDEMC	46	has-mir-320c	dbDEMC, miRdSNP
22	has-mir-151b	**unconfirmed**	47	has-mir-32	dbDEMC, miRdSNP
23	has-mir-130b	dbDEMC, miRCancer	48	has-mir-30c	dbDEMC
24	has-mir-374a	dbDEMC, miRCancer	49	has-mir-516b	dbDEMC
25	has-mir-302b	dbDEMC	50	has-mir-98	dbDEMC, miRdSNP

4.4   Case Studies: Lung Neoplasm and Breast Neoplasm

Case studies of two diseases, lung neoplasms and breast neoplasms, are analyzed for further evaluation of the ability of our method to predict potential miRNA-disease associations. The top 30 predictive results are then confirmed using the following three databases: dbDEMC [25], miRCancer [26] and miRdSNP [27]. The results are shown in Tables 3 and 4. For lung neoplasms, associations with the three miRNA candidates, hsa-mir-199a-2, and hsa-mir-92a-2 do not appear in the above three databases. 29 out of top-30 predicted miRNAs have been verified to be associated with Breast neoplasms in the three databases. We conclude that these case studies demonstrate that our method is powerful and effective for predicting miRNA-disease associations with a high level of reliability.

Table 3. The top 30 Lung neoplasms - related candidates.

Rank	miRNAs	Evidence	Rank	miRNAs	Evidence
1	has-mir-1236	dbDEMC	16	has-mir-195	dbDEMC, miRCancer
2	has-mir-218	dbDEMC, miRCancer	17	has-mir-92a-2	**unconfirmed**
3	has-mir-451a	dbDEMC	18	has-mir-141	dbDEMC, miRCancer
4	has-mir-125b-2	dbDEMC, miRCancer	19	has-mir-302c	dbDEMC
5	has-mir-199a-2	**unconfirmed**	20	has-mir-130a	dbDEMC
6	has-mir-429	dbDEMC, miRCancer	21	has-mir-373	dbDEMC
7	has-mir-16	miRCancer	22	has-mir-15a	dbDEMC, miRCancer
8	has-mir-16-2	miRCance	23	has-mir-708	dbDEMC
9	has-mir-24	dbDEMC, miRCancer	24	has-mir-99a	dbDEMC
10	has-mir-15b	dbDEMC, miRCancer	25	has-mir-367	dbDEMC
11	has-mir-106b	dbDEMC	26	has-mir-128-1	dbDEMC
12	has-mir-151a	dbDEMC	27	has-mir-625	dbDEMC
13	has-mir-193b	dbDEMC	28	has-mir-296	dbDEMC
14	has-mir-149	dbDEMC, miRdSNP	29	has-mir-378a	dbDEMC
15	has-mir-302b	dbDEMC, miRCancer	30	has-mir-367	dbDEMC

Table 4. The top 30 Breast neoplasms - related candidates.

Rank	miRNAs	Evidence	Rank	miRNAs	Evidence
1	has-mir-1293	dbDEMC	16	has-mir-130b	dbDEMC, miRdSNP
2	has-mir-185	dbDEMC, miRCancer	17	has-mir-19b-2	dbDEMC
3	has-mir-330	dbDEMC, miRCancer	18	has-mir-138-2	**unconfirmed**
4	has-mir-186	dbDEMC, miRdSNP	19	has-mir-138	dbDEMC, miRCancer
5	has-mir-95	dbDEMC	20	has-mir-192	dbDEMC, miRdSNP
6	has-mir-219-2	dbDEMC	21	has-mir-370	dbDEMC, miRCancer
7	has-mir-449b	miRCance, miRdSNPr	22	has-mir-92b	dbDEMC, miRdSNP r
8	has-mir-99a	dbDEMC, miRCancer	23	has-mir-542	dbDEMC
9	has-mir-130a	dbDEMC, miRCancer	24	has-mir-574	miRCancer
10	has-mir-32	dbDEMC, miRCancer	25	has-mir-212	dbDEMC, miRCancer
11	has-mir-372	dbDEMC, miRdSNP	26	has-mir-517a	dbDEMC
12	has-mir-449a	dbDEMC, miRCancer	27	has-mir-15b	dbDEMC
13	has-mir-153	dbDEMC, miRCancer	28	has-mir-378a	dbDEMC
14	has-mir-142	miRCancer	29	has-mir-494	dbDEMC
15	has-mir-106a	dbDEMC, miRdSNP	30	has-mir-503	dbDEMC

5   Conclusion

With the rapid development of biotechnology, many studies have shown that miRNAs play a critical role in multiple biological processes as well as the developments and progressions of various human diseases. We proposed a novel approach based on a meta-path similarity to predict potential disease-related miRNAs. We constructed miRNA-disease heterogeneous network by three kinds of experimental validated datasets, miRNA-miRNA similarity network, disease-disease similarity network, miRNA-disease association network. The comprehensive data are beneficial in enhancing performance. We proposed MPSMDA by considering network topological information and meta-paths semantic information. The results of experiments by comparing with other methods confirms MPSMDA's superior performance. MPSMDA is also successfully applied to diseases without any known miRNAs. The case studies on two diseases further show that MPSMDA has a strong ability to discover candidate disease miRNAs.

There are still some potentials to further improve the performance of MPSMDA. First, with the update of biological datasets, more and more miRNA-disease relationships have been verified, and the latest biological data can influence the performance of MPSMDA. Second, the constant α we use to combine the similarity score of different paths is fixed, and it limit the improvement of our method. In the future, we can learn the parameters via a machine learning approach to further improve prediction performance.

Acknowledgement. The authors acknowledge the financial support from the following foundations: National Natural Science Foundation of China (U1802271, 61562091), Natural Science Foundation of Yunnan Province (2016FB110), Program for Excellent Young Talents of Yunnan University (WX173602), and Natural Science Foundation of Yunnan University (2017YDJQ06).

References

1. Meister, G., Tuschl, T.: Mechanisms of gene silencing by double-stranded RNA. Nature **43** (7006), 343–349 (2004)
2. Bartel, D.P.: MicroRNAs: genomics, biogenesis, mechanism, and function. Cell **116**, 281–297 (2004)
3. Ambros, V.: microRNAs: tiny regulators with great potential. Cell **107**(7), 823–826 (2001)
4. Ambros, V.: The functions of animal microRNAs. Nature **431**, 350 (2004)
5. Xu, Y., Guo, M., Liu, X., Wang, C., Liu, Y., Liu, G.: Identify bilayer modules via pseudo-3D clustering: applications to miRNA-gene bilayer networks. Nucleic Acids Res. **44**, e152 (2016)
6. Calin, G.A., Croce, C.M.: MicroRNA-cancer connection: the beginning of a new tale. Cancer Res. **66**(15), 7390–7394 (2006)
7. Meola, N., Gennarino, V.A., Banfi, S.: MicroRNAs and genetic diseases. Pathogenetics **2**(1), 7 (2009)
8. Sayed, D., Abdellatif, M.: MicroRNAs in development and disease. Physiol. Rev. **91**(3), 827–887 (2011)
9. Lu, L., Zhou, T.: Link prediction in complex networks: a survey. Phys. A **390**(6), 1150–1170 (2010)
10. Xuan, P., Han, K., Guo, M.: Prediction of microRNAs associated with human diseases based on weighted k most similar neighbors. PLoS ONE **8**(8), e70204 (2013)
11. Luo, J., Ding, P., Liang, C.: Collective prediction of disease-associated miRNAs based on transduction learning. IEEE/ACM Trans. Comput. Biol. Bioinform. **PP**(99), 1 (2016)
12. Chen, X., Liu, M., Yan, G.: RWRMDA: predicting novel human microRNA–disease associations. Mol. BioSyst. **8**(10), 2792–2798 (2012)

13. Liu, Y., Zeng, X., He, Z., Zou, Q.: Inferring microRNA-disease associations by random walk on a heterogeneous network with multiple data sources. IEEE/ACM Trans. Comput. Boil. Bioinform. **14**, 905–915 (2017)
14. Chen, M., Liao, B., Li, Z.: Global similarity method based on a two-tier random walk for the prediction of microRNA–disease association. Sci. Rep. **8**(1), 6481 (2018)
15. Chen, X., Yan, G.: Semi-supervised learning for potential human microRNA-disease associations inference. Sci. Rep. **4**, 5501 (2014)
16. Chen, X., Yan, C.C., Zhang, X., Li, Z., Deng, L.: RBMMMDA: predicting multiple types of disease microRNA associations. Sci Rep. **5**(1), 13877 (2015)
17. You, Z.H., Huang, Z.A., Zhu, Z.: PBMDA: a novel and effective path-based computational model for miRNA-disease association prediction. PLoS Comput. Boil. **13**(3), e1005455 (2017)
18. Zhang, X., Zou, Q., Rodriguez-Paton, A.: Meta-path methods for prioritizing candidate disease miRNAs. IEEE/ACM Trans. Comput. Boil. Bioinform. **PP**(99), 1 (2017)
19. Chen, X., Yan, C.C., Zhang, X.: HGIMDA: heterogeneous graph inference for miRNA-disease association prediction. Oncotarget **7**(40), 65–69 (2016)
20. Chen, X., Yan, C.C., Zhang, X.: WBSMDA: within and between score for MiRNA–disease association prediction. Sci. Rep. **6**(1), 21106 (2016)
21. Li, Y., Qiu, C., Tu, J., et al.: HMDD v2.0: a database for experimentally supported human microRNA and disease associations. Nucleic Acids Res. **42**(Database issue) (2013)
22. Jiang, Q., Wang, Y., Hao, Y., et al.: miR2Disease: a manually curated database for microRNA deregulation in human disease. Nucleic Acids Res. **37**(Database), D98–D104 (2009)
23. Wang, D., Wang, J., Lu, M., et al.: Inferring the human microRNA functional similarity and functional network based on microRNA-associated diseases. Bioinformatics **26**(13), 1644–1650 (2010)
24. Bauer-Mehren, A., Rautschka, M., Sanz, F., et al.: DisGeNET: a cytoscape plugin to visualize, integrate, search and analyze gene-disease networks. Bioinformatics **26**(22), 2924–2926 (2010)
25. Yang, Z., Ren, F., Liu, C., et al.: dbDEMC: a database of differentially expressed miRNAs in human cancers. BMC Genomics, **11**(4 Suppl.) (2010)
26. Xie, B., Ding, Q., Han, H., et al.: miRCancer: a microRNA-cancer association database constructed by text mining on literature. Bioinformatics **29**(5), 638–644 (2013)
27. Bruno, A.E., Li, L., Kalabus, J.L., et al.: miRdSNP: a database of disease-associated SNPs and microRNA target sites on 3'UTRs of human genes. BMC Genomics **13**(1), 44 (2012)

Medical Question Retrieval Based on Siamese Neural Network and Transfer Learning Method

Kun Wang[1], Bite Yang[2], Guohai Xu[1], and Xiaofeng He[1(✉)]

[1] School of Computer Science and Software Engineering,
East China Normal University, Shanghai, China
51164500119@stu.ecnu.edu.cn, guohai.explorer@gmail.com,
xfhe@sei.ecnu.edu.cn
[2] DXY, Hangzhou, Zhejiang, China
yangbt@dxy.cn
http://www.dxy.com

Abstract. The online medical community websites have attracted an increase number of users in China. Patients post their questions on these sites and wait for professional answers from registered doctors. Most of these websites provide medical QA information related to the newly posted question by retrieval system. Previous researches regard such problem as question matching task: given a pair of questions, the supervised models learn question representation and predict it similar or not. In addition, there does not exist a finely annotated question pairs dataset in Chinese medical domain. In this paper, we declare two generation approaches to build large similar question datasets in Chinese health care domain. We propose a novel deep learning based architecture Siamese Text Matching Transformer model (STMT) to predict the similarity of two medical questions. It utilizes modified Transformer as encoder to learn question representation and interaction without extra manual lexical and syntactic resource. We design a data-driven transfer strategy to pre-train encoders and fine-tune models on different datasets. The experimental results show that the proposed model is capable of question matching task on both classification and ranking metrics.

Keywords: Health care · Question matching · Transfer learning

1 Introduction

With the great improvement of public health consciousness, it is hard for traditional offline medical services to meet the rapidly increasing demands. To satisfy public health care demands, some medical communities, such as xywy.com and www.dxy.com, offer abundant medical knowledge, question answering and other online services. As shown that the number of users in online medical communities has increased to 192 million at the end of 2017[1].

---

[1] https://www.qianzhan.com/analyst/detail/220/181210-db903bba.html.

© Springer Nature Switzerland AG 2019
G. Li et al. (Eds.): DASFAA 2019, LNCS 11448, pp. 49–64, 2019.
https://doi.org/10.1007/978-3-030-18590-9_4

Usually, users can not get a timely answer from online doctors, and instead, they have to wait hours even days for response from online doctors. Actually, the medical communities have a large number of solved medical QA pairs in database. Most of these websites exploit retrieval systems to search related questions for new posted queries, and return topK QA results by computing the word correlation scores. The major challenge of this kind of retrieval system is lexical gap between different expressions of patients. For instance, for the query "Do hypertension patients require long-term medication?", similar question would be "When can hypertension patients stop taking medicine?", however, they have very low similarity score because they have few overlap words. On the other hand, the dissimilar question "Do hyperlipidemia patients require long-term medication?" has a high word overlap with the query. Patients may have different ways to express the same medical contents, which is observed from the collected question corpus. For example, users with medical background may take the professional word "hypoimmunity (免疫力低下)" as a query, while most patients would use "poor immunity (免疫力差)".

In previous works, researchers have proposed many methods to relieve semantic gap of sentence similarity problem, like translation models [9,25], topic model [10,26] and supervised neural networks [21,27]. The above methods have already been applied in general domain, while there are still several challenges when import these approaches into public health domain. Firstly, there is no open source collection of similar question pairs for model training in Chinese health care domain. The sample generation method of question pairs has great influence on the generalization performance of models. Secondly, linguistic resources for Chinese medical NLP tasks are scarce, such as medical encyclopedia and taxonomy, far from complete. Thirdly, patients have different expression ways, which is difficult to be learned only based on a small volume of training data. In addition, typical Siamese neural networks have two encoders with shared weights to process questions separately, without considering the interaction and alignment of information in both questions.

In this paper, we design a Siamese Text Matching Transformer (STMT) neural network to learn question representation and interaction in Chinese health care domain, which fully utilizes the context information and interactive information of question pairs. Specially, the key contributions of this paper are as follow:

- We propose a modified Transformer neural network namely STMT to incorporate context information and interactive information of question pairs to alleviate the problem of ignoring interaction information of typical Siamese neural network. The results on a large Chinese question pairs dataset demonstrate the effectiveness of the proposed model.
- For the lack of linguistic resource in Chinese health care domain, we build a large QA corpus and a huge health care terminology crawled from several well-known Chinese medical websites. We take two different approaches for training dataset annotation. The semantic relatedness between words in medical questions is captured by the pre-trained word embedding.

– A multi-class classification task is applied as a transfer strategy to pre-train encoders of Siamese neural networks, which helps the encoders to learn more general representation of similar questions and distinguish different concepts in medical semantic spaces. The experimental results show the advantages of transfer strategy than random initialization.

The rest of this paper is organized as follows: Sect. 2 discuss related research. Section 3 declares the details of proposed method. Section 4 describes the dataset and experiments. We analyze the results in Sect. 5 and conclude our work in Sect. 6.

2   Related Work

The task of question matching (QM) in community question answering (CQA) systems, has attracted increasing attention in recent years. This task is closely related to paraphrase identification (PI) problem [11]. The goal of PI task is to determine whether or not two sentences have the same meaning, but sentences that are non-paraphrase can still be semantic similar. Nevertheless, the methods share commonness. In addition, methods in QA and natural language inference (NLI) can also be lead into dealing with this problem.

At the early stage, retrieval based methods are widely used to find similar questions [14,16]. However, such approaches only consider lexical keywords, the detailed semantic information in the questions can not be captured. Another unsupervised methods is to train a topic model with unlabeled text corpus, and medical questions can be represented as vectors in latent topic space, and then, the similarity of questions is measured by the similarity of these mapped questions' vectors [10,26]. Some researchers regard this problem as a translation task [5,9,25], which incorporates lexical similarity and semantic similarity into a unified structure. The traditional translation models focus on translation probability of words or phrases among questions. The similarity of questions is the probability of translating one question to another [4]. [3] proposes an end-to-end neural network for question similarity learning, which utilizes RNN as encoder and decoder. The feature engineering based studies utilizes word overlap [8], linguistic features [20] and other hand-crafted statistic information to measure the lexical and phrase level similarity.

More recently, deep learning based methods are widely used to solve this problem [21,27]. Convolutional neural networks (CNN) are applied to extract multiple granularity features for similarity comparison [15,24]. [1,2] use two LSTM to compose two sentences and calculate the similarities between the encoded sentence vectors. These neural networks are called Siamese neural network [7], which process two sentences in parallel. In medical QA domain, [18] proposes an interactive attention based LSTM model to evaluate the similarity of question pairs, which also takes LSTM as question encoder and only focuses on unidirectional attention interactivity. [4] combines translation model and Siamese CNN model to learning question-question similarity and question-answer relations, which is better than traditional retrieval algorithms, like BM25

[16]. BiMPM is design to learning sentence matching from multiple perspectives [23]. [6] propose ESIM model to solve NLI task, which utilize stacked LSTMs and attention mechanism to capture interactive similarity information of sentence pairs. The above two models are the state-of-art method and achieve good performance on different datasets [11,23].

3   Methodology

In this section, we first define the question matching task in Chinese health care domain. Then, we introduce sentence encoders for learning question representations, which are the core component of Siamese neural networks. Next, we describe the similarity measurements and architecture of our proposed model. Further, we explain how we address the problem of dataset generation in Chinese medical domain. Finally, we discuss about a transfer strategy to pre-train encoders to improve the performance of question matching models.

3.1   Task Definition

We first define the task formally and declare some notations that we used in this paper. Given an unsolved medical question $Q = [w_1, w_2, \ldots, w_n]$ containing n words. We embed each word of question as distributed word vector, then $Q = [q_1, q_2, \ldots, q_n]$, where $q_i \in R^{d_e}$ is the word embedding of the i-th word in this question. Define a set of relevant candidate questions $C = \{C_1, C_2, \ldots, C_m\}$, retrieved from large solved medical question corpus. We need to determine whether or not each candidate C_i is similar to Q and rank them by similarity scores to query Q. We describe the workflow of finding similar medical question in Fig. 1.

3.2   Encoder

In previous research, Siamese RNN structure consists of two share-weights encoders, such as LSTM or GRU. Each encoder process one question in the given question pair [1,2]. We first introduce bidirectional LSTM encoder that we used in baseline and then present the modified Transformer encoder designed in this section.

BiLSTM. We take bidirectional LSTM encoder to learn sentence level representation of medical questions. The unidirectional LSTM model process word sequence from left to right, when LSTM cell goes through the whole question, the information of previous words can be transmitted and accumulated into its memory cell, and the output of last hidden state is used as vector representation of question. The bidirectional LSTM consists of a forward LSTM and a backward LSTM, at each time step t, forward LSTM computes a representation $\overrightarrow{h_t}$ with the left context of word x_t, and the backward LSTM computes a representation

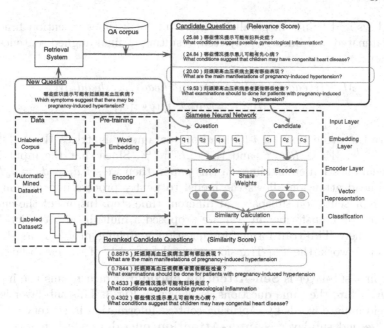

Fig. 1. The workflow of similar medical question retrieval.

$\overleftarrow{h_t}$ of the same sequence in reverse order. Then the representation of word x_t is obtained by concatenating its left and right context representations, namely $h_t = [\overrightarrow{h_t}, \overleftarrow{h_t}]$. Thus, the output of BiLSTM would capture the abundant context information of medical questions at sentence level. The concatenated final hidden state $h_n = [\overrightarrow{h_n}, \overleftarrow{h_n}]$ of BiLSTM is used as the semantic vector representation of questions, thus, we take v_q to represent the question vector, then $v_q = h_n$.

TM-Transformer. Transformer is a neural network architecture proposed by Google for machine translation [19], which adopts multi-head attention to encode sentence instead of RNN models. The basic attention mechanism in Transformer model is scaled dot-product attention, which is described formally as follow,

$$Attention(Q, K, V) = softmax(\frac{QK^T}{\sqrt{d_k}})V \qquad (1)$$

Where $Q \in R^{n \times d_k}$, $K \in R^{m \times d_k}$ and $V \in R^{m \times d_v}$ are sentence embedding matrices. In translation task of [19], the K and V are the same representations of a sentence and Q is the other one of an aligned bilingual sentence pair. The output of such attention is called aligned embeddings.

To enhance the efficiency and effectiveness, instead of taking d_e-dimensional Q, K, V to perform a single attention function, the multi-head attention linearly projects the Q, K and V matrices h times to d_k, d_k and d_v dimensions, respectively. The scaled dot-product attention is performed on each of these projected queries, keys and values in parallel. The results of h times attention functions are concatenated and linearly projected, resulting in final aligned sentence representations.

$$MultiHead(Q, K, V) = Concat(head_1, \ldots, head_h)W_o \tag{2}$$

$$head_i = Attention(QW_i^q, KW_i^k, VW_i^v) \tag{3}$$

Where the projections of i-th head are parameter matrices $W_q \in R^{d_e \times d_k}$, $W_k \in R^{d_e \times d_k}$, $W_v \in R^{d_e \times d_k}$ and $W_o \in R^{hd_v \times d_e}$. Since that the input is d_e-dimensional question embeddings in this paper, then we set $d_k = d_v = d_e/h$.

Multi-head attention allows the model to jointly process information from different representation subspaces at different time step. Inspire of the encoder and decoder of Transformer in [19], we proposed a multi-head attention based encoder namely TM-Transformer for question matching problem, as show in Fig. 2. In this work, the TM-Transformer model consists of three sub-layers.

(1) The first sub-layer is **Self-Attention encoder**, which means the input of Q, K, V are the same questions, as shown in Fig. 3. This sub-layer is used to reformulate words by capturing context information in sentences.
(2) The second sub-layer is **Inter-Attention encoder**, which has the same structure as the first sub-layer. However, the input of this sub-layer is different that Q is different from K and V. If Q is the embedding of an unsolved medical question, the K and V represent the same candidate questions, or vice versa. In this sub-layer, the inter-attention is used to encode question with the aligned context information from the other question in question pair.
(3) The third sub-layer is **Feed-Forward layer**, which consists of two convolutions with kernel size 1 as linear transformations. The first convolution has dimensionality $d_f = 2 * d_e$ and the dimensionality of the second one is $d_e = 100$.

A global max pooling layer is used to retain important information of the new representation of questions along the words dimension. At last, questions can be represented as final semantic vectors. The number of the TM-Transformer layer is set as $N = 6$ in this paper, which is as same as [19].

Similarity Calculation. For a given question pair $<Q, C_i>$, with the pretrained encoder model, we map Q and C_i into semantic vectors v_q and v_{c_i}, respectively. In most neural network architectures [6,18,23], a multilayer perceptron (MLP) layer is used for label prediction. We put the concatenated vector $v = [v_q, v_{c_i}]$ into a final MLP classifier in our experiments. The MLP has a hidden layer with *relu* activation and a *softmax* output layer. The final score range in $[0, 1]$, where 1 represents similar and 0 is dissimilar. In this work, we take binary-class entropy loss for training and the entire model is trained end-to-end. Figure 2 shows the proposed neural network architecture.

3.3   Data Generation

In this section, we first introduce the acquisition of the medical QA corpus, and how we build a large medical terminology. Then, we declare how we generate the training and test dataset with two different automated ways.

Medical QA Corpus. In order to satisfy the demands of generating the word embedding and the training datasets, we require a large medical QA corpus that covers most of the medical questions and terms. We have crawled 4 online medical QA websites and acquired privacy-free QA data from cooperative company www. dxy.com. The statistics of obtained QA pairs is shown in Table 1.

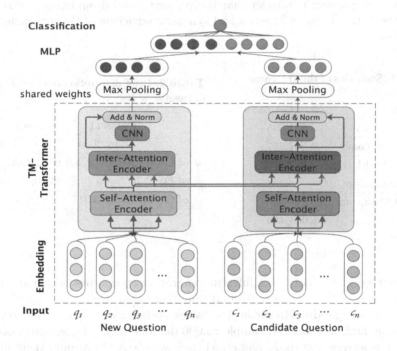

Fig. 2. The framework of Siamese TM-Transformer neural network.

Medical Terminology. According to the analysis of questions in medical QA corpus, we have found that the medical terms in patients queries are usually not exactly equivalent to professional terminology used by medical experts. So in our work, we have collected the medical terms from these sources: (1) the Chinese medical terms of www.dxy.com, which is compiled by professional medical editors, referring to a large number of *medical encyclopedia entries* and *general expressions* of users. (2) the common disease names and drug names crawled from China Public Health Website[2], and the drug names from China Food and

---

[2] http://www.chealth.org.cn/.

Drug Administration[3]. (3) the disease names, symptom names, and their common aliases and other entities on the medical community websites. Finally, we have built a medical terminology that contains 26,619 diseases, 9,904 symptoms, 1,735 medicines, 8,734 examinations, 7,395 operations, 5,472 body parts and other medical entities.

Training Datasets. In previous research [12], the dataset is generated by sampling approach. The negative samples are generated by randomly sampling medical question pairs that share no common medical entities, which directly neglects dissimilar question pairs with common entities. They randomly replace and drop words in medical questions to generate positive samples. In fact, this sampling approach doesn't consider that the replacing and dropping of words may break the semantic characteristics and syntactic structure of the new generated questions.

Table 1. Statistics of the QA pairs

Websites	# of QA pairs
www.dxy.com	623,395
www.xywy.com	1,357,507
www.39health.com	997,717
www.muzhi.com	6,442
www.120ask.com	68,045
Total	3,053,106

Table 2. Brief description of the generated datasets

	Dataset1		Dataset2	
	Train	Test	Train	Test
# of Q	28,045	14,269	165,103	36,094
# of Q&C	-	-	264,685	46,800
# of Group	5,000	3,317	-	-

In our work, we adopt two different generation approaches, which are adapt to different training scenarios. We find that there are some similar questions on each QA web page edited by medical websites, such as xywy.com. In the first generation method, we randomly sample 10,000 different medical questions from the QA corpus as question seeds, and crawl their web pages to acquire their similar questions provided by websites. We remove probably dissimilar questions that have different disease categories with the seed. Finally, about 60,000 questions are automated categorized into 8,317 synonymous sentence groups after filtering out the seeds without similar questions. Several generated question groups are shown in Table 3. In our work, this dataset namly **Dataset1** is used for transfer strategy introduced in Sect. 3.4.

Another method generates question pairs with the help of opensource retrieval system. We randomly sample 80 thousand different questions as queries and search their related questions from the medical QA corpus with Solr[4]. The

---

[3] http://eng.sfda.gov.cn/WS03/CL0755/.
[4] http://lucene.apache.org/solr/.

Table 3. Synonymy question group sample of Dataset1

Group	Medical Questions
0	慢性荨麻疹怎么治疗?(How to treat chronic urticaria?)
0	慢性荨麻疹如何调理?(How to regulate chronic urticaria?)
0	怎样才能根治慢性荨麻疹?(How can I cure chronic urticaria?)
0	得了慢性荨麻疹怎么办?(I have chronic urticaria. What should I do?)
1	高血压伴发的症状.(Symptoms associated with hypertension.)
1	高血压的常见症状有哪些?(What are the common symptoms of hypertension?)
1	高血压的临床表现有什么?(What are the clinical features of hypertension?)

BM25 algorithm built in Solr rank the candidate questions by their *relevance scores* with query. We select the top 20 candidates of each query to form question pairs and represent each question as a mean vector with word embeddings of all words in question. We then calculate *cosine score* between mean vectors of each query and candidate pair. The average of *cosine score* of mean vectors and *relevance score* of BM25 is treated as the final *similarity score* of each question pair, which ranges from 0 to 1. We assume that question pairs with similarity score in a high threshold interval $[0.9, 1)$ are the positive samples and those in a low threshold interval $[0, 0.6]$ as negative samples. Then we randomly sample the positive and negative samples with an approximate ratio of 1:3. Finally, we invite five domain experts of *dxy.com* to validate and remark the auto-tagged labels of samples. We name this dataset as *Dataset2*.

3.4 Pre-training

Word Embedding. Word Embedding is popular in almost NLP tasks since the Word2vec[5] has been proposed by Google. The distributed representation of words learned by the neural networks would capture the semantic and syntactic information from unlabeled corpus. So that we preprocess the crawled medical QA data as training corpus and take CBOW model in Word2vec to train word embedding, and the dimensionality of word vectors is set to $d_e = 100$.

Transfer Learning Method. The pre-trained word vectors are usually used as the initialized weights of embedding layer in neural networks. It is commonly regarded as a transfer learning approach, which performs better than randomly initialization. Besides to the pre-trained word vectors, we can also transfer weights from pre-trained question encoders of other models to initialize corresponding layers in Siamese neural networks.

Since the process of face feature extraction and comparison of face recognization problem is similar to sentence matching task, in this paper, we adopt a multi-class classification neural network to pre-train encoders mentioned in Sect. 3.2. The encoder of multi-class classification model would help to learn

---

[5] https://code.google.archive/p/word2vec/.

more generalized representation among different expressions of similar questions. We take **Dataset1** that medical questions are categorized into different groups as training corpus for multi-class question classification, which is as same as the process in face recognition models [17]. The extracted features of pre-trained encoder will be used as question representations for similarity measurement.

The classification models would capture reasonable features of most samples, but usually confuse samples near group margins. The reason is that the classification constraint is not able to distinguish the close samples in different categories. The loss function of most multi-class classification tasks is softmax categorical cross entropy. If we set Q as input of encoder, and the cross-entropy loss of general multi-class classification can be described formally as follow,

$$z = Encoder(Q) \tag{4}$$

$$f = softmax(zW)$$
$$= softmax((z \cdot w_1), (z \cdot w_2), , (z \cdot w_n)) \tag{5}$$

$$L_{softmax} = -\sum_t log \frac{e^{(z \cdot w_t)}}{\sum_{i=1}^n e^{(z \cdot w_i)}} \tag{6}$$

where, W is the weight of final linear layer, $W = (w_1, w_2, ... w_n)$, $(z \cdot w_i)$ is the dot product between z and w_i, f is the target probability distribution of Q and t is the label of Q.

The margin softmax function perform better than softmax function for feature ranking problem. It has beed proved more effective in field of face recognization. There are many angular margin softmax functions, like A-softmax [13] and AM-softmax [22]. We take AM-softmax to improve classification performance in our work. [22] proposes the AM-softmax for learning large-margin features with small intra-class variation and large inter-class difference. In the design of this function, z and w_i are normalized with $l2$ normalization, and the dot product in Eq. 5 is transformed to cosine function. Then a positive number m is used as a margin to tighten the cosine score of f_t and a positive number s is taken as scale rate to the tightened score. Formally, the AM-softmax loss function is shown as follow,

$$L_{AMS} = -\sum_t log \frac{e^{s \cdot (cos\theta_t - m)}}{e^{s \cdot (cos\theta_t - m)} + \sum_{i \neq t} e^{s \cdot (cos\theta_i)}} \tag{7}$$

where $cos\theta_i$ is the cosine score of z and w_i. In this paper, we choose the best hyper-parameters $m = 0.35$ and $s = 20$ in experiments. We attempt the Bi-LSTM and the *Self-Attention encoder* of TM-Transformer as the basic encoder in multi-class classification model respectively. Figure 3 is the pre-training model with Self-Attention encoder.

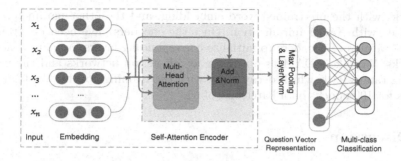

Fig. 3. The architecture of pre-training model with Self-Attention Encoder.

4 Experiments

In this section, we evaluate the proposed model architecture and the transfer learning strategy on large scale QA dataset in the following aspects: (1) We compare the results of Siamese Bi-LSTM and STMT model on the *Dataset2* to validate the effectiveness of our proposed architecture. (2) We pre-train the encoders with the transfer method declared in Sect. 3.4. Then we use pre-trained encoders to initialize corresponding layers of Siamese Bi-LSTM and STMT models and fine-tune model weights on finely labeled dataset.

4.1 Dataset

As mentioned in Sect. 3.3, the *Dataset1* which contains 8,317 similar question groups is used for pre-training encoders. The *Dataset2* that consists of about 300 thousand question pairs is applied to train and fine-tune the Bi-LSTM and TM-Transformer models. Table 2 lists the statistics of the datasets that we used in our experiments.

4.2 Experiment Setup

We randomly choose 5000 question groups from the *Dataset1* as training data for pre-training. In other words, we pre-train a classifier with 5000 target classes. We split the *Dataset2* into two subsets, we randomly select 60% question pairs as train set, 20% as dev set and the rest as test set. For all the experiments, we train the model on train set and tune parameters on dev set and pick the parameters which works the best on dev set. Finally, we train models with all the data in train set and dev set by fixing the best hyper-parameters. We calculate the Precision, Recall and F1 score for evaluating similarity prediction ability and normalized discounted cumulative gain (NDCG@3) as ranking score of these models.

We set the hidden size as 100 in Bi-LSTM architecture. We set head $h = 10$ of the multi-head attention in all the sub-layers of TM-Transformer. In the pre-training stage, we use the same hyper-parameters for above encoders in multi-class classification models. We initialize the embedding layer of all the neural networks with the pre-trained word embedding, and the other parameters are initialized with Xavier initialization. Once the encoders have been pre-trained, we take weights of encoders to initialize corresponding layers in Siamese neural networks. We take ADAM as optimizer for all neural networks and the initial learning rate is 0.01, which decay along with the training epochs. We train the models for 100 epochs on a batch size of 64.

5   Discussion

5.1   The Effectiveness of TM-Transformer

In this part, we discuss about the effectiveness of our proposed architecture on medical question matching task. For comparison, we use the following four methods as baselines: (1) The logistic regression (LR) model takes the average of all the word vectors of a question as its representation, namely LR-vec. (2) The logistic regression model takes the TFIDF based bag-of-words as question representation, namely LR-tfidf. (3) The Siamese Bi-LSTM model utilize Bi-LSTM as sentence encoder. (4) The traditional Transformer model ($N = 6$ layers) for translation task declared in [19].

The results of these models are shown in Table 4. From the results, we can find that Siamese TM-Transformer model performs better on both classification and ranking metrics than all baseline models. We observe that the F1 score of STMT model exceeds about 2.61% than traditional Transformer, while NDCG score gets 0.89% improvement. It means that the proposed architecture can capture more semantical similarity information between question pairs. Since two attention based sub-layers of TM-Transformer focus on learning question representation and question interaction respectively. The simple sentence representation methods in SVM-vec and SVM-tfidf can not capture question context information and the interactive information between question pairs.

To validate the effect of these sub-layers, we train the TM-Transformer by removing the first sub-layer or the second one and the number of TM-Transformer layers $N = 6$. As shown in Table 4, either of two sub-layers has good performance, and the combination of them achieve a great improvement. In other words, stacked

Table 4. Performance of Siamese neural networks on *Dataset2*

Model	Accuracy	Precision	Recall	F1	NDCG@3
LR-tfidf	78.02	61.68	36.14	45.58	93.13
LR-vec	74.68	62.92	55.62	59.05	93.43
Siamese Bi-LSTM	80.98	60.55	73.57	65.05	96.64
Transformer [19]	85.64	65.92	**90.25**	75.19	97.07
STMT (only Self-Attention Encoder)	82.27	60.91	85.63	70.00	97.33
STMT (only Inter-Attention Encoder)	84.88	70.22	73.46	71.80	97.20
STMT	**88.93**	**76.77**	81.36	**77.80**	**97.96**

multi-head attention layers would help us to capture more semantic information and enhance the fitting and generalization ability of model.

5.2   The Effectiveness of Transfer Learning Method

In this part, we verify the effectiveness of transfer learning method we declared for model pre-training. The baseline are Siamese Bi-LSTM and STMT models, which we initialize model weights with Xavier initialization method in above experiments. We share the same encoder layers between baseline models with the multi-class classification model. And then we take weights of the pre-trained encoders as initialization of baseline models and fine-tune model weights on *Dataset2*. The result is shown in Table 5. There are great improvements of all the basic models with the transfer strategy. We can observe that the F1 score of STMT has a 2.81% increase while Bi-LSTM gets a improvement of 6.41%. For the ranking metrics, the transfer learning approach gives a little bit improvement over baseline, even though the baseline models have achieved good ranking scores already. The NDCG score of Bi-LSTM has a raise of 0.43%, as for TM-Transformer, the improvement is 0.35%. To further verify the effect of transfer strategy, we compare the training loss of TM-Transformer and TM-Transformer with transfer strategy. As shown in Fig. 4(b), the transfer strategy accelerates the model training and achieves better convergence. From the results in Fig. 4(a), it shows that the transfer learning method would increase the performance of different models more or less on question matching task.

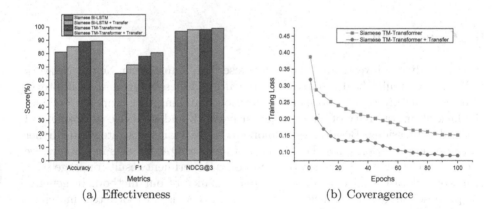

(a) Effectiveness   (b) Coveragence

Fig. 4. Comparison between plain Siamese neural network and Siamese neural network with transfer strategy.

We also compare the performance of our proposed method with other existing method on our datasets, such as ESIM in [6,11] and BiMPM model in [23]. As shown in Table 5, we find that our STMT model with transfer strategy gets a 6.55% F1 score higher than BiMPM model and 8.85% higher than ESIM model. We can conclude that our proposed method can relieve question matching problem in medical domain, which perform better than the state-of-art model on other datasets. In the process of generating *Dataset2* based on the crawled medical QA corpus, we can find similar candidates for almost all queries. We also restrict the ratio of randomly selected negative samples. That is why the ranking scores of each model are relatively higher than classification metrics. At the pre-training stage, the feature extraction of multi-class classification is related to the feature ranking of sentence matching but they are not equivalent to each other. In other words, the features extracted by classifier may not work well at similarity ranking stage. So good loss functions, like AM-softmax or even better ones, are really helpful to relieve the gap between feature extraction and feature ranking.

Table 5. Performance of transfer learning method and other existing models

Model	Transfer	Accuracy	Precision	Recall	F1	NDCG@3
Siamese Bi-LSTM	No	80.98	60.55	73.57	65.05	96.64
	Yes	85.20	70.22	72.75	71.46	97.07
STMT	No	88.93	**76.77**	81.36	77.80	97.96
	Yes	**89.30**	74.91	87.25	**80.61**	**98.31**
ESIM [6]	No	85.88	73.13	70.45	71.76	97.18
BiMPM [23]	No	83.70	61.49	**96.33**	74.06	98.09

6   Conclusion

In this study, we investigate a novel Siamese TM-Transformer Neural Network (STMT) for similar health question retrieval in Chinese. Our method improves internal structure of Transformer for question matching task, which overcomes the lack of interactivity of Siamese neural network and the issue of diversity of medical expressions. Besides, we explore to use a transfer strategy to further enhance model performance. For the lack of medical QA datasets in Chinese, we declare two different data generation methods. Experiment results on large scale real world datasets have validates the performance of our method. In general, the proposed model and transfer strategy can also be used to solve text matching tasks in other domain.

Acknowledgment. This work is supported by the National Key Research and Development Program of China under Grant No. 2016YFB1000904.

References

1. Aditya, T.: Siamese recurrent architectures for learning sentence similarity. In: Thirtieth AAAI Conference on Artificial Intelligence, pp. 2786–2792 (2016)
2. Baziotis, C., Pelekis, N., Doulkeridis, C.: Datastories at semeval-2017 task 6: Siamese LSTM with attention for humorous text comparison. In: Proceedings of the 11th International Workshop on Semantic Evaluation, SemEval@ACL 2017, pp. 390–395 (2017)
3. Borui, Y., Guangyu, F., Anqi, C., Ming, L.: Learning question similarity with recurrent neural networks. In: IEEE International Conference on Big Knowledge, pp. 111–118 (2017)
4. Cai, H., Yan, C., Yin, A., Zhao, X.: Question recommendation in medical community-based question answering. In: Liu, D., Xie, S., Li, Y., Zhao, D., El-Alfy, E.-S.M. (eds.) ICONIP 2017. LNCS, vol. 10638, pp. 228–236. Springer, Cham (2017). https://doi.org/10.1007/978-3-319-70139-4_23
5. Cao, X., Cong, G., Cui, B., Jensen, C.S., Zhang, C.: The use of categorization information in language models for question retrieval. In: Proceedings of the 18th ACM Conference on Information and Knowledge Management, pp. 265–274 (2009)
6. Chen, Q., Zhu, X., Ling, Z., Wei, S., Jiang, H., Inkpen, D.: Enhanced LSTM for natural language inference. In: Proceedings of the 55th Annual Meeting of the Association for Computational Linguistics, ACL, pp. 1657–1668 (2017)
7. Das, A., Yenala, H., Chinnakotla, M.K., Shrivastava, M.: Together we stand: Siamese networks for similar question retrieval. In: Proceedings of the 54th Annual Meeting of the Association for Computational Linguistics, ACL (2016)
8. Eyecioglu, A., Keller, B.: Twitter paraphrase identification with simple overlap features and SVMs. In: Proceedings of the 9th International Workshop on Semantic Evaluation, pp. 64–69 (2015)
9. Jeon, J., Croft, W.B., Lee, J.H.: Finding similar questions in large question and answer archives. In: Proceedings of the 2005 ACM CIKM International Conference on Information and Knowledge Management, pp. 84–90 (2005)
10. Ji, Z., Xu, F., Wang, B., He, B.: Question-answer topic model for question retrieval in community question answering. In: 21st ACM International Conference on Information and Knowledge Management, CIKM 2012, pp. 2471–2474 (2012)
11. Lan, W., Xu, W.: Neural network models for paraphrase identification, semantic textual similarity, natural language inference, and question answering. In: Proceedings of the 27th International Conference on Computational Linguistics, COLING 2018, pp. 3890–3902 (2018)
12. Li, Y., et al.: Finding similar medical questions from question answering websites. CoRR abs/1810.05983 (2018)
13. Liu, W., Wen, Y., Yu, Z., Li, M., Raj, B., Song, L.: Sphereface: deep hypersphere embedding for face recognition. In: 2017 IEEE Conference on Computer Vision and Pattern Recognition, CVPR 2017, pp. 6738–6746 (2017)
14. Ponte, J.M., Croft, W.B.: A language modeling approach to information retrieval. SIGIR Forum 51(2), 202–208 (2017)
15. Qiu, X., Huang, X.: Convolutional neural tensor network architecture for community-based question answering. In: Proceedings of the Twenty-Fourth International Joint Conference on Artificial Intelligence, IJCAI, pp. 1305–1311 (2015)
16. Robertson, S.E., Jones, K.S.: Relevance Weighting of Search Terms. Taylor Graham Publishing (1988)

17. Taigman, Y., Yang, M., Ranzato, M., Wolf, L.: Deepface: closing the gap to human-level performance in face verification. In: 2014 IEEE Conference on Computer Vision and Pattern Recognition, CVPR 2014, pp. 1701–1708 (2014)
18. Tang, G., Ni, Y., Xie, G., Fan, X., Shi, Y.: A deep learning-based method for similar patient question retrieval in chinese. In: MEDINFO 2017: Precision Healthcare through Informatics - Proceedings of the 16th World Congress on Medical and Health Informatics, pp. 604–608 (2017)
19. Vaswani, A., et al.: Attention is all you need. In: Advances in Neural Information Processing Systems 30: Annual Conference on Neural Information Processing Systems, pp. 6000–6010 (2017)
20. Vo, N.P.A., Magnolini, S., Popescu, O.: FBK-HLT: an effective system for paraphrase identification and semantic similarity in Twitter. In: Proceedings of the 9th International Workshop on Semantic Evaluation, pp. 29–33 (2015)
21. Wan, S., Lan, Y., Guo, J., Xu, J., Pang, L., Cheng, X.: A deep architecture for semantic matching with multiple positional sentence representations. In: Proceedings of the Thirtieth AAAI Conference on Artificial Intelligence, pp. 2835–2841 (2016)
22. Wang, F., Cheng, J., Liu, W., Liu, H.: Additive margin softmax for face verification. IEEE Signal Process. Lett. **25**(7), 926–930 (2018)
23. Wang, Z., Hamza, W., Florian, R.: Bilateral multi-perspective matching for natural language sentences. In: Proceedings of the Twenty-Sixth International Joint Conference on Artificial Intelligence, IJCAI 2017, 19–25 2017, pp. 4144–4150 (2017)
24. Wang, Z., Mi, H., Ittycheriah, A.: Sentence similarity learning by lexical decomposition and composition. arXiv:1602.07019 (2016)
25. Xue, X., Jeon, J., Croft, W.B.: Retrieval models for question and answer archives. In: Proceedings of the 31st Annual International ACM SIGIR Conference on Research and Development in Information Retrieval, SIGIR 2008, pp. 475–482 (2008)
26. Zhang, K., Wu, W., Wu, H., Li, Z., Zhou, M.: Question retrieval with high quality answers in community question answering. In: Proceedings of the 23rd ACM International Conference on Conference on Information and Knowledge Management, CIKM 2014, pp. 371–380 (2014)
27. Zhou, G., Zhou, Y., He, T., Wu, W.: Learning semantic representation with neural networks for community question answering retrieval. Knowl.-Based Syst. **93**, 75–83 (2016)

An Adaptive Kalman Filter Based Ocean Wave Prediction Model Using Motion Reference Unit Data

Yan Tang[1], Zequan Guo[1(✉)], and Yin Wu[2]

[1] College of Computer and Information, Hohai University, Nanjing, China
{tangyan,gzq}@hhu.edu.cn
[2] Transocean Inc., Houston, USA
yin.wu@deepwater.com

Abstract. Fleets like the ocean drilling platforms need to remain stationary relative to the bottom of the ocean, therefore the ship or platform need to pay close attention to the fluctuation of ocean currents. The analysis of the ocean waves is of great significance to the stability of the ocean operation platforms and the safety of the staffs onboard. An effective ocean current prediction model is helpful both economically and ecologically. The fluctuations in the ocean waves can be seen as a series of sinusoidal time series data with different frequencies and is usually captured by the sensors known as MRU (Motion Reference Unit). The study aims to analyze and accurately predict the movement of the ocean in the future time based on the historical movement of the ocean currents collected by MRU. All of these data have a fixed high resolution acquisition frequency. This research focuses on how to effectively fill in the missing values in the time series of MRU data. We also aim to accurately predict the future ocean wave. Therefore, an novel ARIMA (Autoregressive Integrated Moving Average) Model based missing data completion method is proposed to fill the data by artificial approximation of missing data. More importantly, an novel adaptive Kalman filter based Ocean Wave Prediction model is proposed to predict the ocean current in the near future by leveraging dynamic wave length. Experiment results validates the correctness of the ARIMA model based missing data completion method. The adaptive Kalman filter based Ocean Wave Prediction model is also shown to be effective by outperforming three base line prediction models.

Keywords: ARIMA model · Missing data completion model ·
Kalman filter · Dynamic wave length · Ocean wave prediction model

1 Introduction

Fleets like the ocean drilling platforms need to remain stationary relative to the bottom of the ocean. Take oil rig for example, the riser of the drilling platform is connected to the oil well at the bottom of the ocean. The abrupt movement

© Springer Nature Switzerland AG 2019
G. Li et al. (Eds.): DASFAA 2019, LNCS 11448, pp. 65–79, 2019.
https://doi.org/10.1007/978-3-030-18590-9_5

of the drilling platform brings immediate damage to the riser. When the ocean wave abruptly pushes the platform vertically, it is sometimes necessary to disconnect the riser and the Blow-Out Preventer (BOP) in time to prevent disasters like the leakage of oil and gas resources. If the riser is damaged, the drilling and oil company will suffer losses of 1 to 2 million dollars per day and the ecological environment near the drilling platform will be destroyed, and the life safety of the staff on the platform is also at a high risk. Therefore, it is mission critical to predict the ocean wave in advance, and to give a warning signal for the oceanic platforms/fleets to act in advance before the future ocean wave becomes a threat. To solve this challenging problem, in this paper, we propose an ARIMA (AutoRegressive Integrated Moving Average) model based missing data completion model and an adaptive Kalman filter based ocean wave prediction model. These two novel models work together to accurately predict future ocean wave and raise warning signals to prevent damage and disaster in the future.

The existing ocean wave prediction methods are not accurate enough [1,2]. In recent year, with the fast development of Industrial Internet of Things (IOT) technology, sensor units like Motion Reference Unit (MRU) are being widely deployed to measure ocean wave and platform moments. This gives opportunity to capture ocean wave at high resolution and design new big data driven ocean wave prediction methods and models. The MRU is a part of Dynamic Position System Logger. The MRU device collects the real-time working state of the drilling platform. The data set used in this study is collected from the MRU devices on deep ocean oil rigs.

This paper mainly aims to solve the following problems using the MRU ocean wave datasets:

- How to fill the missing values in the data set?
- How to accurately predict the ocean wave form in the future by analyzing the current and previous ocean wave time series data?

For addressing the above problems, this paper proposes two novel models: an ARIMA based missing data completion model that leverages Kalman filter to complete missing values after the transformation of the state space in the ARIMA process and an Kalman filter based prediction model based on adaptive calculation of the Dynamic Wave Length (DWL) to accomplish the prediction of future wave form. Dynamic Wave Length is the duration of the next ocean wave cycle from peak to peak.

Four different data sets are used in this paper. Each data set is the ocean wave data at different time. Datasets are used to verify the validate the proposed models and methods. In summary, the contributions of this paper are as follows:

- A novel ARIMA Model based Missing Data Completion model is designed to effectively complete missing values in the original data set.
- An adaptive Kalman filter based Ocean Wave Prediction Model is proposed to predict the future ocean wave by leveraging the Dynamic Wave Length and adjusting the coefficients of the time and measurement update equations.

– Extensive experiment are carried out to validate the proposed models, our
Ocean Wave Prediction model outperforms three baseline prediction models
by up to 60% in prediction accuracy.

The remainder of the paper is organized as follows. Section 2 presents related
works. Section 3 gives the overview of the ocean wave prediction process. Sections 4 and 5 describes our models in detail. Section 6 shows experiment results
and discusses findings. Lastly, the paper is concluded in Sect. 7.

2   Background and Related Works

In this paper, key attributes in data from the Motion Reference Unit (MRU) are
selected as features. The main output of MRU is roll, pitch and yaw, and mag-
netic northward looking, as well as relative dynamic wave fluctuations, undulat-
ing and rocking measurements [3]. Figure 1 is the work model diagram of MRU.

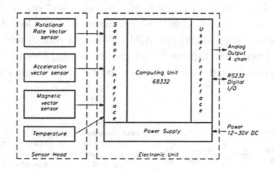

Fig. 1. MRU workflow diagram

The lack of data, especially time-based data, often lead to unreliable data
results in the analysis of data. Many methods have been put forward to deal
with missing values in data. Weerasinghe [4] claims that the minimum mean
square error (MMSE) prediction and MMSE reverse prediction are useful in the
time series for improving ocean waves analysis and forecasting. Others, such as
Stefanakos and Athanassoulis, have proposed using simulated values to complete
the missing values [5].

In this article, we assume that the data is Missing completely at random
(MCAR). ARIMA Based Missing Data Completion Model is based on the transformation of state space, the missing data values are estimates using Kalman
filter after the ARIMA state space transformation.

In the prediction of waves, scientists have proposed many methods. In 1999,
scientists proposed a SWAN model [6] to monitor the sea conditions of the
Western Iberian Peninsula. In recent years, many methods are proposed for
predicting ocean current. James et al. realized the prediction of ocean currents

through deep learning [7]. Shi et al. use heave prediction model based on support vector machine for regression (SVR) to reduce the wind and wave impact [8]. However, these models lack clarity and visibility in the prediction process. It is difficult to explain the prediction process in a reliable way.

The WAM model [9] is the third generation of operational wind forecasting systems that operate on grid nodes with interlaced longitude and latitude. It integrated the law of conservation of energy to obtain the two-dimensional wave spectrum. The model is mainly for the near shore ocean. The SWAN model [10] is a prediction system for predicting the coastal currents combining with the WAM model. It is also a third-generation forecasting system that combines the generation of random winds. This system is designed to predict coastal and wave frequencies in shallow water too. Both WAM and SWAN models are not very effective for deep ocean wave prediction.

3   Approach Overview

The flow chart in Fig. 2 is the overview of the proposed ocean wave prediction process. The process begins by using the observed ocean wave data as input. If there is missing ocean wave data, the ARIMA based model will fill the missing ocean wave data. Then the adaptive Kalman filter based Ocean Wave Prediction model will use current ocean wave data to predict wave heights and wave forms in the near future.

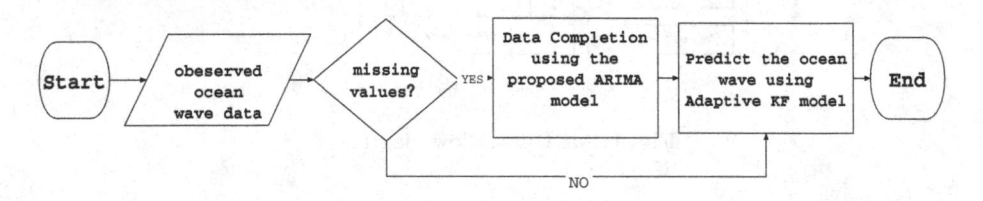

Fig. 2. The flow chart of the proposed approach and ARIMA based missing data completion model

4   ARIMA Model Based Missing Data Completion Method

The model uses Kalman filter to check the effect of filling missing data by ARIMA. It is worth mentioning that during the process of handling the wave data, it is necessary to select a set of suitable ARIMA model parameters to obtain accurate results. Then, the time update equation of Kalman filter calculates the values for the missing data. Finally, the Kalman gain is used to obtain the confidence interval of data completion.

4.1   Theoretical Basis

In many time series, data are often measured in an irregular manner and always missing. This situation will make it difficult to analyze this data sequence. Here, the state space model needs to be introduced to deal with data missing problem.

Suppose t is a given time, if y_t is the $q \times 1$ observation vector, X_t is a $p \times 1$ state vector and V_t is the observation noise at time t. Then a partition $y_t = (y^{(1)'}, y^{(2)'})'$ is defined, where $y^{(1)}$ is $q_{1,t} \times 1$ vector, value, and $y^{(2)}$ is $q_{2,t} \times 1$ vector and unobserved value [11]. By the above formula, we can obtain $q_{1,t} + q_{2,t} = q$. This partition extends naturally to A_t and V_t. The partition observation equation is Eq. 1.

$$\begin{pmatrix} y^{(1)} \\ y^{(2)} \end{pmatrix} = \begin{pmatrix} A_t^{(1)} \\ A_t^{(2)} \end{pmatrix} X_t + \begin{pmatrix} V_t^{(1)} \\ V_t^{(2)} \end{pmatrix} \tag{1}$$

It can be seen that $A_t^{(1)}$ is $q_{1,t} \times P$ matrix corresponding to the observation value. $A_t^{(2)}$ is $q_{2,t} \times P$ matrix corresponds to the unobserved value. The covariance matrix of the measurement errors equation is Eq. 2.

$$cov \begin{pmatrix} V_t^{(1)} \\ V_t^{(2)} \end{pmatrix} = \begin{bmatrix} R_{11,t} & R_{11,t} \\ R_{21,t} & R_{22,t} \end{bmatrix} \tag{2}$$

In time-domain analysis, the random process of generating time series data is divided into two classes according to whether the characteristics of the statistical laws change over time. If the characteristics of the random process change over time, the process is unstable. On the contrary, if the characteristics of the random process do not change over time, the time series of temperature components are similar and this process is stable. The stability of stochastic process determines the basic form of the model to be selected. The ocean wave is an unstable system, so the ARIMA model is chosen to deal with it [12].

The ARIMA model mainly uses P, D and Q to decide the state space characteristics. P represents the lag of the time series data in the prediction model. D represents time series data, which needs to be differentiated by several orders. Q represents the lag of the prediction error used in the prediction model. In the ARIMA model, X_t is represented as a function of X past values or past errors. When we predict a value over the end of the series, on the right side of the equation, we may need values from the observed sequence.

In the model, $X_t = \delta + \Phi_1 X_{t-1} + \Phi_2 X_{t-2} + W_t$, X_t is a linear function of the previous two X. W_t is the noise. Suppose we have observed n data values and want to use observation data and the estimated model to predict the value of X_{n+1} and X_{n+2}. The equations for the two values are as follows:

$$X_{n+1} = \delta + \Phi_1 X_n + \Phi_2 X_{n-1} + W_{n+1} \tag{3}$$

$$X_{n+2} = \delta + \Phi_1 X_{n+1} + \Phi_2 X_n + W_{n+2} \tag{4}$$

In order to use the first equation, The observations of X_n and X_{n-1} is required and W_{n+1} is replaced with the expected value of 0 (the assumed mean value of error). There are problems in the second equations of time $n + 2$ for predicting value. It requires the unobserved values of x_{n+1}. The solution is to use a predicted value to approximate x_{n+1}.

4.2 Detailed Description of the Model

When dealing with ocean wave data in ARIMA, The appropriate model parameters need to be selected to describe the characteristics of the model. We will add noise to ARIMA model to fit the Kalman filter equation. In simulation, as in the previous hypothesis, the wave data contain a random measurement error of V_k, which satisfies the characteristics of white Gaussian noise. At the same time, we also need to select the order and parameters of ARIMA.

The construction of the whole model is used to analyze the time prediction values of the missing fragments, the standard deviation between the prediction and the observed data and the 95% confidence interval that can be obtained by the standard deviation.

The detailed missing data completion process is as follows in four steps:

- **Step 1**: Select a suitable ARIMA model. The ARIMA model is related to the historical data of the previous time and the prediction data error of the next time. It is AR and MA. This paper gives a assumed coefficient to describe the model. It is known that the two coefficients are used to describe the coefficient of error covariance for P and Q.
- **Step 2**: According to the model parameters set up above, create the normal state spatial model and transform the data for processing in the state space.
- **Step 3**: The Kalman filter time update equation is applied to the transformed data that satisfies the spatial condition. The core formula is the update equation of the provisional measurement mentioned in Eq. 7. Since the error covariance of Kalman gain has also changed. The core equation is transformed into a vector form (Eqs. 1 and 2).
- **Step 4**: Estimation of parameters: Use maximum likelihood estimation method to estimate parameters. From the formula, $X'_k = AX'_{k-1} + \epsilon_t$, Maximum likelihood estimation can be applied to process error ϵ_t in Eq. 5.

$$- log L_Y(\Theta) = \frac{1}{2} \sum_{t=1}^{n} log \left| \sum_t (\Theta) \right| + \frac{1}{2} \sum_{t=1}^{n} \epsilon'_t \sum_t (\Theta)^{-1} \epsilon_t \tag{5}$$

In the Eq. 5, it can be seen that $\sum_t (\Theta)$ is the P'_k in the Eq. 8. In order to solve the equation and achieve the minimum sum of Θ, the data that has not been known is assigned to 0. The parameters which is related to the equation in the latter half of the prediction model is assigned to 0. Then the equation can be simplified to Eq. 6:

$$y_{(t)} = \begin{pmatrix} y_t^{(1)} \\ 0 \end{pmatrix} \quad A_t = \begin{bmatrix} A_t^{(1)} \\ 0 \end{bmatrix} \quad R_{(t)} = \begin{bmatrix} R_{11,t} & 0 \\ 0 & R_{22,t} \end{bmatrix} \tag{6}$$

5   The Proposed Approach

In this chapter, a novel Adaptive Kalman Filter and Dynamic Wave Length (DWL) based Ocean Wave Prediction model will be introduced. The key novelty is in the concept of Dynamic Wave Length, since the ocean wave length is not a fixed value, we need to estimate the duration of the future wave cycle denoted as DWL based on the duration of the current wave cycle. Then, the prediction model will uses the time series data until time T to predict the future wave form in time $[T, T + DWL]$.

Definition 1: Dynamic Wave Length

Dynamic Wave Length is the duration of the future ocean wave cycle from peak to peak.

This model is concerned with attributes related to wave height. From 982 different attributes of MRU records, three wave attributes containing heave keywords are extracted and analyzed. The vale of these attributes form three time series. The input is the wave height Z_t of the preceding DWL seconds, and the wave height Z_{t+DWL} of the next DWL seconds is obtained by using adaptive Kalman filter measurement update equation and time update equation.

5.1   Theoretical Basis

Kalman filter is a recursive algorithm for linearly filtering discrete data by minimizing the covariance of error. Its purpose is to track the moving target and to predict the next state through the measurement state at the moment. Kalman filter contains a series of equations which can be used as predictors and error correctors.

The Kalman filter consists of two main steps: time update and measurement update [13]. In the following equations, A is a matrix of $n \times n$ to preserve the previous state of the system, while B is the effect of the input data on the output of the result. The variables Q and R are random variables, representing the errors in the process and measurement. These two errors, Q and R, are not related to each other, but satisfy Gaussian white noise. In essence, Gaussian white noise is a random noise. Its probability density function is Gaussian distribution and directly superimposed on the original signal, which can be interpreted as a fixed, random noise layer on the spectrum. X_k' and X_k are predicted values at time k. P_k' and P_k are both covariance matrix. K represents Kalman gain. H converts sensor data to the data we need.

The time update equation contains the computation, the current state of the transmission system and the covariance of the error. These values are used to estimate the next time segment. The following two equations can be regarded as predictive equations. The most important equation is [14]:

$$X_k' = AX_{k-1}' + Bu_k \tag{7}$$

$$P_k' = AP_{k-1}'A^T + Q \tag{8}$$

The measurement update equation is used to add a new measurement value, which is obtained from the feedback mechanism, to the value obtained in the last step to get a more accurate result. The following is the error correction equation. The main equation is:

$$X_k = X_k^{'} + K_k(Z_k - HX_k^{'}) \tag{9}$$

$$P_k = (1 - K_kH)P_k^{'} \tag{10}$$

$$K_k = P_k^{'}H^T(HP_k^{'}H^T + R)^{-1} \tag{11}$$

5.2   An Adaptive Kalman Filter Based Ocean Wave Prediction Model

There are a large number of models for data prediction, which are based on Kalman filter. Because the essence of Kalman filter is to divide a time series into two parts. One part is signal and the other is noise [15]. Therefore, Kalman filter is often applied to deal with long waves with white Gaussian noise, such as the tracking of submarine missiles. In the application of Kalman filter, the model must be able to calculate the correct data and the data can affect the subsequent data. In the use of Kalman filter we have to have an estimate of the impact of its input on the output. When the model uses this method, the model needs to satisfy another premise that the original time series is a smooth curve after eliminating the noise.

This model attempts to use the observed data to predict the wave data of the next DWL seconds. We design the following algorithm to calculate DWL.

Algorithm 1. Dynamic Wave Length Calculation

Input:
  D: Ocean Wave Data until time T
  K: number of complete wave cycles;
Output:
  DWL: Dynamic Wave Length;
  1: The start point is selected as the minimum point of wave data D from T back to K wave cycles, recorded as D_{start};
  2: The Kth minimum point after D_{start} is chosen as the end point, denoted as D_{end};
  3: Obtain K complete wave cycles from D_{start} to D_{end}, obtain the wave length WL_k of each complete wave as an vector denoted as $\mathbf{WL} = [WL_1, WL_2, ..., WL_k]$
  4: Apply linear regression model to estimate the next wave length WL_{k+1}
  5: **return** DWL as WL_{k+1}

When dealing with the DWL seconds ocean wave data in this paper, it can be assumed that the impact of the wave in a short period of time is to satisfy the white noise. Regardless of the influence of the other extended disturbance, the wave in different sea areas have their own inherent wave characteristics and meet the smoothness characteristics.

Although the model has a high accuracy in predicting the smooth data, it can not cope with the sudden change in direction. It is because the discrete Kalman filter model is the prediction model that maintains a smooth trend and can only predict the sudden change of the future data. For the sea wave data, it is impossible to predict a sudden change of sea wave height.

This model uses the time update equation and the measurement update equation of Kalman filter to complete the estimation of the next DWL seconds data. The time update equations and the measurement error equation are mentioned in Sect. 5.1. According to the above five core equations, we can see that the core of Kalman filter prediction is recursive. The wave height at the next segment is derived from the value of the wave height at the previous moment. Prediction can be divided into different step lengths, depending on the time interval unit of the predicted signal data. The core flow of the whole model is iterative between the time renewal equation and the measurement error estimation. Algorithm 2 describes the prediction process in detail.

Algorithm 2. An Adaptive Kalman filter based Ocean Wave Prediction Model

---

Input:
    $Z1$: DWL seconds observed ocean wave data by Algorithm 1;
Output:
    X_k: predicted ocean wave data by Kalman filter;
 1: According to the data of $Z1$, the mean and the mean square deviation are calculated and stored in X_1 and P_1 respectively;
 2: Choose suitable coefficients A, B;
 3: According to the previous prediction value X_{k-1}, $Z1_k$ and $Z1_{k-1}$, X_k' is calculated. The equation used is Equation 7;
 4: Update the Kalman gain K_k by updating the Kalman gain K_{k-1} at the previous moment. It converges to a fixed value after a certain step. The equation used is Equation 11;
 5: The covariance P_k of the update error is calculated by the Kalman gain and the last P_{k-1}. The equation used is Equation 10;
 6: The final measured value X_k is obtained through the temporary prediction value X_k' and the Kalman gain K_k calculated in the previous step. The equation used is Equation 9;
 7: **return** X_k

5.3 Triple Exponential Smoothing ($Holt - Winters$)

The essence of Triple Exponential Smoothing ($Holt-Winters$) [16] is the exponential smoothing method. A new parameter p is added to indicate the trend of smoothing. Triple exponential smoothing retains seasonal information on the basis of double exponential smoothing, so that it can predict time series with seasonality. The triple exponential smoothing adds a new parameter p to indicate the trend after smoothing. The triple exponential smoothing has two methods:

accumulation and multiplication. what follows is the cumulative triple exponential smoothing core Equations [16]. L is DWL, which represents the length of wave data we selected.

$$s_i = \alpha(x_i - p_{i-L}) + (1 - \alpha)(s_{i-1} + t_{i-1}) \tag{12}$$

$$t_i = \beta(s_i - s_{i-1}) + (1 - \beta)t_{i-1} \tag{13}$$

$$p_i = \gamma(x_i - si) + (1 - \gamma)p_{i-L} \tag{14}$$

The prediction Equation of triple exponential smoothing [16]. m is the prediction step.

$$x_{i+m} = s_i + mt_i + p_{i-L+1+(m-1) \bmod L} \tag{15}$$

s_1 and t_1 is calculated in Eq. 16

$$s_1 = \sum_{t=1}^{L} Z1 \qquad t_1 = \frac{1}{L}\left(\frac{Z1_{L+1} - Z1_1}{L} + \frac{Z1_{L+2} - Z1_2}{L} + ... + \frac{Z1_{L+L} - Z1_L}{L}\right) \tag{16}$$

p_i is calculated by Eq. 17. N is the number of complete cycles present in our data

$$p_i = \frac{1}{N} \sum_{j=1}^{N} \frac{x_{L(j-1)+i}}{A_j} \quad i = 1, 2, ... L \qquad A_j = \frac{\sum_{i=1}^{L} x_{L(j-1)+i}}{L} \quad j = 1, 2, ..., N \tag{17}$$

Algorithm 3 describes the prediction process in detail.

Algorithm 3. Triple Exponential Smoothing($Holt - Winters$)

Input:
   $Z1$: $N \times L$ seconds observed ocean wave data;
Output:
   X_{i+m}: predicted ocean wave heights in time series;
1: According to the data of $Z1$, s_1 and t_1 are calculated in Equation 16 respectively, p_i is calculated by Equation 17;
2: Choose suitable coefficients α=0.2, β=0.3, γ=0.6 ;
3: Smoothness index s(i) is calculated in Equation 12;
4: Trend index t(i) is calculated in Equation 13;
5: Seasonal index p(i) is calculated in Equation 17;
6: The final predicted value of X_{i+m} is calculated in Equation 15;
7: **return** X_{i+m}

5.4   Moving Average and Double Moving Average Model

In this paper, the moving average (MA) and the double moving average (DMA) model proposed here are for comparison with Kalman filter. In time series analysis, the moving average model is a common approach for modeling univariate time series. The moving average model specifies that the output variable depends linearly on the current and various past values of a stochastic term. DMA is based on MA. The core of MA and DMA is using different window size K to calculate different ocean wave fragments.

6 Experiment and Data Analysis

In this chapter, In order to analyze the effectiveness of the these models and methods, different data are used. Because there is no obvious missing data in the MRU data, the validity of the model is verified by generating similar data and the accuracy of the ocean wave height prediction model is verified by the ocean wave data of the 2015-04-27 based on the MRU record.

This paper selects 3 attributes (heave1, heave2, heave3) with the keyword Heave and 14400 rows of data during 6:00–9:00 at night. The time of a complete wavelength of ocean wave (WL) from a 10 s in the past reduces to 7–8 s. But the wave length of each wave is dynamic. The adaptive Kalman filter based Ocean Wave Prediction Model will select three complete dynamic wave length (DWL) for analysis according to Algorithm 1.

6.1 Evaluation Criteria for Wave Prediction Value

In this paper, RMSE is used as an evaluation criterion for ocean wave prediction and to express the accuracy of prediction. The RMSE of predicted values $\hat{y}_t$ for times t of a regression's dependent variable y_t, with variables observed over T times, is computed for T different predictions as the square root of the mean of the squares of the deviations:

$$\text{RMSE} = \sqrt{\frac{\sum_{t=1}^{T}(\hat{y}_t - y_t)^2}{T}}. \tag{18}$$

6.2 Experiment of Continuous Missing Data Completion

In order to generate data similar to MRU, ARIMA $(2, 1, 2)$ is selected for analysis and simulation. As mentioned in the previous content, the simulated ocean wave data are consistent with the first order difference and are related to the historical data at the previous moment and the prediction data error at the next moment. Therefore, the measurement error $Q = 1$ and the system error $R = 1$ need to be selected for the whole simulation system.

This paper will simulate the time length of n = 100, while the missing data is in the selected interval and the selected data are independent of the other values in the time series. Try to select 5, 30 consecutive missing values. Four consecutive missing fragments are randomly selected to simulate the data. The black ring represents observed data, while the green ring represents the missing data points. The red line represent the complete line. The black line represents the confidence interval. Figure 3(a) and (b) represents the missing data state of continuous missing values K = 5, 30, respectively. The completion of the missing data points and the 95% confidence interval for completion are shown in the figure. The horizontal axis in the picture is second. While the longitudinal axis in following figures is the wave height data magnified 100 times. The unit is cm.

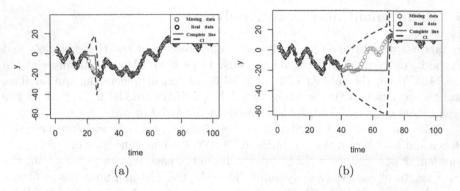

(a)   (b)

Fig. 3. Continuous missing data filling experiment (Color figure online)

By comparing the Fig. 3(a) and (b), it can be seen that as the number of continuous missing data increases and the confidence interval increases, the effect of Kalman filter prediction is gradually getting worse. It can be seen that the effect of Kalman filter depends on the number of missing data. This method is more suitable for filling on missing fragments of short time fragments

6.3   Experiment on Ocean Wave Prediction

In this experiment, The data set recorded by MRU will be used. This data set is a wave data that records in 2015-4-27 by MRU, with 982 attributes and 170 thousand rows of data. The starting time node chosen in the Fig. 4(a) and (c) is 1600 s and in the Fig. 4(b) and (d) is 4197 s. The DWL are 25 and 26 s respectively. The two nodes were chosen because the wave had larger fluctuations during these two periods. This section attempts to modify the value of A, B to reduce the dependence of the model on historical data. But this experiment is mostly done by using error corrector to predict the value. This chapter gives the comparison between A = 0.7, B = 3.1 and A = 0.85, B = 0.75. The decrease of B can reduce the dependence of the model on historical data.

In Fig. 4(a) and (b), A = 0.7, B = 3.1 is chosen. It can predict the trend of the model well, but it has the disadvantage of low accuracy. In Fig. 4(c) and (d), it can be seen that the degree of dependence on historical data has changed by changing A to 0.85 and B to 0.75, the accuracy of the model has been improved.

In the comparison with three baseline models, from Fig. 5, the Double Moving Average model is more stable in subsequent prediction. Because the Double Moving Average model asks for a single average through the temporary prediction value of the preorder sequence. So it will become stable during the second

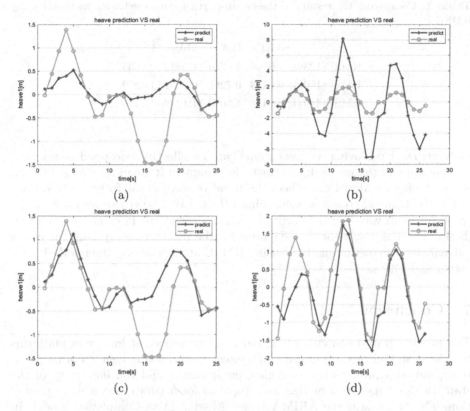

Fig. 4. Prediction and observed ocean wave comparison when A = 0.7, B = 3.1 and A = 0.85 B = 0.75

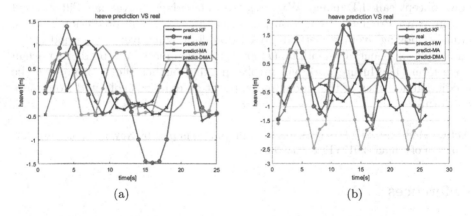

Fig. 5. Comparison of moving average model, double moving average model, triple exponential smoothing (*Holt − Winters*) and Adaptive Kalman filter based Ocean Wave Prediction Model

Table 1. Comparing the results of the ocean current data prediction methods using RMSE

	AKF	HW	MA	DMA
RMSE(1230)	1.0005	1.6931	1.5932	2.2035
RMSE(1600)	0.6133	0.7269	0.7491	1.1259
RMSE(4197)	0.6446	1.3001	1.1259	1.7415

forecast. Double Moving Average model can not effectively respond to the subsequent state of the waves. In contrast, the Adaptive Kalman filter based Ocean Wave Prediction Model can reflect the trend of wave in the future better. Similarly, the triple exponential smoothing $(Holt - Winters)$ can also reflect the trend of ocean waves in the early prediction. But in the later period, the prediction effect is very poor. Table 1 provides the results of comparing the ocean current data prediction methods using RMSE at the starting time node 1230 s, 1600 s and 4197 s.

7   Conclusion

The prediction of ocean currents is particularly important for ocean platforms like fleets and oil rigs. It not only guarantees the economic benefits of the company, but also protects the ecological environment and the life safety of the staff. In this paper, two models are proposed for accurate ocean wave prediction. The first one is the ARIMA Based Missing Data Completion model. In this model, Kalman filter is used to complete the filling of missing data after the transformation of the state space in the ARIMA process. The second model is an adaptive Kalman Filter based Ocean Wave Prediction model based on a new concept called Dynamic Wave Length. The adaptive Kalman Filter based Ocean Wave Prediction model outperforms three base line prediction models, namely, moving average method (MA), double moving average method (DM) and triple exponential smoothing $(Holt - Winters)$ method in predicting ocean wave form. The future work will involve making further effort to improve the prediction accuracy of the model and apply it on data sets collected from more platforms.

Acknowledgments. This work has been supported in part by Key Technologies R&D Program of China (2017YFC0405805-04).

References

1. Kumar, N.K., Savitha, R., Mamun, A.A.: Ocean wave height prediction using ensemble of extreme learning machine. Neurocomputing **277**, 12–20 (2017)
2. Özger, M.: Prediction of ocean wave energy from meteorological variables by fuzzy logic modeling. Expert Syst. Appl. **38**(5), 6269–6274 (2011)

3. Emmanouil, G., Galanis, G., Kallos, G.: Combination of statistical Kalman filters and data assimilation for improving ocean waves analysis and forecasting. Ocean Model. **59–60**(12), 11–23 (2012)
4. Weerasinghe, S.: A missing values imputation method for time series data: an efficient method to investigate the health effects of sulphur dioxide levels. Environmetrics **21**(2), 162–172 (2010)
5. Stefanakos, C.N., Athanassoulis, G.A.: A unified methodology for the analysis, completion and simulation of nonstationary time series with missing values, with application to wave data. Appl. Ocean Res. **23**(4), 207–220 (2001)
6. Booij, N., Ris, R.C., Holthuijsen, L.H.: A third-generation wave model for coastal regions: 1. Model description and validation. J. Geophys. Res. Oceans **104**(C4), 7649–7666 (1999)
7. James, S.C., O'Donncha, F., Zhang, Y.: An ensemble-based approach that combines machine learning and numerical models to improve forecasts of wave conditions. In: Oceans (2017)
8. Shi, B.H., Xian, L., Wu, Q.P., Zhang, Y.L.: Active heave compensation prediction research for deep sea homework crane based on KPSO-SVR. In: Control Conference, pp. 7637–7642 (2014)
9. The WAMDI Group: The WAM model–a third generation ocean wave prediction model. J. Phys. Ocean. **18**(12), 1775–1810 (1988)
10. Sips, H.J.: SWAN (Simulating Waves Nearshore) (2012)
11. Miao, D., Qin, X., Wang, W.: The periodic data traffic modeling based on multiplicative seasonal ARIMA model. In: Sixth International Conference on Wireless Communications and Signal Processing, pp. 1–5 (2014)
12. Hodge, B.M., Zeiler, A., Brooks, D., Blau, G., Pekny, J., Reklatis, G.: Improved wind power forecasting with ARIMA models. Comput. Aided Chem. Eng. **29**, 1789–1793 (2011)
13. Evensen, G.: The ensemble Kalman filter: theoretical formulation and practical implementation. Ocean Dyn. **53**(4), 343–367 (2003)
14. Welch, G., Bishop, G.: An introduction to the Kalman filter, vol. 8, no. 7, pp. 127–132 (1995)
15. Faragher, R.: Understanding the basis of the Kalman filter via a simple and intuitive derivation [lecture notes]. IEEE Signal Process. Mag. **29**(5), 128–132 (2012)
16. Fried, R., George, A.C.: Exponential and holt-winters smoothing. In: Lovric, M. (ed.) International Encyclopedia of Statistical Science, pp. 488–490. Springer, Heidelberg (2011). https://doi.org/10.1007/978-3-642-04898-2_244

ASLM: Adaptive Single Layer Model for Learned Index

Xin Li, Jingdong Li, and Xiaoling Wang[⊠]

Shanghai Key Laboratory of Trustworthy Computing,
MOE International Joint Lab of Trustworthy Software,
East China Normal University, Shanghai, China
xinli@stu.ecnu.edu.cn, lljjdd567@gmail.com, xlwang@sei.ecnu.edu.cn

Abstract. Index structures such as B-trees are important tools that DBAs use to enhance the performance of data access. However, with the approaching of the big data era, the amount of data generated in different domains have exploded. A recent study has shown that indexes consume about 55% of total memory in a state-of-the-art in-memory DBMS. Building indexes in traditional ways have encountered a bottleneck. Recent work proposes to use neural network models to replace B-tree and many other indexes. However, the proposed model is heavy, inaccuracy, and has failed to consider model updating. In this paper, a novel, simple learned index called adaptive single layer model is proposed to replace the B-tree index. The proposed model, using two data partition methods, is well-organized and can be applied to different workloads. Updating is also taken into consideration. The proposed model incorporates two data partition methods is evaluated in two datasets. The results show that the prediction error is reduced by around 50% and demonstrate that the proposed model is more accurate, stable and effective than the currently existing model.

Keywords: B-tree · Neural networks · Model · Updating

1 Introduction

When talking about efficient data access, we are talking about data access through indexes. Indexes have been through thorough development during the past decades due to their importance for different applications [6–9]. Among all kinds of indexes, B-tree is the one most used and supported by different kinds of database systems. In the big data era, however, a large amount of main memory is consumed by B-Tree or its different variants. This would reduce the amount of valuable space, which ought to be used to store some important intermediates and to process existing data. Meanwhile, with the increase of tree structure, a large part of it would be held up in disks. For online transaction processing intensive workloads, frequently I/Os are unavoidable, and the cost is completely unacceptable. To summarize, traditional indexes have encountered bottlenecks in both space and time when storing and processing large-scale data.

G. Li et al. (Eds.): DASFAA 2019, LNCS 11448, pp. 80–95, 2019.
https://doi.org/10.1007/978-3-030-18590-9_6

To reduce the storage overhead of B-trees, various compression techniques [6,10,18,22,23] have been introduced to remove the redundancy between keys or the size of keys. These techniques would cause a higher runtime cost for that decompressing must be done to search for a data point. These compressed indexes still need a significant amount of memory if they are many distinct keys for that one entry (key and pointer) ought to be contained for each different value. Based on this observation, A-Tree [4] proposed to index a fraction of keys and to use interpolation search between keys. Linear functions are used to approximate the data in leaf nodes. However, linear functions would fail to approximate other data distribution patterns other than linear distribution.

Trying to solve the current problems, former work [1] inspired by the mixture of experts work [2] proposes a kind of recursive regression model to implement B-tree. The main idea of this staged model is to use neural networks (or other machine learning models) to learn the distribution of data and give predictions based on it. In this work, each stage of the staged model can be seen as one layer of B-tree, and the selection of the next stage model depends on the output of the previous stage. However, the staged model is heavy, and the segmenting method adopted by this model has not considered the correlation between data, which would cause an increase in prediction error. The model cannot deal with updating also, which is impractical.

In this paper, we propose an adaptive single layer model (ASLM). Compared with the state-of-the-art work, the proposed model is light-weighted and more intelligent in partitioning workloads. The design of ASLM improves the querying time and prediction accuracy effectively. At the same time, we also propose updating strategies based on ASLM, both insertion and deletion are considered.

The main contributions of this work can be summarized as follows:

(1) We propose ASLM, and the memory occupation and querying time of the proposed model is much smaller than existing learned index structure.
(2) Compared with the previous model, we propose better data partitioning strategies, which significantly reduce the prediction error by around 50%.
(3) We propose update strategies, including insert and delete, which were not found in previous works.
(4) We perform experiments on two existing datasets. Results show that our model is more accurate than the current model and remains stable, effective during the process of updating.

The remainder of the paper is organized as follows. The next section introduces existing work related to this topic and its main drawbacks. In Sect. 3, we present an overview of ASLM with two workload partitioning strategies in detail. In Sect. 4, model updating is introduced (insertion and deletion are included). In Sect. 5 we introduce experiment settings and two datasets used in our experiment. Afterward, we discuss the experiment results and give brief explanations of the results in this section. In Sect. 6 we conclude this paper.

2 Related Work

2.1 B-Trees and Variants

B-tree has experienced great development, and to reduce the cost of storage, many of its variants have been proposed. B+-Trees [3] were proposed for disk-based systems and several caches conscious B+-tree variants were proposed, such as CSB+-tree [11]. T-trees [12] or balanced/red-black trees [13,14] were proposed for in-memory systems. For read-heavy workloads, some tree structures such as FAST [15] proposed to make use of hardware features (e.g., SIMD) or even GPUs [15–17].

However, with the rapid expansion of data volume, the space occupied by indexes increases, and the capacity of the memory is limited. As a result, most of the index will be held up on disk, and for online transaction processing intensive workloads, frequently I/O operations are required, resulting in inefficient data access.

A-trees [4], BF-trees [18] and B-tree interpolation search [19] are the ones closely related to our work. A-trees use linear functions, which is piece-wise to reduce the number of leaf nodes in traditional B-trees. Similar to A-tree, BF-tree uses a B+-tree to store information about a region of the dataset. However, BF-tree uses bloom filters as leaf nodes instead of linear functions. BF-tree does not consider the distribution of data either when segmenting workloads. Work [19] proposed to use interpolation search within a B-tree page.

2.2 Learned Index Structure

In work [1], the author suggests that almost all indexes used by database systems can be highly optimized as long as the exact data distribution is known. Machine learning, NN especially, can learn the patterns and correlations in data and thus enable the automatic synthesis of specialized index structures, termed learned indexes, with lower engineering cost. Taking B-tree as an example, a hierarchical structure model called recursive model index (RMI) is proposed (See Figs. 1 and 2). The division of data for the second stage of models depends

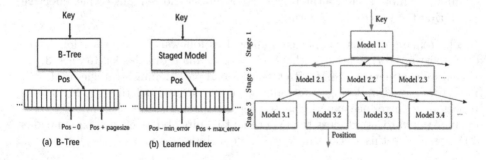

Fig. 1. B-tree index is a model   Fig. 2. Staged model [1]

on the prediction from the previous stage. The first stage gives a prediction in the range from 0 to N. If the RMI has K models in the second stage, then any points with a prediction in the range [0, N/K] go to model 1, points with predictions in the range [N/K + 1, 2 * N/K] go to model 2, etc. Similar to B-tree, each stage of this staged model further narrows the search region. The selected model at each stage takes the same key as an input and selects another model in the next stage based on the output of this model until the final stage gives the position prediction.

3 Proposed Model

3.1 Motivation

The partitioning strategy adopted by RMI does not adequately consider the similarity between data separated in the same model. As can be seen from Fig. 3, two data points are far apart and are divided into the same model, however. It will be easier for NN to fit similar data rather than disordered data. Meanwhile, the proposed RMI cannot deal with updating properly. Taking insertion as an example. When new data is inserted into RMI, retraining is needed to fit these newly inserted data. Not only should the model in the last stage that the data finally inserted in be retrained, but those models above the last stage concerning the inserted data should also be retrained. It is a recursive retraining process that cannot be avoided, and the cost of such updating strategy is unacceptable.

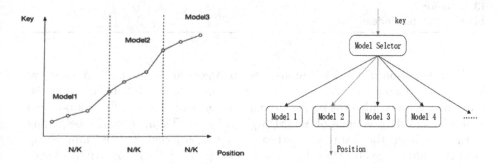

Fig. 3. Partition method of RMI   **Fig. 4.** Single layer model (SLM)

3.2 Initial Solution

Based on the above observations, we propose a better data partitioning algorithm and a single layer model (See Fig. 4).

3.2.1 Data Partition Method

The partition method adopted by RMI would eventually result in the inaccuracy of the whole model. Two factors have impacts on the accuracy of the model: volume and distribution of data. Based on the greedy strategy, data partition method tries to achieve a balance between these two factors. If the data trained in the same model is in great volume, then they should be well distributed, which means the distance between two adjacent data points should be as small as possible. If the data trained in the same model is in chaos, then the data volume should be low.

Algorithm 1. Data Partition Algorithm

Input: workload, K
Output: split_index
1: SUM = 0, split_index = [];
2: **for** $i = 0$; $i < workload.length$; $i + +$ **do**
3:     SUM = SUM + (workload[i+1]- workload[i]);
4: **end for**
5: V = SUM/K;
6: temp_Sum = 0
7: **for** $i = 0$; $i < workload.length$; $i + +$ **do**
8:     temp_Sum = temp_Sum + (workload[i+1]- workload[i]);
9:     **if** $temp\_Sum > V$ **then**
10:         split_index.append(i)
11:         temp_Sum = 0
12:     **end if**
13: **end for**
14: **return** split_index;

Data partition method is summarized in Algorithm 1. In the beginning, we have to specify how many sub-models the SLM has (Value of K). Then all the distance between two adjacent keys is summed up as SUM (Line 3). Each sub-model would have a data volume of V = SUM/K (Line 5). Then, from the beginning of the workload, the distance between every two adjacent keys is summed up until the value is greater than V, break at that point (Line 7–13). After all this, we finally have K pieces of data. Such a division method guarantees that data points are separated into the same part if their keys are similar to each other and data points are split into different pieces if their keys are different. Models can achieve better accuracy if all the data points are close to each other even if the data volume is great, and if the data in the same model is different from each other, the model's accuracy can also be ensured due to the small volume of data. This method is an improved version of the method of equal separation, and it works better than the method taken by RMI.

3.2.2   Single Layer Model (SLM)

Based on the data partition method, we proposed a single layer model (See Fig. 4). It is a simple model which only has one layer of K sub-models. This SLM can deal with updating well for that insertion (or deletion) only happens in one layer, and only one or two sub-models are involved in this layer.

3.3   Optimized Solution

The method introduced above only considers the distance between keys and leaves positions aside. We also find that the setting of K is a problem when partitioning an unknown workload.

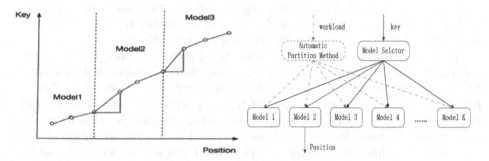

Fig. 5. Partition method of RMI   **Fig. 6.** Adaptive Single Layer Model (ASLM)

3.3.1   Automatic Data Partition Method

To solve the problem of the setting of K, in this method, the distance (Both keys and positions are considered) between data points is calculated and used as a feature to partition workloads automatically. One heuristic idea is that the area of the triangle (See Fig. 5) between two adjacent data points can reflect the similarity between data points.

Automatic data partition method is summarized in Algorithm 2. First, the areas of all triangles are calculated and sorted in descending order (Line 1–5). Second, the top k (different from K in method 1) areas in that descending order are selected as the splitting points (Line 6). The selected split indexes should be through the boundary check process (Line 7). Boundary setting is used to put a constraint on the data volume of the sub-model. Training too much data in one sub-model would cause a long training time and inaccuracy of the model. Training fewer data in one sub-model would cause an increase in the total sub-model number and the storage cost. Thus, for each sub-model, the upper bound and lower bound are set. After many experiments, we find that when the upper bound is set to 10k, the results would be better. Due to space limitations, it will not be described in detail here. The setting of k depends on the requirement for

---

Algorithm 2. Automatic data partition method

Input: workload
Output: split_index
 1: area = [];
 2: **for** $i = 0$; $i < workload.length$; $i + +$ **do**
 3:    area.append(Area_of_Triangle(workload[i], workload[i+1]));
 4: **end for**
 5: area.sort();
 6: *split_index ← top k of area*;
 7: Boundary Check;
 8: **return** split_index;

---

accuracy and efficiency. Setting a large number to k would cause wasting of time during the boundary check process. Setting a small number to k would result in inaccuracy of the model for that the workload is not well divided.

The method introduced above is a heuristic idea after all, and we have to give a normalized definition of how to measure the similarity between data points. Euclidean distance is usually used to define the similarity between data. Therefore, we propose to replace the areas of triangles with Euclidean distance, and the rest operations remain the same as the heuristic method for that there is a proportional relationship between them. One most significant advantage of this method is that this method helps to decide how many sub-models (Value of K) the SLM should have for different workloads and the decision is far smarter than our intuitive experience. Using automatic data partition method along with the SLM, the final model: Adaptive Single Layer Model is proposed next.

3.3.2   Adaptive Single Layer Model (ASLM)

As can be seen from Fig. 6, first to do is workload distribution analysis. Choosing the area of the triangle or the Euclidean distance as the feature to divide the workload in this step, and after automatic data partition method, the workload is partitioned into K pieces and K sub-models are created accordingly. The same network structure can be applied to all the sub-models or different structures for different sub-models. We finally use the same structure for two reasons. For one thing, the data is sorted according to the key, and there is not much difference between different pieces of data, the same structure will be fine for all sub-models. For another, when the training of one sub-model is finished, its weights could be used to initialize the sub-models next. The training process would be faster than starting training from zero.

The K pieces of data are distributed to K sub-models, and the next step is model training. As all the data have been sorted and the model does not need to be trained from zero except for the first one, the training process would be quick. At the same time, the min-error and max-error of each sub-model are remembered during training to provide the same semantic guarantees that B-tree provides.

The model is designed to be adaptive, and it has to support model updating. However, the model cannot be updated every time one data point is inserted into it, which means the model retraining. The cost is unacceptable, especially when there is a vast amount of data to be inserted. Considering all these, a buffer is maintained for each sub-model. The buffer in each sub-model is organized like a hash table in order to answer related queries quickly, which means if some data is inserted into the buffer but has yet to be updated into the model, queries about these data can be answered promptly and precisely.

Supposing that the proposed ASLM has been well trained and has experienced so many operations like queries, inserts, etc. For the most general circumstances, model querying should follow the following steps: model selector takes the query key as an input and decides which sub-model should be selected below; once the sub-model is selected, first thing of all is to check whether the query key is in this sub-model's buffer, and if so return the position immediately; if the key is not in the sub-model's buffer, the model should take this key as an input and output the position prediction. To find the precise position of the query key, binary search around the predicted position in the range from min-error to max-error will do. The query speed is rapid, for there is no I/O cost and only a few mathematical multiplications and additions operations are needed compared with traditional B-tree.

The first data partition method can divide the workload in linear time and achieve better results, however, some problems exist in this method. The time complexity of the optimized method is O(nlogn). Compared with the first data partition method, the partitioning precision is further improved at the cost of the slightly increased of partitioning time. In actual scenario, data segmentation occurs in the data storage phase, and the efficiency of querying and updating would not be affected. The second partition method also enables the model to be adaptive to any workloads without specifying the exact sub-model number. Thus, the model in the following of this paper refers to ASLM.

4 Model Maintenance

4.1 Data Insertion

Concerning inserting, two circumstances should be taken into consideration: general inserting in the middle and appending to the end. Appending to the end is easier for ASLM, a new sub-model is created directly at the end. The new sub-model created for appending is like any other sub-models, a buffer is maintained to avoid being frequently retrained. At the beginning of appending, records should all be inserted in the buffer and no training is involved until the buffer is full. The weights of the last sub-model can be used to initialize the new sub-model. When model training is finished, the buffered should be cleared for new inserts. The appending process is terminated if the data volume of this sub-model exceeds the upper bound and a new sub-model is created.

Inserting in the middle is more complicated than appending to the end. The algorithm of inserting in the middle is summarized in Algorithm 3. At line 1, the

ASLM selects a sub-model according to the given key and gets the average error and the buffer of the selected sub-model at lines 2 and 3. Then the sub-model gives a position prediction according to the given key, and thus the prediction error of the model can be calculated for the key. If $|Position - predict\_pos| <=$ $average\_error$, this data record can be inserted directly into the selected sub-model without training (Line 5–6) for that the inserted record will not decrease the accuracy of the sub-model. Once some data is inserted into one of the sub-models, whether the amount of this sub-model's data exceeds the data upper bound should be checked (Line 7). The model splitting process is meant to be triggered if the condition is satisfied. Before model splitting, however, we have to check whether the buffer of the sub-model is empty, if some data is inside then we have to update these into the sub-model (Line 8–11).

Algorithm 3. Insertion of ASLM

Input: (Key, Position)
1: $model \leftarrow model\_selector(Key)$;
2: $average\_error \leftarrow model.average\_err$;
3: $Buffer \leftarrow model.Buffer$;
4: $predict\_pos \leftarrow model.predict(Key)$;
5: **if** $abs(Position - predict\_pos) <= average\_error$ **then**
6:     insert into model.data directly
7:     **if** $Size(model.data) > Upper\_bound$ **then**
8:         **if** $Size(Buffer) > 0$ **then**
9:             Update Buffer.data into model.data
10:        **end if**
11:        model.split()
12:    **end if**
13: **else**
14:    Buffer.insert(Key, Position)
15:    **if** $Size(Buffer) > Threshold$ **then**
16:        Update Buffer.data into model.data
17:        Buffer.clear()
18:        **if** $Size(model.data) > Upper_bound$ **then**
19:            model.split()
20:        **end if**
21:    **end if**
22: **end if**

If $|Position - predict\_pos| > average\_error$, this record should be inserted into the buffer (Line 14) for that inserting this record into this sub-model would reduce the model's accuracy. The buffer also has a threshold, and if the size of the current sub-model's buffer exceeds the threshold, the data in the buffer should be updated into the sub-model, which means the sub-model should be retrained (Line 15–16). In the retraining process, the sub-model's average error, min-error and max-error should be updated accordingly, and the buffer of this sub-model should be cleared after this procedure (Line 17). Each sub-model has a data volume upper bound, and when the data covered by this sub-model exceeds its upper bound, the model split process is meant to be triggered (Line 18–19).

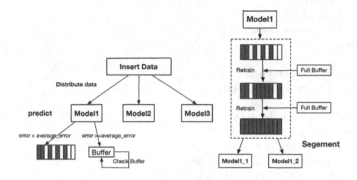

Fig. 7. Model insertion

Model splitting is like node splitting in B-tree index; however, the model splitting is much easier. Compared to the complex splitting strategy of B-tree, the splitting of our model only involves the current sub-model. As is shown in Fig. 7, after so many inserts, data volume is likely to exceed the sub-model's upper bound, and if so a new sub-model should be created and can be initialized with the previous sub-model's weights. The breakpoint can be found using the automatic data partition method. The two sub-models should be trained separately based on the data they cover. This process is quick, for all the data originally belongs to one sub-model, and that sub-model has already been trained to fit these data. The weights of the original sub-model can be used to train the new model in order to reduce the training time.

4.2 Data Deletion

The first thing to do is querying when deleting a data point. Querying is to ensure that the record does exist in the model and to make sure which sub-model it lies in. The algorithm of deletion is summarized in Algorithm 4.

After a bulk of insertions, there must be some data in the buffer of each sub-model. If the record to be deleted lies in one sub-model's buffer, then we can remove it directly from the buffer without any impacts on the sub-model's error (Line 7–8).

If the record to be deleted lies in one of the sub-models, removing it directly from the model does not affect the model accuracy actually for the model itself has not changed. At the same time, the counter, used to count how many records have been deleted from this sub-model, should be auto incremented (Line 10–11). The amount of the remaining data should be calculated to check whether it is less than lower bound (Line 12). When the data volume is smaller than the lower bound (Line 13), merging sub-models should be considered. The merging

---

Algorithm 4. Deletion of ASLM

Input: Key

 1: $model \leftarrow model\_selector(Key)$;
 2: $counter \leftarrow model.counter$;
 3: $Buffer \leftarrow model.Buffer$;
 4: **if** $Key\ not\ in\ model$ **then**
 5:     **return**
 6: **else**
 7:     **if** $Key\ in\ Buffer$ **then**
 8:         delete from Buffer directly
 9:     **else**
10:         delete from model.data
11:         counter++
12:         $residual \leftarrow Upper\_bound\text{-}counter$;
13:         **if** $residual < Lower\_bound$ **then**
14:             **if** $Size(Buffer) > 0$ **then**
15:                 **continue**
16:             **end if**
17:             left_data_len = Len(model_left.data)
18:             right_data_len = Len(model_right.data)
19:             **if** $residual + left\_data\_len$ $<$ $Upper\_bound$ **or** $residual +$ right_data_len $< Upper\_bound$ **then**
20:                 model_converge()
21:             **end if**
22:         **end if**
23:     **end if**
24: **end if**

---

of models is not necessary. Actually, data can always be deleted from one sub-model until there is no data left and the sub-model is removed. However, there is no need to maintain a sub-model if the data volume left is too small. The merging of sub-models may bring up additional overhead, but to free up some of the memory, the merging is worth consideration. Before merging, whether the buffer of this sub-model contains data points should be considered. If there exists some data in the buffer (Line 14), updating these data into this sub-model must trigger the model retraining process and bring up additional cost. In this case, merging between two sub-models may not happen in order to be efficient at the expense of additional memory occupation (Line 15). The merging of models involves two sub-models next to this sub-model, which means this sub-model may be merged into its left sub-model or its right one. Into which sub-model should the current sub-model be merged depends on the amount of data of its two adjacent sub-models (Line 17–20). If sub-models on both sides do not meet the data volume requirements, no merging would happen and leave it be. All the above is the whole implementation of model deletion.

5   Experiment

5.1   Experiment Setup

Dataset: Two datasets are used in our experiments: Map_Data and TPC-H. Map_Data is a small part of our extraction from the OpenStreetMap [20]. It has approximately 2 million records. TPC-H is extracted from TPC-H [21] benchmark and it has approximately 14 million records. It is divided into two datasets equally (TPC-H1 and TPC-H2), one for model training and the other for model update experiments (insertion data). Each of these two datasets has approximately 7 million records.

Settings: To be fair, in the model comparison experiment, the original model design in work [1] is followed. For RMI, a two-layer fully-connected neural network with 32 neurons per model is trained using ReLU activation function. The same settings are used in ASLM, which means each sub-model has 32 neurons and uses ReLU activation function. Due to the adaptability of the proposed model, experiment on the proposed ASLM is carried out first and gets the final model number. Then the number of models is assigned to K, which is the number of models in RMI's second layer and the number of sub-models in SLM with the first proposed data partition method.

Environment: All experiments are conducted on Windows 10 platform with Intel(R) Core(TM) i5 CPU @3.2 GHz and 16 GB RAM. All models are implemented under Keras framework using TensorFlow [5] as the backend and Python as the front-end, and all the other experimental environments are the same.

5.2   Model Comparison

We evaluate the performance of our model versus RMI from the perspective of prediction accuracy and time consumption. The predicting and time consumption results are shown in Table 1 and Fig. 8.

In the table, the third column represents the average error of prediction and the fourth, fifth and sixth columns indicate the proportion of the error less than five, ten and fifty respectively. It can be seen from the results that our model achieves better accuracy with the three dataset partitioning methods we propose. There are only subtle differences between the results using the area of the triangle and Euclidean distance, which may be caused by the training process or some other tinny factors.

As is shown in Fig. 8, we have done querying 100 times on different sub-models, and the results show that the average time consumed by RMI is almost two times than that of SLM. The proposed model performs better than RMI for that the proposed ASLM only has one layer and RMI has at least two layers, which means that the consumption of time will at least be doubled.

The time consumption between B-tree and ASLM is not compared for specific reasons. RMI in work [1] is implemented under the learning index framework (LIF). It is a framework designed for index configuration and optimization

Table 1. Prediction error of the proposed model and RMI

Map_Data	Model	Average error	Error < 5	Error < 10	Error < 50
	RMI	6.07	74.4%	83.5%	97.9%
	SLM (data partition method)	2.06	87.94%	95.90%	99.99%
	ASLM (area of triangle)	1.74	91.71%	97.94%	99.99%
	ASLM (Euclidean distance)	1.60	92.99%	97.94%	99.99%
TPC-H1	Model	Average error	Error < 5	Error < 10	Error < 50
	RMI	51.58	17.57%	36.37%	94.92%
	SLM (data partition method)	16.79	18.08%	36.75%	97.56%
	ASLM (area of triangle)	13.93	19.37%	39.40%	99.71%
	ASLM (Euclidean distance)	15.16	18.41%	37.50%	98.49%

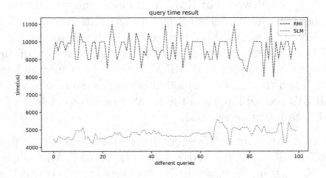

Fig. 8. Query time comparison

and can be seen as an acceleration framework that optimizes some unnecessary overhead and simplifies the implementation of the model. Unfortunately, the framework is currently not open-sourced, and thus we cannot implement ASLM under LIF. However, as is stated in [1], the querying of RMI implemented under LIF is faster than traditional B-tree. We argue that our model is faster than the stage model both in training time and in giving prediction time (See Fig. 8). The proposed model's time consumption will be much less than B-tree if it is implemented under LIF.

5.3   Model Updating

To test the performance of ASLM during insertion, we trained ASLM using TPC-H1 and TPC-H2 is used as external data for insertion. The first 1 million records in TPC-H1 are used as model training data in our experiment. 300,000 pieces of data are randomly selected from the first 1 million pieces of data in TPC-H2 as insertion data. The sub-models being inserted are evaluated from the perspective of the average error, which represents the accuracy of the model prediction. The original ASLM has 155 sub-models after training, and after insertion, it has 178 sub-models. The additional 23 sub-models are caused by model splitting or model creating during the process of insertion. The average error before and after insertion of the original 155 sub-model is shown in Fig. 9(a). The average error of the additional 23 sub-models is at the same level as the original 155 sub-models.

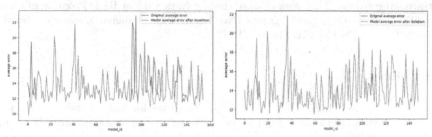

(a) Model average error before and after   (b) Model average error before and after
model insertion   model deletion

Fig. 9. Model average error

If the proposed model is stable and effective, the model accuracy should remain unchanged or at the same level in the process of or after inserting data. As can be seen from Fig. 9(a), the model average error remains at the same level or reduces a lot after insertion, compared with the original model average error, which means the stability and effectiveness of our model are ensured.

To test the performance of the proposed model during deletion, the first 1 million records of TPC-H1 are used to train ASLM and 50,000 records of the training data are randomly selected as the data to be deleted from the model. The original ASLM has 155 models after training, and after deletion, it has 148 models. The merging of models causes the missing 7 models during deletion. The average errors before and after deletion of the remaining 148 models are shown in Fig. 9(b).

As can be seen from Fig. 9(b), the average error of most models remains unchanged, this is because the data been deleted is limited and is not sufficient to cause so many merges. However, there are still some merging operations around the model 10, 35, 78 and 110. We can see that the average error before and after deletion remains at the same level as the original error. Some model's average error is slightly higher than before, but it remains at the same level as its original average error.

6   Conclusion

In this paper, we present ASLM a new learned index structure that incorporates two workloads partition strategies to replace B-tree. The adaptiveness of the model is reflected in the ability to be adapted to different workloads and to support model updates well. We evaluate ASLM using two datasets and show that it can achieve better accuracy than the state-of-the-art learned index RMI. Meanwhile, experiment results show that the proposed model is relatively stable when updating, and the accuracy of the model does not change too much during the progress of updating.

Acknowledgement. This work is supported by National Key R&D Program of China (No. 2017YFC0803700), NSFC grants (No. 61532021), Shanghai Knowledge Service Platform Project (No. ZF1213) and SHEITC.

References

1. Kraska, T., Beutel, A., Chi, E.H., et al.: The case for learned index structures. In: Proceedings of the 2018 International Conference on Management of Data, pp. 489–504. ACM (2018)
2. Shazeer, N., Mirhoseini, A., Maziarz, K., et al. Outrageously large neural networks: the sparsely-gated mixture-of-experts layer. arXiv preprint arXiv:1701.06538 (2017)
3. Bayer, R., McCreight, E.: Organization and maintenance of large ordered indexes. In: Broy, M., Denert, E. (eds.) Software Pioneers, pp. 245–262. Springer, Heidelberg (2002). https://doi.org/10.1007/978-3-642-59412-0_15
4. Galakatos, A, Markovitch, M, Binnig, C, et al.: A-tree: a bounded approximate index structure. arXiv preprint arXiv:1801.10207 (2018)
5. Abadi, M., Barham, P., Chen, J., et al.: TensorFlow: a system for large-scale machine learning. In: OSDI 2016, pp. 265–283 (2016)
6. Graefe, G., Larson, P.A.: B-tree indexes and CPU caches. In: 2001 Proceedings of the 17th International Conference on Data Engineering, pp. 349–358. IEEE (2001)
7. Richter, S., Alvarez, V., Dittrich, J.: A seven-dimensional analysis of hashing methods and its implications on query processing. Proc. VLDB Endow. **9**(3), 96–107 (2015)
8. Fan, B., Andersen, D.G., Kaminsky, M., et al.: Cuckoo filter: practically better than bloom. In: Proceedings of the 10th ACM International on Conference on Emerging Networking Experiments and Technologies, pp. 75–88. ACM (2014)
9. Alexiou, K., Kossmann, D., Larson, P.Å.: Adaptive range filters for cold data: avoiding trips to Siberia. Proc. VLDB Endow. **6**(14), 1714–1725 (2013)
10. Zhang, H., Andersen, D.G., Pavlo, A., et al.: Reducing the storage overhead of main-memory OLTP databases with hybrid indexes. In: Proceedings of the 2016 International Conference on Management of Data, pp. 1567–1581. ACM (2016)
11. Rao, J., Ross, K.A.: Making B$^+$-trees cache conscious in main memory. In: ACM SIGMOD Record, vol. 29, no. 2, pp. 475–486. ACM (2000)
12. Lehman, T.J., Carey, M.J.: A study of index structures for main memory database management systems. In: Proceedings of the VLDB, p. 1 (1986)
13. Bayer, R.: Symmetric binary B-trees: data structure and maintenance algorithms. Acta Inform. **1**(4), 290–306 (1972)
14. Boyar, J., Larsen, K.S.: Efficient rebalancing of chromatic search trees. J. Comput. Syst. Sci. **49**(3), 667–682 (1994)
15. Kim, C., Chhugani, J., Satish, N., et al.: FAST: fast architecture sensitive tree search on modern CPUs and GPUs. In: Proceedings of the 2010 ACM SIGMOD International Conference on Management of Data, pp. 339–350. ACM (2010)

16. Shahvarani, A., Jacobsen, H.A.: A hybrid B$^+$-tree as solution for in-memory index-ing on CPU-GPU heterogeneous computing platforms. In: Proceedings of the 2016 International Conference on Management of Data, pp. 1523–1538. ACM (2016)

17. Kaczmarski, K.: B$^+$-tree optimized for GPGPU. In: Meersman, R., et al. (eds.) OTM 2012. LNCS, vol. 7566, pp. 843–854. Springer, Heidelberg (2012). https://doi.org/10.1007/978-3-642-33615-7_27

18. Athanassoulis, M., Ailamaki, A.: BF-tree: approximate tree indexing. Proc. VLDB Endow. **7**(14), 1881–1892 (2014)

19. Graefe, G.: B-tree indexes, interpolation search, and skew. In: Proceedings of the 2nd International Workshop on Data Management on New Hardware, p. 5. ACM (2006)

20. OpenStreetMap. https://www.openstreetmap.org

21. Transaction Processing Performance Council. http://www.tpc.org/tpch/

22. Bayer, R., Unterauer, K.: Prefix B-trees. ACM Trans. Database Syst. (TODS) **2**(1), 11–26 (1977)

23. Zukowski, M., Heman, S., Nes, N., et al.: Super-scalar RAM-CPU cache compres-sion. In: 2006 Proceedings of the 22nd International Conference on Data Engineer-ing, ICDE 2006, p. 59. IEEE (2006)

SparseMAAC: Sparse Attention for Multi-agent Reinforcement Learning

Wenhao Li, Bo Jin$^{(\boxtimes)}$, and Xiangfeng Wang

Shanghai Key Lab for Trustworthy Computing,
School of Computer Science and Software Engineering,
East China Normal University, Shanghai, China
{bjin,xfwang}@sei.ecnu.edu.cn

Abstract. In multi-agent scenario, each agent needs to aware other agents' information as well as the environment to improve the performance of reinforcement learning methods. However, as the increasing of the agent number, this procedure becomes significantly complicated and ambitious in order to prominently improve efficiency. We introduce the sparse attention mechanism into multi-agent reinforcement learning framework and propose a novel Multi-Agent Sparse Attention Actor Critic (SparseMAAC) algorithm. Our algorithm framework enables the ability to efficiently select and focus on those critical impact agents in early training stages, while eliminates data noise simultaneously. The experimental results show that the proposed SparseMAAC algorithm not only exceeds those baseline algorithms in the reward performance, but also is superior to them significantly in the convergence speed.

Keywords: Multi-agent deep reinforcement learning ·
Sparse attention mechanism · Actor-attention-critic

1 Introduction

1.1 Multi-agent Deep Reinforcement Learning

Reinforcement learning is one of the most popular research area in academia and industry. Its goal is to maximize the cumulative feedback while the agent interacts with the environment, as a result the agent can guarantee an optimal strategy. The biggest limitation of traditional reinforcement learning is the demand to manually design features to model the state of environment, which significantly increases the difficulty of extending to sophisticated task. With the development of deep learning, representational learning in complex tasks has become achievable, and deep reinforcement learning has recently achieved remarkable results in games [19], robotics [5], and autonomous driving [2].

Our work focuses on the multi-agent reinforcement learning (MARL) [1]. Actually in many practical problems, there are multiple agents that need to controlled more intelligently, e.g., multi-robot control [13], multiplayer games

G. Li et al. (Eds.): DASFAA 2019, LNCS 11448, pp. 96–110, 2019.
https://doi.org/10.1007/978-3-030-18590-9_7

[20] and etc. These agents always affect each other while completing their own targets at the same time, so that the single agent reinforcement learning algorithms can not be directly applied to multi-agent scenario [22]. The process of state transition in multi-agent system can be principally described as: At a certain time step, the environment receives the joint actions of all agents, further moves to the next state with reward returned from each agent. Multi-agent problems can be divided into two categories according to the relationship between agents, i.e., collaboration problem and competition problem [1]. In the competition problem, the agents of different teams are independent of each other, that is, they have their own reward functions. The goal of each team is to only maximize the cumulative rewards belong to themselves. On the contrary, in collaboration problem, the goal of all agents is no longer to maximize the cumulative rewards they receive individually, but to maximize whole return of all agents in groups. Furthermore in general, all agents share the same value function (objective) in the collaboration problem.

1.2 Attention Mechanism in Multi-agent Reinforcement Learning

The attention mechanism automatically extract the semantic information with respect to each task prior information through an end-to-end manner. This prominent advantage of attention mechanism has been greatly extracted in recent years and it has many successful applications in the field of computer vision [27], natural language processing [11,25] and even reinforcement learning [6,7,18].

The Multiple Actor-Attention-Critic algorithm (MAAC) [6] is a typical attention mechanism based multi-agent reinforcement learning method. MAAC learns the multi-agent system through one centralized critic and many separated decentralized actors. With the purpose to solve the limitations of both traditional value function method and strategy gradient method on multi-agent problems, MAAC borrowed the basic idea of Multi-Agent Deep Deterministic Policy Gradient method (MADDPG) [10], which allows introducing extra agent information in the training process. Further in details, the critic is augmented with extra information about the policies of other agents. As the increasing of agents number, the information of each agent which needs to process or communicate in order to make decision is significantly growing. If each agent considers the behaviors of all other agents, the decision-making learning procedure becomes extremely more difficult. For a large-scale multi-agent system, each agent's decision is not affected by all other agents. Furthermore, it will inevitably make the useful signal submerged in the background noise once all other agents are equally taken into account. The MAAC algorithm introduced the attention mechanism to address this problem, by sharing an attention mechanism which selects relevant information for each agent at every timestep.

However, MAAC still take advantages of all the agents although the utilization level is considered to increase accuracy and efficiency. In practical applications, the tasks performed by different agents maybe quite different. Therefore,

in the process of learning, each agent needs to select the related agents while filter out the independent ones. This selection scheme should not only focus on the
degree judgement but more importantly on agent picking. In order to solve this
problem, we combine the sparse selection technique with classical MAAC algorithm and propose a multi-agent sparse attention actor critic algorithm (Sparse-
MAAC) for multi-agent reinforcement learning problem. Our main contributions
are summarized as follows:

1. We introduce the sparse attention mechanism into the multi-agent reinforcement learning combined with MAAC algorithm, which enables our algorithm
   to quickly filter out the useful parts from the received information in the
   complex environment and eliminate the noise data. This allows the algorithm
   to have a faster convergence or the ability to jump out of a local optimal
   solution. Through the guaranteed sparse attention weights, the established
   system can be more interpretable.
2. Based on the related work [16], we introduce the hyperparameters of the
   control algorithm sparsity of sparsemax [12], as a result prior environment
   information can be easily introduced to our algorithm framework in order to
   acclimate varied complexity environments.
3. To better verify the effects of sparse attention, we design a more complex
   collaborative environment called *Grouped Cooperative Treasure Collection*
   (GCTC), and conduct related and significative experiments performance.

The rest of the paper is organized as follows. In Sects. 2 and 3, we discuss
related work and backgrounds, followed by a detailed description of our approach
in Sect. 4. We report experimental results in Sect. 5 and conclude our paper in
Sect. 6.

2   Related Work

2.1   Sparse Attention Mechanism

The attention mechanism has been widely applied in deep learning in recent years
because of its automatically output driven learning performance. The attention
mechanism was first successfully applied in [27] which automatically generates a
description that matches the content of the pictures. The images are extracted
by convolutional neural network (CNN) and further combined with the implicit
state of long short-term memory (LSTM). A multi-layer perceptron is introduced
to calculate the similarity, and the softmax function is used to compute the
attention weights. The authors also put forward the concept of hard attention
and soft attention which are significantly different.

Then [11] extended the attention mechanism into the machine translation
application. By considering the partial relationship between output word and
input words, they proposed both global attention and local attention schemes.
The attention mechanism is regularly and prosperously combined with recurrent
neural networks (RNN). However because of the recurrent nature of RNN, it

is not conducive to parallel computing, so that usually the training procedure always be time-consuming. [25] uses a self-attention mechanism instead of a circular neural network, while both scaled dot-product attention and multi-head attention scchemes are introduced.

The softmax function are usually introduced to calculate the attention weights, which is dense without any possibility to select agents. Although both [27] and [11] attempt to introduce sparsity by adding special structures, like hard attention or local attention, they ignore introducing sparse constraints on activation function. Reference [12] introduced sparsity into the attention mechanism by seeking a sparse activation function for the first time. The core idea is to take advantage of the fact that the Euclidean projection of any input vector to a simplex is sparse. Reference [16] found that it also falls into the above situation by calculating the max function subgradient. By considering the discontinuity of this projection, the authors introduced a strongly convex regularizer on the dual problem of the max operator with the purpose to guarantee efficient training. The sparsemax algorithm in [12] can be included into the algorithm framework of [16] In our paper, we will utilize the generalized γ-sparsemax algorithm in [16].

2.2 Attention Mechanism in Reinforcement Learning

Attention mechanism is also popularly applied to reinforcement learning methods. Reference [17] used reinforcement learning to conduct brain activity research and introduced attention mechanism to screen input signals in order to accelerate the training process. Reference [18] considered the memory structure in the reinforcement learning algorithm to preserve the knowledge learned by the agent. In decision-making procedure, the attention mechanism is introduced to select relevant information from the memory with the purepose to assist decision-making. A soft attention mechanism combined with the Deep Q-Network (DQN) [14] model is proposed in [15] to highlight task-relevant locations of input frames.

In the field of multi-agent reinforcement learning, [6] combines the attention mechanism with the actor-critic algorithm to train a centralized critic that automatically selects relevant information. [7] also proposed an attention-based actor-critic algorithm. Their main concerns are on learning attention models for sharing information between policies. Our work in this paper is basiclly based on [6], but firstly and originally designs a sparse attentional mechanism for multi-agent reinforcement learning, which strengthens the algorithm stability for complicated and noisy environments and guarantee better interpretability.

3 Preliminaries and Backgrounds

3.1 Markov Decision Process and Markov Game

Reinforcement learning models the process in which an agent learns in constant interaction with the environment as a Markov decision process (MDP) [21]. An MDP is defined by a quintuple $\langle S, A, R, P, \gamma \rangle$ where S and A represent a limited state space and a limited action space respectively. The probability transfer matrix $\mathcal{P}$ is a three-dimensional tensor, where each element represents the probability that the agent will move to the next state s' after executing action a in state s. The reward function $\mathcal{R} : \mathcal{S} \times \mathcal{A} \rightarrow \mathbb{R}$ defines the instant feedback made by the environment after the agent executes the action a in the state s and this feedback can be stochastic. Finally, $\gamma \in [0,1]$ represents the discount factor that defines the trade-off between the immediate feedback and the accumulated future feedbacks during the agent learning process

Since our work focuses on multi-agent situation, at the beginning we present a formal definition of this problem. A Markov game $\langle \mathcal{S}, \mathcal{N}, \mathcal{A}, \mathcal{R}_{\mathcal{N}}, \mathcal{P}, \gamma \rangle$ denotes an extension of the Markov decision process in multiple agent scenarios [9], while $\mathcal{N}$ represents the number of agents. The key differences with traditional MDP exist on the reward functions, i.e., $\mathcal{R}_{\mathcal{N}}$, while the probability transition matrix $\mathcal{P}$ and the policy π both depend on the actions of all agents in the environment.

3.2   Deep Q-Network

DQN is the most famous deep reinforcement learning method in recent years, which learns the optimal action value function corresponding to the optimal policy by minimizing the mean square bellman error (MSBE):

$$\mathcal{L}(\theta, \mathcal{D}) = \mathbb{E}_{(s,a,r,s') \sim \mathcal{D}} \left[Q(s, a; \theta) - Q_{\text{targ}} \right]^2, \tag{1}$$

where $s, s' \in \mathcal{S}$, $a \in \mathcal{A}$ and $r \in \mathcal{R}$ denote the state, action and reward respectively. $Q(s, a; \theta)$ denotes the action value function which is parameterized with θ. $Q_{\text{targ}} = r(s, a) + \gamma \max_{a'} \bar{Q}(s', a')$. Since DQN approximates the action value function Q with neural network, its convergence can not be guaranteed theoretically until now. In order to ensure the stability of the training procedure, traditionally the target function $\bar{Q}$ and the experience replay buffer $\mathcal{D}$ are introduced, where the parameters of $\bar{Q}$ are periodically updated with the behavior function Q's parameters.

3.3   Vanilla Policy Gradient and Actor Critic

The policy gradient-type method is another typical reinforcement learning algorithm framework. Its core idea is to learn policy directly rather than indirectly establish through learning value functions. More specifically, policy gradient method directly adjust parameters θ of the policy in order to maximize the objective function $J(\theta) = \mathbb{E}_{\tau} [R_{\tau}]$ by gradient ascent scheme with $\nabla_{\theta} J(\theta)$. In details the gradient takes the follow formulation [24]:

$$\nabla_{\theta} J(\theta) = \mathbb{E}_{s_t \sim p^{\pi}, a_t \sim \pi_{\theta}} \left[\nabla_{\theta} \log \pi_{\theta}(a_t | s_t) \sum_{t}^{T} r(s_t, a_t) \right]. \tag{2}$$

where $s_t \in \mathcal{S}$, $a_t \in \mathcal{A}$ denote the state and action at time t, $p^{\pi} = P(s_t | s_{t-1}, a_{t-1}) \times \pi(a_{t-1} | s_{t-1})$ and $\pi_{\theta} = \pi(a_t | s_t)$. The resulting method with the above formula is called vanilla policy gradient method (or REINFORCE method [26]). However, this method used a real reward in order to directly estimate the gradient, as a result the variance would be very large because of error accumulation. Lots of variant algorithms are proposed with the purpose to solve this shortcoming. Their core idea is to estimate the cumulative rewards with another function. This function is usually called *critic*, while the resulting algorithm framework is called *actor-critic* algorithm [23]. The advantage function $A(s,a) = Q(s,a) - V(s)$ ($V(s)$ denotes the value function at state s) is a popularly used critic function, which measures the pros and cons of one action relative to the average performance.

3.4   Soft Actor Critic

Soft Actor Critic (SAC) [4] is an algorithm of training stochastic policy through off-policy way. The entropy regularization plays an important role in the algorithm framework. The objective of SAC is to maximize a tradeoff between cumulative rewards together with the entropy of the policy distribution. Each agent is encouraged to pay more attention to exploration in the early stage of training while also avoids converging to a local optimal solution. In SAC algorithm framework, the goal becomes to maximize the following objective function, i.e.,

$$\pi^* = \arg\max_{\pi} \mathbf{E}_{\tau \sim \pi} \left[\sum_{t=0}^{\infty} \gamma^t (r(s_t, a_t) + \alpha H(\pi(\cdot|s_t))) \right], \tag{3}$$

where $s_t \in \mathcal{S}, a_t \in \mathcal{A}$ denote the state and action at time t, $\gamma \in (0,1]$ denote the discount factor and $\alpha > 0$ is the trade-off coefficient. Then state value function and action value function are calculated as

$$V_\pi(s) = \mathbf{E}_{\tau \sim \pi} \left[\sum_{t=0}^{\infty} \gamma^t (r(s_t, a_t) + \alpha H(\pi(\cdot|s)))|s_0 = s \right], \tag{4}$$

$$Q_\pi(s,a) = \mathbf{E}_{\tau \sim \pi} \left[\sum_{t=0}^{\infty} \gamma^t r(s_t, a_t) + \alpha \sum_{t=1}^{\infty} \gamma^t H(\pi(\cdot|s_t))|s_0 = s, a_0 = a \right]. \tag{5}$$

With these definitions, the corresponding MSBE's target term and policy gradient respectively become

$$Q_{\text{targ}}(s_t, a_t) = r(s_t, a_t) + \mathbf{E}_{s_{t+1} \sim p^\pi, a_{t+1} \sim \pi_\theta} \left[\bar{Q}(s_{t+1}, a_{t+1}) - \alpha \log(\pi_\theta(a_{t+1}|s_{t+1})) \right],$$

$$\nabla_\theta J(\theta) = \mathbf{E}_{s_t \sim p^\pi, a_t \sim \pi_\theta} \left[\nabla_\theta \log \pi_\theta(a_t|s_t)(\alpha \log \pi_\theta(a_t|s_t) - Q(s_t, a_t) + b(s_t)) \right],$$

where $b(s)$ is a state-dependent baseline for variance reduction [3]. Under the giving two formulations, the SAC algorithm alternately iteratively updates the *actor* and *critic* like the traditional *actor-critic* algorithm.

4   Algorithm Framework: SparseMAAC

In this section, we introduce our proposed **Multi-Agent Sparse Attention Actor Critic** (SparseMAAC) algorithm. As discussed above, the main motivation for our algorithm is the partial relationship between agents in a practical multi-agent system. Especially only a limited number of other agents have impacts on each agent's own decisions. Excessive consideration of other agents that have less influence will not only reduce the sampling efficiency, but also cause the convergence to a "worse" local optimal solution. The sparse attention mechanism introduced to solve these problems also leads to strong interpretability.

In the following, we introduce our sparse attention mechanism together with its variants, and then discuss our proposed SparseMAAC algorithm in details.

4.1   Sparsemax

The softmax function is commonly used for attention mechanism designing. Suppose the input of the softmax function is a K-dimensional vector $z \in \mathbb{R}^K$, we can guarantee the K-dimensional output vector $p \in \mathbb{R}^K$ which satisfies $\|p\|_1 = 1$. It is obviously that p belongs to a $K - 1$ dimension simplex set $\triangle^{K-1}$. The function $f : \mathbb{R}^K \to \triangle^{K-1}$ denotes the mapping to calculate attention weights.

Further in order to guarantee sparse attention weights, [12] proposed the sparsemax attention mechanism by introducing the projection operation to the target simplex set, i.e.,

$$\text{sparsemax}(z) = \arg \min_{p \in \triangle^{K-1}} \|p - z\|_2^2. \tag{6}$$

The optimal solution of problem (6) can be calculated in closed form [12], which greatly simplifies the process of extending sparsemax attention mechanism to multi-agent reinforcement learning algorithms.

4.2 γ-Sparsemax

Although sparse attention weights are obtained through the sparsemax mechanism, but it cannot handle the degree of sparseness. For multi-agent reinforcement learning algorithms, we can properly take advantages of the prior information on the degree of mutual influence between the agents, which can significantly accelerate the training process.

We introduce the general γ-sparsemax attention mechanism proposed in [16] to our multi-agent sparse attention actor critic algorithm framework. The corresponding optimization problem is defined as

$$\max(z) = \max_{i \in \{1,2,\cdots,K\}} (z_i) = \sup_{p \in \triangle^{K-1}} (p^T z), \tag{7}$$

where the subgradient $\partial \max(z) = \{e, e_i = 1, e_{-i} = 0 | i \in \arg \max_{i \in \{1,2,\cdots,K\}} z_i\}$. The subgradient $\partial \max(z)$ maps z to the $K - 1$ dimension simplex set. However this mapping can not be directly used due to its discontinuity, so that a strongly convex regularizer can be introduced. We can define the new mapping Π_Ω : $\mathbb{R}^K \rightarrow \triangle^{K-1}$ by solving the following optimization problem:

$$\Pi_\Omega(z) = \arg \max_{p \in \triangle^{K-1}} p^T z - \gamma \Omega(p), \tag{8}$$

where $\Omega(p)$ denotes a strongly convex function. When $\Omega(p) = \frac{1}{2}\|p\|_2^2$, the γ-sparsemax attention mechanism can be calculated by

$$\gamma\text{-sparsemax}(z) = \Pi_\Omega(z) = \arg \max_{p \in \triangle^{K-1}} \left\| p - \frac{z}{\gamma} \right\|_2^2. \tag{9}$$

The coefficient γ can be considered as the sparsity control hyperparameter. Smaller γ leads to more sparsity attention weights. This should be the first work to introduce self-adjusted sparsity machanism into multi-agent reinforcement learning algorithm, while a priori information about the degree of influence between agents can be introduced to guidance adjusting γ.

4.3   Multi-agent Sparse Attention Actor Critic

The MAAC algorithm modifies the framework of MADDPG [10] by replacing the deep deterministic policy gradient (DDPG) [8] with the soft actor critic (SAC). The Q-value function of each agent not only depends on its own state and actions, but also the strategies of all other agents which are utilized as additional information, i.e.,

$$Q_i^\pi(s, a) = Q_i(s, a_1, \cdots, a_n)|_{a_i = \pi_i(s_i)}. \tag{10}$$

For the agent i, all other agents have the same influence on each agent's decision-making procedure. The attention mechanism for constructing special structured Q-valued function plays the main role in MAAC, which indicates that different weights are used for different agents in the training procedure.

The attention distribution are usually continuous in MAAC, which means that every agent keeps attention on all other agents. However, in many practical applications, the agent's attention has some selective manner. For example, in soccer game, the striker doesn't need to pay attention to his own goal keeper while attacking. Therefore, the attention mechanism needs to extended with sparsity. In this paper, to introduce the sparse attention tactics, we propose the multi-agent sparse attention actor critic (SparseMAAC) algorithm. The main procedure is presented in Algorithm 1, and the key idea of SparseMAAC is shown in Fig. 1(a).

To calculate the attention weight of the remaining agents to the agent i, the attention module receives the observations $s = \{s_1, s_2, \cdots, s_n\}$ and actions $a = \{a_1, a_2, \cdots, a_n\}$ for all agent. Then all the data is divided into two parts, which belong to the agent i and not belong to the agent i, and the latter is denoted as $-i$, and j is used to index in $-i$.

Since the observation of the agent in reinforcement learning often contains a lot of noise, we, like the MAAC algorithm, first encode the agent's observation-action pair, then pass a bilinear mapping, and the final scaled output is used as input of the γ-sparsemax algorithm:

$$\alpha_j = \gamma\text{-sparsemax} \left(\frac{(W_k E_{-i})^T (W_q e_i)}{\sqrt{d}} \right)_j, \tag{11}$$

where d is the dimension of the encoding of tuple (s, a), $W_k, W_q \in \mathbb{R}^{n \times d}$, $E_{-i} \in \mathbb{R}^{d \times (n-1)}$. The j-th column $e_j \in \mathbb{R}^{d \times 1}$ of E_{-i} is the encoding of tuple (s_j, a_j) and $e_i \in \mathbb{R}^{d \times 1}$ is the encoding of tuple (s_i, a_i). We also have used the multiple attention heads to explicitly cluster the coding on the semantic level, so that the learned attention weight is more representative.

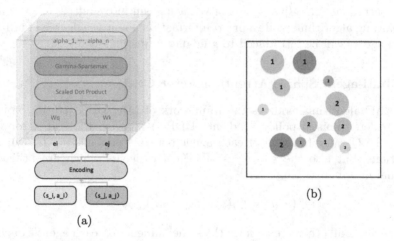

(a)   (b)

Fig. 1. (a) Multiple sparse attention head in SparseMAAC. (b) Grouped Cooperative Treasure Collection. The number in the circle represents group.

5   Experiments

5.1   Environments

We evaluation the performance of the proposed algorithm for two aspects. Firstly, we demonstrate the sparse attention can improve the final performance of the algorithm and make the results more interpretable in a multi-agent environment. Second, the affects of different attention sparsity are analyzed.

In the multi-agent literature, Cooperative Treasure Collection and Rover-Tower are usually used for evaluation But, only the latter one is specifically designed to verify the effectiveness of the attention mechanism. Therefore, we choose two environment in our experiments, i.e. Rover-Tower and a modified Cooperative Treasure Collection.

Algorithm 1. SparseMAAC

1: Input: initialize policy parameters θ_i, target policy parameters $\bar{\theta}_i$ centralized Q-function paramaters ϕ and centralized target Q-function paramaters $\bar{\phi}$ for every agent, initialize replay buffer $\mathcal{D}$ and E parallel environments

2: **for** $m = 1$ to n_{episodes} **do**

3:    Reset all environments and get $s_{i,1}^m$ for each agent i in each environment

4:    **for** $t = 1$ to $\min(T_{\text{done}}, T_{\text{max\_per\_episode}})$ **do**

5:       Select action $a_{i,t}^m \sim \pi_{\theta_i}(\cdot | s_{i,t}^m)$ for each agent i in each environment

6:       Each agent i excutes action $a_{i,t}^m$ in each environment

7:       Each agent i observes $s_{i,t+1}^m$ and get $r_{i,t+1}^m$ in each environment

8:       Store transition $(s_t^m, a_t^m, r_{t+1}^m, s_{t+1}^m)$ for all environments in $\mathcal{D}$

9:       Randomly sample a batch of transion $(s_t, a_t, r_{t+1}, s_{t+1})$ from $\mathcal{D}$

10:      **for** each agent i **do**

11:         Calculate Q_{targ}^i by 3.4

12:         Calculate critic loss $\mathcal{L}_i(\phi) = \frac{1}{m} \sum_{j=1}^m (Q_{\text{targ}}^i - Q^i(s_j, a_j^1, \cdots, a_j^N))^2$

13:      Calculate overall critic loss $\mathcal{L}(\phi) = \frac{1}{N} \sum_{i=1}^N \mathcal{L}_i(\phi)$ and update critic

14:      **for** each agent i **do**

15:         Update actor using sampled policy gradient 3.4

16:      Update target critic network parameters $\bar{\phi}$

17:      Update target policy network parameters $\bar{\theta}_i$ and for each agent i

18: Output: Trained policy parameters θ_i for each agent i

Rover-Tower. The configuration is the same as the experiment in MAAC [6], where each two agents are teamed up and only one agent in each team can move For the agent which can move, the immovable agent in other teams has little influence on its policy. Therefore, this environment can also be used to verify the effectiveness of sparse attention.

Grouped Cooperative Treasure Collection. We design a new collaborative environment show as Fig. 1(b), which consists of two four-agent groups. For each group (1 or 2), there are two deposits (biggest circles), two collectors (gray circles) and two treasures (smallest circles). Each group can be seen as a mini version of the original environment. All entities are moveable except for the treasures. The collectors' task is to collect the treasure and place it in the deposit of the same color. And the task of deposits is to store as many treasures as possible. The agent can not only observe the agent from the same group, but also see different groups. To verify the effectiveness of sparse attention, the tasks of different groups are independent. Therefore, the optimal strategy should be with sparse attention. The complexity of the environment is the same as the original environment.

5.2   Settings

To verify the effectiveness of sparse attention, we compare the proposed Sparse-MAAC algorithm with the MAAC. For fairness, we keep all the hyperparameters in the original paper to obtain its best performance. Besides, in order to demonstrate the influence of different sparsity, we further compare the sparsemax with different sparsity ($\gamma = 1, 0.1, 0.01$) in our SparseMAAC, as well as the softmax in MAAC.

5.3   Results and Analysis

Firstly, we demonstrate that the proposed γ-sparsemax function can produce the sparse attention weight. Figure 2 shows the attention weight of agent-0 in Rover-Tower and Grouped Cooperative Treasure Collection by MAAC (softmax) and SparseMAAC (sparsemax, $\gamma = 0.01, 0.1, 1$). The x axis (0–1) stands for the value of attention, the y axis (0–120000) stands for the iteration number, and the height stands for the attention of agent-0 on other agents. It can be seen that, the output of MAAC (softmax) is all greater than 0 and is concentrated around a certain value (about 0.15 in Fig. 2(a)). So, the agents in MAAC have similar attention for every other agents. But, the attention value of SparseMAAC (sparsemax) are partially 0. The smaller γ is, the more sparse the attention distribution is. Therefore, the agents can learn a sparse policy via SparseMAAC, and treat other agents selectively.

Impact of Sparse Attention. Figure 3(a) shows the performance of the proposed SparseMAAC algorithm with MAAC in the Rover-Tower environment by the mean episode rewards. Obviously, the proposed SparseMAAC algorithm is significantly superior to MAAC, not only in terms of mean rewards, but also the convergence speed. The phenomenon proofs that the introduced attention sparsity can help the agent quickly filter out useful information in the early stage of training. Thus, it speeds up the training process with a better results.

Figure 3(b) shows the performance of the proposed SparseMAAC algorithm with MAAC in the Grouped Cooperative Treasure environment by the mean episode rewards. Because each agent can observe all other agents, the performances of compared algorithms are nearly the same. But the convergence speed of the proposed SparseMAAC is still faster than the MAAC algorithm. Sparse-MAAC can select useful agents with the sparse attention, so it achieve similar mean episode reward with shorter training time and fewer computations.

Impact of the Sparsity. In this part, we delve into the effectiveness of different sparsity on the performance of the proposed algorithm. We choose the Rover-Tower environment for discussion, where the agents are paired in pairs. For an agent, there is only one agent which has the greatest impact on it. Figure 4 visualizes the attention results of the paired relationship in Rover-Tower. We first train all the algorithms for 50k epochs, and then let each agent randomly sample an observation from the environment and calculate the corresponding attention weight. MAAC and SparseMAAC both adopt the multi-head attention mechanism, and 4 attention heads are totally used in our experiments. The first column in Fig. 4 stands for the ground truth, the 2–5

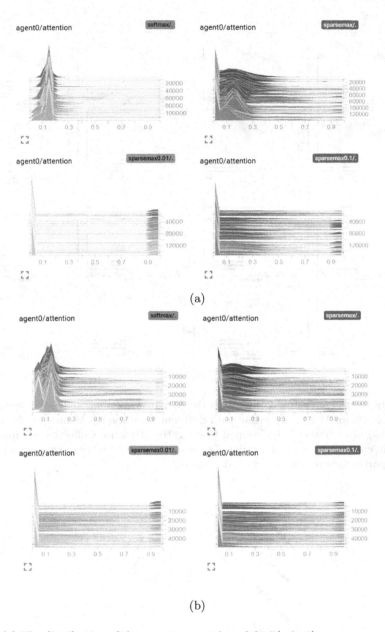

Fig. 2. (a) The distribution of the attention weights of the 7(=8−1) agents outputted in the Grouped Cooperative Treasure Collection environment with the number of training iterations. (b) The distribution of the attention weights of the 7(=8−1) agents outputted in the Rover-Tower environment with the number of training iterations.

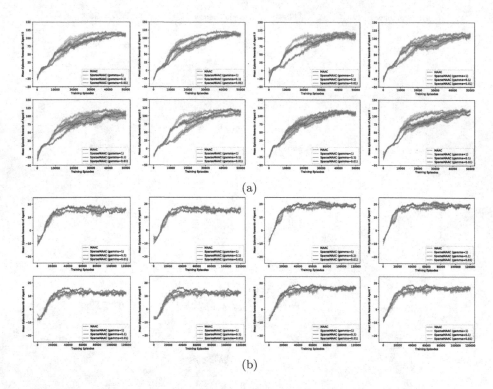

Fig. 3. In the legend, softmax represents the MAAC algorithm, and γ-sparsemax represents the SparseMAAC algorithm with different γ values. (a) Mean episode rewards (of 5 runs) of each agent on Grouped Cooperative Treasure Collection environment. (b) Mean episode rewards (of 3 runs) of each agent on Rover-Tower environment. The shaded part represents the standard deviation.

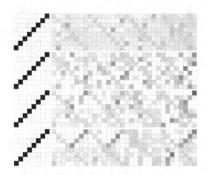

Fig. 4. Attention thermodynamic map in *Rover-Tower* environment. Attention weight is calculated by randomly sampling an observation. The darker the color, the greater the attention weight. Since each agent does not calculate the attention with itself, the main diagonal of the squares in color columns is filled with twill. (Color figure online)

columns stand for different head attention, and the 1–4 row stand for MAAC (softmax), SparseMAAC($\gamma = 0.01, 0.1, 1$).

From Fig. 4, it can be seen that the smaller the γ value, the more the number of white grids. That that the obtained attention weight is more sparse. This also coincides with the results from Fig. 2. We can see that our SparseMAAC algorithm has the trend to concentrate the true diagonal elements. It is most noticeable when the γ is 1. It is worth noting that the third attention head differs greatly from the other three modes in the fourth column. This is because different attention heads learn different level of environment, due to the advantages of the multi-head attention mechanism. In our experiments, the third attention head represents the overall information more than other heads.

6   Conclusion

In order to adjust the agent's selective attention and reduce the complexity of relationship between agents, we introduce the sparse attention mechanism into multi-agent reinforcement learning via a sparsemax function, and propose the Sparse Multiple Attention Actor Critic (SparseMAAC) algorithm. The proposed SparseMAAC can learn a selective policy with sparse attention, and detect helpful agent on the early stage of training procedure. This guarantees our Sparse-MAAC with a better optimization routine and lower complexity. We also introduce an adjustable scheme via a hyperparameter to control the sparseness of attention. The proposed algorithm under different sparsity level is evaluated in two different environments with MAAC. The results demonstrate that Sparse-MAAC can achieve sparse attention distribution effectively. The experiment also show that our SparseMAAC can not only exceed or be equal to the MAAC algorithm, but also converge significantly faster than MAAC.

Acknowledgment. This work is supported by the National Natural Science Foundation of China (Grant No. 61702188, No. U1609220, No. U1509219 and No. 61672231).

References

1. Busoniu, L., Babuska, R., De-Schutter, B.: A comprehensive survey of multi-agent reinforcement learning. IEEE Trans. Syst. Man Cybern. Part C (Appl. Rev.) **38**(2), 156–172 (2008)
2. Chen, C., Seff, A., Kornhauser, A., Xiao, J.: Deepdriving: learning affordance for direct perception in autonomous driving. In: IEEE International Conference on Computer Vision, pp. 2722–2730 (2015)
3. Foerster, J., Farquhar, G., Afouras, T., Nardelli, N., Whiteson, S.: Counterfactual multi-agent policy gradients. In: 32nd AAAI Conference on Artificial Intelligence (2018)
4. Haarnoja, T., Zhou, A., Abbeel, P., Levine, S.: Soft actor-critic: off-policy maximum entropy deep reinforcement learning with a stochastic actor. arXiv preprint arXiv:1801.01290 (2018)

5. Haarnoja, T., et al.: Soft actor-critic algorithms and applications. arXiv preprint arXiv:1812.05905 (2018)
6. Iqbal, S., Sha, F.: Actor-attention-critic for multi-agent reinforcement learning. arXiv preprint arXiv:1810.02912 (2018)
7. Jiang, J., Lu, Z.: Learning attentional communication for multi-agent cooperation. arXiv preprint arXiv:1805.07733 (2018)
8. Lillicrap, T., et al.: Continuous control with deep reinforcement learning. arXiv preprint arXiv:1509.02971 (2015)
9. Littman, M.: Markov games as a framework for multi-agent reinforcement learning. In: International Conference on Machine Learning, pp. 157–163 (1994)
10. Lowe, R., Wu, Y., Tamar, A., Harb, J., Abbeel, O.P., Mordatch, I.: Multi-agent actor-critic for mixed cooperative-competitive environments. In: Advances in Neural Information Processing Systems, pp. 6379–6390 (2017)
11. Luong, M.T., Pham, H., Manning, C.: Effective approaches to attention-based neural machine translation. arXiv preprint arXiv:1508.04025 (2015)
12. Martins, A., Astudillo, R.: From softmax to sparsemax: a sparse model of attention and multi-label classification. In: International Conference on Machine Learning, pp. 1614–1623 (2016)
13. Matignon, L., Jeanpierre, L., Mouaddib, A.I.: Coordinated multi-robot exploration under communication constraints using decentralized Markov decision processes. In: 26th AAAI Conference on Artificial Intelligence (2012)
14. Mnih, V., et al.: Human-level control through deep reinforcement learning. Nature **518**(7540), 529 (2015)
15. Mousavi, S., Schukat, M., Howley, E., Borji, A., Mozayani, N.: Learning to predict where to look in interactive environments using deep recurrent Q-learning. arXiv preprint arXiv:1612.05753 (2016)
16. Niculae, V., Blondel, M.: A regularized framework for sparse and structured neural attention. In: Advances in Neural Information Processing Systems, pp. 3338–3348 (2017)
17. Niv, Y., Daniel, R., Geana, A., Gershman, S., Leong, Y., Radulescu, A., Wilson, R.: Reinforcement learning in multidimensional environments relies on attention mechanisms. J. Neurosci. **35**(21), 8145–8157 (2015)
18. Oh, J., Chockalingam, V., Singh, S., Lee, H.: Control of memory, active perception, and action in minecraft. arXiv preprint arXiv:1605.09128 (2016)
19. OpenAI: Openai Five (2018). https://blog.openai.com/openai-five/
20. Peng, P., et al.: Multiagent bidirectionally-coordinated nets for learning to play starcraft combat games. arXiv preprint arXiv:1703.10069 (2017)
21. Puterman, M.: Markov Decision Processes: Discrete Stochastic Dynamic Programming. Wiley, Hoboken (2014)
22. Shoham, Y., Powers, R., Grenager, T.: If multi-agent learning is the answer, what is the question? Artif. Intell. **171**(7), 365–377 (2007)
23. Sutton, R., Barto, A.: Reinforcement Learning: An Introduction (2018)
24. Sutton, R.S., McAllester, D.A., Singh, S.P., Mansour, Y.: Policy gradient methods for reinforcement learning with function approximation. In: Advances in Neural Information Processing Systems, pp. 1057–1063 (2000)
25. Vaswani, A., et al.: Attention is all you need. In: Advances in Neural Information Processing Systems, pp. 5998–6008 (2017)
26. Williams, R.: Simple statistical gradient-following algorithms for connectionist reinforcement learning. Mach. Learn. **8**(3–4), 229–256 (1992)
27. Xu, K., et al.: Show, attend and tell: Neural image caption generation with visual attention. In: International Conference on Machine Learning, pp. 2048–2057 (2015)

The 4th International Workshop on Big Data Quality Management (BDQM 2019)

Identifying Reference Relationship of Desktop Files Based on Access Logs

Yukun Li[1,2,3(✉)], Xun Zhang[1], Jie Li[1], Yuan Wang[1], and Degan Zhang[1,2]

[1] Tianjin University of Technology, Tianjin 300384, China
liyukun_tjut@163.com, zx_knight@163.com, 18507913520@163.com,
www_wyuan@163.com, gandegande@126.com
[2] Key Laboratory of Intelligence Computing and Novel Software Technology,
Tianjin, China
[3] Key Laboratory of Computer Vision and System,
Ministry of Education, Tianjin, China

Abstract. When writing a document, people sometimes refer to other files' information such as a picture, a phone number, an email address, a table, a document and so on, therefore reference becomes a natural relationship among desktop files which can be utilized to help people re-finding personal information or identifying information linage. Therefore how to identify reference relationship is an interesting and valuable topic. In this paper, we propose an access log-based method to identify the relationship. Firstly we propose a method to generate access logs by monitoring user desktop operations, and implement a prototype based on the method, and collect several persons' access logs by running it in personal computers. Then we propose an access logs-based method to identify reference relationship of desktop files. The experimental results verify the effectiveness and efficiency of our methods.

Keywords: Reference relationship · Desktop files · Access logs

1 Introduction

Studies show that most accesses to desktop information are re-accesses [1], so re-finding is a popular desktop operation. With increment of personal desktop files, people have to pay much time for re-finding. The popularly-used desktop search tools are based on keyword search technology, whose assumption is that people can remember some words included in the expected files. Because of the limitation of human memory and the difference of individuals, the keyword-based method cannot work well in all situations. Users sometimes meet the following situations: (1) When searching an Excel Stylesheet document which stores the experimental data and is referenced by a paper draft, and she/he can only remember the filename of the draft instead of the filename of the Excel stylesheet document; (2) When a person wants to re-find a picture made before and she/he cannot remember the filename, but she/he remembers the picture

© Springer Nature Switzerland AG 2019
G. Li et al. (Eds.): DASFAA 2019, LNCS 11448, pp. 113–127, 2019.
https://doi.org/10.1007/978-3-030-18590-9_8

has been inserted into a document that she/he can remember. In these scenarios the reference relationship can help people to re-find the expected files efficiently. To the best of our knowledge, there are few works about identifying reference relationship of desktop files.

1.1 Related Work

Desktop search tools are popularly utilized for desktop re-findings, but they fail to work well when people cannot recall proper keywords. Chirita and Nejdl [2] proposed to connect semantically related desktop items by exploiting usage analysis information about sequences of accesses. They also considered the local resources organization structure as features for ranking desktop items. Li et al. [3] proposed a method to exploit the potential of desktop context to refine the search returned list. Peery et al. [5] presented a multi-dimension query method in personal dataspace, which individually grades each dimension(content, structure and metadata), then combines the three dimensional scores into a meaningful unified score. Some researchers studied about the methods of managing personal data set, which involves personal dataspace model [6,7], pay-as-you-go integration [8,9], index [10] and query. Wang and Tian [11] proposed a framework for automatically building a concise resource summary based on users' interests in personal dataspace management systems. Some interesting prototype systems were developed such as iMemes [4], Semex [8], MyLifeBit [12], HyStack [13], Orientspace [14] and so on. Dumais et al. [15] designed a system called Stuff I've Seen(SIS), that facilitates the reuse of on personal computers by indexing the documents people used.

Although there has been some works about searching desktop files, few of them paid attention to the role of reference relationship in helping users re-find desktop files. There are also some works about reference relationship, which include evaluating and ranking patents using weighted references [16], recommending users more references by considering the papers' reference relationships [17]. These works are obviously different from our work: These works assume that the reference relationships are known, and try to propose methods about recommendation or computing similarity based on the reference relationships. We aim to identify the reference relationships.

1.2 Challenges and Contributions

As for identifying reference relationships among desktop files, people can easily think of the following naive methods. (1) By identifying special operations. Obviously some operations can reflect reference relationships, such as "copy-paste", "cut-paste", "save a", "download from", "insert from" and so on, and by detecting these operations, some reference relationships can be captured. This method sounds effective, but it has the following limitations: There are a great number of application softwares (MS Office, Photoshop, LaTeX, web page explorer, email management tools, and etc.), and few of them are open source, therefore it is challenging to monitor so many different applications. (2) By computing the content similarity of any two files. This method has the following disadvantages. Firstly, to find the same string of two documents is often of high cost; Secondly, it is infeasible to compare the similarity of two files with different types. For example, if a text file refers to a picture, this method fails to work well.

Pointing to challenges above, being inspired by Allen's interval algebra [18], which defines possible relations between time intervals for reasoning about temporal descriptions of events, we propose to compute referencing relationship of two desktop files based on people access logs to personal computer. The main contributions of this work can be summarized as below:

- Propose a conceptual model of reference relationship of personal desktop files. including some new concepts like desktop activity list, time inclusive relationship and so on, and propose a method to generate personal sequential access logs in computer desktop by monitoring user's operations.
- Propose an access log-based method to identify file reference relationship based on sequential inclusive relationship of desktop files. Based on the collected access logs, we create an experimental data set and a baseline for evaluation. A prototype system has been implemented based on the method.

The rest of this paper is organized as follows: In Sect. 2, we introduce the reference relationship model. In Sect. 3, we describe methods on identifying reference relationship. Section 4 evaluates our methods. Section 5 concludes this paper.

2   Access-Based Reference Model

How to model reference relationship is the first problem. In this section we will define reference relationship, and propose some concepts based on user access logs for identifying reference relationship.

Definition 2.1 [Reference Relationship]. We denote it as $Rc(f_1, f_2)$, where f_1 and f_2 are personal desktop files, and f_1 has been created or modified by user, it means "When creating or modifying f_1, the user has referred to content of f_2". The first parameter f_1 is called referencing item, and the second parameter is called referenced item.

For example, if there is a reference relationship ("D:\project proposal.do", "E:\Department information.doc"), it means the user referenced information of the file "E:\Department information.doc" when editing document "D:\project proposal.doc". Here "D:\project proposal.doc" is the referencing item, "E:\Department information.doc" is the referenced item. Obviously, the common feature of the referencing items is that they are certainly modified by user in the operation, as to the referenced item, it must be accessed but not certainly be modified.

By analysis we have the following observations about user behaviors on operation. In most cases, when people do a reference operation, usually the referencing file has been opened, then the user open the referenced file and copy the referenced content from the referenced file into the referencing file. From the behavior pattern, we can see in a reference operation, the access time to referencing file and the access time to the referenced file mostly have overlap. The observations provide a possible solution about identifying reference relationship based on user sequential access logs. For doing it, we propose the following concepts.

Definition 2.2 [Desktop Access Activity]. A Desktop Access Activity(DAA) points to a user operation on a desktop file, and is described as a 3-tuple (Fi, Ot, Tp), where Fi represents the file accessed by user. Ot represents the operation type, and is denoted as "read-only" and "modification", Tp means the lasting time of the activity, and is denoted as a 2-tuple (Ts, Te), where T_s means start time of the activity, and T_e means its end time.

For example, the tuple ("D:\project proposal.doc", "Modification", (2013-09-01 08:05:01, 2013-09-01 08:56:10)) represents a user activity, which means the user accessed the desktop file "D:\project proposal.doc" during the time interval (2013-09-01 08:05:01,2013-09-01 08:56:10), and in the period the user modified the file. If we collect desktop access activities and order them by start time, we can get a sequential activity list.

Definition 2.3 [Sequential Activity List]. A Sequential Activity List(SAL) is a list of activity ordered by start time. We denote it as $\Gamma = (a_1, a_2, ..., a_n)$, where a_i is a desktop access activity, $a_i.T_s \leq a_{i+1}.T_s$, and $a_i.F_i <> a_{i+1}.F_i (1 \leq i \leq n-1)$.

From the definition we can see a sequential activity list has two features: (1) It is ordered by start time; (2) The two neighbored files refer to different desktop files. Figure 1 shows an example of activity list $(a_1, a_2, a_3, a_4, a_5, a_6, a_7)$, where the x coordinate axes means access time, and the time unit is minute, the y coordinate axes means time sequential list. Each rectangle represents an activity. In the example there are 7 activities, which illustrate the start time, end time, lasting-time and the operation type. Taking activity a_3 for example, we can see its start time is 6, end time is 9, lasting time is $9-6=3$, the operated object is f3, and the operation type is "Read". From Fig. 1 we can see, some activities have overlap of lasting time, such as a_2 and a_3, a_5 and a_6, and so forth. And we propose a new concept to model this relationship of time overlap between different activities.

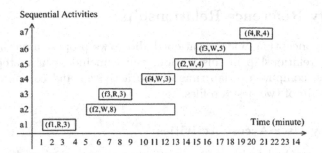

Fig. 1. Sequential activity examples

Definition 2.4 [Sequential Inclusive]. Let Γ be a sequential activity list, α and β be two activities. We think α sequentially includes β if the time of α fully or partially includes the time of β. We denote it as $S_I = \{(\alpha, \beta)|\alpha \in \Gamma, \beta \in \Gamma, \alpha.Tp \cap \beta.Tp \neq \varnothing\}$. In this paper we take the operator $\supset$ to represent the relationship. If α sequential includes β, we denote it as $\alpha \supset \beta$.

From the above definition we can divide the relationship into two classes: (1) Sequential Full Inclusive (SFI). We denote it as $\supset_F$, and $\supset_F (\alpha, \beta)$ means the time of α activity fully includes the time of β; (2) Sequential Partial Inclusive(SPI). We denote it as $\supset_P$, and $\supset_P (\alpha, \beta)$ means the time of α activity partially includes the time of β. Let's take the activity list shown in Fig. 1 for example, $(a1, a2)$, $(a2, a3)$ and $(a5, a6)$ all have sequential inclusive relationship, but the time range of a3 is fully included in the time range of a2, and the time of a6 is partially included in the time of a5. Therefore a2 $\supset_F$ a3, a5 $\supset_P$ a6.

Table 1 shows the sequential inclusive relationship of the example shown by Fig. 1. There are 5 sequential inclusive relationships. Table 2 shows the full sequential inclusive relationship and the partial sequential inclusive relationship of the example. There are 2 full sequential inclusive relationships and 3 partial sequential inclusive relationships. Table 2 also shows the size of the time overlap of activities. For example, $\supset_F (a_2, a_3, 3)$ means the lasting time of a_2 and a_3 are overlapped, and the size of overlap time is 3 min.

Table 1. Examples of SI relations

Activity	a_1	a_2	a_3	a_4	a_5	a_6	a_7
a_1		$\supset$					
a_2			$\supset$	$\supset$			
a_3							
a_4							
a_5						$\supset$	
a_6							$\supset$
a_7							

Table 2. Examples of SFI and SPI

SI type	Relation
$\supset_F$	$(a_2, a_3, 3)$
	$(a_2, a_4, 3)$
$\supset_P$	$(a_1, a_2, 0)$
	$(a_5, a_6, 3)$
	$(a_6, a_7, 1)$

3 Identify Reference Relationship

Based on the conceptual model mentioned above, we propose a method to identify reference relationship in this section, which includes three steps: identify user activities, compute overlap time of two activities and compute the reference relationship of two desktop files.

3.1 Identify User Access Activities

We propose a method to identify user desktop access activities by monitoring the opened windows and the recently-accessed folder of operating systems, and take a 6-ary tuple {*ActivityID, OperatedFile, FileDirectory, StartTime, End-Time, OperationType*} to represent the schema of access logs. As the first step, the recently accessed folder and the opened window list will be checked every one second. When a new opened window is found, a new access record will be generated, and the attributes of OperatedFile, FileDirectory, StartTime can be identified through monitoring the changes of the "RecentAccess" folder provided by operating systems like Windows XP, where the shortcuts of the recent accessed files are placed. On the other hand, when a window is found closed, the activity it refers to will be located and its *EndTime* and *OperationType* will be recorded, and the operation type is worked out by checking if there is a change on the file's last modification time.

We ran the prototype on 8 people's computers in our laboratory and had collected their desktop activities for about one month, and the data sets were stored in a relational table. By analyzing the collected desktop activities, we have the following observations. (1) Re-finding is a popular requirement of people on desktop operation. (2) When a user wants to reference information of file f1 in another file f2, she/he popularly needs to obey the following sequential steps: opening the referencing file, opening the referenced file, getting the information from the referenced file, placing the referenced information in the referencing file. The observations make it possible to identify reference relationship based on sequential inclusion.

3.2 Compute Time Overlap of Activities

To describe our algorithm about computing time overlap of two activities of access logs, we firstly give a new computation symbol, and denote it as $S_{TO}(t_1, t_2)$, where t_1 and t_2 are variable of time interval. S_{TO} is a binary operation and the variable type is time interval. Based on the definition of Desktop Access Activity, it has a parameter Tp, which represents a time interval of the activity lasting. For two given sequential activities a_1 and a_2, we can compute the $S_{TO}(a_1.Tp, a_2.Tp)$ based on the formula (1).

$$S_{TO}(a_1.Tp, a_2.Tp) = min(a_1.Te, a_2.Te) - max(a_2.Ts, a_2.Ts) \qquad (1)$$

As shown in Fig. 1, $S_{TO}(a1, a2) = min(a_1.Te, a_2.Te) - a_2.Ts = min(13, 4) - 4 = 0$. $S_{TO}(a2, a3) = min(a_2.Te, a_3.Te) - a_3.Ts = min(13, 9) - 6 = 3$.

$S_{TO}(a5, a6) = min(a_5.Te, a_6.Te) - a_3.Ts = min(18, 20) - 15 = 3$. From the examples we can see, this formula can be used to compute the size of time overlap.

How to efficiently compute all activities' time overlap is a key problem. The naive method is to scan the activity list two times, and compute every two activities' overlap time. This method is simple but high cost, because its cost is $O(n \times n)$, where n is the activity number of activity list. Based on our analysis on activities we collected, the largest size of activities is about 10000 in about one month. Evaluating based on it there are about 120000 activities collected one year. To this data set, the cost of computation is $O(120000 \times 120000)$, therefore it is an algorithm with high time complexity.

By analysis we have the following observations: If two activities are related to a referencing operation, at least one of them is modification operation. Therefore to a modification activity a', we only need to care about the activities which are close to a' in the list and have overlap time with it. Based on this, we only need to scan the activity list one time and only need to consider the modification activities. Moreover, when finding a modification activity we just need to scan a few activities followed it sequentially. This algorithm improves the performance greatly.

Algorithm 1. Identify Sequential Inclusive Activities

Input: An sequential activity list AL, whose size is S_{AL}.
Output: A set of sequential inclusive activity $SIA(Fa, Fb, OverlapTime)$

```
1:  procedure Identify SI(AL, S_AL, SI)
2:      for (int i = 1, i ≤ S_AL, i++) do
3:          if AL[i].Ot = "Write" then
4:              for (int j = i + 1, j ≤ S_AL ∧ AL[i].Tp.Te > AL[j].Tp.Ts, j++) do
5:                  S_TO = min(AL[i].Tp.Te, Al[j].Tp.Te) − AL[j].Tp.Ts)
6:                  if S_TO >= 0 then
7:                      Add (AL[i], AL[j], S_TO) into SIA
8:                  end if
9:              end for
10:         end if
11:     end for
12: end procedure
```

3.3   Identify Reference Relationship

By analysis we found: for two file f_1 and f_2, more times that f_2 is sequentially included by f_1, more possibly f_2 is referenced by f_1. This observation provides us an idea on computing reference relationship. Considering about two files f_1 and f_2, we firstly define a parameter: overlap times, which means the total times that the two files are handled concurrently, because the overlapping access to the same two files may happen multiple times.

Algorithm 2. Identify Sequential Inclusive files

Input: A set of sequential inclusive activity $SIA(Fa, Fb, S_{TO})$ with size S_{SIA}
Output: A sequential inclusive set of object SIO.
1: **procedure** *Identify SIO(SIA, S_{SIA},SIO)*
2:      Sort SIA to SIA' on Fa and Fb
3:      **for** (int i = 1, i $\leq S_{SIA}$, i++) **do**
4:          **for** (int j = i, j $\leq S_{SIA} \wedge SIA'[j].Fa.Ob = SIA'[i].Fb.Ob$, j++) **do**
5:              **for** (int k = j, k $\leq S_{SIA} \wedge SIA'[k].Fa.Ob = SIA'[j].Fb.Ob$, k++) **do**
6:                  $T_t ime = T.time + SIA'[j].S_{TO}$
7:                  $T_t imes = T.times + 1$
8:              **end for**
9:              Add $(AL'[i].Fa.ID,\ AL'[j].Fa.ID,\ T_t imes, T_t ime)$ into SIO
10:          **end for**
11:      **end for**
12: **end procedure**

Because the method for identifying the reference relationship is based on user activities and user activity often shows randomness, therefore our method shows a good recall, but its precision is not good enough. So a good ranking method becomes more necessary. We propose four methods to compute reference relationship degree for ranking referenced files. (1) Ranking based on full inclusive times(FI-Times), (2) Ranking based on inclusive times(I-Times), (3) Ranking based on inclusive time(I-Time). (4) Ranking based on the ratio of inclusive times by a given object and the inclusive times by all objects(I-Times/AI-Times).

To make the methods clear, we take $R_c(f_2, f_3)$ for example. We compute the four values based on the four methods as shown in Table 3. (1) Because f_2 is only fully includes f_3 once, then FI-Times$(f_2, f_3) = 1$; (2) Because f_2 fully includes f_3 once and partially includes f_3 once, then I-Times$(f_2, f_3) = 2$; (3) Because f_2 fully includes f_3 three minutes and partially includes f_3 three minutes, then I-Time$(f_2, f_3) = 3 + 3 = 6$; (4) Because f_2 fully or partially includes f_3 two times, and f_3 is not included by other objects, then I-Times/AI-Times$(f_2, f_3) = 2/2$ $= 1$. Table 3 overviews reference relationship of the four methods.

Table 3. Overview of reference relationship

CR	FI-Times	I-Times	I-Time	I-Times/AI-times
(f2, f3)	1	2	6	1
(f2, f4)	0	1	3	0.5

4   Evaluation

There is no public data set for evaluation, which is the biggest challenge. We created a data set by capturing personal access activities on desktop computers, and created a baseline for evaluation.

4.1   Create Experimental Data Set

We developed a program to record users' accesses to desktop information, and ran it on the computers of 8 members of our lab. We had collected their desktop access logs for about one month, which are stored by a relational table. Table 4 overviews the participants. We can see the participants include undergraduates, professor and master students. Table 5 overviews the data set we collected, and its column is specified as below: (1) *Time* means the length of time for capturing user accesses; (2) Op-N means the number of user operation; (3) Op(R) means the number of "read" operation; (4) Op(M) means the number of "modify" operation; (5) Op-web means the number of accessed web pages in the collections; (6) Files means the number of accessed desktop files in the collections; (7) File(R) means the number of files which were "read only"; (8) File(M) means the number of files which have been modified; (9) WebPage means the number of web pages which are accessed by the user; (10) Mo-R means the ratio of the number of modified file to total file number.

Table 4. Overview of the participants

User	Age	Gender	Position	Time
user1	23	Female	Undergraduate	23
user2	21	Male	Undergraduate	27
user3	22	Female	Undergraduate	21
user4	24	Female	Master student	22
user5	23	Female	Undergraduate	29
user6	44	Male	Professor	38 ·
user7	24	Male	Master student	38
user8	25	Female	Master student	38

Table 5. Overview of activities of eight users

User	Time	Op-N	Op(R)	Op(M)	Op-Web	Files	File(R)	File(M)	W-Page	Mo-R(%)
1	23	1987	1981	6	1875	415	410	5	408	1.20
2	27	531	527	4	531	174	170	4	174	2.30
3	21	3091	3066	25	3059	621	611	10	617	1.61
4	22	9950	9945	5	9875	1834	1830	4	1830	0.22
5	29	13889	13630	259	13825	1983	1937	46	1979	2.32
6	38	20355	20348	70	20174	3103	3073	30	3078	0.97
7	38	11180	11028	152	10978	2023	1954	69	1999	3.41
8	38	11851	11769	82	11714	2041	2022	19	2032	0.93

By analyzing participants' accesses to their desktop computers, we observed:
(1) The data size of different people show difference. For example, in about the
same time period, the Op-N of user2 is only 531, but that of user6 is 20355. (2)
Mo-R is a small value, which means the number of modified desktop files is much
smaller than the number of accessed files. It means the most user activities are
read operations. When a user handles a modify activity, even it lasts for a long
time, we still regard it as one time. Our evaluation is based on this data set.
Although the time of collecting data is not long, because we focus on identifying
reference relationships, it can satisfy the need.

4.2   Create Baseline for Evaluation

Creating a baseline includes two steps: (1) Select representative files developed
by user as samples of referencing files; (2) Ask the participants to decide the
files referenced by referencing files. Because of the limitation of human memory,
it is hard for people to recall all referencing conditions. The participants need
to review the content of related files for judging if two files have referencing
relationship, so they have to pay much time for creating the baseline. Therefore
we ask each participant to select 3 modified files as referencing files for evaluation.

After selecting referencing files, we ask each user to manually identify referenced files by each file. At this stage we face two challenges: (1) This process
is subjective and hard to manage; (2) It is difficult for users to precisely recall
past event, and it is impossible for them to check each desktop file. To make the
results exact and less subjective, we develop a program to help participants to
make out each one's referenced files, and take the following measures to make
this stage manageable and efficient.

- *Take a greedy method to generate candidate sets.* For a given referencing file,
  we take all files which have time inclusive relationship with it as candidates
  for users to select.
- *Make clear the definition of file reference relationship.* We only consider direct
  references. For example, if A references content of file B, and B references
  content of C, but A does not reference content from C, we think C is not
  referenced by A.
- *Predesign some must-fill forms for participants.* Firstly, we set two options for
  each candidates: referenced and unreferenced. When a user marks a candidate
  file "referenced", she/he is asked to clarify what was referenced.
- *Allow users to add references out of the given candidates.*

To make the process repeatable, we collect not only the results generated by
participants, but also all related source files. Therefore the experiments can be
re-done. In fact, the authors of this paper checked each participants' results, and
corrected some errors by discussing with the corresponding participant.

4.3   Evaluation Measures

In this paper we take two well-accepted information retrieval metrics, namely Precision(P(K)) and normalized Discounted Cumulative Gain (nDCG) [19]. Each metric focuses on a different aspect. (1) P(K) reports the precision of top K results, and it is independent on the sequence of top k items. (2) nDCG is a retrieval measure devised specifically for web search evaluation, and its value depends on not only the top k files, but also their sequence. In this work, each file of ranking results is marked with an integer lable (0 means "Not referenced and 1 means "Referenced). The formulas (2) and (3) show the computation of DCG and nDCG in this paper, where $rel(i)$ means sequence number of the ith file of the results, p means the number of files of the result list, and iDCG means the DCG value of the best ranking list given by the files' owner.

$$DCG_p = \sum_{i=1}^{p} \frac{2^{rel(i)}}{log_2(1+i)} \tag{2}$$

$$nDCG_p = \frac{DCG_p}{iDCG_p} \tag{3}$$

Here we take an example to show how to compute nDCG in this work. Let A be a file for testing, and the top-5 possible referenced files by A generated by our method are F1, F2, F3, F4 and F5, and their referential relationships to A is (1, 0, 0, 0, 1), obviously the best ranking list should be (F1, F5, F2, F3, F4), and its according referential relationship list is (1, 1, 0, 0, 0). We can see the precision of this example is 0.4, $DCG = 2 + 1/log_2 3 + 1/log_2 4 + 1/log_2 5 + 2/log_2 6$, and $iDCG = 2 + 1/log_2 6 + 1/log_2 3 + 1/log_2 4 + 2/log_2 5$, then nDCG = 3.96/4.58=0.86.

As mentioned in Sect. 3.3, we proposed 4 methods to rank referenced files of a given referencing file. In our experiments we compared the results of the 4 different methods, and took top 10 items of the results for evaluation in our experiments. If the number of the results was smaller than 10, we took the actual returned results for computation.

4.4   Experimental Results

As we haven't found studies about identifying reference relationship of personal desktop files, we only compared the four methods we proposed. Table 6 shows P-precision of the four different methods of the eight participants. Table 7 shows the nDCG values of the four methods, and Table 7 considered the top-10 files. Figure 2 shows the average value of the four methods. We can see the method based on access time is the best.

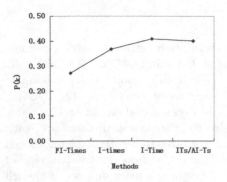
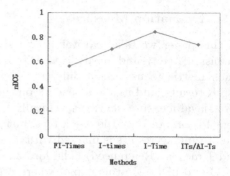

Fig. 2. The evaluation result of P-precision and nDCG

Table 6. P-precision

User	FI-Times	I-Times	I-Time	I-Times/ AI-times
user1	0.22	0.33	0.43	0.40
user2	0.07	0.13	0.23	0.13
user3	0.69	0.73	0.73	0.73
user4	0	0.50	0.50	0.50
user5	0.53	0.67	0.60	0.63
user6	0.40	0.40	0.57	0.57
user7	0.1	0.1	0.10	0.10
user8	0.07	0.7	0.10	0.13

Table 7. nDCG value

User	FI-Times	I-Times	I-Time	I-Times/ AI-times
user1	0.68	0.71	0.87	0.66
user2	0.57	0.79	0.50	0.53
user3	0.80	0.82	0.96	0.99
user4	0	0.84	0.84	0.73
user5	0.93	0.87	0.76	0.81
user6	0.60	0.69	0.99	0.97
user7	0.57	0.55	0.93	0.55
user8	0.37	0.37	0.91	0.68

From the tables, we have following observations: To either P-precision or nDCG, the inclusive time-based method shows the best effectiveness, and the full inclusive time-based method shows the worst. The results are in line with people's experiences. When people do referencing operation, she/he often needs to access the referencing file or the referenced file concurrently for a period of time. We re-analyzed the access logs and found some reasons. When a person is writing a paper, she/he often need to write or design a figure or table with other tools like MS Office Word, MS Office Excel and so forth, then there is often a long concurrent time of the two processes. But if a person references an existing file by inserting operation, instead of generating it by himself/herself, the inclusive time-based method shows poor performance. The I-Times/AI-Times method also has its own advantages and disadvantages. It can filter out the files which are sequentially included by many unrelated files, like some songs the user prefer to listen to when working, but there is no reference relationship between the song and the file being handled.

In general, the different methods have its own working stage, in the future we plan to find a way to combine them to have a method with a high precision in most cases.

4.5   Analysis and Discussion

From the experimental results we have the following observations: (1) There is a sharp distance between the results of Precision and nDCG value. It looks unreasonable but factual. In our method to compute nDCG value, we only take two values 0 and 1 to express whether the two files have referencing relationship, where 1 means yes and 0 means no. Because the precision is low, both DCG and iDCG have a small value, as the ratio of them, the nDCG is certainly not small. (2) The precision is commonly low and averagely no more than 0.5. By analysis we observed there were mainly two reasons. The first, because our method is based on computing the overlap of time of two files' accesses, but people often handle more than one tasks at the same time, therefor some files are taken as referenced file by mistake; The second, in our method we don't consider the operations like "copy and paste", "renaming files" and so forth, or else it will work better. In the future we plan to refine the solution by considering more factors. (3) The participants' experimental results show great difference. In the future, we plan to survey the relationship between the results and people's individuality and operation habits. (4) Sometimes the referencing relationship are transitive. For example, coping data from file A to file B and then the same data from file B to file C. Possibly the relationship will be meaningful for people to understand the prevalence of different relationship types.

We also find some referenced files are not identified by our methods. The main reason is the limitation of our program of collecting access logs. As the first edition, we get access logs by monitoring the open and close action of windows, which cannot find all user operations in some applications. For example, insert a picture into a document edited by a specific tool, like Microsoft Office Word and so on. In this condition the picture don't need to be opened. In the future, we will pay more attention to complete the algorithm on collecting access logs to refine the method.

This method is still preliminary, but it uncovers many interesting research issues. (1) The method should be refined by considering more user patterns; (2) The personalization should be considered and an adaptive method needs to be studied more; (3) How to index personal activities when the number of activity becomes large; (4) How to improve the precision by combining multiple factors (directory, time distance, access pattern, etc.) and so on. All these research issues are meaningful in the field of personal information management.

4.6   Implementation

In order to verify the feasibility of our model and methods, we develop a proto-type system based on the methods we proposed, which demonstrates the usage of our method in re-finding, and Fig. 3 shows the search interface. The top part is an area where a user can input for re-finding, including access times, access time, access type and keywords. The middle part displays the searching results. To a modified file, the user can explore its referenced files by double-clicking it, the bottom of the frame lists all possible referenced files of it. In this prototype, we take a database system to store the user access logs, and utilize the existing technologies of database like indexing to get a good searching performance.

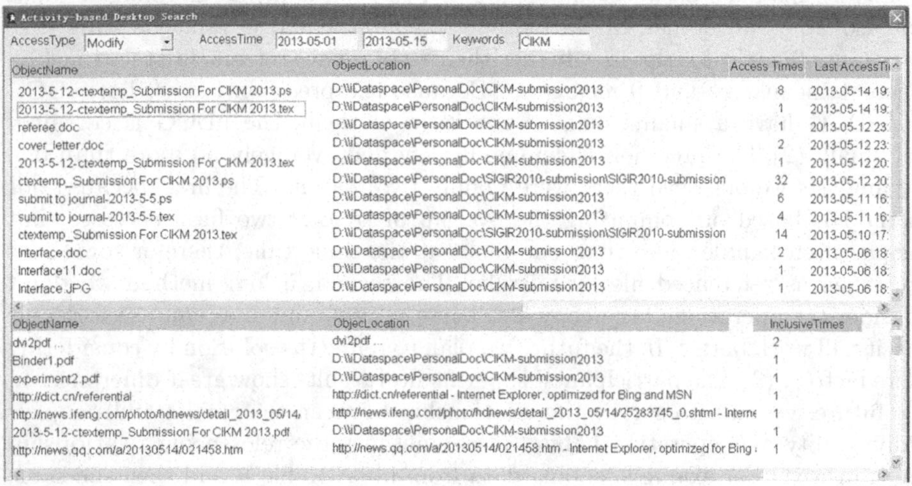

Fig. 3. The main interface of prototype system

5 Conclusion

In this paper, we propose a conceptual activity model by analyzing features of user behaviors on desktop, and present an efficient method to automatically identify activities and identify reference relationship based on desktop collections. This is only a preliminary work, and it discovers some interesting research issues. In the future, we will make effort to improve precision of reference relationship identification method, and to develop a system which have better performance and can be utilized really.

Acknowledgments. This research was supported by Natural Science Foundation of Tianjin (No. 15JCYBJC46500), the Training plan of Tianjin University Innovation Team (No. TD13-5025), the Major Project of Tianjin Smart Manufacturing (No. 15ZXZNCX00050).

References

1. Elsweiler, D., Baillte, M., Ruthven, I.: Exploring memory in email refinding. ACM Trans. Inf. Syst. **26**(4), 21 (2008)
2. Chirita, P.A., Nejdl, W.: Analyzing user behavior to rank desktop items. In: SPIRE, pp. 6–97 (2006)
3. Li, X., Yu, Y., Ouyang, C.: Refine search results based on desktop context. In: Li, J., Ji, H., Zhao, D., Feng, Y. (eds.) NLPCC 2015. LNCS (LNAI), vol. 9362, pp. 209–218. Springer, Cham (2015). https://doi.org/10.1007/978-3-319-25207-0_18
4. Blunschi, L., Dittrich, J.-P., Girard, O.R., Karakashian, S.K., Salles, M.A.V.: A dataspace odyssey: the imemex personal dataspace management system. In: CIDR (2007)
5. Peery, C., Wang, W., Marian, A., Nguyen, T.D.: Multi-dimensional search for personal information management systems. In: EDBT (2008)
6. Franklin, M.J., Halevy, A.Y., Maier, D.: From databases to dataspaces: a new abstraction for information management. SIGMOD Record **34**(4), 27–33 (2005)
7. Dittrich, J.P., Antonio, M., Salles, V.: iDM: a unified and versatile data model for personal dataspace management. In: VLDB, pp. 367–378 (2006)
8. Dong, X., Halevy, A.: A platform for personal information management and integration. In: Online Proceedings CIDR, pp. 119–130 (2005)
9. Vaz Salles, M.A., Dittrich, J.P., Karakashian, S.K., Girard, O.R., Blunschi, L.: iTrails: pay-as-you-go information integration in dataspaces. In: VLDB, pp. 663–674 (2007)
10. Dong, X., Halevy, A.: Indexing dataspaces. In: SIGMOD, pp. 43–54 (2007)
11. Wang, N., Tian, T.: Summarizing personal dataspace based on user Interests. Int. J. Softw. Eng. Knowl. Eng. **26**(5), 691–714 (2016)
12. Gemmell, J., Bell, G., Lueder, R., Drucker, S.M., Wong, C.: MyLifeBits: fulfilling the Memex vision. In: ACM Multimedia, pp. 235–238 (2002)
13. Karger, D.R., Bakshi, K., Huynh, D., Quan, D., Sinha, V.: Haystack: a customizable general-purpose information management tool for end users of semistructured data. In: Proceedings of CIDR, pp. 13–26 (2005)
14. Li, Y., Zhang, X., Meng, X.: Exploring desktop resources based on user activity analysis. In: SIGIR, p. 700 (2010)
15. Dumais, S.T., Cutrell, E., Cadiz, J., Jancke, G., Sarin, R.: Robbins D.: Stuff i've seen: a system for personal information retrieval and re-use. In: SIGIR Forum, vol. 49, no. 2, pp. 28–35 (2015)
16. Oh, S., Lei, Z., Mitra, P., Yen, J.: Evaluating and ranking patents using weighted citations. In: JCDL, pp. 281–284 (2012)
17. Huang, W., Kataria, S., Caragea, C., Mitra, P., Giles, C.L., Rokach, L.: Recommending references: translating papers into references. In: CIKM, pp. 1910–1914 (2012)
18. Allen J.: Maintaining knowledge about temporal intervals. In: Communications of the ACM, ACM Press. pp. 832–843 (1983)
19. Järvelin, K., Kekäläinen, J.: IR evaluation methods for retrieving highly relevant documents. In: SIGIR, pp. 41–48 (2000)

Visualization of Photo Album: Selecting a Representative Photo of a Specific Event

Yukun Li[1,2,3(✉)], Ming Geng[1], Fenglian Liu[1,2], and Degan Zhang[1,2]

[1] Tianjin University of Technology, Tianjin 300384, China
liyukun_tjut@163.com, gengming1124@163.com,
lflian@tjut.edu.cn, gandegande@126.com
[2] Tianjin Key Laboratory of Intelligence Computing and Novel Software
Technology, Tianjin, China
[3] Key Laboratory of Computer Vision and System, Ministry of Education,
Tianjin, China

Abstract. In order to effectively manage photos in personal photo album and improve the efficiency of re-finding photos, the visualization of photo album has received attention. The most popular and reasonable visualization method is to display a representative photo of each photo cluster. We studied the characteristics of representative photos and then proposed a method of selecting the representative photos from a set of photos related to a specific event. The method mainly considered two aspects of photos: aesthetic quality and memorable factor. Aesthetic quality contains the area and location of the salient region and the sharpness of photo; memorable factors contain the salient people and text information. The experimental data sets are real-world personal photo collections, including more than 7,000 photos and more than 2000 specific events. The experimental results show the efficiency and reliability of selecting representative photos to visualization of photo album.

Keywords: Photo album · Visualization · Representative photo

1 Introduction

With development of technology, more and more people like to take photos with mobile phone. How to effectively manage photos in personal mobile phone and improve the efficiency of re-finding photos is an important research issue. The photo classification and visualization of photo clusters are two effective ways to manage photos and increase the speed of re-finding photos. Most researches focus on photo classification, but there are several researches on the visualization of photo album. However, the visualization of photo cluster after the classification of photos is also very important for photo management and re-finding. Our previous work [1] was to study the classification of photos based on specific events, and this paper conducted a research on the visualization of specific events. A specific event can be taken as a series of actions taking place for the same target subject in a relatively continuous period of time and relatively neighboring location. For example, attending a wedding of a friend at noon one day is a specific event.

© Springer Nature Switzerland AG 2019
G. Li et al. (Eds.): DASFAA 2019, LNCS 11448, pp. 128–141, 2019.
https://doi.org/10.1007/978-3-030-18590-9_9

There are some challenges for people to study this problem. Because people have some personalized features, and the content and the number of each person's photos is different. It is not easy to get a general method to classify personal photos. The representative photo is the photo which can stimulate the user to recall the corresponding specific event. This paper has done relevant research and proposed related solutions. The main contributions are summarized as follows:

(1) We collect a number of phone pictures from different persons and summarized the characteristics of the representative photos by analyzing them;
(2) According to the characteristics of the representative photos, we proposed a method to select the representative photo from each photo cluster of specific event;
(3) Based on real-world personal photo collections, the experimental data sets are established. The experimental results verify the efficiency and reliability of the proposed method.

2 Related Work

By reading relevant documents there are four main factors considered in visualization of photo album: aesthetic quality, image near-duplication detection, the correlation between the photo and the event, and memorable factor.

The literature [2] and the literature [3] judge the importance of photos through the aesthetic quality. The literature [4], literature [5] and literature [6] select the representative photos according to the correlation between the photo and the event. The literature [7], literature [8] and literature [9] take into account the near-duplication detection of photos to ensure the diversity. The literature [10] and literature [11] not only consider the aesthetic quality of photos, but also the correlation between the photo and the event. The literature [12] considers the aesthetic quality and the diversity of photos to select representative photos. The literature [13], literature [14] and literature [15] mainly consider three factors: aesthetic quality, near-duplication detection and memorable factors. But the memorable factors only consider the "people" in photos. The photos that contain people are easier to remember and recall.

The aesthetic quality must be considered when selecting a photo to visualization of photo cluster. However, the aesthetic quality only refers to the visual effect of the photo, such as sharpness, brightness etc., which is not enough to recall the event. The current method of identifying the event is mainly by recognizing the object and scene in the photo, then use the relevance between scenes, objects, and events to identify events. Since there are abundant scenes and objects corresponding to different events, and the same scene or object may also correspond to different events, such as the photos of tour and wedding may both contain sky. So the method of using scenes and objects to identity the event is not accurate. The near-duplication of photos can ensure the diversity of photos. But only duplicate photos are filtered out, there are still many photos left in one cluster. Using all the rest of the photos to represent this cluster cannot achieve good visualization. About the memorable factor, this is very important. Although the literature [13], literature [14] and literature [15] have considered the memorable factor, but not enough, it is necessary to find out more memorable factors that affect people's memories. This will be more effective in stimulating human memory of events.

In this paper, we summarize the characteristics of the representative photos. Combining these characteristics, we propose a method to select the representative photo of specific event cluster.

3   Method

3.1   Survey and Findings

We collect 10 real-world personal photo collections, including 2030 specific events. We let the 10 people to choose representative photos from the specific events. Then we summarize the representative photos with the following characteristics through comparing the representative photos selected by users and other photos.

(1) The representative photos contain semantic information, such as the photo of a building with text.
(2) The representative photos contain salient people, and the position of the salient people is relatively centered. It is better to contain more people. This is mentioned in the literature [13], literature [15] and literature [16].
(3) The representative photos contain salient object. The area of the salient object is relatively larger and the position is relatively centered.
(4) The representative photos are relatively clear.

3.2   Solution Framework

We propose a method to grade photos, and select the photo with the highest score as the representative photo of the specific event. Our rating method takes into account two aspects of a photograph: aesthetic quality and memorable factor. Aesthetic quality: we mainly consider the layout and sharpness of the photo. If there is the salient region in the photo, the area and the location of the salient region affect the layout of the photo. Memorable factors: memorable factors refer to the factors that affect human memory and can stimulate people to recall the specific event of the photo. This paper considers two memory factors: people information and semantic information. The memory factor mentioned in the literature [12], literature [13] and literature [14] is the information of the people in the photos. Moreover, the more people there are in the photo, the more it can help people recall the event information, which has also been found through research. After investigation and analysis, another memory factor is also found, that is the semantic information contained in photos. For example, in a photo cluster about the campus landscape photos, a photo of the school gate with the name of the school is more representative than a photo of the teaching building or a photo of the campus lake and so on. In this paper, the semantic information is the text information, because the text itself is the expression of semantic information, and can even convey some unique information. At the same time, more words often convey more semantic information, which can help people recall the event information.

Firstly, the saliency detection of the photos is carried out to filter the useless background noise in the photos. If there is no salient region in the photo, we just consider the sharpness of the photo. If the photos contain salient region, aesthetic quality and memory factors should be considered. Regarding aesthetic quality, we should consider the area of salient region, the location of salient region and the sharpness of the photo. As for memory factors, if the salient region contains people, the number of people should be considered; if the salient region contains text information, the score of text information should be considered. Then the aesthetic quality score and memorableness score are weighted together to get the photo score. Finally, the photos in specific events are sorted according to the photo scores, and the photo with the highest score is selected as the representative photo. The whole process of selecting representative photos in this paper is shown as Fig. 1. The process of algorithm is shown as Algorithm 1. How to identify whether the photo contains text information and people, and how to identify the number of characters and the number of people, we do not research here. The current image recognition technology can be realized.

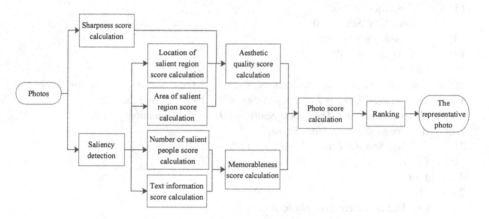

Fig. 1. The whole process of selecting representative photos

3.3 Saliency Detection

In this paper, we use a saliency fusion algorithm based on conditional random field (CRF) [17] to predict the region of salient objects. The Fig. 2 shows the diagram of the saliency detection method.

Under the conditional random field CRF framework, for a photo I(x, y) whose size is W*H, the probability of a labeling configuration A(x, y) can be modeled as a conditional distribution p(A|I), as shown in the formula (1)

$$p(A|I) = \frac{\exp(-E(A|I))}{Z} \tag{1}$$

Algorithm 1. The algorithm of selecting a representative photo from a specific event.

Input: Personal photos of a specific event, $I_1, \ldots, I_n$.
Output: The representative photo R.
1: *Max_Score* = 0; $j = 0$;
2: for each ($i \in [1, n]$)

3:      Saliency detection of photo I_i;
4:      Calculate the sharpness score *Sharpness_Score* of photo I_i;
5:      If (I_i has salient region)
6:          Calculate the *Area_Score* of I_i;
7:          Calculate the *Location_Score* of I_i;
8:          Calculate the *People_Score* of I_i;
9:          Calculate the *Text_Score* of I_i;
10:     Else
11:         *Area_Score* = 0;
12:         *Location_Score* = 0;
13:         *People_Score* = 0;
14:         *Text_Score* = 0;
15:     End If
16:     *Aesthetics_Score* = λ_1 **Sharpness_Score*+λ_2**Area_Score*+λ_3**Location_Score* ;
17:     *Memorable_Score* = λ_4 * *People_Score* + (1 - λ_4) * *Text_Score*;
18:     Photo_Score = λ * *Aesthetics_Score* + (1 - λ) * *Memorable_Score*;
19:     If (*Photo_Score* > *Max_Score*)
20:         *Max_Score* = *Photo_Score*; $j = i$;
21:     End If
22: End for
23: $R = I_j$;
24: **return** The representative photo R.

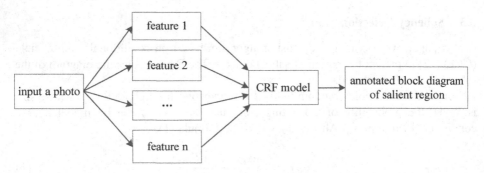

Fig. 2. The diagram of the saliency detection method

Z is a normalized constant and E(A|I) is the energy function. The energy function E (A|I) as a linear combination of multiple static salient features and pairwise features.

$$E(A|I) = \sum_{x=1}^{W}\sum_{y=1}^{H}\left(\sum_{m=1}^{M}\lambda_m * Map_m(x,y) + \sum_{m=1}^{M}Pair_m(x,y)\right) \qquad (2)$$

M is the total number of salient features. The annotation function A(x, y) for the pixel p(x, y) is:

$$\begin{cases} A(x,y) = +1 \\ A(x,y) = -1 \end{cases} \qquad (3)$$

"+1" indicates that the pixel belongs to the salient region, and "−1" indicates that the pixel belongs to the background region. The Map(x, y) is the salient feature function. The Pair(x, y) is the pairwise feature function, which is a penalty term that marks adjacent pixels as different values.

Finally, the salient region is outlined by a square, and the result of the saliency detection is shown in Fig. 3.

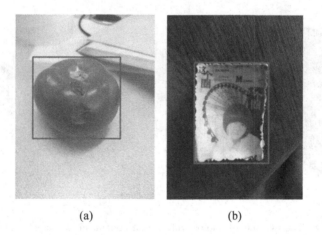

(a)   (b)

Fig. 3. The figure of salient region

3.4   Aesthetic Quality Score Calculation

The photo aesthetic quality score includes the image sharpness score, the location of salient region score and the area of salient region score.

(1) The sharpness score
The Average Gradient calculation method reflects the small detail contrast and texture variation features in the photo, and also reflects the sharpness of the photo. The larger the value of the Average Gradient is, the higher the sharpness of the photo can be. The

size of the photo P is W*H, and the average gradient value of the photo is calculated as the formula (4).

$$\nabla \bar{g} = \sum_{i=0}^{M-1} \sum_{j=0}^{N-1} \frac{\sqrt{(\Delta I_x^2 + \Delta I_y^2)/2}}{M \times N}, \Delta I_x = I(i+1,j) - I(i,j), \Delta I_y = I(i,j+1) - I(i,j)$$

(4)

I(i, j) is the pixel value at (i, j). Figure 4 shows two sets of photographs with similar content in two specific events, and the average gradient is calculated separately. The average gradient value of (a) and (b) are 0.5691 and 1.0899 respectively; the average gradient value of (c) and (d) are 0.9051 and 2.5259 respectively. It can be seen that the larger the average gradient value is, the higher the sharpness of the photo can be.

First find the photo with the largest average gradient value in the specific event, and record the value of average gradient as Max_AG. Then calculate the average gradient value AG of each photo in the specific event. The calculation of sharpness score is shown in formula (5).

$$Sharpness\_Score = 1 * \frac{AG}{MAX\_AG}$$

(5)

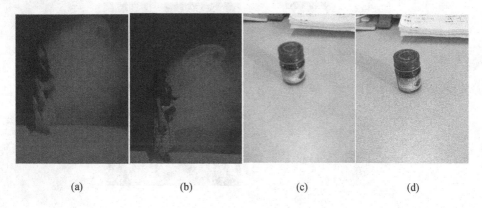

(a)   (b)   (c)   (d)

Fig. 4. The two sets of photos with similar content in two specific events

(2) Area of salient region score

The length of the photo is M, the width is N, and the area is M*N. After performing the saliency detection on the photo, the salient region is outlined by a square. Then the area s of the salient region is calculated, and the area of salient region score is calculated as the formula (6).

$$Area\_Score = \frac{s}{M * N}$$

(6)

(3) Location of salient region score

"Rule of thirds" is the most famous composition guide rule in photography, which can create aesthetic feeling. The rule of thirds is to divide a photo into nine equal parts with two equidistant horizontal lines and two equidistant vertical lines, showing a grid. Important composition elements are placed at four intersections. The scenery at the intersection often attracts people's attention and makes people understand the true meaning of the photo. These four intersections are the center of people's visual focus.

Assuming the length of the photo is M and the width is N, the rule of thirds of the photo is shown in Fig. 5. We find out the central position of the salient region in the photo, then find the intersection of the four intersections that is closest to the center of the salient region. We calculate the distance Min_x between the two points. The location of salient region score is calculated as the formula (7) and formula (8).

$$Location\_Score = 1 - 1 * (Length\_Ratio)^2 \tag{7}$$

$$Length\_Ratio = \frac{Min\_x}{\sqrt{(M/3)^2 + (N/3)^2}} \tag{8}$$

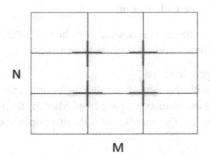

Fig. 5. The photo's rule of thirds

The location of salient region score of the photo varies with the distance from the center of the salient region to the intersection, and the change is shown in the curve in Fig. 6. Compared with the linear scoring method (shown by the straight line in Fig. 6), the photo score of the photos whose center of salient region near the intersection decreased slowly. What's more, the farther the distance is, the faster the score decreases. Our method is more reasonable. Because the photos taken by people do not necessarily conform to the rule of thirds, the focus of the photos may not be exactly at the intersection. After all, there is a certain visual error, which may be near the intersection.

The final aesthetic quality score of the photo is shown in formula (9), where $\lambda_1 + \lambda_2 + \lambda_3 = 1$.

$$Aesthetics\_Score = \lambda_1 * Sharpness\_Score + \lambda_2 * Area\_Score + \lambda_3 * Location\_Score \tag{9}$$

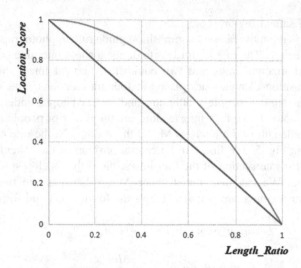

Fig. 6. The location of salient region score of the photo's changes

3.5   Memorableness Score Calculation

The photo memorableness score includes the number of salient people score and the text information score. These scores are all between 0 and 1.

(1) Number of salient people score
First find out the photo with the largest number of people in the salient region of the specific event, and record the number of people as Max_p; then identify the number of salient people in each photo. The number of salient people score is calculated as the formula (10).

$$People\_Score = 1 * \frac{p}{Max\_p} \tag{10}$$

(2) Text information score
First find out the photo with the most words in the specific event, write down the number as Max_t; then identify the number of word in each photo as t. The text information score is calculated as the formula (11).

$$Text\_Score = 1 * \frac{t}{Max\_t} \tag{11}$$

The final Memorableness Score of the photo is shown in formula (12).

$$Memorable\_Score = \lambda_4 * People\_Score + (1 - \lambda_4) * Text\_Score \tag{12}$$

3.6 Select the Representative Photo of a Specific Event

The aesthetic quality and memorable factors may have different effects on the selection of representative photos, because the representative photo we choose are to stimulate people's memory and make it easier for people to recall the specific event of the photo. We suspect that the impact of the aesthetic quality may be less than memory factors. So we set the weight λ in the final calculation formula for the photo score, as shown in the formula (13).

$$Photo\_Score = \lambda * Aesthetics\_Score + (1 - \lambda) * Memorable\_Score \qquad (13)$$

Calculate the photo score of each photo in a specific event, and then sort. Select the photo with the highest score as the representative photo of this specific event, which is the visualization photo.

4 Evaluation

4.1 Data Sets

We use real-world personal photo collections of 10 mobile phone users as the experimental data sets, a total of 7122 photos and 2030 specific events, as shown in Table 1. The reason for not using public photo data sets is that the photos of the public data set are not consecutive photos of specific events taken by mobile phone, and we can not let the photographer choose the representative photo to represent the specific event. The photographer of photos knows more about the photos he has taken, and the selected representative photos are more valuable, which helps us to summarize the features of representative photos. Then classify the photos based on specific event, and let the ten users select the representative photo for each specific event.

Table 1. The data sets' information

Users	Gender	Age	Number of photos	Number of specific events	Time-length of photo cluster
User1	Female	24	491	148	10 months
User2	Female	25	634	178	12 months
User3	Male	24	646	220	7 months
User4	Male	25	209	53	3 months
User5	Female	23	886	265	16 months
User6	Male	25	1260	300	18 months
User7	Female	24	866	202	12 months
User8	Male	25	653	224	16 months
User9	Female	22	524	186	8 months
User10	Male	21	953	254	14 months

4.2   Evaluation Indicators

This paper uses accuracy to analyze experimental results. Accuracy is the most common indicator of evaluation. The calculation method is the ratio of the correct number of samples obtained to the total number of samples. Use N represents the set of representative photos selected from the specific events using our method in this paper, and T represents the set of representative photos selected by the photographer from each specific event, then the accuracy calculation is shown in formula (14).

$$Accuracy = \frac{|N \cap T|}{|T|} \tag{14}$$

4.3   Evaluations

Regarding the parameters λ_1, λ_2, λ_3, λ_4, λ of the method proposed in this paper, we took the parameters from 0 to 1 with the step size 0.1 in the process of calculating the photo score of photos in the training. During the experimental training process, the parameters are continuously adjusted, and the photo score of each photo in the specific event are calculated. The photo with the highest score is selected as the representative photo, and then compared the selected representative photo with the representative photo selected by photographer to determine whether the selected representative photo is correct. Finally, the parameter value corresponding to the highest accuracy is selected, and $\lambda_1 = 0.4$, $\lambda_2 = 0.2$, $\lambda_3 = 0.4$, $\lambda_4 = 0.2$, $\lambda = 0.3$. So the calculation of the aesthetic quality score is Aesthetics_Score = 0.4*Sharpness_Score + 0.2*Area_Score + 0.4*Location_Score, the calculation of the memorableness score is

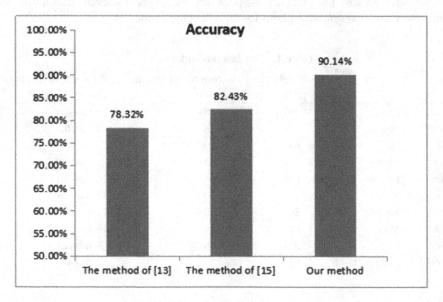

Fig. 7. The accuracy of three methods

Memorable_Score = 0.2*People_Score + 0.8*Text_Score, and the calculation of the photo score is Photo_Score = 0.3* Aesthetics_Score + 0.7*Memorable_Score.

We compared the method proposed in this paper with the methods proposed in [13] and [15]. In each specific event, we used the three methods to select representative photo. The accuracy of the experimental results is shown in Fig. 7.

It can be seen from Fig. 7 that the accuracy of our method is higher when selecting representative photos. The representative photos selected by our method are more suitable as photos for cluster visualization.

We analyzed the reasons why our method is more accurate. On the one hand, there are disadvantages in aesthetic quality scoring of other methods. On the other hand, there are too few memorable factors considered in other methods, which is also the main reason. The memorable factor of literature [13] and literature [15] only consider that the photos that are helpful for people to recall the event are the photos containing people, and the more people in the photo, the easier it is to stimulate people's memories of specific event. But the factors that can affect people's memory are not only this, it need to consider more factors. After survey and research, we found the characteristics of the representative photos. Then combined the characteristics, and proposed a method of selecting representative photos. Therefore, using our method to select representative pictures is more in line with people's memory habits, and the representative photos selected by our method are more suitable for the visualization of photo cluster. The Fig. 8(a) and (b) are two specific events from data set. The photos in (a) were taken by the photographer to commemorate the university after graduation; the photos in (b) were taken by the photographer when she went to YunMeng Mountain in Henan with her boyfriend. The Fig. 8(c) is the representative photos selected from the photo cluster in (a) by three methods; the Fig. 8(d) is the representative photos selected from the photo cluster in (b) by three methods.

However, the representative photos selected by our method are not exactly the same as the representative photos selected by the users, that is, the accuracy rate is less than 100%. We analyzed several reasons as following:

(1) Due to people's personalized characteristics, it is difficult to meet the requirements by a unified standard for all people to recall specific events through the representative photos;

(2) We think that the text itself contains semantic information, so it is believed that the more words the photos contain, the more semantic information there is. In our method, the Text_Score is only measured by the number of words, but the semantic information of the words is not analyzed. Since different words express different semantic information, it is not completely accurate to measure the content of semantic information only by the number of words. It needs further study and improvement.

The method of literature[13]   The method of literature[15]   Our method

(c)

The method of literature[13]   The method of literature[15]   Our method

(d)

Fig. 8. The photo clusters of two specific events and the representative photos selected by three methods

5 Conclusions

By studying the characteristics of representative photos, we put forward a method to select representative photos as the visualization photos of specific events. The representative photos can help users to recall the specific event, so they can help users locate the photo clusters they want to find more quickly, which can greatly improve the re-finding rate. The experiment also prove the validity of our method. However, the accuracy of our method cannot reach 100%, we will do further study about semantic analysis of photo content.

Acknowledgements. This research was supported by Natural Science Foundation of Tianjin (No. 15JCYBJC46500), the Training plan of Tianjin University Innovation Team (No. TD13-5025), the Major Project of Tianjin Smart Manufacturing (No. 15ZXZNCX00050).

References

1. Geng, M., Li, Y., Liu, F.: Classifying personal photo collections: an event-based approach. In: U, L.H., Xie, H. (eds.) APWeb-WAIM 2018. LNCS, vol. 11268, pp. 201–215. Springer, Cham (2018). https://doi.org/10.1007/978-3-030-01298-4_18
2. Yeh, C.H., Ho, Y.C., Barsky, B.A., Ouhyoung, M.: Personalized photograph ranking and selection system. In: Proceedings of MM 2010 (2010)
3. Li, C., Loui, A.C., Chen, T.: Towards aesthetics: a photo quality assessment and photo selection system. In: International Conference on Multimedia (2010)
4. Wang, Y., Lin, Z., Shen, X., et al.: Event-specific image importance. In: Computer Vision and Pattern Recognition, pp. 4810–4819. IEEE (2016)
5. Nakaji, Y., Yanai, K.: Visualization of real-world events with geotagged tweet photos. In: IEEE International Conference on Multimedia and Expo Workshops, pp. 272–277. IEEE Computer Society (2012)
6. Wang, Y., Lin, Z., Shen, X., et al.: Recognizing and Curating Photo Albums via Event-Specific Image Importance (2017)
7. Chu, W.T., Lin, C.H.: Automatic summarization of travel photos using near-duplication detection and feature filtering. In: ACM International Conference on Multimedia, pp. 1129–1130. ACM (2009)
8. Bolanos, M., Mestre, R., Talavera, E., et al.: Visual summary of egocentric photostreams by representative keyframes. In: IEEE International Conference on Multimedia & Expo Workshops, pp. 1–6. IEEE (2015)
9. Karlsson, K., Jiang, W., Zhang, D.Q.: Mobile photo album management with multiscale timeline. In: ACM International Conference on Multimedia, pp. 1061–1064. ACM (2014)
10. Zhang, L., Denney, B., Lu, J.: Sub-event recognition and summarization for structured scenario photos. Multimed. Tools Appl. **75**(15), 1–20 (2016)
11. Nowak, S., Paduschek, R.: Photo summary: automated selection of representative photos from a digital collection. In: ACM International Conference on Multimedia Retrieval, pp. 1–2. ACM (2011)
12. Sinha, P.: Summarization of archived and shared personal photo collections. In: International Conference on World Wide Web, pp. 421–426 (2011)
13. Chun, C., Lee, H., Kim, D., et al.: Visualization of photo album on mobile devices. In: Proceedings of SPIE - The International Society for Optical Engineering (2014)
14. Ceroni, A., Solachidis, V., Papadopoulou, O., et al.: To keep or not to keep: an expectation-oriented photo selection method for personal photo collections. In: ACM on International Conference on Multimedia Retrieval, pp. 187–194. ACM (2015)
15. Kim, J.H., Lee, J.S.: Travel photo album summarization based on aesthetic quality, interestingness, and memorableness. In: Signal and Information Processing Association Summit and Conference, pp. 1–5. IEEE (2017)
16. Isola, P., Xiao, J., Torralba, A., et al.: What makes an image memorable? In: Computer Vision and Pattern Recognition, pp. 145–152. IEEE (2012)
17. Sheng, Q., Zong-Hai, C., Ming-Qiang, L., et al.: Saliency detection based on conditional random field and image segmentation. Acta Automatica Sin. **41**(4), 711–724 (2015)

Data Quality Management in Institutional Research Output Data Center

Xiaohua Shi[1,2](✉) [iD], Zhuoyuan Xing[1] [iD], and Hongtao Lu[2] [iD]

[1] Library, Shanghai Jiaotong University, Shanghai, China
xhshi@sjtu.edu.cn
[2] Department of Computer Science, Shanghai JiaoTong University, Shanghai, China

Abstract. Institutional research output data center will store norma-
tive and convinced scholar's research output data, and it will effectively
support dynamic presentation of research output, reveal institutional
academic publication in multiple dimensions, advance open access, and
provide data support for subject evaluation and discipline development.

In this paper, we propose a data quality management framework to
build institutional research output data center, and put forward relevant
technical solution for different data governance problems, such as department name similarity estimation in data matching, author name disambiguous problem in data merging and security issue in data exchange. We
also introduce some learning algorithms such as text distance and community detection with matrix factorization. Comparing with different
ways, our methods achieve good performance in quality manage processing.

Keywords: Research information system ·
Author name disambiguous · Text distance · Community detection ·
Matrix factorization

1 Introduction

Following disciplinization development in world's universities, it has been widely
recognized that it is urgent to build institutional research output data center
to collect and store unitary and authoritative academic research output data
of their scholars'. With advance of chinese 'double world-class' initiative and
vigorous development of global research data infrastructure, pace of building
research output data center to store and manage employee's output, has been
also significantly accelerated in China.

Main system function of research output data center is to harvest, integrate,
correlate and reposit acdemic achievements data from existing scattered data
source in batches, and aggregate all research information into a 'big data' center.
Research output data center have become a new breakthrough in construction
of research information system and research data application in universities, and
will also promote development of campus informatization.

G. Li et al. (Eds.): DASFAA 2019, LNCS 11448, pp. 142–157, 2019.
https://doi.org/10.1007/978-3-030-18590-9_10

Various campus affiliations and their information platforms will need to acquire and apply scholar's research output data:

- Human Resources Office. For teacher's title appraisal, talents evaluation, and teachers annual performance confirmation, etc.
- Scientific Bureau. For significant paper award and publication statistics in scientific research management system.
- Planning and Development Department. For further discipline construction, subject funds transfer, and interdisciplinary collaboration analysis, etc.
- Graduate school. For examination and assessment of graduate qualification.
- Library. For scholars and institutions results retrieving and subject service, etc.
- Various schools or departments. For personal homepages, discipline evaluation and academic contribution analysis, etc.

Different functional platforms do not adopt uniform standards for collection and storage of research output data, so they cannot share and interoperate with each other. This will bring many problems for collection and utilization of academic results. For example, scholars need to input their publication data frequently between different systems. Dispersing construction of information system for academic output by each subordinate department in university according to its own needs, will lead to different data formats, poor integrity and low accessibility among campus information systems.

Research output data center can effectively support dynamic update of research output, reveal current institutional research situation in multiple dimensions, advance open access and provide data support for academic evaluation and discipline development. Effective data quality control [7, 16] in research output data center will comprehensively guarentee data severity of inconsistency, incompleteness, accuracy, decision and finding missing or unknown data. Main issues in data quality control procedure in research data center may include:

- Before data collection. Accurately integrating various data from different source and format, such as English publications from Web Of Science and Scopus etc, and Chinese publications from CNKI and CSSCI etc.
- During data collection. Data governance with different kinds of technical methods for data matching and data fitting etc, in data center platform.
- After data collection. Data authentication and security policy of data application for campus information systems.

Different phases need different data quality management policies and technologies. In this paper, we will propose a data quality management framework to build the research output data center, and put forward relevant technical solution for different data governance problems, such as department name similar estimation in data integration, author name disambiguous problem in data linkage and security issue in data exchange. We also introduce some effective learning algorithms such as text distance and community detection with matrix learning in our test to achieve good performance in quality manage processing.

2   Related Works

For purpose of systematically collecting current research output of individuals for further assessment and analysis of researchers, higher educations or research institutions are committed to build their institutional research outputs knowledge center or data denter. All academic publications written by their researchers from different journals or conferences will be fully collected in this research information data center or current research information system (CRIS) [4,10]. Berkhoff *et al.* [2] define CRIS system as research output data center that supports all research processes, sharing a common database with other components that are integrated within an integrated higher administration software. In general, data attributes of research output include author, affiliation, paper (journal paper and conference paper), patent and grant of whole university or institution.

The data center need to determine a data governance strategy for capture, collect, process, storage and manage of scholars' publication data, and maintain good data quality to serve different information system in a campus research data lifecycle. The quality of data is determined by factors such as accuracy, completeness, reliability, relevance and how up-to-date it is. As research data has become more intricately linked with different campus systems, data quality has gained greater attention.

In long run, it is undoubtedly an effective way to promote unique identifiers of scholars and require them to provide identifiers when submitting papers. However, at present stage, various institutions have accumulated a large number of publications that have not been clearly linked to scholars. In this case, it is unrealistic to expect all scholars to rely on their own efforts to consciously associate their own scholar identifiers. In order to improve work efficiency, it is necessary to design an application process, propose relevant technical solutions, and identify scholars.

Chang [6] discusses the concept, function and meaning of scholar identification, then deeply analyzes the authority control database of author name and the scholar unique ID system, which are two kinds of the classic solutions for scholar identification. She also brings insight into the mechanism of scholar identification and research on the key technology to identify scholars. Azeroual *et al.* [1] consider that research data is usually not uniformly formatted and structured. These include various source systems with their different data formats, and all data must be constantly synchronized and the results of the data links checked. However, since many fields of scholar information extracted from publications is not strictly structured, most scholars have a variety of input methods in the name abbreviation and affiliation name. In addition, there are errors or irregularities, which make it difficult to effectively attribute and associate the publication information directly with the actual author. In these large amounts of publications, it will also happen that same researcher publishes different papers with different affiliations.

It's a difficult problem for data quality administrator to precisely link these papers to one unique faculty or student. This process is also regarded as Author Name Disambiguation (AND) [13,21]. AND tries to find right person who publish the paper, under the conditions of same people appear different name in various situation, or different people share same name. Identification of scholars is not only crucial in the aggregation process of school research results, but also one of the important basic tasks to ensure data quality of individual or institutional academic output results. Zhang *et al.* [27] has proposed a Bayesian non-exhaustive classification framework for solving online name disambiguation task. Treeratpituk *et al.* [23] describe an algorithm for pairwise disambiguation of author names based on 'random forests' machine learning classification algorithm. Momeni *et al.* [12] focus on homonym names in the computer science bibliography DBLP, and employ community detection method which uses co-authorship networks and analyze the effect of common names on it.

After collection, processing and storage in research output data center, far-reaching value will come from widely application of data. It is true that we may use institutional research output data for data visualization and other technical methods to reveal data for a certain extent and conduct simple statistical analysis. From a global perspective, as a kind of important resources data in campus informational data cycle, it will be much meaningful to share research output data for university administrative or departments systems. This will undoutedly improve research output data utilization and expand its vitality.

3   Conceptual Framework of Data Processing with Data Quality Control

As shown in Fig. 1, we design a conceptual framework of data processing with data quality control enhancement. External data generated from other platform, such as publications, reviews and institutional repository information, will be continuously integrated into research output data center, and a data clearning and processing policy will be applied. We also design and develop a series of data directories and interoperation interface to provide comprehensive inquiry and application services of academic output data for other institutional systems, such as scholar homepage, talent accessment, subject development, and institutional prospect analysis.

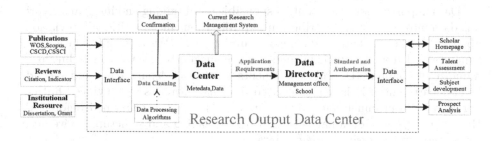

Fig. 1. Research output data center framework.

3.1   Main Data Processing for Quality Control

In institutional common research data management situation, user's standard information, such as name and title, is available from HR System or Student Management System, and these data can easily convert into research output data knowledge base. Basic task of central research output data database is to collect all institutional research output information and as accurately as possible to associate the output to real faculties or students. We successively design data process for quality control, and divide it into eight modules: **data inputing**, **preprocessing**, **merging**, **learning**, **matching**, **fitting**, **manual processing**, and **save and feedback**. Relationship between data and logical transfer among modules is shown in Fig. 2. Solid line is common data transfer, and dotted line is data transfer process in iterative optimization. Through whole processing, scholars' publications are finally divided into three categories:

1. Paper-author matching results. In principle, it is directly related to the actual personnel information corresponding to author, and regularly reminds data administrator to confirm or review, and author can also claim or reject the results. If it is modified as wrong result, it will be turned into no credible result.
2. Author not finding results. Some of results can be combined by using text distance or machine learning algorithms. However, data center may not exist related personnel information of this scholar. Therefore, data administrator may select to add author's information, such as user name, user campus ID and department information.
3. No credible results. Handled by manual discrimination processing.

3.2   Data Cleaning

Main data cleaning in research output data center includes three phases: **data importing**, **data preprocessing** and **data merging**. Data sources are generally divided into three categories: academic publication data, existing supervised information (for purpose of supervision information, such as list of published documents or verification report of confirmed institutional personnel, etc.), and real personnel information.

Data importing stage is mainly responsible for obtaining and integrating various data into the system. Attribution of publications include title, abbreviation, abstract, author affiliation, document number, etc. Personnel information are synchronized from institutional personnel management system, including user name, user Campus ID, and departments (first-level departments and second-level departments). Existing supervised information refers to convinced linkage between real personnel and publications that have been confirmed before. All these three types of data have different data meaning and structure, so we need to do data preprocessing for good data quality.

Data preprocessing is relatively complex, for many data is mainly reflected in its multi-source and multi-hierarchy structure. Due to inconsistency of publication data format from different databases. Generally, data fields related to

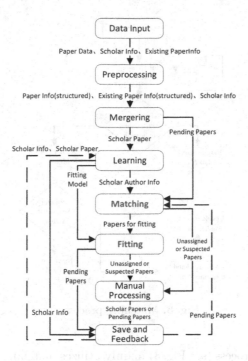

Fig. 2. Main flow of data processing design in research output data center.

authors always contain information of more than one author, and some authors will sign more than two affiliated department.

Data Merging process includes two parts: one process is splitting author data field into multiple records for storage separately, and another process is integrating information of the same author belonging to different affiliations into same record. The newly input result data also needs to be uniquely de-weighted with existing result data in data backend to ensure uniqueness of result data.

3.3   Data Linking

The process of research output data Linking with publications and scholars is divided into three modules: **Learning**, **Matching** and **Fitting**.

Learning process mainly includes general process and special process. The general process mainly fulfills integration of the author's signature information. From basic information of teachers and students, the author's signature information and corresponding information of the first-level department and second-level department in both Chinese and English are established. The special process regularly trains a series of results of a scholar (including published information and published information confirmed by system), obtains characteristics of the scholar's research fields, main quotations, co-authors and other aspects, and uses them for subsequent intelligent fitting analysis.

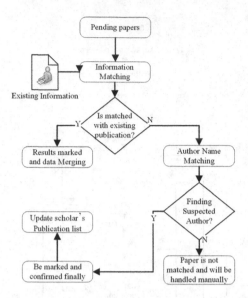

Fig. 3. Matching process.

The matching process (see Fig. 3) mainly utilizes matching between scholar's publication signature information and existing faculties or students background information in the system (mainly matching author name and affiliation department fields). In matching process, text distance matching method can be used to improve data quality for matching accuracy. After matching, three types of intermediate data are generated:

1. Matching successful results. In principle, author's signature and existing scholar's information of the system are exactly matched, and institutional information of the scholar is completely consistent with or similar to institutional HR information of one scholar. The similarity of mechanism is mainly determined by text distance function, and we will introduce it later. In case of certain text similarity, it can be directly correlated for confirmation. However, for sake of perfect data quality, this kind of data may still need to be reviewed manually.
2. The synthesized results contain incomplete information, which can be matched to one or more authors, but it is not possible to directly determine whether results belong to the scholar (for example, publication department information does not absolutely matching with existing information). This kind of problem needs to use further classification fitting algorithm to analyze in the fitting processing and model.
3. No attribution results for completely unmatched results, which are inconsistent with any scholar registered in the system. Such results may be due to spelling errors, and cannot be dealt with computer automatically.

The fitting process is to identify some suspected results after matching processing and try to build classification model with various algorithms. The main algorithms can be machine learning model based on attribute association or community detection based from co-author network. The intermediate data after fitting analysis are mainly divided into two categories:

1. Preliminarily confirm the relevant publication information and the scholars through fitting, and submit it to phase of manual confirmation or personal claim,
2. If a paper is not well fitting, then it is set as unfinished paper.

As shown in Fig. 2, after fitting processing, when a result transfer to manual confirmation. Many feedback and update can be carried out back to learning process to enhance data quality, so as to further expand existing supervised information source. In process of information analysis of scholars' output, it is necessary to effectively match a large number of scholars' and published data, so as to achieve good data link accuracy with automatic integration of papers and scholars.

4   Data Matching with Text Distance Algorithm

In matching process of author and affiliation, an effective text distance function can greatly improve data quality for irregularity of scholars' published papers, and correct misspelling mistake in paper publication procedure.

In many case, we cannot guarantee that author name or background information is same when collecting his/her publication data from different database source, for example:

- Publish papers in english with phonetic inversion, such as Liu, Yang and Yang, Liu.
- Publish papers with name abbreviation, such as Liu, Y.
- Publish papers woith additional symbols, such as Shi, Xiao-Hua.
- when publishing a paper, scholar use different spellings for the department information, such as "Stem Cell Res Ctr" and "Clin Stem Cell Res Ctr".

Therefore, a text distance algorithm is needed to calculate and analyze distance between name characters or affiliation characters. If text similarity distance of some scholar name fields is greater than a certain threshold, they are possible to be same author. For same author name with different department name, if the text similarity distance of department fields is greater than a certain threshold, it will represented with two named departments for same person. Common text distances are Cosine distance, Jaccard distance, and Levenshtein distance, etc.

Levenshtein distance, also known as Edit Distance. It was proposed by Russian scientist Vladimir Levenshtein in 1965. Levenshtein Distance [14] between two strings is determined by minimum number of edits required to convert one string to the other, and the greater the distance, the more different they are.

In computation algorithm[1], the permitted editing operations include replacing one character with another, inserting a character, and deleting a character. Through practical comparison and analysis, Levenshtein distance Algorithm (1) can achieve a better matching effect on the similarity of scholar's department fields.

Levenshtein distance between two strings a,b is given by $\mathbf{lev}_{a,b}(|a|,|b|)$ where

$$\mathbf{lev}_{a,b}(i,j) = \begin{cases} \max(i,j) & if \quad min(i,j) = 0, \\ \min \begin{cases} \mathbf{lev}_{a,b}(i-1,j)+1 \\ \mathbf{lev}_{a,b}(i,j-1)+1 \\ \mathbf{lev}_{a,b}(i-1,j-1)+1_{a_i \neq b_j} \end{cases} & otherwise \end{cases} \quad (1)$$

where $1_{(a_i \neq b_j)}$ is the indicator function equal to 0 when $a_i = b_j$ and equal to 1 otherwise, and $\mathbf{lev}_{a,b}(|a|,|b|)$ is the distance between the first i characters of a and the first j characters of b.

We collect some SCI publications of Shanghai JiaoTong university since 2012 to 2015, after published data for processing, calculation the author's name, same level but not at the same time, secondary institutions to calculate the Levenshtein distance between the text of secondary institutions (we use the transform number divided by the maximum length of the string as a result the output distance). We can see the test results from Table 1, this text distance algorithm can effectively deal with different institutions or non-standard abbreviations, and we may chose thresold in 0.5 to 0.6 to solve the data matching problem.

Table 1. Text distance in secondary department name.

Name	Department1	Department2	Text distance
An, Yuan	Coll Agr & Biol	Sch Agr & Biol	0.733333
An, Yuan	Coll Agr & Biol Sci	Sch Agr & Biol	0.578947
Ao, Huafei	Peoples Hosp 3	Shanghai Peoples Hosp 3	0.608696
Bai, Haitao	Affiliated Peoples Hosp 1	Affiliated Shanghai Peoples Hosp 1	0.735294
Bai, Haitao	Affiliated Peoples Hosp 1	Shanghai Peoples Hosp 1	0.64
Bai, Haitao	Affiliated Shanghai Peoples Hosp 1	Shanghai Peoples Hosp 1	0.676471
Bai, Jing	Sch Chem & Chem Engn	Sch Chem & Chem Technol	0.73913
Bai, Min	Affiliated Peoples Hosp 1	Peoples Hosp 1	0.56
Bai, Min	Affiliated Peoples Hosp 1	Shanghai Peoples Hosp 1	0.64
Bai, Min	Peoples Hosp 1	Shanghai Peoples Hosp 1	0.608696
Bai, Yuehong	Affiliated Peoples Hosp 6	Shanghai Peoples Hosp 6	0.64

In our real test based on Levenshtein distance, we suggest some optimizations according to special bylines of subordinate affliation of one university or institution. For example, in some chinese universities' subordinate hospital, the biggest difference of their affiliations name is just a number, such as 'Affiliated hospital 1' and 'Affiliated hospital 3'. We need to reduce distance value while department name contain 'Affiliated hospital' by reducing distance value. In case

---

[1] https://en.wikipedia.org/wiki/Levenshtein_distance/.

that some affliation names are similar, such as 'Renji Hosp' and 'Ruijin Hosp', these names are also need to be adjust to improve accuracy for good data quality. After comparision, 10–15% of the data of same scholars name with different department name can be directly matched by calculating and merging effective text distance.

5 Author Name Disambiguous with Community Detection

The identification of scholars is a hot topic in fields of research information retrieval and scientific evaluation for solving the confusion of scholars' names. Tang et al. [22] substitut a hidden markov model into supervised learning on basis of previous studies and supplemented relationship between scholars with scholars, scholars with output, and output with output as an important part of compare basis, which greatly improved the accuracy of the discrimination algorithm. Xia et al. [25] focus on algorithms while emphasizing the integrated management and collection of academic data and made a series of attempts to visually present screening results. Sedelnikov et al. [17] calculated the sensitivity parameters in the process of personal identification in the scientific research management system to identify the author names of all kinds of latest academic articles, patents and data sets. Yang et al. [26] discussed the discrimination of Chinese names, and propose a cluster method based on contextual similarity measurement by reducing data sparsity.

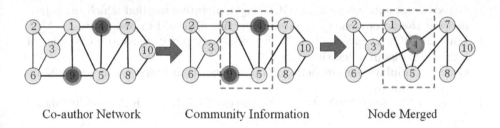

Co-author Network   Community Information   Node Merged

Fig. 4. Author merging with co-author network community detection.

In general, network community detection algorithm comes from the need of Internet user group analysis. However, network community detection algorithm can also be applied to our data fitting process. As shown in Fig. 4, according to scholars' co-author network, we can judge whether two nodes belong to same person, and merge it. We will test a community detection algorithm based on Non-negative Matrix Factorization to accomplish data fitting process.

In this section, we artificially generate a real institutional scientific collaborative network to compare the performance of different methods. Detection community from scientific collaborative network in a specific institution may help us to learn the knowledge of scientific cooperation from all faculties, analysis the difference between communities and predict the trend of interdisciplinary demand.

We download $35,659$ SCI publications[2] stated with Shanghai Jiaotong University (SJTU) from 2012 to 2016, and these publications are written by $1,346,279$ authors without pruning. We decide that authors who write a paper signed with same name and affiliation will be regard as one person. We create a network with $17,908$ nodes and $718,967$ links from whole institutional publication data. We use this real large collaborative network to compare community number (K_C) and modularity within GCD, BGLL, GMO, BNMF and BSNMF, for these five methods can find communities without given initial community number:

In task of network community detection, all network structure can be represented by the diagram, and we can describe the diagram with its adjacency matrix. Therefore, application of matrix factorization learning method can effectively aggregate the nodes from whole network into different communities, and achieves good experimental results. In this section, we use a bayesian symmetric nonnegative matrix factorization algorithm (BSNMF) [19,20] we proposed before, and apply this algorithms in co-author scientific network data. Comparing with relevant community detection methods, our algorithms obtain good experiment and application effects.

We compare our algorithm with other four state-of-the-art community detection methods, and all five algorithms are listed below. Algorithm 4 to 5 are Matrix Learning methods, and all algorithm can capture communities dimension automatically without any given information.

1. Greedy community detection (GCD) agglomerative method which takes into account the heterogeneity of community size observed in real networks [8].
2. Louvain (BGLL) can compute high modularity partitions and hierarchies of large networks in quick time [3].
3. Greedy modularity optimisation (GMO) algorithm based on the multi-scale algorithm but optimised for modularity [11].
4. Bayesian Non-negative Matrix Factorization (BNMF) with Poisson likelihood [5].
5. Bayesian Symmetric Non-negative Matrix Factorization (BSNMF) with Poisson likelihood [19].

The update rule of s_{ik} in BSNMF is:

$$s_{ik} \longleftarrow s_{ik} \left(\frac{X}{SS^T} \right)_{ik} \left(\frac{S}{1S + SB} \right)_{ik} \tag{2}$$

where X is generated suspend co-author network, S is a target symmetric coefficient matrix.

We can see from Table 2 that, Bayesian matrix factorization method we proposed can effectively find much dense community, and achieves best Modularity [15] value in 0.7564 and is highly close to results of BGLL and GMO.

---

[2] http://webofknowledge.com/.

Table 2. Community detection results in SJTU co-author network

Methods	K_C	Modularity
GCD	2057	0.6696
BGLL	1233	0.7478
GMO	1045	0.7506
BNMF	868	0.7235
BSNMF	625	**0.7564**

Table 3. Author name merging from BSNMF results.

Name	Department	Paper ID	Community ID
Liu, Yang	Sch Mat Sci & Engn	2	149
Liu, Yang	Sch Med	9	**183**
Liu, Yang	State Key Lab Oncogenes & Related Genes	1	**183**
Liu, Yang	Stem Cell Res Ctr	1	**183**
Liu, Yang	Clin Stem Cell Res Ctr	1	**183**
Liu, Yang	Ren Ji Hosp	1	**183**
Liu, Yang	Renji Hosp	1	**183**
Liu, Yang	State Key Lab Microbial Metab	1	214
Liu, Yang	Sch Life Sci & Biotechnol	1	214
Liu, Yang	Sch Agr & Biol	3	328
Liu, Yang	Minist Educ	1	604

We find that community detection in scientific co-author network can help to solve AND problem in research output data fitting processing, and we assume that researcher with same name and different affiliation in same community should be same true person. We further analysis the BSNMF results of previous real scientific collaborative network with 17,908 nodes, and find that some nodes with same name and different afflictions are partition into one community. Scholars may share different positions in different academic institutions, like different schools, departments or laboratories. Table 3 show the example of person named **Liu, Yang**. Through actual confirmation, five **Liu, Yang** appeared in our real network community NO. 1683, and they are same person from School of Medicine and a State key laboratory for this six nodes. Two **Liu, Yang** from community 214 is another person in School of Life Science. So we can effectively discriminate authors who sign different affiliation information with same name published. By checking all communities results, our fitting algorithm can prune 1803 nodes from whole network with 17908 nodes, and improve 10.07% data accuracy to aggregate the scholar signed with same name and different affiliations.

6 Data Exchange Policy Design with Quality Control

A sound data quality management framework is also critical to getting the most out of key values of data in data exchange. Shen *et al.* [18] introduce the content and framework of universal data interchange platform in terms of categories of processes and layers of information exchange, and develop a data interchange model to elaborate data exchange between different departments on campus. Wang *et al.* [24] design and realize data exchange platform of Tsinghua University to accomplish the data exchange in large data quantity of heterogeneous database based on ETL technology, and accomplish the real-time exchange in small data quantity of heterogeneous database based on ESB technology.

External data service can be divided into **data output** and **data input**. Data output refers to packaging part of research output data in a certain format and providing for other application systems. Data input means that data has been manually reviewed or confirmed by teachers and flowed back into research output data center as supervised information. Through reflux of data, more completeness and authority of data quality are guaranteed.

In order to control data quality in process of data service, data input and output should be completed with other systems at interface level through API basically. As shown in Fig. 5, in interaction process of interface invocation, security and legality of invocation should be considered first.

OAuth protocol[3] is a common protocol for application authorization [9]. It provides a safe, open and simple standard for user resource authorization. Its protocol has good characteristics: **Simple**, whether OAuth service providers or application developers, are easy to understand and use; **Security**, no user keys and other information, more secure and flexible; **Open**, any service provider can implement OAuth, any software developers can use OAuth. By integrating OAuth protocol into the API architecture, different permissions can be controlled for different API callers.

Based on application of authorization, different users with different data access, such as in "School of Electronics, Information and Electrical Engineering" data access system, the interface should not by batch get research output data of 'School of Naval Architecture, Ocean & Civil Engineering', although these data in research output data support platform is open and visible. Therefore, API interface should also be able to implement row-level and column-level data access control, that is, to control the fields and records accessible to application and only allow users to access part of data records related to it.

At same time, for sensitive data such as ID information, we should design desensitization algorithm for sensitive data and making use of it. The sensitivity of the same data may be different in various system calls. For example, ID information may not belong to sensitive information when university HR system calls, because the record of ID information in HR system is more authoritative.

---

[3]   https://oauth.net/.

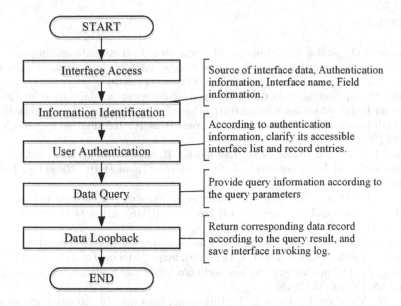

Fig. 5. Data interface security design with quality control policy.

Therefore, desensitization algorithms should be designed or configured for different applications and sensitive data fields to safeguard data quality in data center.

7   Conclusion

This paper discuss data quality issue in building institutional research output data center. We put forward a conceptual framework of data processing with data quality control, and introduce some useful methods, such as text distance, community detection and data exchange security policy. We also compare different community detection algorithm in scientific network for Author Name Disambogous, our matrix learning method achieve good performance than other community detection methods. In long run of research output data center, we think that key ways for promote data quality in research output data center, is to regulate publication scholars' ID, such as ORCID[4]. This is most effective way to solve data quality problem, but requires joint attention and promotion of global publishers, research institutions and authors.

Acknowledgments. This work was supported by NSFC (Grant No. 61772330), the Science and Technology Commission of Shanghai Municipality (Grant No. 16JC1402800), China Next Generation Internet IPv6 project (Grant No. NGII20170609), the Social Science Planning of Shanghai (Grant No. 2018BTQ002), and Arts and Science Cross Special Fund of Shanghai JiaoTong University (Grant No. 15JCMY08).

---

[4]   https://orcid.org/.

References

1. Azeroual, O., Saake, G., Abuosba, M., Schöpfel, J.: Text data mining and data quality management for research information systems in the context of open data and open science. arXiv preprint arXiv:1812.04298 (2018)
2. Berkhoff, K., Ebeling, B., Lübbe, S.: Integrating research information into a software for higher education administration-benefits for data quality and accessibility. In: 11th International Conference on Current Research Information Systems. euroCRIS (2012)
3. Blondel, V.D., Guillaume, J.L., Lambiotte, R., Lefebvre, E.: Fast unfolding of communities in large networks. J. Stat. Mech.: Theor. Exp. **2008**(10), P10008 (2008)
4. Bryant, R., et al.: Practices and patterns in research information management: findings from a global survey. OCLC Research (2018). https://doi.org/10.25333/BGFG-D241
5. Cemgil, A.T.: Bayesian inference for nonnegative matrix factorisation models. Comput. Intell. Neurosci. **2009**, 1–17 (2009). https://doi.org/10.1155/2009/785152
6. Chang, E.: The mechanism and key technology of scholar identification. Libr. Tribune **35**(10), 88–95 (2015)
7. Dai, W., Yoshigoe, K., Parsley, W.: Improving data quality through deep learning and statistical models. In: Latifi, S. (ed.) Information Technology - New Generations. AISC, vol. 558, pp. 515–522. Springer, Cham (2018). https://doi.org/10.1007/978-3-319-54978-1_66
8. Danon, L., Díaz-Guilera, A., Arenas, A.: The effect of size heterogeneity on community identification in complex networks. J. Stat. Mech.: Theor. Exp. **2006**(11), P11010 (2006)
9. Hardt, D.: The oauth 2.0 authorization framework. Technical report (2012)
10. Joint, N.: Current research information systems, open access repositories and libraries: antaeus. Libr. Rev. **57**(8), 570–575 (2008)
11. Le Martelot, E., Hankin, C.: Fast multi-scale detection of relevant communities in large-scale networks. Comput. J. **56**(9), 1136–1150 (2013)
12. Momeni, F., Mayr, P.: Using co-authorship networks for author name disambiguation. In: 2016 IEEE/ACM Joint Conference on Digital Libraries (JCDL), pp. 261–262. IEEE (2016)
13. Müller, M.C., Reitz, F., Roy, N.: Data sets for author name disambiguation: an empirical analysis and a new resource. Scientometrics **11**, 1–34 (2017)
14. Navarro, G.: A guided tour to approximate string matching. ACM Comput. Surv. (CSUR) **33**(1), 31–88 (2001)
15. Newman, M.E.: Modularity and community structure in networks. Proc. Nat. Acad. Sci. **103**(23), 8577–8582 (2006)
16. Olson, J.E.: Data Quality: The Accuracy Dimension. Elsevier, Amsterdam (2003)
17. Sedelnikov, M.S., Gordeev, R.N., Kuzmicheva, A.V., Odulov, A.G.: Disambiguation solution for persons' accounts in research information management systems. Indian J. Sci. Technol. **9**(43), 1–12 (2016)
18. Shen, S.S., Ding, A.X.: Design and establishment of information exchange standard on campus. In: Applied Mechanics and Materials, vol. 513, pp. 1294–1298. Trans Tech Publications (2014)
19. Shi, X., Lu, H.: Community detection in scientific collaborative network with Bayesian matrix learning. Front. Comput. Sci. **13**(1), 212–214 (2019)

20. Shi, X., Lu, H., Jia, G.: Adaptive overlapping community detection with Bayesian nonnegative matrix factorization. In: Candan, S., Chen, L., Pedersen, T.B., Chang, L., Hua, W. (eds.) DASFAA 2017. LNCS, vol. 10178, pp. 339–353. Springer, Cham (2017). https://doi.org/10.1007/978-3-319-55699-4_21
21. Smalheiser, N.R., Torvik, V.I.: Author name disambiguation. Annu. Rev. Inf. Sci. Technol. **43**(1), 1–43 (2009)
22. Tang, J., Fong, A.C., Wang, B., Zhang, J.: A unified probabilistic framework for name disambiguation in digital library. IEEE Trans. Knowl. Data Eng. **24**(6), 975–987 (2012)
23. Treeratpituk, P., Giles, C.L.: Disambiguating authors in academic publications using random forests. In: Proceedings of the 9th ACM/IEEE-CS Joint Conference on Digital Libraries, pp. 39–48. ACM (2009)
24. Wang, Q., Liu, N.J., Cheng, Z.R.: The application research of data exchange technology in digital campus. In: Zhang, Y., Zhou, Z.-H., Zhang, C., Li, Y. (eds.) IScIDE 2011. LNCS, vol. 7202, pp. 607–613. Springer, Heidelberg (2012). https://doi.org/10.1007/978-3-642-31919-8_77
25. Xia, F., Wang, W., Bekele, T.M., Liu, H.: Big scholarly data: a survey. IEEE Trans. Big Data **3**(1), 18–35 (2017)
26. Yang, X., Jin, P., Xiang, W.: Exploring word similarity to improve Chinese personal name disambiguation. In: Proceedings of the 2011 IEEE/WIC/ACM International Conferences on Web Intelligence and Intelligent Agent Technology, vol. 03, pp. 197–200. IEEE Computer Society (2011)
27. Zhang, B., Dundar, M., Al Hasan, M.: Bayesian non-exhaustive classification a case study: online name disambiguation using temporal record streams. In: Proceedings of the 25th ACM International on Conference on Information and Knowledge Management, pp. 1341–1350. ACM (2016)

Generalized Bayesian Structure Learning from Noisy Datasets

Yan Tang$^{(\boxtimes)}$, Yu Chen, and Gaolong Ge

College of Computer and Information, Hohai University, Nanjing 210098, China
{tangyan,chenyu12,gegaolong}@hhu.edu.cn

Abstract. In recent years, with the open data movement around the world, more and more open data sets are available. But, the quality of the datasets poses issues for learning models. This study focuses on learning the Bayesian network structure from data sets containing noise. A novel approach called GBNL (Generalized Bayesian Structure Learning) is proposed. GBNL first uses a greedy algorithm to obtain an appropriate sliding window size for any dataset, then it leverages a difference array-based method to quickly improve the data quality by locating the noisy data sections and removing them. GBNL can not only evaluate the quality of the data set but also effectively reduce the noise in the data. We conduct experiments to evaluate GBNL on five large datasets, the experiment results validate the accuracy and the generalizability of this novel approach.

Keywords: Bayesian network structure learning · Bayesian score · Noise reduction · Data quality

1 Introduction

A Bayesian network (BN) [1,2] is a probabilistic graphical model that represents a set of random variables and their conditional dependencies via a directed acyclic graph. BNs have been broadly applied to modeling and reasoning in many domains [3–6]. With the increasing availability of open datasets in academia, government and business, constructing Bayesian network from domain data are becoming more and more valuable and mission critical, especially when uncertainties are involved [2,7,8].

The sparsity of the data, incomplete and noisy, introduces challenges to the algorithm stability. Small changes in training data may significantly change the models [9]. This requires good generalization ability of the learning model. However, most of the current BN structure learning algorithms perform well given a DAG-faithful dataset with good data quality [10]. A dataset is DAG-faithful if its underlying probabilistic model can be structured as a DAG. This condition makes a dataset suitable for BN learning. However, facing real-world datasets, most of the current BN structure learning algorithms have limitations. First,

© Springer Nature Switzerland AG 2019
G. Li et al. (Eds.): DASFAA 2019, LNCS 11448, pp. 158–169, 2019.
https://doi.org/10.1007/978-3-030-18590-9_11

given any data set, it is hard to know whether it is suitable for Bayesian network learning. Second, even if the data set is suitable for learning, it may still contain noise such as incomplete or missing data. Very few studies are conducted to deal with the presence of noise in the data for BN learning.

For the first limitation, one solution is to estimate the accuracy levels of a dataset by assessing the quality of the data [11]. Several methods are proposed to deal with noise in learning parameters of Bayesian network [12,15] and the sensibility of Bayesian score function [13]. For the second limitation, even though there are numerous algorithms are proposed to learning Bayesian network from data such as [10,14]. However, most learning algorithms operate individually and lack the ability to deal with the data quality issue.

Therefore, we introduce a novel approach called Generalized Bayesian network Learning (GBNL). GBNL is an iterative function that begins with calculating the Data Sliding Window Size (DSWS) from a data set. DSWS is the number of records GBNL reads for each data checking iteration. Then GBNL calculates the Bayesian Network Learning Health Degree ($BNLHD$) for the data slice of size DSWS. Lastly, GBNL scans through the whole data set to find out the data sections containing noise and remove them. Experimental results on five different datasets show that by intelligently selecting the sliding windows size, evaluating the dataset's health degree and locating the noise data. GBNL is a generalized method that leads to significantly improved learning results on real-world datasets for different BN learning algorithms.

To summarize, the main contributions of this paper are as follows:

- This paper proposes a greedy algorithm to obtain an appropriate sliding window size for any dataset. This ensures the precision for locating the data section containing noise.
- This paper proposes a generalized difference array-based method called GBNL to quickly improve the data quality by locating the noisy data sections and removing them. This method can not only detect the noise in the data but also effectively improves the data quality. GBNL is a generalized approach that can be applied to different kinds of BN learning algorithms.
- We further evaluate GBNL on five big datasets, the experiment results validate the accuracy and the generalizability of this novel approach.

The rest of the paper is organized as follows: Sect. 2 is the related work. The proposed method including algorithms is presented in Sect. 3. After giving experimental results and discussion in Sect. 4, we conclude this work in Sect. 5.

2   Related Works

Early in 2009, [11] estimates the accuracy levels of a dataset by assessing the quality of the data. [12] studies noise smoothing in learning parameters of the Bayesian network. The robustness of Bayesian networks learning from non-conjugate sampling is studied in [15]. Ueno et al. [13] describes some asymptotic analyses of BDeu score to explain the reasons for the sensitivity and its effects. Furthermore, this paper presents a proposal for a robust learning score for the equivalent sample size (ESS). But most approach focus on the parameter learning of Bayesian network.

Numerous algorithms are proposed to learning Bayesian network from data, such as Hill Climbing (HC), Tabu Search (Tabu), Three Phase Dependency analysis (TPDA) [10], Inter-IAM [16] and Max-Min Hill-Climbing (MMHC)[14]. However, most algorithms operate individually, our previous work [7,17] achieved higher accuracy of BN structure learning through ensemble methods. But most BN learning algorithms still lack the ability to deal with the data quality.

3   The Proposed Approach

3.1   Overview of the Method

GBNL works as follows (Fig. 1): First, it calculates Data Sliding Window Size ($DSWS$) according to the data set. Then, GBNL calculates the window health threshold called $BNLHD$ to evaluate data quality variation in the data. Lastly, GBNL scans through the whole data set and uses a difference array-based method to locate the data sections containing noise and remove them.

GBNL uses the Bayesian score function in to calculate the score denoted as:

$$bayesianScore = score(B, data) \qquad (1)$$

Where $data$ is the data set and B is the Bayesian network structure learned from the data set. There are many score functions. One of them is the BIC (Bayesian information criterion) score. The BIC score is based on the assumption that the sample satisfies the independent and identical distribution hypothesis, and uses log likelihood to measure the degree of fitness of the structure of the data. The BIC is formally defined as:

$$BIC = \ln(n)k - 2\ln(\hat{L}) \qquad (2)$$

Where $\hat{L}$ is the maximized value of the likelihood function of the model.x is the observed data.n is the number of data points in x, the number of observations, or equivalently, the sample size. k is the number of parameters estimated by the model [18]. One of the BIC scores is the log likelihood of the model, which is the degree to which the metric structure fits the data. Another item is the penalty for model complexity, preventing over-fitting of data and structure [19].

3.2   Data Sliding Widows Size Calculation

Definition 1. *(Data Sliding Widows Size)*
Assuming that the interval of DSWS is [lborder, rborder], we divide [l, r] into n small intervals and take random numbers in small intervals. DSWS is the size of the sub dataset that has the highest score, denoted as:

$$\mathbf{DSWS} = \max_{i=1...n} \mathbf{DSWS}_{l_i - r_i} \qquad (3)$$

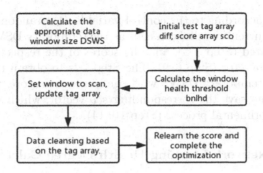

Fig. 1. Overview of the method

In order to get DSWS, we define the following algorithm. The Intervals variable in the algorithm represents the interval between the$[r/lborder, r/rborder]$ intervals divided into k equal parts, r is the number of rows in the dataset. P is a control variable. $len1$ is the random number taken in the cell, that is, the test window size. $stp1$ is the step progress variable. Sco is a scoring vector for each operation. The score is the maximum of all sco 's average values. The algorithm is as follows:

Algorithm 1. CalculateDSWS

Input:
   D: Dataset;
Output:
   $DSWS$
1:   $step=0; score=$MIN$; r=$read size of Dataset;
2:   $intervals=$ Decentralization interval $[lborder, rborder]$;
3: **while** $step<cnt$ **do**
4:    $len1=$ Random numbers from intervals$[step]$;
5:    $stp1=$round$(len1/P)$;
6:    **for** $i=1$ to $r-len1+1$ by $stp1$ **do**
7:     $window=$ Segmentation of Dataset;
8:     $sco=$sco.add(score(structureLearning(window),window));
9:    **end for**
10:   **if** mean$(sco)>score$ **then**
11:    $score=$mean(sco)
12:    $len=len1$
13:    $window sco = sco$
14:   **end if**
15:   $step=step+1$
16: **end while**
17: **return** len

The data size r is read from the data set D, the initialization variable score is the minimum value, and the value range $[lborder, rborder]$ of the DSWS is obtained

(Step 1–2). Then, according to the control variable P, a random number $len1$ is selected from the current ten-division cells as the temporary DSWS, and the step size $stp1$ is calculated (Step 4–5), and the scores of the respective windows are recorded in the $scovector$ (Step 6–9). The Step 7 is to obtain the data forming window. Take the temporary DSWS with the largest sco score vector average as the real DSWS, and save the corresponding sco vector, which is convenient for observing the experimental process (Step 10–14).

3.3   Bayesian Network Learning Health Degree Calculation

In order to judge whether the data window is qualified, we need a judgment standard. Here we define a notion called $BNLHD$ (Bayesian Network Learning Health Degree).

Definition 2. *(Bayesian Network Learning Health Degree)*
$BNLHD$ is the scoring standard for windows. It is the data health threshold corresponding to DSWS. Assuming the number of windows is n, BNLHD can be replaced by the mean of window score, denoted as:

$$\mathbf{BNLHD} = \frac{1}{n} \sum_{i=1..N} \mathbf{score}(DS_i) \tag{4}$$

which is to determine whether the data set is healthy through this health threshold. BNLHD is obtained on the basis of Algorithm 2. By dividing the entire data set into blocks, the score of each block is obtained by the score scoring function, and the scores of larger fluctuations are removed, and the average of the remaining scores is BNLHD. The *structureLearning* algorithm can be any learning algorithm like MMHC, TPDA or HC [20].

The algorithm for calculating $BNLHD$ is as follows:

Algorithm 2. CalculateBNLHD

Input:
   D: Dataset;
Output:
   $BNLHD$
1: $step$=0;$score$=MIN;r=read size of Dataset;
2: $intervals$= Decentralization interval $[lborder,rborder]$;
3: **while** $step<cnt$ **do**
4:    $len1$= Random numbers from intervals[$step$];
5:    $stp1$=round($len1/cnt$);
6:    **for** i=1 to r-$len1$+1 by $stp1$ **do**
7:       D_w= Segmentation of Dataset of size w;
8:       sco= $sco.append(score(structureLearning(D_w), D_w))$;
9:    **end for**
10:   **if** mean(sco)>$score$ **then**
11:      $score$=mean(sco)

12:    **end if**
13:    *step=step+1*
14: **end while**
15: **return** *score*

The algorithm starts to read the data set size r, initializes the score (BNLHD) to the minimum, and obtains the test interval [lborder, rborder] (Steps 1–2). The [lborder, rborder] decimal is divided into random numbers as temporary DSWS, and the corresponding BNLHD (Steps 4–9) is obtained. Idea analogy Algorithms 1, 2 takes the maximum value of all temporary BNLHDs as real BNLHDs (Steps 10–14).

3.4   Noisy Data Positioning and Removal

For a given dataset, DSWS and BNLHD are obtained from Algorithms 1 and 2, and then the dataset is scanned by Algorithm 3. In the scanning process, it is necessary to mark the qualified data window, that is, the current window score is higher than BNLHD, then all data in the data window are marked as qualified. For each piece of data in the data set, it is not marked at the beginning, and the value is marked as 0; then, every time it is marked, the value is added by 1. In order to implement such a marking process, we introduce a differential array diff:

$$Diff[i] = value(i) - value(i - 1) \qquad (5)$$

The data of the current window [l, r] is marked as qualified, that is, [l, r] plus 1, which can be converted into diff [l] + 1, diff [r + 1] − 1. Finally, the whole dataset is traversed, and the value of 0 is the unqualified data, which is removed. So far, the whole process is transformed into interval operation and single point query based on differential array diff. In addition, in order to ensure the accuracy of the scanning, we set the translation step of the window to $1/Q$ (Q = 10 in the experiment) of the window, so that each piece of data can be scanned more times, such as Figure 1, which can ensure the accuracy of noise location (Fig. 2).

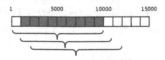

Fig. 2. Scan example

The algorithm is as follows:

Algorithm 3. Remove noise data and score

Input:
   D: Dataset;
   k: Window step coefficient,Generally take 0.1
Output:
   $BICScore$
1: $diff[]=0$;$windowsco[]=0$;
2: **while** $windows<D.size$ **do**
3:     $windowsco=score(structureLearning(D[head, head + DSWS))$;
4:     $head=head+$k$*DSWS$;
5:     **if** $windowsco<BNLHD$ **then**
6:         $value=0$;
7:     **end if**
8:     $value=value+1$
9:     $step=step+1$
10: **end while**
11: $diff(i)=value$(i)$-value$(i-1)
12: **for** $i=0$ to $diff.size$ **do**
13:     **if** $diff(i)==0$ **then**
14:         delete Di
15:     **end if**
16: **end for**
17: $score=$socre(mmhc(D))
18: **return** $score$

The input of Algorithm 3 is the data set. Algorithm 3 first initializes a differential array $diff$ and $windowsco$ to store the sliding window's score (Step 1). Within the scope of the data set, the sub-dataset of the selected range is scored (Step 2–3). The window score is compared with $BNLHD$. If it is higher than BNLHD, the corresponding array value is incremented by one, otherwise, it is unchanged (Step 5–8), and then the data sliding window is moved to the next position to start scanning. After scanning the entire array, the differential array $diff$ (Step 11) is calculated by the value array. If the value of $diff$ is 0, the data is judged to be noise data and deleted (Step 12–14). Finally, the network structure learning score of the optimized data is obtained.

4   Experiments and Discussion

After completing the design of the algorithm, we collect a lot of data sets suitable for Bayesian network structure learning, choose three learning algorithms, record the scores before and after the algorithm optimization, compare the results by calculating the shd of the structure before and after optimization, and analyze and summarize the experimental results. It is found that the quality of the data set has been significantly improved, and GBNL is a generalized approach that could improve the learning accuracy of different BN learning algorithms.

4.1 Method Selection

The scoring criteria for the entire experiment are given by the score() function in the bnlearn package. Considering computer performance issues, we did not choose a large data set in the experiment, so we choose the bic score in the commonly used Bayes score. And the Bayesian network structure learning is performed by the MMHC [14], HC and Tabu. Structural Hamming Distance (SHD) is used to measure the learning accuracy. SHD between two graphs is the number of edge insertions, deletions or flips in order to transform one graph to another graph. The SHD function is denoted as:

$$structureHammingDistance = SHD(learnedStructure, trueStructure) \quad (6)$$

learnedStructure is the Bayesian network learned from data, *trueStructure* is the actual gold standard BN structure.

4.2 Data and Measures

The experimental data include the real data set Asia (20000 data, 8 variables) Cancer (20000 data, 5 variables), Earthquake (20,000 data, 5 variables), Sachs (20,000 data, 10 variables), Survey (20,000 data, 6 variables). For all the datasets, we add 5 %–10 % of noise by randomly assign a continues section of dataset with arbitrary values.

Structure hamming distance (SHD) is a function that compares the difference between two BN structures [14]. When comparing a learned BN structure with the correct BN structure, lower SHD indicates a more accurate learned structure.

4.3 Experimental Result

In the experiment, By calculating the SHD value of the pre-optimized structure and the SHD of the optimized structure by GBNL, we can verify whether the optimization process is successful.

4.3.1 Comparison of Score Results

The following table shows the SHD study in details (Tables 1, 2 and 3):

Table 1. Comparison of scoring results before and after algorithm optimization.mmhc

Dataset	Asia	Cancer	Earthquake	Sachs	Survey
Unoptimized score (BIC)	−104254.1	−79418.03	−27735.49	−221394.8	−108678.8
Optimized score (BIC)	−25885.79	−22051.64	−2748.639	−78116.86	−40814.01
Unoptimized shd	11	9	9	9	10
optimized shd	6	5	5	6	6

Table 2. Comparison of scoring results before and after algorithm optimization.hc

Dataset	Asia	Cancer	Earthquake	Sachs	Survey
Unoptimized score (BIC)	−104254.1	−79418.03	−27735.49	−221394.8	−108678.8
Optimized score (BIC)	−24243.05	−22036.57	−2646.685	−75268.17	−40762.13
Unoptimized shd	14	9	9	10	10
optimized shd	6	5	4	7	6

Table 3. Comparison of scoring results before and after algorithm optimization.tabu

Dataset	Asia	Cancer	Earthquake	Sachs	Survey
Unoptimized score (BIC)	−104254.1	−79418.03	−27735.49	−221394.8	−108678.8
Optimized score (BIC)	−24243.05	−22051.23	−2666.473	−75295.08	−40758.94
Unoptimized shd	14	9	9	10	10
optimized shd	6	4	5	6	6

By comparing the experimental results, we can find that the data obtained by our algorithm has a significant decrease in the structure of the learning algorithm and the SHD value of the structure before optimization. The decline of the BIC score is more obvious, so the expected purpose of the algorithm is achieved, and the dataset can be optimized to a large extent, and the Bayesian network learning structure is improved.

In order to test the versatility of the optimization algorithm for various Bayesian network structure learning algorithms, we select HC, MMHC, and tabu algorithms to learn about five different data sets. After comparing the structure of the three sets of experimental data, we can find that the learning effect of the tabu and HC algorithms on the noisy data set is slightly worse than MMHC, which is different from the learning effect of the algorithm itself without looking at the learning effect of the algorithm itself, we can compare the experimental results of the three learning algorithms, and we can find that no matter which learning algorithm SHD has been significantly declined, and then compare the learned network structure, we can find out that through the optimized learning the network structure is more similar to the correct structure.

4.3.2   Comparison of Structure

Considering the page limit, this section only shows the comparison results of some data sets. The structure learned by MMHC algorithm before and after optimization is compared as shown below (Figs. 3, 4 and 5).

It can be observed that for an algorithm as good as MMHC, noise in the data set can significantly reduce the learning accuracy, resulting in a low-quality BN structure. But after applying GBNL, the optimized network structure is more similar to the correct structure with fewer edges. The same improvement is also observed in other learning algorithms. Therefore, the GBNL algorithm

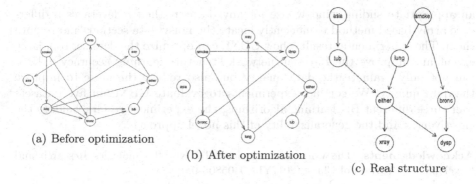

(a) Before optimization

(b) After optimization

(c) Real structure

Fig. 3. Asia data set

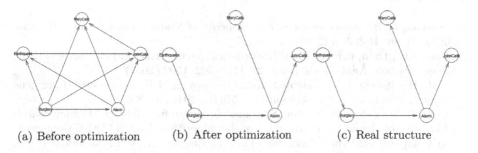

(a) Before optimization   (b) After optimization   (c) Real structure

Fig. 4. Earthquake data set

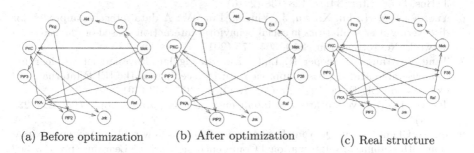

(a) Before optimization   (b) After optimization   (c) Real structure

Fig. 5. Sachs data set

has a good optimization effect on various Bayesian network structure learning algorithms and achieves the expected general algorithm optimization purpose.

5   Conclusion

Aiming to find a generalized method to improve the data quality for Bayesian network learning, this study proposed a novel approach called GBNL (Generalized Bayesian Structure Learning). GBNL first uses a greedy algorithm to obtain an appropriate sliding window size for any dataset, then it leverages a difference array-based method to efficiently locate the noisy data sections and remove them. The experimental results show GBNL can optimize the data set to a large extent and improve the Bayesian network structure learning accuracy. GBNL can not only evaluate the data quality but also reduce the noise to improve the data quality. We conduct experiments to evaluate GBNL on five datasets over three different BN learning algorithms, the experiment results validate the effectiveness and the generalizability of this novel approach.

Acknowledgments. The work was supported by Key Technologies Research and Development Program of China (2017YFC0405805-04).

References

1. Ben-Gal, I.: Bayesian Networks. Encyclopedia of Statistics in Quality and Reliability. Wiley, Hoboken (2007)
2. Jain, A.K., Duin, R.P.W., Mao, J.: Statistical pattern recognition: a review. IEEE Trans. Pattern Anal. Mach. Intell. **27**(11), 1502–1502 (2002)
3. Njah, H., Jamoussi, S.: Weighted ensemble learning of Bayesian network for gene regulatory networks. Neurocomputing **150**(B), 404–416 (2015)
4. Yang, J., Tong, Y., Liu, X., Tan, S.: Causal inference from financial factors: continuous variable based local structure learning algorithm. In: 2014 IEEE Conference on Computational Intelligence for Financial Engineering & Economics (CIFEr), pp. 278–285. IEEE (2014)
5. Giudici, P., Spelta, A.: Graphical network models for international financial flows. J. Bus. Econ. Stat. **34**(1), 128–138 (2016)
6. Yue, K., Wu, H., Fu, X., Xu, J., Yin, Z., Liu, W.: A data-intensive approach for discovering user similarities in social behavioral interactions based on the Bayesian network. Neurocomputing **219**, 364–375 (2017)
7. Tang, Y., Wang, Y., Cooper, K., Li, L.: Towards big data Bayesian network learning - an ensemble learning based approach. In: Proceedings of the IEEE International Congress on Big Data (BigData Congress), pp. 355–357 (2014)
8. Jensen, F.V.: Bayesian artificial intelligence. Pattern Anal. Appl. **7**(2), 221–223 (2004)
9. Li, D., Chen, C., Lv, Q., Yan, J., Shang, L., Chu, S.: Low-rank matrix approximation with stability. In: International Conference on Machine Learning, pp. 295–303 (2016)
10. Cheng, J., Greiner, R., Kelly, J., Bell, D., Liu, W.: Learning Bayesian networks from data: an information-theory based approach. Artif. Intell. **137**(1–2), 43–90 (2002)
11. Sessions, V., Valtorta, M.: Towards a method for data accuracy assessment utilizing a bayesian network learning algorithm. J. Data Inf. Qual. **1**(3), 1–34 (2009)
12. Wang, S.C., Leng, C.P., Rui-Jie, D.U.: Noise smoothing in learning parameters of Bayesian network. J. Syst. Simul. **21**(16), 5046–5053 (2009)
13. Ueno, M.: Robust learning Bayesian networks for prior belief. In: Proceedings of the Twenty-Seventh Conference on Uncertainty in Artificial Intelligence, pp. 698–707. AUAI Press (2011)
14. Tsamardinos, I., Brown, L.E., Aliferis, C.F.: The max-min hill-climbing Bayesian network structure learning algorithm. Mach. Learn. **65**(1), 31–78 (2006)

15. Smith, J.Q., Daneshkhah, A.: On the robustness of Bayesian networks to learning from non-conjugate sampling. Int. J. Approximate Reason. **51**(5), 558–572 (2010)
16. Yaramakala, S., Margaritis, D.: Speculative Markov blanket discovery for optimal feature selection. In: Fifth IEEE International Conference on Data Mining (ICDM 2005), pp. 809–812. IEEE (2005)
17. Wang, J., Yan, T., Mai, N., Altintas, I.: A scalable data science workflow approach for big data Bayesian network learning. In: IEEE/ACM International Symposium on Big Data Computing (2015)
18. Wit, E., Heuvel, E.V.D.: 'All models are wrong...': an introduction to model uncertainty. Statistica Neerlandica **66**(3), 217–236 (2012)
19. Scutari, M.: Bayesian network constraint-based structure learning algorithms: parallel and optimised implementations in the bnlearn R package. J. Stat. Softw. **077** (2017)
20. Ruohai, D., Xiaoguang, G., Zhigao, G.: Parameter learning of discrete Bayesian networks based on monotonic constraints. Syst. Eng. Electron. **36**(2), 272–277 (2014)

The Third International Workshop on Graph Data Management and Analysis (GDMA 2019)

ANDMC: An Algorithm for Author Name Disambiguation Based on Molecular Cross Clustering

Siyang Zhang[1,3(✉)], Xinhua E[2,3], Tao Huang[1], and Fan Yang[1,3]

[1] Beijing Advanced Innovation Center for Future Internet Technology,
Beijing University of Posts and Telecommunications,
No. 10, Xitucheng Road, Haidian District, Beijing, People's Republic of China
{zhangyanrui,htao,yfan}@bupt.edu.cn
[2] Beijing University of Technology,
No. 100, Pingleyuan, Chaoyang District, Beijing, People's Republic of China
517893410@qq.com
[3] Peng Cheng Laboratory, No.2, Xingke first street, Nanshan District, Shenzhen,
Guangdong, People's Republic of China

Abstract. With the rapid development of information technology, the problem of name ambiguity has become one of the main problems in the fields of information retrieval, data mining and scientific measurement, which inevitably affects the accuracy of information calculations, reduces the credibility of the literature retrieval system, and affect the quality of information. To deal with this, name disambiguation technology has been proposed, which maps virtual relational networks to real social networks. However, most existing related work did not consider the problem of name coreference and the inability to correctly match due to the different writing formats between two same strings. This paper mainly proposes an algorithm for Author Name Disambiguation based on Molecular Cross Clustering (ANDMC) considering name coreference. Meanwhile, we explored the string matching algorithm called Improved Levenshtein Distance (ILD), which solves the problem of matching between two same strings with different writing format. The experimental results show that our algorithm outperforms the baseline method. (F1-score 9.48% 21.45% higher than SC and HAC).

Keywords: Name disambiguation · Coreference problem · String matching

1 Introduction

At present, there are several literature retrieval platforms in the world such as China National Knowledge Infrastructure (CNKI), DBLP, CiteSeer, PubMed,

This work is supported by National Natural Science Foundation of China (NSFC) (61702049).

etc. The content and quality of the digital library are seriously affected by the ambiguity of author's name, which is regarded as one of the most difficult issues facing digital library [1]. Therefore, how to reduce the impact due to the name ambiguity, and maximize the effectiveness of the digital library, has become a concern of researchers. The "Name Disambiguation" began to be raised and attracted the attention of a large number of experts and scholars.

Name Disambiguation, also known as Entity Resolution [2,3], Name Identification [4], which mainly solves the problem of name coreference and name ambiguity. The name coreference problem mainly appears in the English digital library. It is common that a single author has multiple names in digital library. For example, a possible form of author names A. Lim is Andrew Lim, Abel Lim, etc. The name ambiguity problem common that different authors may share identical names in the real world. For example, there are 57 papers authored by 2 different "Alok Gupta" in the DBLP database.

A lot of work has been studied for Name Disambiguation. For example, Shen, et al. [5] present a novel visual analytics system called NameClarifier to interactively disambiguate author names in publication. However, NameClarifier still heavily relies on human beings' subjective judgments. Kim, et al. [6] used Random Forest to derive the distance function and obtained a good accuracy rate, but the training set required a lot of manual labeling while the model have poor migration. Lin et al. [7] proposed an approach only use the coauthor and title attributes, but they did not consider the coreference problem. Xu et al. [8] considered that each kind of single feature has very strong fuzziness in the expression and used a similarity algorithm. However, many feature inability to correctly match due to the different writing formats between two same strings.

This paper mainly proposes an algorithm called Author Name Disambiguation based on Molecular Cross Clustering (ANDMC) considering name coreference. Meanwhile, we propose the string matching algorithm called Improved Levenshtein Distance (ILD), which solves the problem of matching between two same strings with different writing format. The experimental results show that our algorithm outperforms the baseline method. (F1 value 9.48% 21.45% higher than SC and HAC).

The structure of this paper is as follows: In Sect. 2, we introduce the related research work of name disambiguation. In Sect. 3, we introduce the core of this article including the similarity calculation method of the author name disambiguation and merging procedure. In Sect. 4, we describe our experiment and verify the proposed method. In Sect. 5, we summarize the method proposed in this paper. This part also addresses the shortcomings of the method and its ideas for future improvement.

2   Related Work

The problem of name ambiguity often appears in the literature retrieval platforms, digital library and other similar systems, which has become one of the main problems in the fields of information retrieval, data mining and scientific measurement. [9] The "Name Disambiguation" which mainly solves the problem of name coreference and name ambiguity began to be raised.

The name coreference problem mainly appears in the English digital library. Newman et al. [10] proposed a heuristic method for complete matching the first letter of the last name and the first name, but some authors is the same as the spelling but different name such as "M. Li", "Min. Li" and "Ming. Li" are merged to reduce the accuracy.

The name ambiguity problem common that difference authors may share identical names in the real world. In general, existing methods for name disambiguation mainly fall into three categories: supervised based [11,12], semi-supervised based [13] and unsupervised based [14–17]. The supervised based method has a high accuracy rate, but the training of massive data requires a lot of manual labeling, which is time-consuming and labor-intensive. What is more, with the advancement of time, the data iteration is rapid. Therefore, the supervised based method has poor portability. Semi-supervised based method use user's feedbacks to get more useful information, but when the amount of data is very large, the user feedbacks information are very difficult to collect and also expend much manpower and material resources in the process of collecting [7]. The biggest advantage of the unsupervised based method is that it does not require a lot of training data and training time. On large-scale data, no method is more feasible and scalable than the unsupervised based method.

The factors that determine the performance of unsupervised based method, not only by the clustering algorithm but also by the calculation of similarity. On the problem of name ambiguity, both the selection of features and how to use these features to calculate similarity are as important as the choice of clustering algorithm. Shin et al. [18], Fan et al. [19] Kang et al. [20] selected coauthor relationships as features, but the author who has not coauthor cannot be distinguish. Lin et al. [7] proposed an approach only use the coauthor and title attributes, but they did not consider the coreference problem. Xu et al. [8] considered that each kind of single feature has very strong fuzziness in the expression and used a similarity algorithm. However, many feature inability to correctly match due to the different writing formats between two same strings.

Based on the previous research results, this paper further studies the Name Disambiguation. The main contributions of this paper can be summarized as follows:

1. Propose the string matching algorithm called Improved Levenshtein Distance (ILD), which solves the problem of matching between two same strings with different writing format. (F1-score 13.08% higher than LD).
2. Propose an algorithm called Author Name Disambiguation based on Molecular Cross Clustering (ANDMC) considering name coreference. (F1-score 21.45% higher than SC, F1-score 9.48% higher than HAC).

3   Proposed Approach

This paper proposes a molecular cross clustering method. The Fig. 1 shows the process of molecular cross clustering. We regard each paper as an atom. Firstly, these papers are classified according to author's name, while keep the associated category records, and perform atom clustering [21] in the same category to form a molecular. Calculate the molecular similarity between molecular according to the associated category records differentiated by the standard segmentation feature values, and finally obtain the classification result. Each time extract the feature of the previous merge result, which could effectively increase the data amount of the corresponding feature and improve the accuracy of the merge.

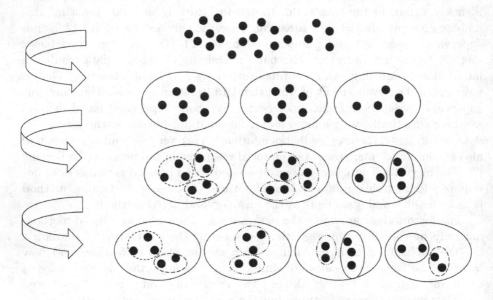

Fig. 1. The process of molecular cross clustering.

The Table 1 lists five records containing the authors of paper, title of paper, and affiliation of paper. It is difficulty for us to make sure that the author "Andrew Lim" is the same person. According to our algorithm, firstly, we can divide this paper into two major categories, Andrew Lim {{1}, {2}, {4}, {5}} and A. Lim {3}. Secondly, it is difficult to directly judge whether Andrew Lim in 1 and 2 is the same person, but 1, 4 have the same collaborator Zhou Xu. After merge 1, 4 we can find that Hu Qin, who is the same collaborator with 2 that means it has a higher probability that 1, 2 are the same person. In the same way, we can easily get the set {1, 2, 4, 5}. At this time, calculate the similarity between the set {1, 2, 4, 5} and {3}, we can find that they have the same collaborator "Fan Wang", the same institution and the similar titles, etc.

The steps of algorithm for Author Name Disambiguation based on Molecular Cross Clustering as follow:

1. Data processing
2. Solve the problem of name ambiguity
   (a) Node relationship division
   (b) Affiliation string matching

Table 1. An example of name disambiguation.

---
1. Author: **Andrew Lim**, Fan Wang, Zhou Xu

Title: A Transportation Problem with Minimum Quantity Commitment

Affiliation: Department of Industrial Engineering and Engineering Management, The Hong Kong University of Science and Technology, Clear Water Bay, Kowloon, Hong Kong

---
2. Author: **Andrew Lim**, Zhenzhen Zhang, Hu Qin

Title: Pickup and Delivery Service with Manpower Planning in Hong Kong Public Hospitals

Affiliation: Department of Industrial and Systems Engineering, National University of Singapore, Singapore 117576;

---
3. Author: **A. Lim**, Fan Wang

Title: Multi-depot vehicle routing problem: a one-stage approach

Organization: Dept. of Ind. Eng. & Logistics Manage., Hong Kong Univ. of Sci. & Technol., China

---
4. Author: **Andrew Lim**, Hu Qin, Zhou Xu

Title: The freight allocation problem with lane cost balancing constraint

Organization: Department of Management Sciences, City University of Hong Kong, Tat Chee Ave, Kowloon Tong, Hong Kong, School of Management, Huazhong University of Science and Technology, No. 1037, Luoyu Road, Wuhan, China

---
5. Author: Lijun Wei, Zhenzhen Zhang, **Andrew Lim**

Title: An Adaptive Variable Neighborhood Search for a Heterogeneous Fleet Vehicle Routing Problem with Three-Dimensional Loading Constraints

Affiliation: School of Information Technology, Jiangxi University of Finance and Economics, Nanchang, 330013 Jiangxi, China

---

3. Solve the problem of name coreference
   (a) Similar name cross match

3.1 Data Processing

Perform pre-processing operations such as integration, cleaning and de-duplication on the data to obtain initial data. Each piece of paper in the initial data as an atom.

Extract the following feature attributes for each paper P:

$$P = (A, \quad T, \quad I) \tag{1}$$

Where A represents the author of the paper, T represents the title of the paper, and I represents the affiliation of the paper.

We treat each paper as a node, let n be a name entity, denoted as n, and for the name n, its variant is denoted as Vn = V1, V2, ..., Vm, where the variant of n include the abbreviated forms, last name and first name rotated form, the change of connection symbol and combinations of them [22]. The set of papers corresponding to the name Vn is denoted by the set Pn = p1, p2, ..., pk, where pi = s1, s2, ..., sk represents a set of all papers containing the author names Vx. Ai = a1, A2, ..., ak represents the author set corresponding to the papers set pi. Ni = n1, n2, ..., nk represents the set of the same name authors corresponding to Ai.

3.2   Node Relation Division (NRD)

In the research of the name disambiguation, the relationship of cooperation between nodes has a strong influence on the correct division of nodes [20]. For two nodes with the same name attribute, if they all have a cooperative relationship with another node, the two nodes have greater similarity.

The set of collaborators of the name Ni can be denoted as:

$$C_i = A_i - N_i = \{a1 - n1, a2, n2, \ldots ak - nk\} = \{c1, c2, \ldots, ck\} \quad (2)$$

Traversal the set N_i, each n_i as a node. Traversal the set C_i, the author in each set c_i generates a node which has a cooperative relationship with the node ni. We use the graph database to generates the author relationship network, and finds the number of connections of the author ni to nj denoted as Num(Lij), according to the Jaccard coefficient similarity function, the similarity between the node ni and the node nj is:

$$sim(n_i, n_j) = \frac{Num(L_{ij})}{|c_i \cup c_j|} \quad (3)$$

When the similarity is greater than the threshold value, ni and nj will be merged.

3.3   Affiliation String Matching (ASM)

The main difficulties in matching affiliation string for English databased is that affiliation write different formats. For example, there have four affiliations as follows: "IBM India Res. Lab, New Delhi", "IBM India Research Laboratory", "IBM India Research Lab, New Delhi, India" and "IBM India Research Lab, New Delhi, India 110 070". It is clearly shown that the above four affiliations belong to the same affiliation, but the writing in different formats which lead the computer cannot match them together correctly. At present, there are many similarity algorithm for string matching, such as Jaccard algorithm, Euclidean Distance, Levenshtein Distance, etc. However, the calculation of the whole affiliation string is not satisfactory. For example, two affiliations as follows:

1. "School of Electrical Engineering & Automation, Henan, Polytechnic University, Jiaozuo, People's Republic of China"

2. "Department of Electrical Engineering and Automation, Tianjin University, Tianjin, People's Republic of China"

If we directly calculate the similarity of the affiliation names, it is likely to judge them as the same affiliation, but they are not the same affiliation actually. There is also a problem with the calculation of Levenshtein Distance. For example, there are two strings include word "Research" and "Res", the Levenshtein distance of two words is 5, and the similarity is 40%. We find that, in reality, these two words actually belong to a same word. In order to solves the problem of matching between two same strings with different writing format while enhance the accurate of similarity calculate. In this paper, we cut each word in the affiliation. We optimize Levenshtein Distance algorithm as ILD (Improved Levenshtein Distance algorithm) to calculate the similarity of each word. For the affiliation X and the affiliation Y, cut through the separator to obtain the set $X = x1, x2, xp$ and set $Y = y1, y2, yq$. Construct the relational matching matrix E with the number of rows p and the number of columns q:

$$E_{pq} = \{sim(i, j)\} \tag{4}$$

For each $xis1, s2, ..., sm, yjs1, s2, ..., sn$ construct the relationship matching matrix LD between xi and yj whose row number is $m+1$ and column number is $n+1$. The first column of the matrix represents X, and the first row represents Y:

$$LD_{(m+1)\times(n+1)} = \{ld_{ij}\} \quad (0 \leq i \leq m, \quad 0 \leq j \leq n) \tag{5}$$

Fill the relationship matching matrix LD according to the following formula:

$$ld_{ij} = \begin{cases} i & j = 0 \\ j & i = 0 \\ min(ld_{i-1j-1}, ld_{i-1j}, ld_{ij-1}) + 1 & i, j > 0, x_i \neq x_j \\ ld_{i-1j-1} & i, j > 0, x_i = x_j \end{cases} \tag{6}$$

After fill the matrix LD, the element dmn is the edit distance between xi and yj, which is recorded as:

$$d(x_i, y_i) = \begin{cases} d_{min(m,n)min(m,n)} & x_i \in y_j \; or \; y_j \in x_i \\ d_{mn} & else \end{cases} \tag{7}$$

The similarity sim(xi, yj) is calculated as:

$$sim(x_i, y_j) = 1 - \frac{d(x_i, y_i)}{max(len(x_i), len(y_j))} \tag{8}$$

Where len(xi) and len(yj) are the lengths of the string xi and the string yj, respectively. When sim(xi, yj) $= 1$, the string xi and yj exactly match. For the matrix Epq, if exist at least one sim(xi, yj) $= 1$ on the p-row or q-column, we think that the affiliation X and the affiliation Y have one word exactly matched which is recorded as:

$$CM(k) = \begin{cases} 1 \; existsim(x, y) = 1 \; in \; link \; k \\ 0 \qquad\qquad\qquad\qquad\quad else \end{cases} \tag{9}$$

The similarity of the word exactly match in the affiliation X and Y as follows:

$$sim(X,Y)_{cm} = \frac{average(\sum_p CM(p), \sum_q CM(q))}{average(p,q)} \qquad (10)$$

The similarity of the word non-exactly match in the affiliation X and Y as follows:

$$sim(X,Y)_{other} = \frac{average(\sum_p max(sim(X,Y)), \sum_p max(sim(X,Y)))}{average(p,q)} \\ - \frac{average(\sum_p CM(p), \sum_q CM(q))}{average(p,q)} \qquad (11)$$

The similarity between the affiliation X and Y is:

$$sim(X,Y) = sim(x,y)_{cm} \times W_1 + sim(X,Y)_{other} \times W_2 \qquad (12)$$

3.4   Similar Name Cross Match (SNCM)

For a name entity n, each variant in Vn = V1, V2, ... Vm has solved the problem of name ambiguity. This part mainly solves problem of name coreference. We need to calculate the similarity between each Ai in Vx, which denoted as "Vx.Ai" and each of Aj in Vy, which denoted as "Vy.Aj". We calculate the corresponding similarity Sx according to the features A, T, I, and set the weight W, respectively. The similarity between Vx.Ai and Vy.Aj are as follows:

$$S(V_x.A_i, V_y, A_j) = \begin{cases} S_A = S_{JACCARD}(V_x.A_i, V_y.A_j) \\ S_T = S_{LD}(V_x.T_i, V_y.T_j) \\ S_I = S_{ILD}(V_x.I_i, V_y.I_j)) \end{cases} \qquad (13)$$

We chose to put similar name cross matching in the last step due to the current similarity calculation does not guarantee 100% accuracy for authors with the same name and a large number of duplicate names. Since each of our mergers is based on the previous step. As a result, we must ensure that the accuracy of the previous merge is as high as possible. If this step is advanced, it will greatly affect the accuracy of the subsequent steps.

4   Experiments

4.1   Data Sets

In our experiments, we perform evaluations on a dataset constructed by Tang et al. [21], which contains the citations collected from the DBLP Website. We downloaded this dataset from the Kaggle. However, the data set is only labelled within the same name range, and the name containing the abbreviation is less. Therefore, we add some real intellectual property disclosure data on the basis of this data set to verify our method. Select some authors as experimental samples.

When evaluating the classification results, we use the author whose name is prone to the same name as a sample. We use manual methods to create standard categories. The process is as follows: For each author name in Table 2, we retrieve all the papers published by the name in the database. Classify the authors of the same name by human annotated, as best as possible to accurately.

Table 2. Evaluation dataset.

Name	Number	Year
Alok Gupta	57	1996–2009
Ming Li	34	2003–2018
M. Li	15	1991–2014
Min Li	30	2001–2018
F. Wang	34	1998–2017
Fan Wang	55	1989–2016
A. Lim	7	1993–2005
Andrew Lim	8	2008–2014
X. Zhang	61	1984–2012
Xin Zhang	46	2002–2018

4.2   Evaluation Indicators

To evaluate and compare the performance of different methods on the Name Disambiguation tasks. In this paper, we use pairwise precision, pairwise recall and pairwise f1-measure to measure the results. We define the measures as follows:

$$PairwisePrecision = \frac{\#PairsCorrectlyPredictedToSameAuthor}{\#TotalPairsPredictedToSameAuthor} \qquad (14)$$

$$PairwiseRecall = \frac{\#PairsCorrectlyPredictedToSameAuthor}{\#TotalPairsToSameAuthor} \qquad (15)$$

$$PairwiseF - Measure = \frac{2 \times PairwiseRecall \times PairPrecision}{PairwiseRecall + PairPrecision} \qquad (16)$$

In the above formula, #PairsCorrectlyPredictedToSameAuthor refers to the number of papers that with the same label predicted by an approach and have the same label in the human annotated data set. #TotalPairsPredictedToSameAuthor refers to the number of papers that with the same label predicted by an approach. #TotalPairsToSameAuthor refers to the number of papers that have the same label in the human annotated data set.

Table 3. Table captions should be placed above the tables.

Author	LD			ILD		
	Precision	Recall	F-Measure	Precision	Recall	F-Measure
Alok Gupta	100.00	100.00	100.00	100.00	90.48	95.00
Ming Li	60.87	70.00	65.12	87.50	70.00	77.78
M. Li	72.73	80.00	76.19	100.00	100.00	100.00
Min Li	80.95	89.47	85.00	100.00	100.00	100.00
F. Wang	50.00	100.00	66.67	100.00	100.00	100.00
Fan Wang	100.00	100.00	100.00	100.00	100.00	100.00
A. Lim	57.14	66.67	61.54	100.00	100.00	100.00
Andrew Lim	100.00	100.00	100.00	100.00	50.00	66.67
X. Zhang	80.56	85.29	82.86	100.00	82.35	90.32
Xin Zhang	40.00	72.73	51.61	100.00	81.82	90.00
Average	74.22	86.42	78.90	**98.75**	**87.46**	**91.98**

4.3 Experimental Results

We considered the baseline methods on LD algorithm. In this step, we only evaluate based on the feature of affiliation, and do not evaluate the results based on other feature. Table 3 shows the results of some examples in our data sets.

Obviously, it can be seen from the experimental results that the ILD algorithm has a better improvement than the LD algorithm in each evaluation value (+17.76% over LD by average F1 score, +24.53% over LD by average Precision). On the other hand, our method has higher precision than baseline methods (+18.3% over SC, +8.51% over HAC by the average Precision value).

According to the name similarity matching, the number of names existing in each name set as follows (Table 4):

Table 4. Evaluation dataset.

Name	Num. authors	Num. records
Alok Gupta	2	57
Ming Li, M. Li, Min Li	44	79
F. Wang, Fan Wang	28	89
A. Lim, Andrew Lim	3	15
X. Zhang, Xin Zhang	72	107

In this paper, we considered several baseline methods based on Hierarchical Agglomerative Clustering (HAC) [24], [23] and single-clustering (SC) [20]. SC only uses the feature of collaborator for disambiguation. HAC uses Jaccard

Similarity and ILD Similarity algorithms with the feature of author's name, affiliation, and collaborator. For a fair comparison, we use the same threshold for the same attribute feature. For each feature, we compare and select the thresholds to ensure that the highest recall rate based on the precision as high as possible. Table 5 gives the threshold values of features.

Table 5. Threshold values of features.

Feature	A	I	T
Thresholds	0.6	0.7	0.5

Table 6 gives the results of some examples in the data set. Obviously, our method outperforms the baseline method in name disambiguation (+21.45% over SC, +9.48% over HAC by average F1 score). On the other hand, our method has higher precision than baseline methods (+18.3% over SC, +8.51% over HAC by the average Precision value).

Table 6. Results of name disambiguation.

Author	SC			HAC			NAS		
	Prec.	Rec.	F1	Prec.	Rec.	F1	Prec.	Rec.	F1
Alok Gupta	36.54	33.33	34.86	80.77	73.68	77.06	80.77	73.68	77.06
A. Lim, Andrew Lim	50.00	33.33	40.00	61.54	53.33	57.14	100.00	93.33	96.55
X. Zhang, Xin Zhang	100.00	87.93	93.58	100.00	89.66	94.55	100.00	93.10	96.43
Ming Li, M. Li, Min Li	93.10	86.27	89.56	86.27	92.16	89.12	92.16	100.00	95.92
F. Wang, Fan Wang	100.00	78.57	88.00	100.00	78.57	88.00	98.21	78.57	87.30
Average	75.93	63.89	69.20	85.72	77.48	81.17	**94.23**	**87.74**	**90.65**

5   Conclusion and Discussion

Name Disambiguation in the digital library is an important task because different authors can share the same name, and an author can have many name variant. This paper mainly proposes an algorithm called Author Name Disambiguation based on Molecular Cross Clustering (ANDMC). We have also explored a string matching algorithm called Improved Levenshtein Distance (ILD). Experimental results indicate that the proposed method significantly outperforms the baseline methods. It's performance in the problem of name coreference is quite satisfying. Meanwhile, we solve the problem of matching between two same strings with different writing format. In the future, we will pay more attention to the speed of the algorithm and improve the efficiency of the algorithm.

References

1. Hussain, I., Asghar, S.: A survey of author name disambiguation techniques. Knowl. Eng. Rev. **32**, 1–24 (2018)
2. Benjelloun, O., Garcia-Molina, H., Menestrina, D., Su, Q., Whang, S.E., Widom, J.: Swoosh: a generic approach to entity resolution. The VLDB J. **18**, 255–276 (2008)
3. Bhattacharya, I., Getoor, L.: Collective entity resolution in relational data. ACM Trans. Knowl. Discov. Data **1** (2007) Article no. 5
4. Li, X., Morie, P., Roth, D.: Identification and tracing of ambiguous names: discriminative and generative approaches. In: Proceedings of 19th National Conference on Artificial Intelligence (AAAI 2004), pp. 419–424 (2004)
5. Shen, Q., Wu, T., Yang, H., Wu, Y., Qu, H., Cui, W.: NameClarifier: a visual analytics system for author name disambiguation. IEEE Trans. Vis. Comput. Graph. **23**(1), 141–150 (2017)
6. Kim, K., Khabsa, M., Giles, C.L.: Random Forest DBSCAN for USPTO inventor name disambiguation, pp. 269–270 (2016)
7. Lin, X., Zhu, J., Tang, Y., Yang, F., Peng, B., Li, W.: A novel approach for author name disambiguation using ranking confidence. In: Bao, Z., Trajcevski, G., Chang, L., Hua, W. (eds.) DASFAA 2017. LNCS, vol. 10179, pp. 169–182. Springer, Cham (2017). https://doi.org/10.1007/978-3-319-55705-2_13
8. Xu, X., Li, Y., Liptrott, M., Bessis, N.: NDFMF: an author name disambiguation algorithm based on the fusion of multiple features. In: IEEE 42nd Annual Computer Software and Applications Conference (COMPSAC), Tokyo 2018, pp. 187–190 (2018)
9. Ferreira, A., Goncalves, M.A., Laender, A.H.: A brief survey of automatic methods for author name disambiguation. ACM Sigmod Rec. **41**(2), 15–26 (2012)
10. Newman, M.E.J., Girvan, M.: Finding and evaluating community structure in networks. Phys. Rev. E **69**, 026113 (2004)
11. Han, H., Giles, L., Zha, H., et al.: Two supervised learning approaches for name disambiguation in author citations. In: Proceedings of JCDL (2004)
12. Huang, J., Ertekin, S., Giles, C.L.: Efficient name disambiguation for large-scale databases. In: Fürnkranz, J., Scheffer, T., Spiliopoulou, M. (eds.) PKDD 2006. LNCS (LNAI), vol. 4213, pp. 536–544. Springer, Heidelberg (2006). https://doi.org/10.1007/11871637_53
13. Quan, L., Bo, W., Yuan, D.U., Wang, X., Yuhua, L.I.: Disambiguating authors by pairwise classification. Tsinghua Sci. Technol. **15**(6), 668–677 (2010)
14. Malin, B.: Unsupervised name disambiguation via social network similarity. In: SIAM SDM Workshop on Link Analysis, Counterterrorism and Security (2005)
15. Pedersen, T., Purandare, A., Kulkarni, A.: Name discrimination by clustering similar contexts. In: Gelbukh, A. (ed.) CICLing 2005. LNCS, vol. 3406, pp. 226–237. Springer, Heidelberg (2005). https://doi.org/10.1007/978-3-540-30586-6_24
16. Cen, L., Dragut, E.C., Si, L., Ouzzani, M.: Author disambiguation by hierarchical agglomerative clustering with adaptive stopping criterion. In: SIGIR 2013, 28 July–1 August 2013
17. Evans, M.D.: A new approach to journal and conference name disambiguation through k-means clustering of internet and document surrogates (2013)
18. Shin, D., Kim, T., Jung, H., et al.: Automatic method for author name disambiguation using social networks. In: IEEE International Conference on Advanced Information NETWORKING and Applications, Aina 2010, Perth, Australia, 20–13 April. DBLP, pp. 1263–1270 (2010)

19. Fan, X., Wang, J., Pu, X., et al.: On graph-based name disambiguation. J. Data Inf. Qual. **2**(2), 10 (2011)
20. Kang, I.-S., et al.: On co-authorship for author disambiguation. Inf. Process. Manag. **45**(1), 84–97 (2009)
21. Tang, J., Fong, A.C.M., Wang, B., Zhang, J.: A unified probabilistic framework for name disambiguation in digital library. IEEE Trans. Knowl. Data Eng. **24**(6), 975–987 (2012)
22. Tang, J., Lu, Q., Wang, T., Wang, J., Li, W.: A bipartite graph based social network splicing method for person name disambiguation. In: Proceedings of the 34th International ACM SIGIR Conference on Research and Development in Information Retrieval (SIGIR 2011). ACM, New York, pp. 1233–1234 (2011)
23. Tan, Y.F., Kan, M.Y., Lee, D.: Search engine driven author disambiguation. In: Proceedings of the ACM/IEEE Joint Conference on Digital Libraries, JCDL, Chapel Hill, NC, USA, 11–15 June, pp. 314–315 (2006)
24. Zepeda-Mendoza, M.L., Resendis-Antonio, O.: Hierarchical agglomerative clustering. Encycl. Syst. Biol. **43**(1), 886–887 (2013)

Graph Based Aspect Extraction and Rating Classification of Customer Review Data

Sung Whan Jeon[1(✉)], Hye Jin Lee[1], Hyeonguk Lee[1],
and Sungzoon Cho[2]

[1] Department of Industrial Engineering, Seoul National University, Seoul, Korea
{sjeon,hyejinlee,hyeonguk21}@dm.snu.ac.kr
[2] Department of Industrial Engineering and Institute for Industrial Systems
Innovation, Seoul National University, Seoul, Korea
zoon@snu.ac.kr

Abstract. This paper introduces graph-based aspect and rating classification, which utilizes multi-modal word co-occurrence network to solve aspect and sentiment classification tasks. Our model consists of three components: (1) word co-occurrence network construction, with aspect and sentiment labels as different modes; (2) dispersion computation for aspects and sentiments, and; (3) feedforward network for classification. Our experiment shows that proposed model outperforms baseline models, Word2Vec and LDA, in both aspect and sentiment classification tasks. Our classification model uses comparatively smaller vector size for representing words and sentences. The proposed model performs better in classifying out of vocabulary contexts.

Keywords: Sentiment classification · Aspect classification · Aspect analysis · Word occurrence network · Dispersion

1 Introduction

With the rise of the World Wide Web, the volume and variety of user-generated text data have been growing exponentially. Amazon, for example, accumulated 142.8 million customer reviews from May 1996 through July 2014 [1]. In the research area, many has employed various text mining techniques to exploit the abundance of online data in an industrial setting. Aspect and ratings analysis, using online customer reviews in particular, constitutes a major branch in such research efforts.

Fig. 1. (a) An example of a Rting.com review; (b) an example of an Amazon review (Color figure online)

© Springer Nature Switzerland AG 2019
G. Li et al. (Eds.): DASFAA 2019, LNCS 11448, pp. 186–199, 2019.
https://doi.org/10.1007/978-3-030-18590-9_13

Given a customer's review, an "aspect" represents a specific feature of the product. A "rating" is a scalar score the reviewer assigns to the product (see Fig. 1). Figure 1(a) shows an example of a product review from Rtings.com, an electronic product review website. Figure 1(a) reviews Samsung's NU8000 by six aspects (in purple box). In the blue box can be found the aspect-level ratings. Such aspect-level, however, may not be always as explicitly presented as the Rtings example. Amazon reviews, as found in Fig. 1(b) only presents the overall rating for the subject product, and the review text is not explicitly labeled with relevant aspect. Ever-rising is the demand for aspect-level analysis, because it may unveil valuable insights about the product based on the customers' aspect-specific sentiments or preferences. For example, a customer may find a certain television product moderately satisfying to his/her taste, yet he/she may have been greatly pleased with the picture quality of the product, while substantially displeased with packaging, hence leading to the moderate satisfaction in overall. In this case, aspect-level analysis of the product review enables a deep dive into the customer's reaction towards the product and allows the sellers to establish sales strategies that are more concisely targeted. At the same time, the producers may develop the next product designed to meet customers' specific needs and wants.

Academic and industry researchers have responded to the growing significance and demand for aspect-level analysis of online review data. Past studies consists of two main components: (1) representing text data in a machine-readable format, and; (2) extracting aspect-specific information from the text representations learned. The most classic example of text representation methods is bag-of-words (BoW), an approach based on the word frequency [2]. It was later modified by adding the inverse term frequency to the measure, TF-IDF, namely, in order to account for specificity of the subject word appearing frequently in a particular document [3]. Recently, Word2Vec [4], a model which learns a distributed vector representation of each word in a given corpus, has gained popularity among other text embedding techniques due to its simplicity and generalizability. Text representations learned from these models are then joined with the conventional machine learning techniques in order to solve a given task. However, the task performance depends largely on the word representations, and these text representation learning are subject to a number of drawbacks, especially in the scope of analyzing tremendously large text corpora such as online reviews. BoW and TF-IDF suffers from the curse of dimensionality, since it uses one-hot coding for each word when recording frequency. While Word2Vec addresses this issue by compressing the dimension of the continuous word vector via utilizing softmax layer of the neural network through which the model learns each word representations, it still demands a great deal of resources in order to learn from ever-growing text corpora. Above all, aforementioned techniques are commonly vulnerable to out-of-vocabulary (oov) problem. BoW or TF-IDF may produce biased results since newly added words are obviously less frequent as compared to the "old" words. One the other hand, Word2Vec and all other Word2Vec alike models need to learn representations for the entire corpus once again upon the arrival of a new word. OOV is a critical problem when dealing with online review data, since new information is streamed in on a real-time basis.

Other than word representation methods, some studies have employed probabilistic graphical models or expectation-maximization algorithms, such as Latent

Dirichlet allocation (LDA) and latent aspect rating analysis (LARA), respectively [5, 6]. These models cluster "topically similar" words into the same group which then represents an aspect. Due to the structural construct of their model framework, LDA and LARA work well with unlabeled data, which is common for industrial datasets, scalability still remains as unresolved issue since learning takes up great resources.

In this paper, we address above issues by presenting a new method for aspect and sentiment classification of online reviews by using multi-model word co-occurrence network. As the measure of the "closeness" between words, we borrow the concept of dispersion as introduced in Backstrom and Kleinberg [9]. Our graph-based aspect ratings classification framework builds word co-occurrence network from a given corpus, defining words as different modes if their source document is labeled with different aspect or sentiment categories. Then, our model computes word-aspect dispersion score and word-rating dispersion score from the network, which are then concatenated and used as input for a feedforward neural network for aspects and ratings classification. The main contributions of the architecture of graph-based classification model are summarized below.

First, our model has better performances. As reported in the experiment section, proposed model performs better than the baseline models (Word2Vec, Doc2Vec, LDA) in both aspect and sentiment classification tasks. Second, our graph-based classification framework is more scalable. By definition, the size of the word-aspect dispersion score and that of the word-rating dispersion are set to the number of aspects and the number of ratings, respectively. Hence, our model has an advantage of significant dimension reduction as compared to the existing word embedding or word frequency models. Last, Proposed classification method is more robust. Once word co-occurrence network is constructed, proposed framework is robust to Out-of-Vocabulary problem, since new word is easily added to the network by connecting with already existing word nodes, with which it occurs in a given document.

The rest of the paper is organized as follows: Sect. 2 surveys through past literature related to our work. In Sect. 3, we provide detailed descriptions of the framework of our model. Section 4 reports the results from our experiment, using Word2Vec and LDA as baseline models. Finally, Sect. 5 concludes the paper.

2 Related Works

In this section, we briefly talk about research from three areas are related to our study: first, distributed representation for text, second, topic modeling mainly used for aspect analysis, and third, dispersion which is a network measure for tie strength and the follow-up studies have adopted dispersion.

As a surge of researches in neural networks, many text embedding models came out, and some of the models received significant attention. Word2Vec [4] is the favorite ways of text embedding to represent each word in a corpus as a vector in the space of N dimensions. Mikolov et al. [4] proposed two approaches of Word2Vec in which are skip-gram and cbow. In skip-gram, the object is to predict the surrounding context words given input as a word. Cbow is merely the reverse of skip-gram. In Cbow, the input is multiple words in a sentence to predict the context word. While Word2vec is the embedding model for each word in a corpus, Doc2Vec [7] is an extension model of Word2Vec to represent documents as a fixed length of vectors. Such distributed representation methods are widely used for text embedding to solve various tasks in Natural Language Processing (NLP).

Online review data is one of the popular data sources used in NLP. The characteristic of online review is that overall rating is the only given label, but a review discusses multiple aspects of the product or service. Therefore, it is not able to know the reviewer's opinion without an explicit rating of each aspect. Topic modeling such as Latent Dirichlet Allocation (LDA) [5] is used frequently on online review data for aspect analysis. LDA automatically generates a set of topics from a document, and it is a fully unsupervised model since it discovers topics based on word distributions. The Latent Aspect Rating Analysis (LARA) [6] is another topic model to detect a reviewer's latent opinion on each aspect and the relative importance of different aspects. A significant drawback of LARA is that the aspects and the initial keywords have to be pre-specified. As a follow-up study, Wang et al. [8] proposed an improved version of LARA which the model does not require a set of seed words, set by the user for every aspect and topic [8].

We borrow the concept of dispersion introduced by Backstrom and Kleinberg [9]. They proposed a new network measure, dispersion to identify romantic partnership given the social network of Facebook data. The meaning of dispersion is the extent to which two individual's mutual friends are not themselves very well connected [9]. In their study, they proved that the accuracy of dispersion measurement almost twice than conventional embeddedness which is tie strength as the number of mutual friends shared by its endpoints embedded in the network [9, 10].

Several studies have used dispersion in their research. Singhal and Pudi [11] used dispersion-based similarity measure to find similar papers in a citation network. They state that dispersed connectivity does capture the inherent structural similarity within a network [11]. Minocha et al. [12] show that dispersion is a crucial structural feature to explain the importance of appropriate legal judgments and landmark decisions [12]. To the best of our knowledge, no study has used dispersion on online review texts to construct a network and feed the dispersion vector into a neural network to classify the online reviews based on its aspect accordingly.

3 Methodology

In this section, we propose a model for classifying aspects and ratings of review texts. The suggested model architecture is below (see Fig. 2).

The model consists of review texts' words co-occurrence network construction (Sect. 3.1), aspects/ratings to words dispersions and weights calculation (Sect. 3.2), and feedforward neural network for aspects and ratings classification (Sect. 3.3).

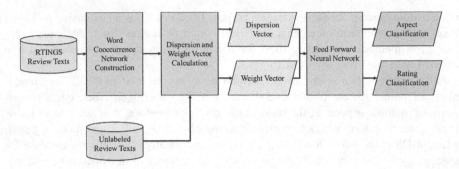

Fig. 2. Model architecture

3.1   Network Construction

RTINGS [13] website has reviews on televisions which each review consists of evaluations on various aspects such as picture quality, sound quality, design, and smart features. Each review also includes ratings score for every aspect. By using the review texts on RTINGS website, we construct word co-occurrence network. First, we crawl the reviews on Samsung, LG, and Sony televisions. Reviews are divided into aspects and aspects have ratings accordingly. We parse the paragraphs in each aspect by sentences. Then, the parsed sentences are tagged by words which this step includes stop words eliminations (see Fig. 3).

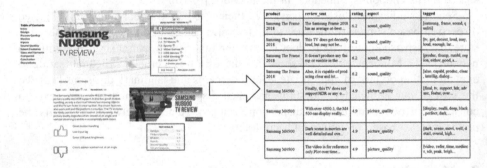

Fig. 3.   RTINGS reviews preprocessing

The word co-occurrence network is constructed based on sentence-wise tagged words. For those words that occurred in same sentence are linked together. If the words are linked already, the link weights are added by one for each co-occurrence. More the words co-occur in same sentence more weights are assigned.

After the word co-occurrence network is generated, we add aspect nodes and rating nodes. This will form multi-mode network with words, aspects, and ratings. There are 16 Aspects in RTINGS reviews which are 'introduction', 'design', 'picture quality', 'motion', 'inputs', 'sound quality', 'smart features', 'mixed usage', 'movies', 'TV shows', 'sports', 'video games', 'HDR movies', 'HDR gaming', 'PC monitor', and 'comparison'. The ratings on reviews are given in range of 0 to 10 with one decimal points. We categorize the ratings by rounding down to the nearest one. There are no ratings with '0s', '1s', '2s', and '10s', which the ratings are categorized into 7 ratings, '3s', '4s', '5s', '6s', '7s', '8s', and '9s'.

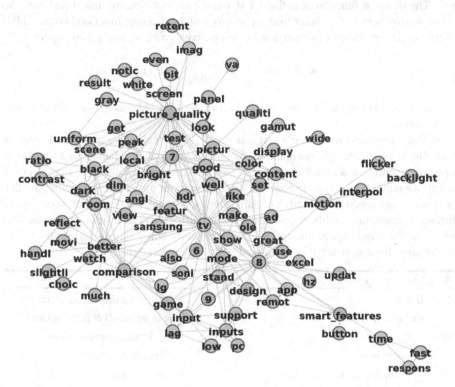

Fig. 4. Select part of the resulting graph (blue: words, orange: ratings, green: aspects) (Color figure online)

All the words that appear in each aspect are linked to each aspect node and all the words that appear in each rating are linked to each rating node. If the more words appear in certain aspect or rating, the weights between word node and aspect/rating node are added by one for each occurrence (see Fig. 4).

3.2   Dispersion and Weight Vector Calculation

It is intuitive that if the certain aspect can be characterized by set of words, the tie between the aspect node and node set of words should be stronger in multi-mode network as constructed above. For example, the word 'speaker' and 'distortion' are mostly used in 'sound quality' aspects. On the other hands, 'TV' or 'television' are widely used throughout the all aspects so these words are not good discriminating factor for aspect classification.

Among many network properties, strong tie between nodes can be calculated by dispersion [9]. Dispersion in our proposed network is calculated by the following steps. First, for given word w_1 and given aspect a_1, let G_{w_1} be the subgraph of w_1 and G_{a_1} be the subgraph of a_1. we take the common subgraph of subgraph $C_{w_1 a_1}$ to be the set of common neighbors of w_1 and a_1. Then, let $d(s, t)$ be the distance function of all nodes in C. The distance function is defined 1 if s and t are not directly linked and have no common neighbors in G_{w_1} other than w_1 and a_1 and the distance function is defined 0 if otherwise. The equation of dispersion between given word w_1 and given aspect a_1 is,

$$disp(w_1, a_1) = \sum\nolimits_{s, t \in C_{w_1, a_1}} d(s, t) \tag{1}$$

The dispersion score between words and ratings are calculated with same procedures above.

In Backstrom and Kleinberg's original paper [9] normalize the absolute dispersion score illustrated above with normalizing factor. The normalizing factor used in original paper is embeddedness which is mutual friends two given nodes share. In our proposed model, we rather normalize the absolute dispersion score by words. Given the fact that some words are specifically used in certain aspects, normalizing dispersion score by the number of common neighbors can dilute the specificity nature of review corpus. The algorithm for word-aspect dispersion score vector is below (see Algorithm 1). The word-rating dispersion score is calculated with same procedure.

Algorithm 1. Dispersion Vector

1:	$D = \{\}$	▷ D is the set dispersion vectors
2:	**For** $w \in N$ **do**	▷ w is the words N is the network
3:	$D_w = \{\}$	▷ D_w is the dispersion vector for w
4:	$Norm = 0$	▷ Normalizing factor
5:	**For** $a \in N$ **do**	▷ a is the aspects
6:	$disp(w, a)$	
7:	$Norm = Norm + disp(w, a)$	
8:	**For** $a \in N$ **do**	
9:	$D_w \leftarrow D_w \cup \{disp(w, a)/Norm\}$	
10:	$D \leftarrow D \cup \{D_w\}$	

The size of word-aspect dispersion score vector is, then, set to number of aspects which is 16 and the size of word-rating dispersion score vector is, then, set to number of rating which is 7.

Word-aspect, word-rating weight vectors are relatively easy to calculate. We search for every word then assess edge weights between given word and every aspect/rating. The weight is also normalized by words to words. The dispersion score presents the relative strength of tie between words and aspects/ratings which also embraces the semantic structure in word co-occurrence network. The simple weight can illustrate the absolute tie between each word and aspects/ratings.

3.3 Feed Forward Neural Network for Aspect/Rating Classification

We use word-aspects/ratings vector for sentence level aspect and rating classification. For aspects and ratings labeled sentences from RTINGS reviews which we preprocess in Sect. 3.1 are used for training and validation of final classification model. We search each tagged word in review sentences in word-aspect/rating vectors. Once the search for tagged words in given review sentence is done, word-aspect/rating dispersion vector are weighted averaged which represents the dispersion vector for given review sentence. Weight vector is also computed with same methodology. Dispersion vector and weighted vector are concatenated as the input for feed forward neural network. We create two feed forward neural network for aspect classification and rating classification. We use ReLU activation function and use softmax function for classification (see Fig. 5).

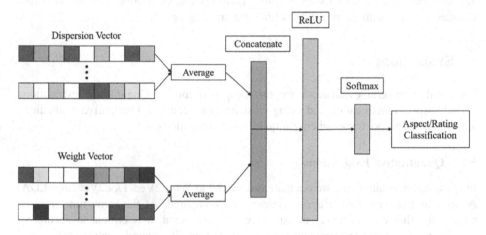

Fig. 5. Feed forward neural network for aspect/rating classification

4 Experiment Settings

4.1 Data

For our evaluation we use RTINGS review data for televisions. Samsung, LG, and Sony television reviews are collected. There are total of 58 television reviews for three companies and each review contains 16 aspects which leads 928 aspect-based reviews. As we parse each review by sentences, there are 10,808 sentences total. After text preprocessing, 2,347 words are tagged. In word co-occurrence network, there are 2,347 nodes with 144,016 weighted edges. After we add aspects and ratings node for multi-mode network construction, there are 157,022 weighted edges.

4.2   Models

Models we use for analysis are Word2Vec [4], Doc2Vec [7], and LDA [5]. We trained each model for Word2Vec and Doc2Vec vector size of 25 to 200 and for LDA topic vector size of 25 to 200. When the vector size is too small trained embedding vectors are not capable of capture all semantic information in text. When the embedding vectors are too large, softmax regression overfits for training data. Therefore, we choose vector size of 100 for all three models. For Word2Vec, words vectors in review sentences are weighted averaged. Doc2Vec is trained as each sentence is a document. Each review sentences for evaluations will be inferred by trained Doc2Vec and LDA models which will result in inferred document embedding vector and inferred topic vector. For our proposed model, concatenated input vectors have size 32 for aspects and 14 for ratings. We use one hidden layer with 512 ReLU activation function and use Adam optimizer. Out of 10,808 sentences in our review corpus, we use 70% of data for training and 30% of data for evaluation. Training and evaluation sets are stratified sampled and we train every model with same training set.

5   Evaluations

We conduct several evaluations for our proposed model. Quantitative evaluations include aspect classification and rating classification accuracy. Qualitative evaluations are aimed for ed post assessment on qualitative evaluations.

5.1   Quantitative Evaluations

In quantitative evaluations, we compare our model to Word2Vec, Doc2Vec, and LDA. Aspects and ratings classification accuracy are compared. At the same time, aspect-wise and rating-wise accuracy measure are also presented. It is critical to accurately classify main aspects in television such as picture quality, sound quality, design and smart features. For ratings it is important to classify bipolar sentiments.

We perform evaluations on classification based on review sentences. For these evaluations can be stricter, we divide 10,808 review sentences in 7:3 ratios. 70% of review sentences will be used in creating word co-occurrence network, training Word2Vec, Doc2Vec, and LDA models. Other 30% of review sentences will be used on evaluations. Out of 2,347 unique words in review dataset, 2,145 words are used in training. Each model will encounter 202 new vocabularies when we perform evaluations. This is about 9% of unique vocabulary in training set.

Aspect Classification. In Table 1, the classification performances of each model are depicted. As mentioned above, there are 16 aspects in this review corpus. Some general sections such as introduction and comparisons are also included in classification evaluation. By simple mathematics, random guess will have accuracy of 1/16 or 6.25% of classification accuracy. The best scoring model is bold faced (see Table 1).

Table 1. Aspect classification performance

Models	Accuracy
Random	6.25%
Word2Vec	72.96%
Doc2Vec	24.42%
LDA	58.93%
Graph based classification	**85.82%**

In general, our model achieved highest accuracy. Doc2Vec has lowest performance. Doc2Vec is more suitable for longer documents. The performance of Doc2Vec cannot be guaranteed with shorter document such as single sentence.

We also included aspect-wise performance of each model. There exist key aspects in television that both consumer and manufacturer are more concentrated. Picture quality, sound quality, design, and smart features are the features are relatively more taken account of. For these four aspects, proposed graph-based classification scored highest. In fact, our model score above 95% accuracy classifying design (97.6%), sound quality (95.2%), and smart features (95.6%), and 89.0% for classifying picture quality (see Fig. 6).

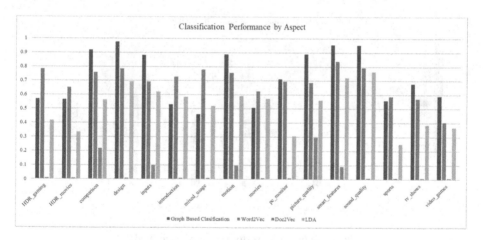

Fig. 6. Aspect-wise performance of each model

Rating Classification. Rating classification is evaluated in the same manner as aspect classification. Since there are 7 ratings in our dataset, the random guess will have accuracy of 14.3%. Review ratings or sentiments predictions are considered more complicated task than aspect classifications. Negations and sarcasms make even difficult for model to detect true sentiment. For more technically focused reviews such as RTINGS reviews are not usually sarcastic. However, there exists many negations that cannot easily be captured with traditional text mining approaches. It is inevitable that general performance for rating classification is lower than aspect classification (see Table 2).

Table 2. Rating classification performance

Models	Accuracy
Random	14.29%
Word2Vec	52.63%
Doc2Vec	38.36%
LDA	48.82%
Graph based classification	**56.60%**

Performance of graph-based classification is the best among other models. However, even the class is decreased to 7, compared to aspect classification, accuracy of all models except Doc2Vec dropped. Sentiment classification is not only relatively harder but also not every sentence in reviews have sentiment. Many sentences from technical reviews remain neutral in sentiment.

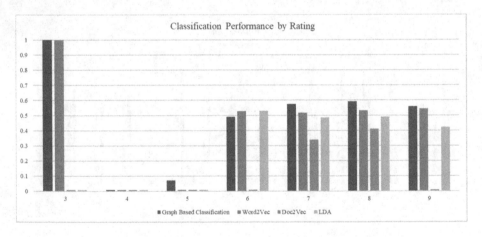

Fig. 7. Rating-wise performance of each model

According to rating-wise performance of each model, Doc2Vec and LDA models tend to classify review sentences in certain ratings range (see Fig. 7). Doc2Vec models have 0 accuracies classifying 3, 4, 5, 6, and 9 ratings. LDA model has trouble classifying low rating reviews. Our model has more stable accuracy than other models which proposed model is the only model which predicted 5 rated reviews. Since detecting bipolarity is more important in sentiment classification, we can conclude that graph-based classification outperform other models.

5.2 Qualitative Evaluations

In qualitative evaluations, we conduct ed post evaluation on quantitative assessments. Since the aspect and rating classifications are done in sentence level, some sentences might be too short and lacks the information for classification. We take sentences that our model is not able to classify correct aspect and rating. By analyzing these sentences with human understanding, some sentences are indeed hard to categorize. At the same time, we can follow logic behind the model's misclassifications that overcoming these errors will be our future works. We selected sentences for both misclassified aspect and rating classifications (see Tables 3 and 4).

Ed Post Evaluation on Aspects. Although our proposed classification framework achieved about 86% of classification accuracy. There are some aspects our model has trouble classifying. We investigate some of evaluation sentences that the model misclassified (see Table 3).

Table 3. Examples of misclassified sentences for aspect

Sentence	True	Predicted
Finally, HDR movies look very good on the Q8C	Picture quality	HDR movies
Viewing angles are ok	TV shows	Picture quality
Sony's new X-Wide viewing angle system delivers wider viewing angles than typically found on VA panels, but unfortunately, they still aren't as good as most IPS TVs, and this comes at the expense of contrast	Introduction	Comparison
Game mode also has motion interpolation, game motion plus, which doesn't look as good as auto motion plus but adds much less input lag, as shown in the input lag box	Motion	Video games
HDR is not supported	HDR gaming	HDR movies

For first sentence in the example above, it explicitly mentioning HDR movies. However, since each aspect is collection of sentences some sentences might seem off the topic when we examine one sentence at a time. It is same for second sentence. For classifying these two sentences with human judgement, we conclude that without background information of adjacent sentences, our model's prediction is reasonable.

The third and fourth sentences are longer. The third sentence compares VA panel to IPS panel which model classified as comparison. In introduction of each review, the topics are heterogeneous. The fourth sentence mainly discuss about game mode of given television model. In video gaming, examination on motion especially input lag is inevitable. Two closely related topics in one sentence makes harder to distinguish.

The last sentence is too short. The discriminating word in this sentence is HDR. There are two HDR related categories in our review dataset, HDR gaming and HDR movies. Given such a short sentence without any further information, the stratification between two aspects are impossible.

Ed Post Evaluation on Ratings. As we mentioned in quantitative analysis, distinguishing sentiment of review is harder task then classifying general aspects in review. We examine the sentences that the difference between true and predicted ratings are more than 2. Misclassification on polarity of sentiment is critical than misplacing sentence with rating 9 to rating 8 (see Table 4).

Table 4. Examples of misclassified sentences for rating

Sentence	True	Predicted
The TV has a 1080p resolution and you won't be able to enjoy 4k content in its native resolution	4	8
Most of the uniformity issues are from the corners being darker than the center of the screen and the top and bottom edges being a bit brighter	4	7
It has a great low input lag perfect for gaming but poor color volume and a limited color gamut, so HDR content doesn't pop like it should	7	9
A high contrast ratio is crucial for good dark scenes performance while watching movies in a dark room	8	4

Many of error arise in negations such as the first sentence in our example. The second sentence is good example of model not correctly recognize the relations between phrases. We see 'uniformity issues' than 'corners being darker than the center' and 'being a bit brighter'. Human can reason that the main context of this sentence is uniformity issue and this television has some areas darker or brighter than other areas which is negative statement. However, only considering the appearance of words, it is hard to understand the semantic relations between phrases in sentence.

When the model encounter sentence with mixed sentiment, the classification accuracy drops. The third sentence starts with positive sentiment because this television has a great low input lag. However, this television has poor color volume and limited color gamut. These mixed sentiments make model harder to assess correct sentiment.

The fourth sentence is the example of sentence without sentiment. This sentence is rather a statement. One cannot easily judge by this sentence whether this television has good dark scenes performance or not.

6 Conclusion

This paper presents graph-based aspect and rating classification, which utilizes dispersion vectors from a multi-modal word co-occurrence network, with aspect and sentiment classes as different modes, to solve aspect and sentiment classification. The use of dispersion scores allows dimension reduction down to the total number of aspect and sentiment labels used in the learning process. Our experiment shows that graph-based aspect and rating classification beats the baseline models, Word2Vec and LDA, in both aspect and sentiment classification tasks. In the future, we plan to expand the model proposed in this paper further by incorporating corpus without aspect labels or sentiment scores. The idea is that, once word co-occurrence network is constructed, any unlabeled data may be "poured on" to the network to expand the size of the network and strengthen the word co-occurrence edges. This will allow aspect-level analysis of data without aspect or sentiment labels.

References

1. Amazon Web Services, Inc. Amazon Customer Reviews Dataset. https://s3.amazonaws.com/amazon-reviews-pds/readme.html. Accessed 27 Jan 2019
2. Luhn, H.P.: A statistical approach to mechanized encoding and searching of literary information. IBM J. Res. Dev. **1**(4), 309–317 (1957)
3. Spärck Jones, K.: A statistical interpretation of term specificity and its application in retrieval. J. Doc. **28**(1), 11–21 (1972)
4. Mikolov, T., et al.: Distributed representations of words and phrases and their compositionality. In: Advances in Neural Information Processing Systems (2013)
5. Blei, D.M., Ng, A.Y., Jordan, M.I.: Latent Dirichlet allocation. J. Mach. Learn. Res. **3**, 993–1022 (2003)
6. Wang, H., Lu, Y., Zhai, C.: Latent aspect rating analysis on review text data: a rating regression approach. In: Proceedings of the 16th ACM SIGKDD International Conference on Knowledge Discovery and Data Mining. ACM (2010)
7. Le, Q., Mikolov, T.: Distributed representations of sentences and documents. In: International Conference on Machine Learning (2014)
8. Wang, H., Lu, Y., Zhai, C.X.: Latent aspect rating analysis without aspect keyword supervision. In: Proceedings of the 17th ACM SIGKDD International Conference on Knowledge Discovery and Data Mining. ACM (2011)
9. Backstrom, L., Kleinberg, J.: Romantic partnerships and the dispersion of social ties: a network analysis of relationship status on facebook. In: Proceedings of the 17th ACM Conference on Computer Supported Cooperative Work and Social Computing. ACM (2014)
10. Marsden, P.V., Campbell, K.E.: Measuring tie strength. Soc. Forces **63**(2), 482–501 (1984)
11. Singhal, S., Pudi, V.: Dispersion based similarity for mining similar papers in citation network. In: 2015 IEEE International Conference on Data Mining Workshop (ICDMW). IEEE (2015)
12. Minocha, A., Singh, N., Srivastava, A.: Finding relevant Indian judgments using dispersion of citation network. In: Proceedings of the 24th International Conference on World Wide Web. ACM (2015)
13. RTINGS Homepage. https://www.rtings.com/

Streaming Massive Electric Power Data Analysis Based on Spark Streaming

Xudong Zhang[1], Zhongwen Qian[1], Siqi Shen[1], Jia Shi[2], and Shujun Wang[3]([⊠])

[1] Zhejiang Electric Power Company, Ltd., Hangzhou 310007, China
[2] Zhejiang Huayun Information Technology Company, Ltd., Hangzhou 310012, China
[3] College of Intelligence and Computing, Tianjin University, Tianjin 300350, China
shujunwang@tju.edu.cn

Abstract. Electric power user classification is one of the most important methods to realize the optimal allocation of power resources. Through the analysis of users'needs, behavior and habits, Countries and enterprises can offer different incentives for different users. In this way, people are more willing to use green and clean Electric power resources. In the analysis of user clustering, there is a need for real-time processing of massive and high-speed data. In this paper we propose a novel distributed user data stream clustering method based on Spark streaming, improved clusStream algorithm and improved K-means algorithm named "DStreamEPK". In the final experimental evaluation, we first tested the clustering effectiveness of DStreamEPK on UCI datasets, the results show that the proposed DStreamEPK is better than the traditional K-means clustering algorithm. At the same time, it is found that DStreamEPK can cluster user's electricity data quickly and efficiently through testing on user's real data sets.

Keywords: Spark streaming · ClusStream · K-means · Electric power

1 Introduction

In recent years, people all over the world are increasingly demanding to protect the environment and achieve sustainable development. In this context, how to make electricity consumption behavior intelligent has become a very important research topic. A great deal of basic electricity consumption data have been accumulated [3]. These data are huge and high Frequency. At the same time, the user's electricity data is constantly generated. The newly generated electricity data can better reflect the user's electricity characteristics. Distributed clustering of the user's electricity data can provide different incentives for different users.

This way can help grid companies to understand the user's consumption habits and provide personalized and differentiated services for users. Furthermore, it helps companies to further expand the depth and breadth of their services, and provides data support for the formulation of future power demand

G. Li et al. (Eds.): DASFAA 2019, LNCS 11448, pp. 200–212, 2019.
https://doi.org/10.1007/978-3-030-18590-9_14

response policies. At the same time, the company will timely feedback the residential power consumption data and residential power consumption to users, so that users can understand their own electricity consumption information and contribute to low-carbon environmental protection.

Cluster analysis is a classical method in the field of data mining. Wang [9] proposed a short-term load forecasting method for power system. Zhao [12] proposed an improved K-means based clustering method for power load curve.

In the meantime, many stream clustering methods have been proposed. Birch [11] algorithm is a hierarchical clustering algorithm proposed by Zhang et al. in 1996. Aggarwal et al. proposed ClusStream [4], a classical two-tier data stream processing framework in 2003. Rakthanmanon et al. proposed E-Stream [8] clustering algorithm in 2007 to improve clusStream algorithm's poor clustering performance for high-bit data. Assent et al. [7] proposed Clustree algorithm in 2011 to cluster data points with arbitrary shape distribution effectively. Marcel et al. proposed StreamKM++ algorithm [1] in 2012.

There exists a lot of clustering algorithms and Data Stream Clustering Algorithms, such as k-means [6] algorithm, improved version algorithm based on k-means, ClusStream, StreamKM++ algorithm et. However, these algorithms can not be directly and efficiently applied to distributed storage and computing environments. How to integrate these algorithms into the current mainstream big data processing frameworks such as Hadoop and spark is a very worthwhile problem.

However, there are few research results on data flow and distributed computing, which are still in the initial stage of exploration. In this paper, the traditional stream clustering algorithm and distributed stream processing platform are introduced firstly, and then the problems of current stream clustering algorithm are analyzed. Based on the original stream clustering algorithm, a stream clustering algorithm DStreamEPK based on SparkStreaming is proposed. DStreamEPK uses a typical two-tier clustering method to maintain the outline information of data stream in online part using X* tree; canopy is used to solve the initial K value selection problem of K-means algorithm in offline part, and an efficient distributed parallel k-means algorithm is designed to cluster power data offline.

Our major contributions can be summarized as follows:

1. A parallel k-means algorithm based on spark is designed.
2. Using canopy algorithm to solve k-value selection problem of K-means algorithm.
3. A clustering method based on spark streaming for user power data is proposed.
4. New stream clustering algorithm DStreamEPK is proposed.

2 Preliminary

2.1 Spark Streaming

Spark is a research product created by APMLab Laboratory at the University of California, Berkeley. It was officially open source in 2010, became an Apache Foundation project in 2013, and became the top project of the Apache Foundation in 2014.

Spark [10] mainly solves the problem of slow computation in hadoop, which is caused by repeated read and write disks.

Resilient Distribute Datasets (RDD) is the core of Spark and the key to Spark's fault recovery and data dependence. RDD model uses Lineage mechanism to solve the dependence between data, and ensures good fault tolerance. It can store intermediate results in memory and minimize disk reading and writing of data, which greatly improves the computing speed. Especially in iterative computing, the computing speed is increased by an order of magnitude.

Data does not exist in the original form in RDD, but is included in RDD in the form of the specific location of the data. New RDD is obtained through different transformations in RDD, and the real calculation is not performed until the action is executed to get the final desired result. Spark Streaming is used for real-time computing in Spark ecosystem.

The essence of Spark Streaming stream processing is to merge data in a short period of time and then do micro batch processing instead of real-time processing each data separately. This is also the biggest difference from other stream processing systems, so it is not real-time processing, but microbatch processing with lower latency.

2.2 Stream Clustering Algorithm

C.C. Aggarwal et al. first proposed a two-layer flow clustering framework in clusStream algorithm, which regards the process of data stream clustering as a process of change, instead of computing and preserving all data with the same granularity. The algorithm can respond to users'queries and clustering requests in time, and return clustering results with different granularity.

For the first time, the algorithm divides the flow clustering process into two parts: offline clustering and online clustering. In the micro-clustering stage, micro-clusters are used to represent the clustering information in the original data. Because micro-clusters are additive, the updating of data is incremental. In the macro-clustering stage, the off-line algorithm can get the final results by clustering all the micro-clusters generated in the online stage. The algorithm introduces a pyramidal inclined time window to deal with the importance of time in different time periods. That is to say, the closer the data is to the current time, the more important it is. The clusStream algorithm uses finer granularity to save new data points in time dimension and coarser granularity to save old data points. The final data snapshot is similar to the inverted pyramid shape, as shown in the figure.

CluStream algorithm uses the clustering feature vector (CF) of BIRCH algorithm to store the outline information of data stream. CF, which first appeared in BIRCH algorithm, is a triple feature.

Vector $\langle$n, LS, SS$\rangle$, n is the number of data points, LS is the linear sum of n points, SS is the square sum of n points. LS and SS are vectors of the same dimension as the original data points. Because CF is additive, updates to micro clusters can be obtained by vector operations. For example, when a new data point is added to a micro-cluster, the general information of the micro-cluster can be updated by adding vectors.

In CluStream algorithm, besides Timestamp T, CF of micro-cluster also maintains several other time labels - LST is the sum of time stamps, SST is the sum of time stamps. CluStream determines whether to create a new cluster or merge it into an existing one by determining the distance between the data point and its nearest cluster and the distance threshold. In order to maintain the relative stability of the number of micro-clusters, when a new micro-cluster is generated, two micro-clusters are selected from the existing micro-cluster set to merge or discard an older micro-cluster. In this way, the algorithm can keep the newer data in the micro-cluster and reduce the memory consumption.

Although the algorithm has good running efficiency and better clustering accuracy, because the Euclidean distance is used to evaluate the similarity, the algorithm can only cluster data streams with spherical distribution. For data streams with other types of distribution, the algorithm can not cluster effectively, and the algorithm can not deal with outliers well.

2.3 K-means Canopy and R Tree

K-means. K-means is a partition-based clustering algorithm, which is simple and efficient.

Because of its strong expansibility, it has been widely used in various fields. K-means algorithm usually uses Euclidean distance between two samples as a measure of similarity.

The calculating steps of K-means algorithm are as follows: firstly, K initial clustering centers are selected artificially in DataSet, then Euclidean distances from the remaining sample data to the initial center are calculated, then each sample is classified into the corresponding cluster centers according to the principle of minimum distance, and then the average distances of all samples of each class are calculated and updated to the new cluster centers of the class. Until the sum of squares of errors function is stable at the minimum value.

Canopy. Although Canopy algorithm [5] does not need to set the number of clusters K, it needs to set the distance range Db and Ds, in which Db refers to the maximum distance threshold and Ds refers to the minimum distance threshold. The relationship between them is $Db > Ds$. The number of canopy subsets in clustering results and their data points can be affected by the setting of distance threshold. However, if the Db is too large, many points will belong to multiple Canopy at the same time, which may eventually lead to a smaller difference between the central points of each cluster, and the difference between the clusters is not obvious. If Ds is too large, it may lead to too few clusters, but if Ds is too small, it will lead to too many clusters and greatly increase the clustering time. So we must set the values of Db and Ds reasonably.

The dataSet R is read into memory, and then the distance threshold parameters Db and Ds are set.

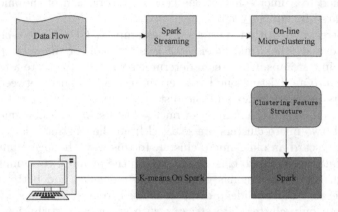

Fig. 1. Architecture

A data point D is randomly selected from R and the distance from D to all canopy subsets is calculated. If there is no canopy subset at the beginning, we need to treat data point D as a new canopy list, and at the same time remove d from R. If the distance between data D and the center of a canopy is less than or equal to Db, then D is written to the canopy, but the data is not deleted from R.

If the distance between the data point$D$ and the center of a canopy is not greater than Ds, D is written to the canopy and the node is deleted from R.

Repeat step 2, 3 until R becomes an empty data set.

R Tree. R-tree [2] is a balanced tree used to store high-dimensional data. It solves the search problem in high-dimensional space very well. R-tree extends the idea of B-tree to multi-dimensional space, uses the idea of B-tree partitioning space, and uses the method of merging and decomposing nodes when adding and deleting operations to ensure the balance of the tree.

R-tree is an extension of B-tree in high-dimensional space and a balanced tree. The leaf nodes of each R tree contain multiple pointers to different data, which can be stored on hard disk or in memory. According to the data structure of R-tree, when we need a high-dimensional spatial query, we only need to traverse the pointers contained in a few leaf nodes to see whether the data pointed by these pointers meet the requirements. This way we can get the answer without traversing all the data, and the efficiency is greatly improved.

3   Architecture of Analysis Model

ClustesStream algorithm points out several problems that need to be solved when designing the framework of stream clustering algorithm. These problems can be summarized as follows:

1. How to Store Data Summary Information Quickly and Efficiently in Continuous Data Stream?
2. At what moments in time should the summary information be stored away on disk?
3. How can the periodic summary statistics be used to provide clustering and evolution insights over user-specified time horizons?

Referring to the algorithm framework model proposed by clusStream algorithm, we propose a framework used in DStreamEPK algorithm.

Theorem 1. *data frame D consists of a set of multi-dimensional records X_1, X_2, X_3 arriving at Timestamps T_1, T_2, T_k...Each X_i is a multi-dimensional record containing d dimensions.*

Theorem 2. *We optimize micro-cluster based on clusStream algorithm. A micro-cluster for a set of d-dimensional points $X_{i1}...X_{in}$ with timestamps $T_{i1}....T_{in}$ is defined as the (2*d+3) tuple (CF2x,CF1x,n,tl,id), wherein CF2x and CF1x ech correspond to a vector of d entries. The definition of each of these entries is as follows:*

- *There are d elements in CF2x, which is the sum of squares for each dimension of data.*
- *For each dimension, the sum of the data values is maintained in CF1x.therefore,CF1x includes d values.*
- *n represents the number of elements in a microcluster.*
- *tl is the last update time of data in micro-cluster.*
- *id is the only symbol of micro-cluster.*

4   On-line Streaming Clustering Algorithm

R* tree is mainly used to perform similarity search in multidimensional data. SS tree is an improved index tree based on R* tree. The original R* tree uses the form of hypercube to partition the multidimensional space, which results in a large number of overlapping nodes in the tree index. However, SS tree uses hypersphere method to divide space. SS tree does not need any additional data besides the central point and radius of subspace, so it can save space and improve the speed of search.

The algorithm based on SS-Tree is executed as follows:

1. Initialization of SS tree structure
2. Pre-clustering the received data to generate several micro-clusters

3. When the new data point X arrives, according to whether the distance from X to the center of each micro-cluster is greater than the RMS deviation (root mean square error) of the micro-cluster, and greater than the new micro-cluster with an independent ID for the q-point, otherwise it will be added to the nearest existing micro-cluster (using the additive property of the eigenvector group).

4. Once a new micro-cluster is established, it is necessary to delete an original micro-cluster. In theory, it is usually determined to delete the micro-cluster according to the recent time stamp formed by the M points that have recently arrived at each micro-cluster. In practice, the mean and standard deviation of arrival time of each data point can be obtained according to the time statistics in the micro-cluster, because the default micro-cluster satisfies the normal distribution. So extracting the time information relevance time of m/(2*n) is compared with the preset threshold delta. If the minimum relevance time is less than delta, the corresponding micro-cluster can be deleted.

5. If all relevance time values are larger than delta, two nearest micro-clusters need to be merged, and the corresponding ID is formed as an idlist.

4.1 Updating of Micro-clusters

Experience shows that the importance of data with timestamps is different, and the latest data has a greater impact on users. That is to say, data flow information has timeliness. When a part of the data exists for more than a certain period of time, it is likely that this part of the data will be de-valued. ClsStream algorithm uses pyramid time window to store micro-clusters in different time periods. In our research, we use a simpler time-decay technique to update and delete micro-clusters periodically. Time attenuation technology can make data of different time show different importance according to the set function. Setting up a process in the algorithm and updating and deleting all data at a fixed time interval can reduce the impact of historical data.

The time weighting function is as follows:

$$w(\triangle t) = 2^{-\lambda \triangle t} \tag{1}$$

In the above formula, A is the attenuation factor. The value of a reflects the role of historical data in clustering. The larger the value, the smaller the impact of historical data on the algorithm.

The time decay function of micro-clusters is calculated as follows:

$$n^{(t)} = \sum_{i=1}^{n} w(t - tp_i)$$
$$LS^{(t)} = \sum_{i=1}^{n} w(t - tp_i) \cdot x_i$$
$$SS^{(t)} = \sum_{i=1}^{n} w(t - tp_i) \cdot x_i^2$$

5 Off-line Streaming Clustering Algorithm

5.1 Preprocessing of Electricity Data

In order to improve the efficiency of the algorithm in massive residential power data mining Rate, data need to be preprocessed, as shown in Fig. 1.

Data Filtering. In the original residential electricity data, there may be a user's electricity information data at a certain time is repeatedly recorded, or is divided into multiple electricity information for recording. For the duplicate records, the method of direct filtering and deletion is adopted. For the latter, the user number can be extracted, and the electricity information can be superimposed and merged into a single data record. In addition, there may be some missing values in a user's data. In this case, a threshold of the number of missing values can be set beforehand. When the threshold is exceeded, the record can be deleted directly. On the contrary, only the missing values can be filtered out.

Data Filling. The method to deal with the missing values is to select the average value of the two adjacent load values of the missing values as the corresponding filling values. If the neighborhood value is also empty, the next non-empty load value is found forward or backward. If there is no non-empty load value, it is filled with zero value.

Normalization of Features. In the original data, after extracting the relevant user features, different eigenvalues may have different ranges. The influence of larger eigenvalues on the global matrix is greater than that of smaller ones, which weakens the role of smaller features. Therefore, it is necessary to normalize the features. In this paper, the interval normalization method is used for eigenvalue matrix $X = X_1 X_2..$ The maximum Max (X_i) and minimum min (X_i) of eigenvalues in the eigenvalue matrix are calculated. According to the formula (8), the eigenvalue fields are normalized to intervals $[0, 1]$, and a set of normalized matrices $V = V_1 V_2$ are obtained.

5.2 K-means Clustering Algorithm Combined with Canopy

When the number of nodes in a distributed cluster is not limited, the performance of the traditional clustering algorithm can be improved with the increase of computing nodes after it is parallelized in Park Streaming. However, when the machine is insufficient and the operation efficiency of the algorithm can not meet the needs of the application, it is also a good solution to optimize the design idea and execution logic of the algorithm itself.

In this paper, the optimization of stream clustering algorithm is mainly embodied in the selection of initial centers of K-Means clustering algorithm, the setting of K value of cluster number and the execution method of K-means algorithm on Spark cluster.

Firstly, this paper chooses Canopy rough clustering algorithm to realize the internal optimization of stream clustering algorithm. The clustering results of Canopy algorithm provide the initial clustering centers and K values for K-Means, which can effectively avoid the randomness of K-Means calculation results and reduce the running time, thus improving the processing performance of K-Means for large-scale data.

By using Map(), Combine() and Reduce() three functions, the spark cluster is implemented efficiently.

Canopy algorithm does not need to specify the number of clusters manually. It can cluster actively according to its own iteration. It only needs to set the threshold Db and Ds in the clustering process, where Db > Ds. Therefore, Canopy rough clustering algorithm and K-Means algorithm can be combined, and the output of Canopy algorithm can be used as the input of K-Means. In this way, the subjectivity of K selection in K-Means and the randomness of initial cluster center selection can be avoided to a certain extent, so that the efficiency and accuracy of clustering can be improved by reducing the number of iterations in K-Means clustering algorithm. Based on the above analysis, the implementation steps of K-Means optimization algorithm based on Canopy are as follows:

Step 1: After data set R is preprocessed, a list of data R' is written to memory, and appropriate distance thresholds are selected: $Db, Ds(Db > Ds)$.

Step 2: According to the Canopy method, all data in R' is divided into several Canopy.

Step 3: Use the number of Canopy generated in Step 2 as the K value of the K-means algorithm.

Step 4: Execute the K-means algorithm.

5.3   Parallel K-means on Spark

Map. Firstly, the pre-processed data set R'is read into memory and stored in the form of <key, value> in which key is the offset of the current sample relative to the starting point of the input data file, and value is a string of coordinate values of the current sample. Firstly, the value of each dimension of the current sample is resolved from value; then the distance between the current sample and K centers is calculated to find the subscript of the nearest cluster; finally, <key', value'> where key'is the subscript of the nearest cluster and value' is the string of coordinates of the current sample. The pseudocode of the function is:

In order to reduce the amount of data and communication cost in the iteration process, after Map operation, we design a Combine operation, which merges the output data after each Map function is processed locally. Because the data output after each Map operation is always stored in the local node, each Combine operation is executed locally, and the communication cost is very small.

Combine. In the pair of key, Value input by Combine function, key is the subscript of cluster, and V is a string list of coordinate values of each sample assigned to cluster with subscript key. Firstly, the coordinate values of each sample are sequentially resolved from the list of strings, and the coordinate values corresponding to each dimension are added up separately. At the same time, the chain is recorded.

The total number of samples in the table. The output key', value 'is the subscript of the cluster to the middle key value is a string, which includes two parts of information: the sum and composition of the total number of samples and the coordinate values of each dimension. The function pseudocode is:

Reduce. In Reduce function input key, V key is the subscript of cluster, V is the intermediate result of transmission from each Combine function. In Reduce function, the number of samples processed from each Combine and the cumulative values of coordinates of the corresponding nodes in each dimension are firstly analyzed, and then the corresponding cumulative values of each dimension are added respectively, and then divided by the total number of samples, the new coordinates of the central points are obtained. The function pseudocode is:

According to the output of Reduce, the new center coordinates are obtained and updated. To the file on HDFS, and then the next iteration, until the algorithm converges.

6   Experiments and Evaluations

Based on the spark platform and the improved K-means algorithm, we have completed the following experiments.

6.1   Clustering Validity

In order to verify the effectiveness of the improved K-means clustering algorithm, we selected some data sets provided by UCI website, and compared the

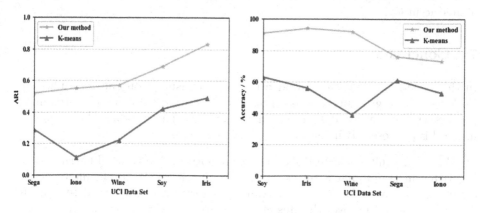

Fig. 2. Cluster result   Fig. 3. Accuracy

traditional K-means algorithm with the improved algorithm in this paper. The clustering results were measured by AdjustRand Index and clustering accuracy. It is easy to draw the following conclusions from Figs. 1 and 2. Compared with the traditional K-means algorithm, the K-means algorithm combined with canopy has better clustering effect and higher accuracy. Therefore, it is very meaningful to use the improved k-means clustering algorithm in the offline part of DStreamEPK algorithm (Figs. 3, 4 and 5).

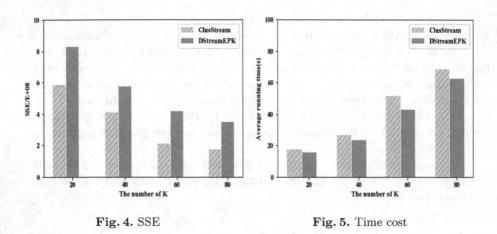

Fig. 4. SSE   Fig. 5. Time cost

6.2   Stream Clustering Verification

Compared with DStream EPK algorithm, CluStream algorithm runs shorter time on data set, because CluStream's online clustering and offline clustering algorithms are implemented by simple K-Means, and it controls the number of iterations. The proposed DStreamEPK algorithm needs to run canopy algorithm once in the offline phase, so it will consume more time. Overall, however, the time consumed is not too much. It is worthwhile to sacrifice part of the time for higher accuracy.

7   Conclusions

In this paper, the analysis of residential power consumption data is studied on the basis of a large number of user power consumption data. A spark streaming processing and improved K-means algorithm based user power data analysis method is proposed. It includes the following aspects:

(1) There are initial clustering centers and optimal K-means clustering algorithm in traditional K-means clustering algorithm. It is difficult to determine the value. In this paper, canopy algorithm is used to determine K value, which improves the clustering accuracy.

(2) An improved K-means algorithm based on spark Streaming is proposed. According to the analysis method. On-line phase can quickly collect micro-cluster information and statistical summary information. In the offline stage, the characteristics of peak-time power consumption rate, load rate, valley load coefficient and flat-section power consumption percentage of each user are extracted by pre-processing of user's power consumption data, and the data vector dimension is established. Then, the improved K-means algorithm is used to cluster the data, and the parallel algorithm is realized on the basis of spark. The user's electricity consumption behavior is analyzed and the characteristics of each type of user are extracted. The experimental results show that the proposed method is stable, efficient and reliable. A massive algorithm based on spark streaming and improved K-means is proposed.

Using data flow analysis method to mine valuable information in power consumption data and analyze users Electricity consumption behavior has important guidance for power dispatching and pricing mechanism formulation. Sexual meaning. Next, combined with the user clustering results of the analysis model, for each category Users conduct research on short-term load forecasting.

References

1. Ackermann, M.R., Märtens, M., Raupach, C., Swierkot, K., Lammersen, C., Sohler, C.: Streamkm++: a clustering algorithm for data streams. ACM J. Exp. Algorithmics **17**(1), 2–4 (2012)
2. Bogojeska, J., Alexa, A., Altmann, A., Lengauer, T., Rahnenführer, J.: Rtreemix: an R package for estimating evolutionary pathways and genetic progression scores. Bioinformatics **24**(20), 2391–2392 (2008)
3. Chen, W., Zhou, K., Yang, S., Cheng, W.: Data quality of electricity consumption data in a smart grid environment. Renew. Sustain. Energy Rev. **75**, 98–105 (2016)
4. Freytag, J.C., Lockemann, P.C., Abiteboul, S., Carey, M.J., Selinger, P.G., Heuer, A. (eds.): VLDB 2003, Proceedings of 29th International Conference on Very Large Data Bases, 9–12 September 2003, Berlin, Germany. Morgan Kaufmann (2003)
5. Goldbergs, G., Maier, S.W., Levick, S.R., Edwards, A.: Limitations of high resolution satellite stereo imagery for estimating canopy height in Australian tropical savannas. Int. J. Appl. Earth Obs. Geoinf. **75**, 83–95 (2019)
6. Hartigan, J.A., Wong, M.A.: Algorithm as 136: a k-means clustering algorithm. J. R. Stat. Soc. **28**(1), 100–108 (1979)
7. Kranen, P., Assent, I., Baldauf, C., Seidl, T.: The clustree: indexing micro-clusters for anytime stream mining. Knowl. Inf. Syst. **29**(2), 249–272 (2011)
8. Udommanetanakit, K., Rakthanmanon, T., Waiyamai, K.: E-Stream: evolution-based technique for stream clustering. In: Alhajj, R., Gao, H., Li, J., Li, X., Zaïane, O.R. (eds.) ADMA 2007. LNCS (LNAI), vol. 4632, pp. 605–615. Springer, Heidelberg (2007). https://doi.org/10.1007/978-3-540-73871-8_58
9. Wang, H.Z., Liu, K., Zhou, J., Wang, Y.F.: Pretreatment of short-term load forecasting based on k-means clustering algorithm. Computer Simulation (2016)

10. Zaharia, M., Chowdhury, M., Franklin, M.J., Shenker, S., Stoica, I.: Spark: cluster computing with working sets. In: Usenix Conference on Hot Topics in Cloud Computing (2010)
11. Zhang, T., Ramakrishnan, R., Livny, M.: BIRCH: an efficient data clustering method for very large databases. SIGMOD Rec. **25**(2), 103–114 (1996)
12. Zhao, W., Gong, Y.: Load curve clustering based on kernel k-means. Electr. Power Autom. Equip. (2016)

Posters

Deletion-Robust k-Coverage Queries

Xingnan Huang[1] and Jiping Zheng[1,2,3]([✉])

[1] College of Computer Science and Technology,
Nanjing University of Aeronautics and Astronautics, Nanjing, China
{huangxingnan,jzh}@nuaa.edu.cn
[2] Collaborative Innovation Center of Novel Software Technology and
Industrialization, Nanjing, China
[3] Department of Computer Science and Technology,
Nanjing University, Nanjing, China

Abstract. The k-coverage query is an ideal solution for representative queries with almost known nice characteristics, such as stability, scale-invariance, traversal efficiency and so on. In this paper, we propose deletion-robust k-coverage queries. First, we calculate a *coreset* from the whole dataset with a *sieving* procedure by various thresholds to make k-coverage queries robust under deletion of arbitrary number of data points. Then our k-coverage queries can be carried out efficiently on the small *coreset* instead of the whole skyline set. Experiments on both synthetic and real datasets verify the effectiveness and efficiency of our proposed method.

Keywords: k-coverage queries · Robust coreset ·
Representative skyline

1 Introduction

Helping end users to identify a small subset with a manageable size from a large dataset is an important functionality in many applications. Considering the deficiencies of top-k and skyline, various k representative queries from different evaluation aspects are raised [1–6], which offer a tradeoff of these two queries. However, most of them are not simultaneously *stable*, *scale-invariant* and *traversal efficient*. Fortunately, k-coverage queries first introduced in [5], returning a solution set that collectively maximizes the coverage area, have good representativeness furnished with these nice characteristics. In addition, none of these besides k-coverage queries is *deletion-robust* when skyline points are deleted.

Take car sales system for example as illustrated in Fig. 1(a) where a car has two attributes HP (horse power) and MPG (miles per gallon). A solution set returned by the greedy algorithm [5] for a 2-coverage query is $S = \{p_1, p_4\}$ with coverage value $f(S) = 0.47$ as shown in Fig. 1(b). In some cases, mercenary salesmen will insert some non-real cars' information in the database, trying to influence the recommendation results. Assume that there is one faked point p_4,

© Springer Nature Switzerland AG 2019
G. Li et al. (Eds.): DASFAA 2019, LNCS 11448, pp. 215–219, 2019.
https://doi.org/10.1007/978-3-030-18590-9_15

Car	HP	MPG	$f(p)$
p_1	0.3	0.9	0.27
p_2	0.4	0.7	0.28
p_3	0.5	0.6	0.3
p_4	0.7	0.5	0.35
p_5	0.8	0.3	0.24
p_6	0.9	0.2	0.18

(a) Car sales database

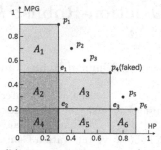

(b) 2-coverage queries w/o p_4

Fig. 1. Car sales system example

after deleting p_4, the recalculated solution set $S' = \{p_1, p_3\}$ and $f(S') = 0.39$. We have to recalculate the solution set after deleting the faked car information, otherwise the results with faked information make no sense. Motivated by this, in this paper, we propose deletion-robust k-coverage queries where query answering is performed on a *coreset* instead of on the whole dataset.

2 Coreset-Based Deletion-Robust k-Coverage Queries

The goal of a k-coverage query is to find a subset S of dataset D containing at most k points such that the coverage value is maximized. However, a subset of D may be removed unexpectedly, we need to execute the k-coverage query again after the removal of the data points. Therefore, our problem is to maximize the coverage function after deletion of any subset $R \subseteq D, |R| = r$. The most straightforward method is to answer k-coverage queries on the dataset $D \setminus R$ for each deletion, but it is too time-consuming for a large data set. Our method preselects a small subset from the dataset D, called *coreset*, where we can still efficiently find a set $S \subseteq coreset \setminus R$ of size k which provides a good coverage value. We show that our coverage function f is a monotone submodular function, thus the *sieving* procedure [7,8] can be exploited to solve our problem.

A Robust-Coreset Algorithm. Based on the submodularity, the optimal value OPT of selected k representatives with largest coverage is in the range $[m, km]$, where m is the largest coverage value in set $\{f(\{p\})|p \in D\}$. The exact value of m is not known due to the deletion, so we get an expanded range $[m_l, km_l]$, m_l is the $(l+1)$th maximum coverage value in the set $\{f\{p\}|p \in D\}$. Similar to [7,8], we apply a discretization of the range of threshold τ, and for each τ, we construct two sets, Q_τ and P_τ. The set Q_τ sieves the points with similar marginal coverage value while the set P_τ picks the points from Q_τ when the size of Q_τ is large enough to make the deletion robust. Algorithm 1 shows the pseudocode of Robust-Coreset algorithm. Note that the *coreset* is determined by three parameters k, l and ϵ. For the same k, we just need to set l and ϵ once to

---

Algorithm 1. Robust-Coreset

Input: A set of n d-dimensional points $D = \{p_1, p_2, ...p_n\}$, positive integer k, the size of tentative deletion set l, a small error ϵ

Output: sets $\{P_\tau\}$, Q

1 Initially, $m_l =$ the $(l+1)$-th largest coverage value of the set $\{f(p)|p \in D\}$;
2 $D_l =$ the top $(l+1)$ points with largest $f(p), p \in D$;
3 $D' = D \setminus D_l$;
4 $T = \{(1+\epsilon)^i | \frac{m_l}{(1+\epsilon)^k} \leq (1+\epsilon)^i \leq m_l\}$;
5 **for** each $\tau \in T$ from high to low **do**
6    $\quad P_\tau = \emptyset$;
7    $\quad$**while** $|Q_\tau| \geq \frac{l}{\epsilon}$ for $Q_\tau = \{p|p \in D' : \tau \leq \Delta_f(p| \cup_{\tau' \geq \tau} P_{\tau'}) < \tau(1+\epsilon)\}$ and $|\cup_{\tau' \geq \tau} P_{\tau'}| < k$ **do**
8    $\quad\quad$ randomly select a point p from Q_τ to P_τ, $P_\tau = \{P_\tau \cup p\}$;
9    $\quad D' = D' \setminus (P_\tau \cup Q_\tau)$;
10 $Q = \{\cup Q_\tau\} \cup D_l$;
11 **return** $\{P_\tau\}$, Q

---

compute a *coreset* as a preselected set, which greatly improves the efficiency. The complexity of Robust-Coreset is $\mathcal{O}((k + \frac{\log k}{\epsilon})n)$ coverage function evaluations.

When the *coreset* is computed, we select the deletion set R from the *coreset* according to a deletion strategy. After the deletion of R, the remaining *coreset* $\{P'_\tau\} \cup Q'$ is used to find k representative points with maximum coverage value. For each threshold τ, we initially put all the points in $P_{\tau'}(\tau' \geq \tau)$ into S_τ. Then, we continue to sieve the points from Q' until S_τ meets the cardinality constraint. Among all the S_τ, the set with maximum coverage value is the final solution.

3   Experimental Results

In this section, we verify the robustness and efficiency of our algorithm. A 15-dimensional anti-correlated dataset generated by dataset generator [9] is exploited. For real data set, we adopt NBA dataset. We compare our RobusT algorithm (RT for short) with two methods studied in previous researches, Greedy [5] and δ-Greedy [6]. We conduct experiments on a machine with 3.40GHz CPU and 8G RAM. All programs are implemented in C++. The preselected sets of Greedy and δ-Greedy ($\delta = 0.1$) are skyline sets, while for our RobusT algorithm, we perform a preprocessing step (*i.e.* Algorithm 1) to get a small and

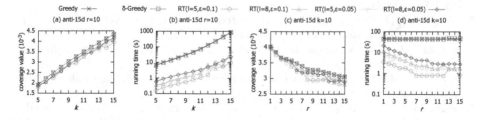

Fig. 2. Performance of the algorithms on the 15-dimensional anti-correlated dataset

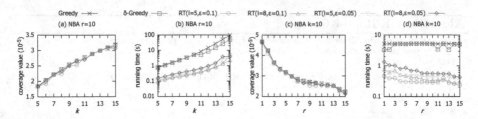

Fig. 3. Performance of the algorithms on the NBA dataset

robust *coreset* which is determined by k, l, ϵ. Figure 2(a–b) show the coverage values of all algorithms which are close to each other with the increase of k, while RobusTs are very fast in practice, but Greedy and δ-Greedy have longer running times, almost 10 times those of RobusTs. In Fig. 2(c–d), we vary the value of r. RobusT algorithms achieve extremely similar coverage values with less running time compared with other algorithms. The experiments on the real dataset have similar results as shown in Fig. 3.

4 Conclusion

In this paper, we propose deletion-robust k-coverage queries. Our method to answer k-coverage queries on precomputed coresets is efficient and effective under deletion of arbitrary subset from the whole dataset. Our future work includes extending our method to streaming environment and for various k representative queries.

Acknowledgment. This work is partially supported by the National Natural Science Foundation of China under grants U1733112, 61702260.

References

1. Chan, C.-Y., Jagadish, H.V., Tan, K.-L., Tung, A.K.H., Zhang, Z.: Finding k-dominant skylines in high dimensional space. In: SIGMOD, pp. 503–514 (2006)
2. Lin, X., Yuan, Y., Zhang, Q., Zhang, Y.: Selecting stars: the k most representative skyline operator. In: ICDE, pp. 86–95 (2007)
3. Tao, Y., Ding, L., Lin, X., Pei, J.: Distance-based representative skyline. In: ICDE, pp. 892–903 (2009)
4. Nanongkai, D., Sarma, A.D., Lall, A., Lipton, R.J., Xu, J.: Regret-minimizing representative databases. In: VLDB, pp. 1114–1124 (2010)
5. Søholm, M., Chester, S., Assent, I.: Maximum coverage representative skyline. In: EDBT, pp. 702–703 (2016)
6. Bai, M., et al.: Discovering the k representative skyline over a sliding window. TKDE **28**(8), 2041–2056 (2016)
7. Badanidiyuru, A., Mirzasoleiman, B., Karbasi, A., Krause, A.: Streaming submodular maximization: massive data summarization on the fly. In: SIGKDD, pp. 671–680 (2014)

8. Kazemi, E., Zadimoghaddam, M., Karbasi, A.: Scalable deletion-robust submodular maximization: data summarization with privacy and fairness constraints. In: ICML, pp. 2549–2558 (2018)
9. Börzsöny, S., Kossmann, D., Stocker, K.: The skyline operator. In: ICDE, pp. 421–430 (2001)

Episodic Memory Network with Self-attention for Emotion Detection

Jiangping Huang[1]([✉]), Zhong Lin[2], and Xin Liu[1]

[1] School of Software Engineering,
Chongqing University of Posts and Communications,
Chongqing 400065, China
{huangjp,liuxin}@cqupt.edu.cn
[2] School of Languages, Chang'an University, Xi'an 710064, China
linz73105@126.com

Abstract. Accurate perception of emotion from natural language text is key factors to the success of understanding what a person is expressing. In this paper, we propose an episodic memory network model with self-attention mechanism, which is expected to reflect an aspect, or component of the emotion sementics for given sentence. The self-attention allows extracting different aspects of the input text into multiple vector representation and the episodic memory aims to retrieve the information to answer the emotion category. We evaluate our approach on emotion detection and obtains state-of-the-art results comparison with baselines on pre-trained word embeddings without external knowledge.

1 Introduction

We use language to communicate not only semantic information but also emotion and the intensity of affect. For example, our utterances can convey that we are very angry, slightly sad, etc. Hence, detecting the type of emotion and measuring the intensity refers to the degree or amounts of an emotion such as anger or sadness are essential for downstream applications in breaking news detection and public opinion analysis [2].

We should focus on the basic set composed by four primary emotions: *anger*, *joy*, *fear* and *sadness*, which are the more common amongst the many proposals for basic emotions [7]. Existing approaches to emotion measurement mostly rely on methods typically used in text classification, and a neural network architecture [1] is designed to obtain the best performance on the shared task of emotion intensity [6]. A joint model which combines convolutional neural networks (CNN) and long short-term memory (LSTM) for extracting both lexical and word feature is proposed [5]. However, these methods commonly consider learning the sentence representation with the whole input text which commonly neglect what words or phrases expressed emotion aspects.

In this paper, we proposed an episodic memory network with self-attention mechanism for capturing the emotional content in input text and computing

© Springer Nature Switzerland AG 2019
G. Li et al. (Eds.): DASFAA 2019, LNCS 11448, pp. 220–224, 2019.
https://doi.org/10.1007/978-3-030-18590-9_16

the score which quantify the contribution to overall emotion of given content. The self-attention mechanism allows extracting different aspects of sentence into multiple vector representations and the episodic memory module is comparised of an internal memory, an attention mechanism and a recurrent network to update its memory. For emotion detection, it is beneficial for episodic memory module to take multiple passes over the input sentence. The proposed model can capture the emotion words or phrases which are related to the type of emotion. We evaluate the proposed approach on emotion detection and the results show that our model obtained a significant performance compared with baselines.

2   Methodology

We should describe the proposed episodic memory network, the model consists of four main parts. The first part is the self-attention mechanism and the second part is an episodic memory structure. In addition, there are also a question module and a answer module, which are used to computes a vector representation $\mathbf{q}$ of the question and generate the model's predicted answer with both $\mathbf{q}$ and episodic memory $\mathbf{em}$ respectively.

Self-attention Mechanism: In order to encode a variable length sentence into a fixed size embedding, a bidirectional LSTM (BiLSTM) is used to produce a forward hidden state $\overrightarrow{h_t}$ and a backward $\overleftarrow{h_t}$ for the input text. We concatenate $\overrightarrow{h_t}$ with $\overleftarrow{h_t}$ as $\mathbf{h}_t$, and all n $\mathbf{h}_t$ is $\mathbf{H}$. Computing the linear combination requires the self-attention mechanism, and the mechanism takes the $\mathbf{H}$ as input, and outputs a vector of weights $\mathbf{a}$:

$$\mathbf{a} = softmax(\mathbf{w_{s2}} \tanh(\mathbf{W_{s1}} \mathbf{H}^T)) \tag{1}$$

Here $\mathbf{W}_{s1}$ is a weight matrix with a shape of $d_a \times u$. And $\mathbf{w}_{s2}$ is a vector of parameters with size d_a, where d_a is a hyperparameter we can set arbitrarily, and the annotation vector $\mathbf{a}$ will have a size n. We need r different parts to be extracted from the sentence. With regard to this, we extend the $\mathbf{w}_{s2}$ into a $r \times d_a$ matrix, note it as $\mathbf{W}_{s2}$, and the resulting annotation vector $\mathbf{a}$ becomes annotation matrix $\mathbf{A}$. Formally,

$$\mathbf{A} = softmax(\mathbf{W_{s2}} \tanh(\mathbf{W_{s1}} \mathbf{H}^T)) \tag{2}$$

Here the $softmax(\cdot)$ is performed along the second dimension of its input. We can deem the Eq. 2 as a two layers MLP without bias, whose hidden unit numbers is d_a, and the parameters are $\{\mathbf{W}_{s1}, \mathbf{W}_{s2}\}$. The embedding vector $\mathbf{m}$ then becomes an $r \times u$ matrix $\mathbf{M}$. We compute the r weight sums up multiplying the annotation $\mathbf{A}$ and LSTM hidden states $\mathbf{H}$, the resulting matrix is the sentence embedding, which means $\mathbf{M} = \mathbf{AH}$, and $\mathbf{M} \in \mathbf{R}^{r \times u}$. And the $\mathbf{M}$ is the representation of the given sentence.

Episodic Memory Module: The episodic memory module is comprised of an internal memory, an attention mechanism and a recurrent network to update its memory. During each iteration, the attention mechanism attends over the fact representations **c** by using a gating function which takes into consideration the question representation **q** and the previous episode memory em_{i-1} to produce an episode e_i. The episode is then used alongside the previous memories em_{i-1}, which is used to update the episodic memory $em_i = LSTM(e_i, em_{i-1})$. The initial state of this LSTM is initialized to the question vector with $em_0 = q$. It is beneficial for episodic memory module to take multiple passes over the input from self-attention module to obtain semantic information.

3   Experiments

3.1   Datasets and Baselines

There is an emotion detection dataset [7]. We implemented it for emotion classification evaluation which has four emotion categories, including *anger*, *fear*, *joy* and *sadness*. For the task, we choose three baselines, including CNN for sentence classification [3], tree-structured long short term memory (TLSTM) for sentence embedding [10] and recursive neural networks (RecNN) for sentence classification [9]. The model is trained using Adam [4] with hyperparameters selected on development set. A two-layer BiLSTM is chose for obtaining hidden state as the input of self-attention module. The glove vectors were used for the word embeddings [8]. Out-of-vocabulary words were randomly initialized with range [−0.01, 0.01]. We crop and pad the input tweet to a fixed length. The models were regularized by using dropouts and a l_2 weight decay.

3.2   Experimental Results

The emotion classification results for each emotion type are shown in Table 1. And the results show that the significant performance of proposed model than three baselines. The EMN obtained 64.0 F_1 value, and EMN+SA has better performance than EMN, which means the self-attention is helpful to EMN for emotion classification as well. For the results of different-level emotion intensity classification are shown in Table 2. The results indicate that our model have obtained significant performance on emotion intensity classification than other three baselines except the *fear* type. The Precision (*Prec*), Recall (*Rec*) and F-measure (F_1) which is computed with $2 \times Prec \times Rec/ (Prec + Rec)$ are used as evaluation metric with percentage.

The experimental results shown in Tables 1 and 2 can be considered as coarse-grained and fine-grained emotion detection respectively. We can find that the episodic memory network obtained the significant performance on two emotion classification. For coarse-grained four emotional categories, there is 1.3% improvement on CNN model. The performance on fine-grained emotion detection is also satisfying exception *fear* type. When the episodic memory network with

Table 1. The experimental results of emotion classification for four emotion categories.

Methods	Prec (%)	Rec (%)	F_1 (%)
CNN	64.3	61.2	62.7
TLSTM	61.7	60.9	61.3
RecNN	57.4	60.1	58.7
EMN	65.2	62.8	64.0
EMN + SA	**65.8**	**63.5**	**64.6**

Table 2. The emotion classification in different-level emotion intensity for each emotion.

Models	Anger			Fear			Joy			Sadness		
	Prec	Rec	F_1	Prec	Rec	F_1	Prec	Rec	F_1	Prec	Rec	F_1
CNN	46.7	51.2	48.8	50.7	**55.0**	**52.8**	47.4	48.9	48.1	45.5	42.1	43.7
TLSTM	44.2	49.9	46.9	48.7	50.2	49.4	42.1	44.3	43.2	40.0	43.3	41.6
RecNN	41.8	39.7	40.7	43.5	40.9	42.2	39.8	43.2	41.4	41.9	40.4	41.1
EMN	50.6	52.1	51.3	51.2	53.0	52.1	50.7	51.9	51.3	49.7	48.8	49.2
EMN + SA	**51.3**	**52.8**	**52.0**	**51.7**	53.8	52.7	**51.3**	**52.0**	**51.6**	**50.7**	**49.5**	**50.1**

self-attention, the performance is improved further on different emotion granularity. The experimental results show that the proposed model is a very effective model for modeling the emotion semantic information and the self-attention is helpful for extracting the emotion aspects of sentence.

4   Conclusion

In this paper, we provide a novel neural network architecture which consists of an episodic memory network and self-attention module to model the semantics of emotion text for emotion detection. The proposed model can model the emotion semantic information from different-level components of input text. We conducted two experiments for evaluating the proposed approach and the results show the effectiveness of our model for emotion measurement.

References

1. Abdul-Mageed, M., Ungar, L.: Emonet: fine-grained emotion detection with gated recurrent neural networks. In: Proceedings of the ACL, pp. 718–728 (2017)
2. Hsiao, S.W., Chen, S.K., Lee, C.H.: Methodology for stage lighting control based on music emotions. Inf. Sci. **412**, 14–35 (2017)
3. Kim, Y.: Convolutional neural networks for sentence classification. In: Proceedings of the EMNLP, pp. 1746–1751 (2014)
4. Kingma, D.P., Ba, J.: Adam: a method for stochastic optimization. CoRR abs/1412.6980 (2014)

5. Köper, M., Kim, E., Klinger, R.: IMS at Emoint-2017: emotion intensity prediction with affective norms, automatically extended resources and deep learning. In: Proceedings of the 8th Workshop on Computational Approaches to Subjectivity, Sentiment and Social Media Analysis, pp. 50–57 (2017)
6. Mohammad, S., Bravo-Marquez, F.: Emotion intensities in tweets. In: Proceedings of the 6th Joint Conference on Lexical and Computational Semantics, pp. 65–77 (2017)
7. Mohammad, S.M., Bravo-Marquez, F., Salameh, M., Kiritchenko, S.: Semeval-2018 task 1: affect in tweets. In: Proceedings of International Workshop on Semantic Evaluation (2018)
8. Pennington, J., Socher, R., Manning, C.: Glove: global vectors for word representation. In: Proceedings of the EMNLP, pp. 1532–1543 (2014)
9. Socher, R., et al.: Recursive deep models for semantic compositionality over a sentiment treebank. In: Proceedings of the EMNLP, pp. 1631–1642 (2013)
10. Tai, K.S., Socher, R., Manning, C.D.: Improved semantic representations from tree-structured long short-term memory networks. In: Proceedings of the ACL, pp. 1556–1566 (2015)

Detecting Suicidal Ideation with Data Protection in Online Communities

Shaoxiong Ji[1]([✉])[iD], Guodong Long[2][iD], Shirui Pan[3][iD], Tianqing Zhu[2][iD],
Jing Jiang[2][iD], and Sen Wang[1][iD]

[1] School of ITEE, The University of Queensland, St. Lucia, QLD 4072, Australia
{shaoxiong.ji,sen.wang}@uq.edu.au
[2] Centre for Artificial Intelligence, University of Technology Sydney,
Ultimo, NSW 2006, Australia
{guodong.long,tianqing.zhu,jing.jiang}@uts.edu.au
[3] Faculty of Information Technology, Monash University,
Clayton, VIC 3800, Australia
shirui.pan@monash.edu

Abstract. Recent advances in Artificial Intelligence empower proactive social services that use virtual intelligent agents to automatically detect people's suicidal ideation. Conventional machine learning methods require a large amount of individual data to be collected from users' Internet activities, smart phones and wearable healthcare devices, to amass them in a central location. The centralized setting arises significant privacy and data misuse concerns, especially where vulnerable people are concerned. To address this problem, we propose a novel data-protecting solution to learn a model. Instead of asking users to share all their personal data, our solution is to train a local data-preserving model for each user which only shares their own model's parameters with the server rather than their personal information. To optimize the model's learning capability, we have developed a novel updating algorithm, called average difference descent, to aggregate parameters from different client models. An experimental study using real-world online social community datasets has been included to mimic the scenario of private communities for suicide discussion. The results of experiments demonstrate the effectiveness of our technology solution and paves the way for mental health service providers to apply this technology to real applications.

1 Introduction

Early detection on suicidal ideation is one of the most effective methods to prevent suicide. Potential victims with suicidal ideation may express the thoughts of committing suicide such as fleeting thoughts, suicide plan and role playing. Suicidal ideation detection is to find out these dangerous thoughts before the tragedy happens. Mental health service practice is labour intensive, requiring social workers to actively and frequently engage with targeted users. Machine learning can help to achieve the goal of early detection; however, the learning

© Springer Nature Switzerland AG 2019
G. Li et al. (Eds.): DASFAA 2019, LNCS 11448, pp. 225–229, 2019.
https://doi.org/10.1007/978-3-030-18590-9_17

process requires a great amount of sensitive information to be obtained from vulnerable people and raises intense privacy concerns when a central server is used to maintain the data. As these data typically comprise sensitive information, people are concerned that their personal data might be misused.

Online communities provide good sources for studying suicidal ideation [4]. De Choudhury et al. applied a statistical methodology to discover the transition from mental health issues to suicidality [2]. Kumar et al. examined posting activity following celebrity suicides and proposed a method to prevent high-profile suicides [6]. A recent technique called federated learning protects the privacy of data by learning a shared model that uses distributed training on local users without sending the data to a central server [1].

The challenge in developing our method is to detect possible suicidal ideation using an individual's data in online communities while preserving their privacy. Balancing learning and data protection requires a trade-off between the accuracy of prediction and the efficiency of protection. One possible way to tackle this challenge is to collect only the aggregated information of users, rather than accurate user information. Based on this intuition, this paper learns a prediction model by exploiting user information with data protection. Our contributions are summarized as: (1) This paper supports early detection of suicidal ideation so a centralized model can be trained while at the same time protecting the data of users. (2) We develop an advanced optimization strategy, called the average difference descent for learning with data protection (AvgDiffLDP for short). (3) Real-world social media datasets are used to mimic the scenario of private discussion communities, then to demonstrate the effectiveness of our model.

2   Proposed Method

This paper proposes an advanced optimization scheme for the data protection learning framework, called `AvgDiffLDP`. The learning framework learns the global model by aggregating distributed local models without collecting the user data. This procedure consists of four steps: (1) Model initialization. Users download the initial model from the server. In this paper, two popular models for text classification, i.e., CNN [5] and LSTM [3], are used. (2) Local training. Users update the model weights on their local devices. If the user data is stored in a centralized server, we can treat each user's personal data as a sandbox with a data protection interface so that authorized third-party applications can only access the data in accordance with data protection rules. (3) Models upload. Users upload their trained models to the server. (4) Aggregation. The server is optimized to produce a new global model by aggregating the local models.

A simple method of aggregating the model is to calculate the average models from all the users. It can be defined as the finite-sum objective $\min_{\theta^k \in \mathbb{R}^d} \frac{1}{n} \sum_{k=1}^{n} F_k(\theta^k)$, where n is the number of users, k as the index of the user, and t as the time stamp of the communication. In the federated learning [1], the federated averaging algorithm uses a weighted average on local weights to update the server model. The global parameters of the next timestamp are updated in the form of $\theta_{t+1} \leftarrow \sum_{k=1}^{n} \frac{n_k}{m} \theta_{t+1}^k$, where k is the index of users, n_k is the number of samples in the k-th user, and $m = \sum_{k=1}^{n} n_k$. After a certain number of communications between the server and the users, the learning procedure gains ensemble parameters.

Our improved algorithm uses the average differences of the server's parameters in the previous time stamp and the updated users' parameters in the current time stamp as the "gradient". This optimization can help global model better represent all the local users. For each iteration, we adopt a similar sampling way of federated averaging by sampling a fraction of users [1]. The selected users then preform the local training on their own data and devices. After the local users have finished the local training, the parameters on the server side are updated using the average difference descent as $\theta_{t+1} \leftarrow \theta_t - \epsilon \frac{1}{n} \sum_{k=1}^{n} (\theta_t - \theta_{t+1}^k)$, similar to gradient descent. By using this method, the update on the server side minimizes the overall objective to some extent. The central model on the server side can also converge to a model that is close to the optimal model for each user. The secure local training procedure runs on the local device of a single user index, denoted as k. Users receive the model and parameters from the central server and then perform local training on their local devices. We use stochastic gradient descent as the optimizer. We also add the momentum term during optimization on the user side during the user update to avoid problems with a local minimum.

3 Experiments

3.1 Datasets and Baselines

The first dataset referred to as Reddit I was collected from the website Reddit. We chose one subreddit related to suicide, "SuicideWatch", and two other subreddits not related to suicide. A total of 39,600 posts were collected. In addition, we collected 9,052 posts from a selected total of 260 users in the Reddit community, referred to as Reddit II. The third dataset was collected from the social website Twitter. A total of 10,200 tweets were collected. We partitioned the data collected from the various subreddits and users on the Reddit and Twitter sites under independently identical distribution. The Reddit dataset and the Twitter dataset have 99 users and 102 users respectively.

Three baselines are: (1) `SimpleLDP` which trains separate local data-preserving models on different devices for each user without sharing data and parameters; (2) `FullbatchLDP` which uses only a single gradient descent step on each local device and assembles an overall aggregation on the full batch of all users; (3) `AverageLDP` which samples a fraction of users to aggregate the weighted average.

3.2 Results

Accuracy-Privacy Balance. First, we conducted three methods of suicidal ideation detection experiments, i.e., `SimpleLDP`, our proposed `AvgDiffLDP`, and

Table 1. Accuracy-privacy balance and performance comparison

(a) CNN as the classifier

Methods	Reddit I	Reddit II	Twitter
SimpleLDP	58.33	53.91	50.28
AvgDiffLDP	88.39	77.06	92.65
NonLDP	91.76	87.55	94.23

(b) LSTM as the classifier

Methods	Reddit I	Reddit II	Twitter
SimpleLDP	50.03	53.03	49.94
AvgDiffLDP	88.83	80.29	88.59
NonLDP	94.55	88.06	95.30

(c) CNN as the classifier

Methods	Reddit I		Reddit II		Twitter	
	Acc.	AUC	Acc.	AUC	Acc.	AUC
FullbatchLDP	85.98	92.88	74.65	60.00	87.01	74.31
AverageLDP	87.67	93.23	72.93	60.09	87.99	72.44
AvgDiffLDP	**88.39**	**94.38**	**77.06**	**61.53**	**92.65**	**76.00**

(d) LSTM as the classifier

Methods	Reddit I		Reddit II		Twitter	
	Acc.	AUC	Acc.	AUC	Acc.	AUC
FullbatchLDP	87.60	91.75	69.45	59.38	81.42	83.37
AverageLDP	88.54	91.85	68.35	59.96	84.66	70.30
AvgDiffLDP	**88.83**	**92.15**	**80.29**	**60.41**	**88.59**	**87.82**

the centralized NonLDP. The average testing accuracy is shown in Table 1a and b. The performance of LSTM was slightly better than CNN using these three datasets. Our proposed method greatly outperformed SimpleLDP on all three datasets when using CNN and LSTM as the classifier. When compared to NonLDP, the results of the methods with data protection were worse than their counterpart. The centralized NonLDP has an advantage over data protection methods because it trains on the entire dataset, but it also violates user privacy and breaks the data protection setting. Our proposed method achieves a balance between preserving privacy and accurate detection.

Comparison Between LDPs with Model Aggregation. The results of the average testing accuracy and the average of area under the receiver operation curve (AUC) of FullbatchLDP, AverageLDP and our AvgDiffLDP are shown in Table 1c and d. Their hyperparamter settings were all the same. AverageLDP and our AvgDiffLDP selected 10% of users. Our algorithm produced better results than the two baselines in terms of both testing accuracy and AUC on all three datasets.

4   Conclusion

Early detection of suicidal ideation is an important and effective way to prevent suicide. By learning a local data-preserving model for each local user and using a global data-free model in the server, our proposed framework can be applied to effective detection in private communities and protect user privacy without sharing the user data. We have improved the optimization of the global data-free model with a novel average difference descent strategy. Experiments on both I.I.D. and real-world datasets from mainstream social media platforms mimic the private discussion communities and prove the effectiveness of our proposed method.

References

1. McMahan, H.B., et al.: Communication-efficient learning of deep networks from decentralized data. In: Artificial Intelligence and Statistics, pp. 1273–1282 (2017)
2. De Choudhury, M., et al.: Discovering shifts to suicidal ideation from mental health content in social media. In: Proceedings of CHI, pp. 2098–2110. ACM (2016)
3. Hochreiter, S., Schmidhuber, J.: Long short-term memory. Neural Comput. **9**(8), 1735–1780 (1997)
4. Ji, S., et al.: Supervised learning for suicidal ideation detection in onlineuser content. Complexity **2018** (2018)
5. Kim, Y.: Convolutional neural networks for sentence classification. In: Proceedings of EMNLP, pp. 1746–1751 (2014)
6. Kumar, M., et al.: Detecting changes in suicide content manifested in social media following celebrity suicides. In: Proceedings of ACM HT, pp. 85–94. ACM (2015)

Hierarchical Conceptual Labeling

Haiyun Jiang[1], Cengguang Zhang[1], Deqing Yang[1], Yanghua Xiao[1,2(✉)],
Jingping Liu[1], Jindong Chen[1], Chao Wang[1], Chenguang Li[1], Jiaqing Liang[1],
Bin Liang[1], and Wei Wang[1]

[1] School of Computer Science, Fudan University, Shanghai, China
{jianghy16,cgzhang15,yangdeqing,shawyh,liujp17,chenjd17,cwang17,
cgli17,liangbin,weiwang1}@fudan.edu.cn, l.j.q.light@gmail.com
[2] Shanghai Institute of Intelligent Electronics and Systems, Shanghai, China

Abstract. The bag-of-words model is widely used in many AI applications. In this paper, we propose the task of hierarchical conceptual labeling (HCL), which aims to generate a set of conceptual labels with a hierarchy to represent the semantics of a bag of words. To achieve it, we first propose a denoising algorithm to filter out the noise in a bag of words in advance. Then the hierarchical conceptual labels are generated for a clean word bag based on the clustering algorithm of Bayesian rose tree. The experiments demonstrate the high performance of our proposed framework.

1 Introduction

The bag-of-words model is widely used in many natural language processing tasks. There are lots of mature technologies to generate a bag of words (BoW) [4]. However, a BoW is just a collection of scattered words and it is difficult to be understood by machines or human beings without explicit semantic analysis. The conceptualization-based methods, i.e., *conceptual labeling* (CL), aim to generate conceptual labels for a BoW to explicitly represent its semantics. In [3,5,6], a BoW is first divided into multiple groups according to their semantic relevance and then each group is labeled with a concept that can specifically summarize the explicit semantics. We present two examples as follows.

In this paper we propose the task of *hierarchical conceptual labeling* (HCL), which represents the semantics of a BoW by *hierarchical* conceptual labels (i.e., a label set with different granularities). For example, given a BoW {China,Japan,France,Germany,Russia}, the hierarchical conceptual labels can be {*Asian country, EU State*} and {*country*}. In general, the hierarchical labels contain more information, which allows real applications to select labels with different abstractness according to their real requirements.

We consider the hierarchical cluster algorithm: Bayesian rose tree (BRT) [1] as our framework to generate hierarchical conceptual labels, where the candidate concepts are derived from the knowledge base: Microsoft concept graph

This paper was supported by National Natural Science Foundation of China under No. 61732004.

(MCG) [8]. Besides, we also propose a simple but effective method to delete the noise before the conceptualization operation.

2   Framework

We first present how to filter out the noise words in a BoW. Then we elaborate on the generation process of hierarchical conceptual labels.

2.1   Filtering Out Noise

The basic idea is: *if a word in a BoW is hard to be semantically clustered with any other word, i.e., difficult to be tagged with the same conceptual label as any other word, then we take it as noise and remove it from the BoW.*

Specifically, let $\mathcal{D}$ be the input BoW, and d_i (d_j) be the i-th (j-th) instance[1] in $\mathcal{D}$. We take $p(c|d_i, d_j)$ to measure how well the concept c conceptualizes the semantics of two instances d_i, d_j. We use Bayesian rule to compute $p(c|d_i, d_j)$ as follows:

$$p(c|d_i, d_j) = \frac{p(d_i, d_j|c)p(c)}{p(d_i, d_j)} = \frac{p(d_i|c)p(d_j|c)p(c)}{p(d_i)p(d_j)} \tag{1}$$

Then $p(c|d_i, d_j) = \frac{1}{p^2}p(d_i|c)p(d_j|c)p(c)$. The prior probability $p(c)$ measures the popularity of c. Intuitively, a larger $p(c|d_i, d_j)$ indicates c can summarize d_i and d_j well, so d_i and d_j have strong semantic relevance. $p(d_k|c)$ and $p(c)$ are estimated using knowledge in MCG [8]. Let $\mathcal{C}_i$ and $\mathcal{C}_j$ be the concept sets of d_i and d_j in MCG, respectively. $\mathcal{C}_{i,j} = \mathcal{C}_i \cap \mathcal{C}_j$ denotes the *shared concept set* of d_i and d_j. We describe the denoising algorithm as follows.

Consider the word $d_i \in \mathcal{D}$, for any other word $d_j \in \mathcal{D}$ ($d_j \neq d_i$), if we cannot find an appropriate concept in $\mathcal{C}_{i,j}$ to conceptualize d_i and d_j, i.e.,

$$\max_{d_j \in \mathcal{D}, c \in \mathcal{C}_{i,j}} p(c|d_i, d_j) < \delta \tag{2}$$

then d_i is treated as noise. δ is a hyperparameter.

2.2   Hierarchical Conceptual Labeling

Next, we describe how to generate hierarchical conceptual labels for a BoW. The basic idea is: *clustering a BoW $\mathcal{D}$ hierarchically based on BRT [1], and for each cluster $\mathcal{D}_m$ an appropriate conceptual label will be generated.* We present the pseudo code in Algorithm 1.

Estimation of $f(\mathcal{D}_m)$ and $p(D_m|T_m)$. $f(\mathcal{D}_m)$ qualifies the probability that all the words in $\mathcal{D}_m$ belong to the same cluster and it further helps us to estimate $p(D_m|T_m)$. Similar to [7], we consider that $\mathcal{D}_m$ with more shared concepts in MCG is more inclined to belong to the same cluster. For each $c \in \mathcal{C}_m$ (the

---

[1] In this paper, the words in BoWs are also called instances.

Algorithm 1. Hierarchical conceptual labeling based on the Bayesian rose tree.

Input: data $\mathcal{D} = \{d_1, d_2, \cdots, d_N\}$
Output: hierarchical conceptual labels
1: Initialize: *LabelSet*= {}, number of clusters $k = N$, $\mathcal{D}_i = \{d_i\}$, $T_i = \{d_i\}$, $p(\mathcal{D}_i|T_i) = 1$ $(i = 1, \cdots, N)$ and $L(T_m) = \gamma_0$
2: **while** $k > 1$ **and** $L(T_m) > \gamma$ **do**
3:　　Find the pair of trees T_i and T_j and the merge operation that can maximize the likelihood ratio:
$$L(T_m) = \frac{p(\mathcal{D}_m|T_m)}{p(\mathcal{D}_i|T_i)\,p(\mathcal{D}_j|T_j)} \tag{3}$$
4:　　Select the conceptual label c_m^*:
$$c_m^* = \arg\max_{c\in\mathcal{C}_m} p(c|\mathcal{D}_m) \tag{4}$$
5:　　Merge T_i and T_j into T_m by the selected merge operation; $\mathcal{D}_m \leftarrow \mathcal{D}_i \cup \mathcal{D}_j$; Add c_m to *LabelSet*; Delete T_i and T_j, $k \leftarrow k - 1$
6: **end while**

---

shared concepts of $\mathcal{D}_m$), the probability that $\mathcal{D}_m$ belongs to the same cluster is computed as
$$p(\mathcal{D}_m|ct.) = \prod_{d_i\in\mathcal{D}_m} p(d_i|c.) \tag{5}$$
When considering all the concepts in $\mathcal{C}_m$, $f(\mathcal{D}_m)$ is computed by $f(\mathcal{D}_m) = \sum_{c\in\mathcal{C}_m} p(c)\,p(\mathcal{D}_m|c)$. Based on $f(\mathcal{D}_m)$, the probability $p(\mathcal{D}_m|T_m)$ can be recursively calculated by $p(\mathcal{D}_m|T_m) = \pi_m f(\mathcal{D}_m) + (1 - \pi_m)\prod_{T_k\in\text{ch}(T_m)} p(D_k|T_k)$.

Estimation of π_m. π_m is a hyperparameter denoting the prior probability that the leaves under T_m are kept in one cluster rather than subdivided by the recursive partitioning process. We simply set $\pi_m = 0.5$ in this paper.

Label Generation. To generate hierarchical conceptual labels for a BoW, we need to select an appropriate conceptual label to well conceptualize each cluster $\mathcal{D}_m$. The following criterion is used to select the most appropriate conceptual label:
$$c_m^* = \arg\max_{c\in\mathcal{C}_m} p(c|\mathcal{D}_m) = \arg\max_{c\in\mathcal{C}_m} p(\mathcal{D}_m|c)p(c) \tag{6}$$

Likelihood Ratio γ. In most cases, a BoW is hard to be semantically merged into *one* cluster, so the cluster operation should be stopped when there is no appropriate label. We take a likelihood ratio γ, and stop clustering when $L(T_m) < \gamma$.

3　Experiments

We evaluate the generated hierarchical conceptual labels. In all experiments, $\delta = 5 \times 10^{-8}$ and $\gamma = 0.8$ are used.

Dataset. The dataset in [7] is used, which contains two subsets: Flickr and Wikipedia. We sample $b = 500$ BoWs from each dataset for evaluation.

Baselines. To the best of our knowledge, there is no work to deal with the task of HCL, so we present two strong baselines constructed by ourselves. *(1) Bayesian hierarchical clustering-based model (BHC).* We first cluster a BoW using Bayesian hierarchical clustering [2]. Each node in the hierarchy is equipped with a concept to conceptualize the corresponding cluster, where the candidate concepts are also from MCG. *(2) Maximal clique segmentation-based model (MCS).* We first construct a semantic graph for a BoW, where the vertex corresponds to a word. Then we take the maximal clique segmentation [5] to split the graph into several subgraphs given a similarity threshold. Finally, we select one conceptual label for each graph, thus generating a flat conceptual label set for a BoW. Furthermore, when considering multiple similarity thresholds, we will get the multiple label sets with different granularities for a BoW.

Metric. We evaluate the models and consider the two cases: with (without) denoising algorithm. We recruit $v = 5$ volunteers to evaluate the labeling results by scoring ($0 \leq score \leq 3$), where the scoring criteria are motivated by [7]. The average score is computed by $\frac{1}{bv} \sum_{i=1}^{v} \sum_{j=1}^{b} s_{i,j}$, where $s_{i,j}$ is the score of the j-th BoW by volunteer i, b is number of BoWs in each dataset and v is the number of volunteers.

Table 1. Average scores on Flickr and Wikipedia data.

Model	Flickr	Wikipedia	Model	Flickr	Wikipedia
BHC	0.228	0.233	BHC + Denoising	0.247	0.261
MCS	0.240	0.245	MCS + Denoising	0.266	0.271
BRT	**0.251**	**0.264**	BRT + Denoising	**0.273**	**0.282**

Results and Analysis. The results are presented in Table 1. We conclude that (1) the scores with the denoising algorithm are higher than these without it for all models, which proves the effectiveness of the denoising method. (2) The proposed model outperforms the other two baselines. In particular, BHC only considers the binary branching structures in the hierarchy and cannot generate multi-branching structures that frequently appear in the BoW clustering. MCS only clusters BoWs into multi-level label sets without hierarchy.

4   Conclusion

This paper first proposes the task of HCL, which aims to generate conceptual labels with different granularities for BoWs. To achieve it, we propose the BRT-based approach with high performance. Besides, we also propose a denoising algorithm to effectively filter out the noise in advance.

References

1. Blundell, C., Teh, Y.W., Heller, K.A.: Bayesian rose trees. In: UAI (2010)
2. Heller, K.A., Ghahramani, Z.: Bayesian hierarchical clustering. In: ICML 21 (2005)
3. Hua, W., Wang, Z., Wang, H., Zheng, K.: Short text understanding through lexical-semantic analysis. In: IEEE International Conference on Data Engineering, pp. 495–506 (2015)
4. Pay, T.: Totally automated keyword extraction. In: 2016 IEEE International Conference on Big Data (Big Data), pp. 3859–3863 (2016)
5. Song, Y., Wang, H., Wang, H.: Open domain short text conceptualization: a generative + descriptive modeling approach. In: International Conference on Artificial Intelligence, pp. 3820–3826 (2015)
6. Song, Y., Wang, H., Wang, Z., Li, H., Chen, W.: Short text conceptualization using a probabilistic knowledge base. In: IJCAI, pp. 2330–2336 (2011)
7. Sun, X., Xiao, Y., Wangy, H., Wang, W.: On conceptual labeling of a bag of words. In: IJCAI, pp. 1326–1332 (2015)
8. Wu, W., Li, H., Wang, H., Zhu, K.Q.: Probase: a probabilistic taxonomy for text understanding. In: SIGMOD, pp. 481–492 (2012)

Anomaly Detection in Time-Evolving Attributed Networks

Luguo Xue[1], Minnan Luo[1], Zhen Peng[1], Jundong Li[2], Yan Chen[1(✉)], and Jun Liu[1,3]

[1] School of Electronic and Information Engineering,
Xi'an Jiaotong University, Xi'an, China
luguoxuecx@gmail.com, zhenpeng27@outlook.com,
{minnluo,chenyan,liukeen}@xjtu.edu.cn
[2] Computer Science and Engineering, Arizona State University, Tempe, USA
jundongl@asu.edu
[3] National Engineering Lab for Big Data Analytics,
Xi'an Jiaotong University, Xi'an, China

Abstract. Recently, there is a surge of research interests in finding anomalous nodes upon attributed networks. However, a vast majority of existing methods fail to capture the evolution of the networks properly, as they regard them as static. Meanwhile, they treat all the attributes and the instances equally, ignoring the existence of noisy. To tackle these problems, we propose a novel dynamic anomaly detection framework based on residual analysis, namely AMAD. It leverages the small smooth disturbance between time stamps to characterize the evolution of networks for incrementally update. Experiments conducted on several datasets show the superiority of AMAD in detecting anomalies.

1 Introduction

Recently, there is a surge of research focusing on anomaly detection on attributed networks, and the task is to identify the anomalous nodes whose patterns deviate from the other majority nodes in the network [4]. Particularly with the increasing use of advanced sensors and social media platforms, an increasingly amount of time-evolving data regarding attributed networks can be collected in real time. It provides us an additional dimension (*i.e.* temporal information) to analyze the evolving patterns of anomalies in attributed networks.

To this end, we study the novel problem of anomaly detection in attributed networks within a dynamic environment. Nevertheless, the problem is nontrivial to solve due to the following three challenges. First of all, as anomalous patterns may evolve in a dynamic environment, it is necessary to continuously update the previously built model in an online fashion. Secondly, since a small disturbance of network might cause a ripple effect to the derived patterns, methods need to characterize the underlying evolution mechanisms of the networks. Third, the structurally irrelevant attributes can impede us to accurately spot anomalies. Hence, identifying and filtering out these attributes is necessary.

© Springer Nature Switzerland AG 2019
G. Li et al. (Eds.): DASFAA 2019, LNCS 11448, pp. 235–239, 2019.
https://doi.org/10.1007/978-3-030-18590-9_19

In this paper, we propose a novel dynamic framework for anomaly detection based on residual analysis, namely AMAD. Under the assumption of temporal smoothness property [1], AMAD leverages the small evolutionary disturbance to characterize the evolution patterns of networks, and therefore update the previously results incrementally. Meanwhile, the incorporation of feature selection ensures the robustness of AMAD against the noisy features in the data. The main contributions of this work are summarized as: (1) Exploring a principled way to characterize the evolutionary patterns of networks to spot anomalies in an online fashion; (2) Formally propose a novel dynamic anomaly detection framework AMAD based on residual analysis and attribute selection; (3) Conducting experiments on several datasets and the results show the superiority of our method.

2   The Proposed Framework - AMAD

We first define the problem of anomaly detection on time-evolving networks and then elaborate the developed anomaly detection framework AMAD. All the notations are summarized in Table 1.

Definition: *Anomaly detection on time-evolving attributed networks.* *Give a time-evolving attributed network $\mathcal{G}(t) = \{\mathcal{V}_t, \mathcal{A}_t, \mathbf{X}_t\}$ over a series of time stamps $t, t+1, t+2, \cdots, t+m$ $(m = 0, 1, 2, \cdots)$, the task of anomaly detection on time-evolving attributed networks is to find a set of nodes at each time stamp that are rare and differ significantly from the majority reference nodes in the attributed network.*

Table 1. Notation definition.

Notation	Definition	Notation	Definition
$\mathbf{A}_t \in \mathbb{R}^{n \times n}$	Adjacency matrix	$\mathbf{W}_t \in \mathbb{R}^{n \times d}$	Wight matrix
$\mathbf{X}_t \in \mathbb{R}^{d \times n}$	Attribute matrix	$\mathbf{R}_t \in \mathbb{R}^{d \times n}$	Residual matrix
$\mathcal{V}_t$	Node set	$\mathbf{L}_t \in \mathbb{R}^{n \times n}$	Laplacian matrix of $\mathbf{A}_t$
$\mathcal{G}(t)$	Attributed network	$\alpha, \beta, \gamma, \varphi$	Trade-off parameters

Modeling Formulation: From the residual analysis perspective, anomalies often have a large residual value [4] and cannot well be reconstructed from the other instances in the data. According to the problem formulation of the residual analysis based anomaly detection [5], the objective function at time stamp t can be formulated as:

$$\min_{\mathbf{W}_t, \mathbf{R}_t} \mathcal{L}(\mathbf{W}_t, \mathbf{R}_t; \mathbf{X}_t) + \Omega(\mathbf{W}_t, \alpha, \beta) + \Psi(\mathbf{R}_t, \gamma, \varphi), \tag{1}$$

where the loss $\mathcal{L}(\mathbf{W}_t, \mathbf{R}_t; \mathbf{X}_t) = \|\mathbf{X}_t - \mathbf{X}_t \mathbf{W}_t \mathbf{X}_t - \mathbf{R}_t\|_F^2$. The fist regularization term on $\mathbf{W}_t$ is $\Omega(\mathbf{W}_t, \alpha, \beta) = \alpha \|\mathbf{W}_t\|_{2,1} + \beta \|\mathbf{W}_t^\top\|_{2,1}$, which is used to control the sparsity of relevant nodes and attributes. The second regularization term on $\mathbf{R}_t$ is $\Psi(\mathbf{R}_t, \gamma, \varphi) = \gamma\|\mathbf{R}_t^{\top}\|_{2,1} + \varphi tr(\mathbf{R}_t \mathbf{L}_t \mathbf{R}_t^{\top})$, where the first term controls the sparsity of anomalies, while the second term follows the Homophily assumption that two connected nodes will be similar.

To fit the dynamic setting, we follow the temporal smoothness property and assume the optimal variables of optimization problem (1) between t and $t+1$ satisfying: $\mathbf{R}_{t+1} = \mathbf{R}_t + \Delta\mathbf{R}, \mathbf{W}_{t+1} = \mathbf{W}_t + \Delta\mathbf{W}, \mathbf{X}_{t+1} = \mathbf{X}_t + \Delta\mathbf{X}$ and $\mathbf{A}_{t+1} = \mathbf{A}_t + \Delta\mathbf{A}$, where Δ denote the small changes variables. As a result, the objective function at $t+1$ is:

$$\min_{\mathbf{W}_t + \Delta\mathbf{W}, \mathbf{R}_t + \Delta\mathbf{R}} \mathcal{L}(\mathbf{W}_t + \Delta\mathbf{W}, \mathbf{R}_t + \Delta\mathbf{R}; \mathbf{X}_t + \Delta\mathbf{X})$$
$$+ \Omega(\mathbf{W}_t + \Delta\mathbf{W}, \alpha, \beta) + \Psi(\mathbf{R}_t + \Delta\mathbf{R}, \gamma, \varphi). \tag{2}$$

When the optimal $\mathbf{W}_t$ and $\mathbf{R}_t$ have been learned at t, we only need to consider the terms which containing $\Delta\mathbf{W}$ and $\Delta\mathbf{R}$. Finally, according to the triangle inequality of norms, we have:

$$\min_{\Delta\mathbf{W}, \Delta\mathbf{R}} \|\Delta\mathbf{X} - (\mathbf{X}_{t+1}\mathbf{W}_{t+1}\mathbf{X}_{t+1} - \mathbf{X}_t\mathbf{W}_t\mathbf{X}_t) - \Delta\mathbf{R}\|_F^2 + \alpha\|\Delta\mathbf{W}_t\|_{2,1}$$
$$+ \beta\|\Delta\mathbf{W}_t^{\top}\|_{2,1} + \gamma\|\Delta\mathbf{R}^{\top}\|_{2,1} + \varphi tr\left((\mathbf{R}_t + \Delta\mathbf{R})\mathbf{L}_{t+1}(\mathbf{R}_t^{\top} + \Delta\mathbf{R}^{\top})\right). \tag{3}$$

To solve the problem, we employ an alternating optimization algorithm to recursively update the optimal variables. Through fixing one variable and updating another, the optimal variables $\Delta\mathbf{R}, \Delta\mathbf{W}$ can be solved by following equations:

$$\Delta\mathbf{R} = (\Delta\mathbf{X} - \varphi\mathbf{R}_t\Delta\mathbf{L} - \varphi\mathbf{R}_t\mathbf{L}_t - \mathbf{X}_{t+1}\Delta\mathbf{W}_t\mathbf{X}_{t+1}) * (\mathbf{I} + \gamma\mathbf{D}_R + \varphi\mathbf{L}_t + \varphi\Delta\mathbf{L})^{-1},$$
$$\alpha\mathbf{D}_{W1}\Delta\mathbf{W} + \beta\Delta\mathbf{W}\mathbf{D}_{W2} + \sum_{M\in\mathcal{M}}\sum_{N\in\mathcal{N}} M\Delta\mathbf{W}N = \mathbf{H}, \tag{4}$$

where $\mathbf{D}_R(k,k) = \frac{1}{2\|\Delta\mathbf{R}^{\top}(k,:)\|_2}, (k = 1, 2, \cdots, n)$, $\mathbf{D}_{W1}(k,k) = \frac{1}{2\|\Delta\mathbf{W}(k,:)\|_2}$ and $\mathbf{D}_{W2}(k,k) = \frac{1}{2\|\Delta\mathbf{W}^{\top}(k,:)\|_2}$ $(k = 1, 2, \cdots, n)$. The sets $\mathcal{M} = \{\mathbf{X}^{\top}\mathbf{X}, \mathbf{X}^{\top}\Delta\mathbf{X}, \Delta\mathbf{X}^{\top}\mathbf{X}, \Delta\mathbf{X}^{\top}\Delta\mathbf{X}\}$ and $\mathcal{N} = \{\mathbf{X}\mathbf{X}^{\top}, \mathbf{X}\Delta\mathbf{X}^{\top}, \Delta\mathbf{X}\mathbf{X}^{\top}, \Delta\mathbf{X}\Delta\mathbf{X}^{\top}\}$. And $\mathbf{H}$ represents the terms which not contain $\Delta\mathbf{W}$. We employ gradient descent method to solve the second equation of the problem (4).

3   Experiments

We compare AMAD with five anomaly detection methods. LOF [3], Radar [4], and ANOMALOUS [5] are static methods, while MTHL [6] and COMPREX [2] are dynamic methods. The information of datasets are listed in Table 2. And we generate time-evolving networks with anomalies by perturb their nodes.

238   L. Xue et al.

Performance Evaluation: The anomaly detection performance is shown in
Fig. 1 and we adopt AUC value as metrics. We have the following observations
from the figure: (1) our method achieves the best performance in majority of the
time stamps, as we characterize the evolution patterns of networks and find the
most relevant attributes; (2) ANOMALOUS and Radar are slightly inferior to
our method as they ignore the evolutionary information of the underlying network for anomaly detection; (3) MTHL and COMPREX obtain the worst results,
though they fit to dynamic setting. It emphasis the importance of instance and
attribute selection.

Additionally, we compare AMAD with ANOMALOUS to demonstrate its
efficiency. As shown in Table 3, AMAD can be converged faster than ANOMA-
LOUS. It can be ascribed to that AMAD could greatly reduce the amount of
computation by leveraging the sparse evolution matrices ΔX and ΔA.

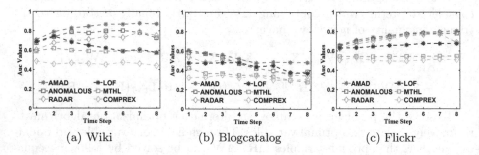

(a) Wiki   (b) Blogcatalog   (c) Flickr

Fig. 1. Time-evolving anomaly detection performance of different approaches

Table 2. Information of datasets

	Nodes	Edges	Attributes
Wiki	2,405	10,976	4,973
Blogcata	4,654	148,372	8,189
Flickr	7,000	203,834	12,047

Table 3. Average running time

	ANOMALOUS	AMAD
Wiki	579.20(s)	**147.30(s)**
Blogcata	885.31(s)	**686.92(s)**
Flickr	5136.81(s)	**2426.04(s)**

4 Conclusions

In this paper, we propose a novel dynamic anomaly detection framework AMAD
and experiments corroborate the effectiveness of AMAD. Additionally, future
work can be focused on detecting group anomaly in a dynamic setting.

Acknowledgements. This work is supported by National Key Research and Devel-
opment Program of China (2016YFB1000903), National Nature Science Foundation of
China (61872287, 61532015 and 61672418), Innovative Research Group of the National
Natural Science Foundation of China (61721002), Innovation Research Team of Min-
istry of Education (IRT_17R86), Project of China Knowledge Center for Engineering
Science and Technology.

References

1. Aggarwal, C., Subbian, K.: Evolutionary network analysis: a survey. ACM Comput. Surv. (CSUR) **47**(1), 10 (2014)
2. Akoglu, L., Tong, H., Vreeken, J., Faloutsos, C.: Fast and reliable anomaly detection in categorical data. In: CIKM (2012)
3. Breunig, M.M., Kriegel, H.P., Ng, R.T., Sander, J.: LOF: identifying density-based local outliers. In: ACM SIGMOD Record, vol. 29, pp. 93–104. ACM (2000)
4. Li, J., Dani, H., Hu, X., Liu, H.: Radar: residual analysis for anomaly detection in attributed networks. In: IJCAI (2017)
5. Peng, Z., Luo, M., Li, J., Liu, H., Zheng, Q.: ANOMALOUS: a joint modeling approach for anomaly detection on attributed networks. In: IJCAI (2018)
6. Teng, X., Lin, Y.R., Wen, X.: Anomaly detection in dynamic networks using multi-view time-series hypersphere learning. In: CIKM (2017)

A Multi-task Learning Framework for Automatic Early Detection of Alzheimer's

Nan Xu[1,2], Yanyan Shen[2], and Yanmin Zhu[1,2](✉)

[1] Shanghai Engineering Research Center of Digital Education Equipment, Shanghai, China
[2] Department of Computer Science and Engineering, Shanghai Jiao Tong University, Shanghai, China
{xunannancy,shenyy,yzhu}@sjtu.edu.cn

Abstract. Alzheimer's disease is a degenerative brain disease which threatens individuals' living and even lives. In this paper, we develop a simple and inexpensive solution to perform early detection of Alzheimer's, based on the individual's background and behavioral data. To alleviate the data sparsity and feature misguidance problems, we propose a novel multi-task learning framework and a pairwise analysis strategy. Extensive experiments show that the proposed framework outperforms the state-of-the-art methods with higher prediction accuracy.

Keywords: Multi-task learning · Neural networks · Early detection

1 Introduction

Alzheimer's disease has become the fifth-leading cause of death. It is well recognized that early detection of Alzheimer's for dementia patients has remarkable benefits. Substantial work on early detection of Alzheimer's adopted clinical strategies or machine learning approaches [3] to analyze multimodal data obtained from extremely complex medical evaluations. However, it is these expensive and time-consuming assessments that delay the diagnoses in reality.

In this paper, we leverage only the individual's demographic and behavioral data to predict his current cognitive status among three categories: cognitively normal (NL), mild cognitive impairment (MCI), and Alzheimer's disease(AD).

There are two key technical challenges in performing automatic Alzheimer's early detection. Firstly, data from the individuals' answers to questionnaires can be insufficient and sometimes misleading. Substantial work has found that the information of the progression and conversion, which is obtained from subsequent examinations, is inevitably critical to verify the correctness of a previous diagnosis [1,2]. Therefore, we construct a multi-task learning neural network (MTN) framework, in which a shared feature representation is learned to improve the generalization performance of multiple tasks. The main *Detection* task shapes its own perceptions about individuals with the general knowledge learned together

© Springer Nature Switzerland AG 2019
G. Li et al. (Eds.): DASFAA 2019, LNCS 11448, pp. 240–243, 2019.
https://doi.org/10.1007/978-3-030-18590-9_20

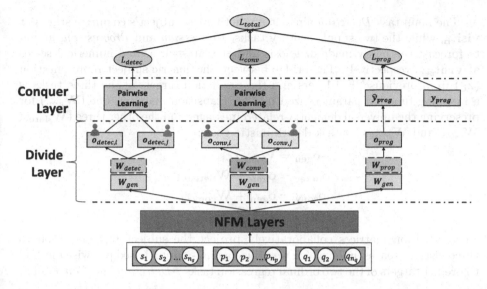

Fig. 1. Our multi-task learning neural network framework MTN.

with two auxiliary tasks: *Conversion* and *Progression*. Secondly, the subjects assigned with different classes (i.e., NL, MCI and AD) may share many symptoms in common and the semantic distances among the three classes are apparently not identical. To better depict the similarities and distinctions of subjects labeled with identical or disparate conditions, we propose to treat *Detection* and *Conversion* as two ordinal regression tasks and introduce an optimizable dual-margin based loss function with two predefined margin parameters as thresholds.

To summarize, this paper has made the following key contributions: (1) We propose a novel and inexpensive multi-task learning framework for early detection of Alzheimer's with only demographic and behavioral data. (2) We treat *Detection* and *Conversion* as ordinal regression tasks, where a dual-margin based loss function is introduced to explore the symptom similarities and discrepancies. (3) Experiments show that our proposed method for early detection of Alzheimer's outperforms several strong baselines in various evaluation metrics.

2   Methodology

Our multi-task neural network is illustrated in Fig. 1. We first extract records of subjects who have paid both the baseline (**visit$_{bl}$**) and the second visit (**visit$_{sec}$**) to the study center. From **visit$_{bl}$**, we select the static information, dynamic numerical data, dynamic categorical data and the current diagnosis accordingly. From **visit$_{sec}$**, we draw out only the dynamic numerical data as target of the *Progression* task and the re-diagnosis data for the *Conversion* task. The concatenated feature vector is quite sparse, hence we employ the Neural Factorization Machine (NFM) layers to embed high-dimensional features into a low-dimensional latent space.

The main task *Detection* aims at predicting the subject's cognitive status at **visit$_{bl}$** while the two supplementary tasks, *Conversion* and *Progression*, intend to forecast the newly-made diagnosis and the latest-measured numerical scales at **visit$_{sec}$**, respectively. The bridge between the generic subject representation ($\mathbf{z_L}$) from previous NFM layers and the prediction targets of the three tasks is a mapping function parameterized by four transition matrices, one ($\mathbf{W_{gen}}$) for preserving the universal dementia-related messages and the other three ($\mathbf{W_{detec}}$, $\mathbf{W_{conv}}$ and $\mathbf{W_{prog}}$) for task-deterministic goals:

$$\mathbf{o_{gen}} = \mathbf{W_{gen}z_L},$$
$$\mathbf{o_{detec}} = \mathbf{o_{gen}} + \mathbf{W_{detec}z_L},$$
$$\mathbf{o_{conv}} = \mathbf{o_{gen}} + \mathbf{W_{conv}z_L},$$
$$\mathbf{o_{prog}} = \mathbf{o_{gen}} + \mathbf{W_{prog}z_L}. \tag{1}$$

The transition matrices collaboratively project the subject representation to three distinct spaces, upon which we utilize a dual-margin based pairwise method to predict targets of the two ordinal regression tasks—*Detection* and *Conversion*:

$$\mathrm{sim}_{detec}(\mathbf{x_i}, \mathbf{x_j}) = \langle \mathbf{o_{detec,i}}, \mathbf{o_{detec,j}} \rangle^2, \tag{2}$$
$$\mathrm{sim}_{conv}(\mathbf{x_i}, \mathbf{x_j}) = \langle \mathbf{o_{conv,i}}, \mathbf{o_{conv,j}} \rangle^2, \tag{3}$$

where $\mathbf{o_{detec,i}}, \mathbf{o_{detec,j}}, \mathbf{o_{conv,i}}, \mathbf{o_{conv,j}}$ are the i-th and j-th subject's representations for task *Detection* and *Conversion*, respectively, $\langle \cdot, \cdot \rangle$ is the dot product function. We predefine two margins λ_i and λ_o , where λ_i is responsible for punishing the too small similarities in pairs of instances coming from the same category, and λ_o controls the penalty for too small discrepancies of pairs containing samples labeled with different types:

$$l_D(c,c) = \frac{1}{N_{D,c}^2} \sum_{\mathbf{x_i},\mathbf{x_j} \in \mathbf{X_{D,c}}} \max\{0, \lambda_i - \mathrm{sim}_{detec}(\mathbf{x_i}, \mathbf{x_j})\}, \tag{4}$$

$$l_D(c,c') = \frac{\frac{1}{N_{D,c}}}{N_{D,c'}} \sum_{\substack{\mathbf{x_i} \in \mathbf{X_{D,c}} \\ \mathbf{x_j} \in \mathbf{X_{D,c'}}}} \max\{0, \mathrm{sim}_{detec}(\mathbf{x_i}, \mathbf{x_j}) - \lambda_o\}, \tag{5}$$

where $X_{D,c}$ and $X_{D,c'}$ are the sample set labeled with class c and class c', $N_{D,c}$, $N_{D,c'}$ are the number of training samples diagnosed with class c and class c' at **visit$_{bl}$**, respectively. For class imbalance consideration, an average loss is adopted here. Similarly, $L_C(c,c)$ and $L_C(c,c')$ are calculated for the intra-class and inter-class loss for *Conversion* task, respectively.

3   Experiments

For performance evaluation, we use the publicly available dataset NACC [1] (19,526 samples in total, 15,526 selected to train). As listed in Table 1, the proposed MTN outperforms all the other methods consistently measured with distinct metrics.

---

[1] https://www.alz.washington.edu/.

Table 1. Model performance. No result for STN on *Progression* as it is specific to ordinal regression. (ACC: accuracy, AUC: Area Under the Receiver Operating Characteristic Curve, RMSE: root mean squared error, MLP^s: Single-task MLP, MLP^m: Multi-task MLP.)

Model	Detection			Conversion			Progression
	ACC	AUC	F1	ACC	AUC	F1	RMSE
RF	0.777 ± 0.02	0.911 ± 0.02	0.697 ± 0.03	0.781 ± 0.03	0.909 ± 0.03	0.663 ± 0.04	10.832 ± 0.81
SVM	0.785 ± 0.02	0.921 ± 0.02	0.685 ± 0.02	0.784 ± 0.02	0.914 ± 0.03	0.633 ± 0.03	10.914 ± 0.80
Bayes	0.768 ± 0.03	0.906 ± 0.02	0.720 ± 0.03	0.761 ± 0.04	0.896 ± 0.03	0.689 ± 0.04	**9.841 ± 0.76**
XGB	0.792 ± 0.02	0.922 ± 0.02	0.721 ± 0.04	0.776 ± 0.03	0.917 ± 0.02	0.649 ± 0.04	10.181 ± 0.71
MLP^s	0.795 ± 0.02	0.931 ± 0.02	0.723 ± 0.04	0.791 ± 0.03	0.927 ± 0.03	0.666 ± 0.03	10.287 ± 0.79
MLP^m	0.797 ± 0.03	0.932 ± 0.02	0.725 ± 0.04	0.793 ± 0.03	0.926 ± 0.03	0.664 ± 0.05	10.622 ± 0.80
STN	0.807 ± 0.01	0.930 ± 0.01	0.723 ± 0.01	0.809 ± 0.01	0.919 ± 0.02	**0.719 ± 0.03**	-
MTN	**0.825 ± 0.01**	**0.936 ± 0.01**	**0.749 ± 0.02**	**0.811 ± 0.01**	**0.934 ± 0.01**	0.663 ± 0.02	11.904 ± 0.21

4   Conclusion

In this paper, we propose an automatic early detection method with a multi-task learning framework for feature learning and a novel dual-margin based loss function to explore symptom similarities. Experimental results demonstrate that the proposed dual-margin based loss function and the joint learning with two relevant tasks are effective to boost prediction performance.

Acknowledgments. This research is supported in part by NSFC (No. 61772341, 61472254) and STSCM (No. 18511103002). This work is also supported by the Program for Changjiang Young Scholars in University of China, the Program for China Top Young Talents, the Program for Shanghai Top Young Talents, and Shanghai Engineering Research Center of Digital Education Equipment.

The NACC database is funded by NIA/NIH Grant U01 AG016976.

References

1. Ferreira, D., et al.: Distinct subtypes of Alzheimer's disease based on patterns of brain atrophy: longitudinal trajectories and clinical applications. Sci. Rep. **7**, 46263 (2017)
2. Risacher, S.L., Saykin, A.J., Wes, J.D., Shen, L., Firpi, H.A., McDonald, B.C.: Baseline MRI predictors of conversion from MCI to probable AD in the ADNI cohort. Curr. Alzheimer Res. **6**(4), 347–361 (2009)
3. Suk, H.I., Lee, S.W., Shen, D., Initiative, A.D.N., et al.: Hierarchical feature representation and multimodal fusion with deep learning for AD/MCI diagnosis. NeuroImage **101**, 569–582 (2014)

Top-k Spatial Keyword Quer
with Typicality and Semantics

Xiangfu Meng[1]([✉]), Xiaoyan Zhang[1], Lin Li[2], Quangui Zhang[1], and Pan Li[1]

[1] Liaoning Technical University, Huludao 125105, China
marxi@126.com
[2] Wuhan University of Technology, Wuhan 430070, China
cathylilin@whut.edu.cn

Abstract. This paper proposes a top-k spatial keyword querying approach which can expeditiously provide top-k typical and semantically
related spatial objects to the given query. The location-semantic relationships between spatial objects are first measured and then the Gaussian
probabilistic density-based estimation method is leveraged to find a few
representative objects from the dataset. Next, the order of remaining
objects in the dataset can be generated corresponding to each representative object according to the location-semantic relationships. The
online processing step computes the spatial proximity and semantic relevancy between query and each representative object, and then the orders
can be used to facilitate top-k selection by using the threshold algorithm. Results of preliminary experiments showed the effectiveness of
our method.

Keywords: Spatial keyword query · Location-semantic relationship ·
Typicality · Top-k selection

1 Introduction

With the rapid development of GPS and universal use of mobile internet, more
and more geo-textual objects are becoming available on the web that represent
Point of Interests (POIs) such as restaurant, hotels, etc. Since spatial database
usually contains a large size of data, too many answer problem often occurs when
a user issues a non-selective spatial keyword query.

To deal with the problem above, several approaches have been proposed to
deal with the issue of spatial keyword query over spatial databases [1,2,4,5,7,8].
According to [8], these approaches can be divided into four categories based on
their scoring functions, that are Boolean Range Queries, Boolean kNN Queries
[2], Top-k Range Queries [6], and Top-k kNN Queries [8]. The last type is the
most popular of spatial keyword queries in the literature, which retrieve the k
spatial objects that have both the high spatial proximity and text relevancy to
the given query as the answer [1,2,4,8]. To quickly retrieve the matching query
results, some hybrid index structures (such as IR-tree [4], quad-tree [7], S2I [1],

© Springer Nature Switzerland AG 2019
G. Li et al. (Eds.): DASFAA 2019, LNCS 11448, pp. 244–248, 2019.
https://doi.org/10.1007/978-3-030-18590-9_21

etc.) are developed to assist the online query processing. It should be pointed out that, however, these queries are mainly confronted with two shortcomings. Firstly, they rarely consider the semantic relevancy between the query keywords and textual descriptions associated to spatial objects. Secondly, the top-k answer objects obtained by existing approaches are usually too similar with each other, which are not benefit for users to recognize the features of whole dataset. Different from the existing query models and indexes, this paper proposes a semantic and typicality query model and TA-based index for top-k result selection.

2   Location-Semantic Relationship Measuring

The location information of a spatial object is usually denoted by a pair of latitude and longitude. We use the *Euclidean* distance to measure the location distance between a pair of spatial objects according to their geo-locations.

The semantic relevancy between a pair of spatial objects can be reflected by their document semantic similarity. The measuring method consists of two steps.

Step 1. Keyword Coupling Relationship Measuring. Given a pair of keyword t_i and t_j, the keyword *intra-correlation* is measured by the frequency of co-occurrence of t_i and t_j appearing in the same documents of the document set associated to the spatial objects. The keyword t_i and t_j are inter-related if there is at least one keyword linked with them. The *inter-correlation* between t_i and t_j via their linked keyword t_c is defined as the minimum value of intra-correlations between $\langle t_i, t_c \rangle$ and $\langle t_j, t_c \rangle$. The linear combination of intra- and inter-correlation between a pair of keywords is called the *keyword coupling relationship*.

Step 2. Document Semantic Similarity Measuring. Based on the keyword coupling relationships, we use a kernel-based cosine similarity method to compute the semantic similarity between a pair of documents. The document d_i associated to each spatial object is first converted into a vector representation $\boldsymbol{d_i}$, and then we can compute the kernel-based cosine similarity based on the matrix of keyword coupling relationships, that is,

$$Sim_{Doc}(d_i, d_j) = cos_{ker}(\boldsymbol{d_i}, \boldsymbol{d_j}) = \frac{k'(d_i, d_j)}{\sqrt{k'(d_i, d_i)}\sqrt{k'(d_j, d_j)}} \qquad (1)$$

where, $k'(d_i, d_j) = \boldsymbol{d_i}(M \cdot M^T)\boldsymbol{d_j}$, M the matrix of coupling relationships between keywords extracted from the compared documents d_i and d_j.

Based on the location similarity and document semantic similarity, the location-semantic relationship for each pair of spatial objects can be then obtained, i.e.,

$$Sim_{LD}(o_i, o_j) = (1 - \lambda)(Sim_{Loc}(o_i, o_j)) + \lambda(Sim_{Doc}(o_i, o_j)) \qquad (2)$$

where, $\lambda \in [0, 1]$ is a weighting parameter.

3   Top-k Typical Relevance Object Selection

The objective of our problem is to find a set of number k objects that are typical and semantically related closely as possible to the given spatial keyword query.

Step 1. Find Representative Spatial Objects. Based on the location - semantic relationships between spatial objects, we provide a density-based typicality estimation algorithm to find the representative objects. Given a set of spatial objects $D = (o_1, o_2, ..., o_n)$, the probability density function $f(o)$ is as follows,

$$f(o) = \frac{1}{n} \sum_{i=1}^{n} G_h(o, o_i) = \frac{1}{n\sqrt{2\pi}} \sum_{i=1}^{n} e^{-\frac{d(o,o_i)^2}{2h^2}} \tag{3}$$

where $d(o, o_i)^2$ is the location-semantic distance between spatial objects o and o_i. According to Eq. (3), the representative objects can be found from the given spatial object dataset. Due to the limited space, we omit the processing procedure and optimal algorithm for probabilistic density estimation.

Step 2. Create Orders for Representative Objects. For each representative object $\bar{o}_i$, create an order τ_i of all remaining objects in D (except $\bar{o}_i$) in descending order, according to their location-semantic relationships to $\bar{o}_i$. Each object o_j has a score that is associated with the position of o_j in each τ_i. The score of o_j in τ_i that corresponds to $\bar{o}_i$ is: $s(o_j|\bar{o}_i) = n - \tau_i(o_j) + 1$.

Step 3. Select Top-k Typical Relevant Object. For a given spatial keyword query q, using the output of step 2, this step computes the set $q_k(D) \subseteq (D)$ with $|q_k(D)| = k$, such that $\forall q_j \in q_k(D)$ and $q_j' \in D - q_k(D)$ it holds that $score(o_j, q) > score(o_j', q)$, with $score(o_j, q) = \sum_{i=1}^{l} (Sim_{LD}(q, \bar{o}_i) \cdot s(o_j|\bar{o}_i))$.

The Threshold Algorithm (TA) is employed to quickly evaluate the top-k objects for a given query [3]. The score of o_j found in each order τ_i to q is computed by: $Sim_{LD}(q, \bar{o}_i) \cdot s(o_j|\bar{o}_i)$. The score of o_j in every other order can be found by using random access mode, and all these scores are summed, resulting in the final score of o_j for q. The termination criterion of TA guarantees that no more retrieving object operations will be needed on any of the orders.

4   Experiments

The experiments are conducted on a computer running Windows 2010 with Intel i5-6300HQ 2.30 GHz CPU, and 8 GB of RAM. All algorithms are developed by Java. We setup a real dataset containing 50,000 POIs extracted from Yelp.

Experiment 1. This experiment aims to test the precision of the top-k answer objects obtained by our TA-based top-k selection algorithm. We generated 10 test spatial keyword queries by extracting the location information and keyword information from the spatial objects in the dataset. Figure 1(a) showed the precision corresponding to different numbers of l (the number of orders for representative objects having the highest location-semantic relationships to the

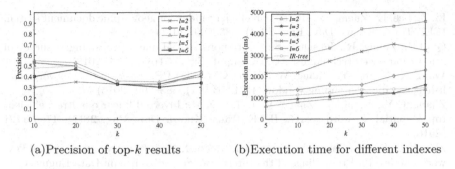

(a)Precision of top-k results   (b)Execution time for different indexes

Fig. 1. The precision and performance of our top-k selection method.

given query) are nearly identical when $l = \{3, 4, 5\}$, and the precision of top-10 results achieves 55% when $l = 4$, which means that when only a small number of orders are used to find the top-10 results, the precision is acceptable.

Experiment 2. This experiment aims to compare the typicality of the top-k answer objects returned by using our method and IR-tree-based top-k selection method, respectively. The measuring criterion is to compute the sum of typicality values of all objects in the answer set by using the Eq. (3). Experimental results showed that the typicality of top-10 results returned by our method steadily outperforms the IR-tree index for the 10 test queries, and the averaged normalized typicality of top-10 results obtained by our method and IR-tree based method are 66% and 46%, respectively.

Experiment 3. This experiment aims to verify the performance of our TA-based index top-k selection compared to IR-tree index. Figure 1(b) showed our method runs faster when $l \leq 3$ and $k \leq 40$ than IR-tree index. Although our method runs a little slower than IR-tree index when $l = 4$, as mentioned above, we can get a high typicality of top-k answers in this situation.

By integrating the experimental results above, we conclude that our method can achieve high typicality with a relative high precision and good performance.

Acknowledgement. This work is supported by the National Natural Science Foundation of China (No. 61772249), and partly by the Natural Science Foundation of Liaoning Province, China (20170540418, 20180550995).

References

1. Cong, G., Jensen, C.S.: Querying geo-textual data: spatial keyword queries and beyond. In: Proceedings of the ACM SIGMOD Conference, pp. 2207–2212 (2016)
2. De Felipe, I., Hristidis, V., Rishe, N.: Keyword search on spatial databases. In: Proceedings of the International Conference on Data Engineering, pp. 656–665 (2008)
3. Fagin, R., Lotem, A., Naor, M.: Optimal aggregation algorithms for middleware. J. Comput. Syst. Sci. **66**(4), 614–656 (2003)

4. Li, Z., Lee, K., Zheng, B.: IR-tree: an efficient index for geographic document search. IEEE Trans. Knowl. Data Eng. **23**(4), 585–599 (2011)
5. Qi, J.H., Zhang, R., Jensen, C.S.: Continuous spatial query processing: a survey of safe region based techniques. ACM Comput. Surv. **51**(3), 1–39 (2018)
6. Wang, X., Zhang, Y., Zhang, W.J., Lin, X.M.: SKYPE: top-k spatial-keyword publish/subscribe over sliding window. PVLDB **9**(7), 588–599 (2016)
7. Zhang, C.Y., Zhang, Y., Zhang, W.J., Lin, X.M.: Inverted linear quadtree: efficient top k spatial keyword search. IEEE Trans. Knowl. Data Eng. **28**(7), 1706–1721 (2016)
8. Zheng, K., Su, H., Zheng, B.L., Liu, J.J., Zhou, X.F.: Interactive top-k spatial keyword queries. In: Proceedings of the International Conference on Data Engineering, pp. 423–434 (2015)

Align Reviews with Topics in Attention Network for Rating Prediction

Yile Liang, Tieyun Qian$^{(\boxtimes)}$, and Huilin Yu

School of Computer Science, Wuhan University, Wuhan, Hubei, China
{liangyile,qty,huilin_yu}@whu.edu.cn

Abstract. Rating prediction has long been a hot research topic in recommendation systems. Latent factor models, in particular, matrix factorization (MF), are the most prevalent techniques for rating prediction. However, MF based methods suffer from the problem of data sparsity and lack of explanation. In this paper, we present a novel model to address these problems by integrating ratings and topic-level review information into a deep neural framework. Our model can capture the varying attentions that a review contributes to a user/item at the topic level. We conduct extensive experiments on three datasets from Amazon. Results demonstrate our proposed method consistently outperforms the state-of-the-art recommendation approaches.

Keywords: Rating prediction · Topic model · Deep learning

1   Introduction

Rating prediction aims to predict the user's ratings on unrated items which may reflect the user's potential interests towards the item. Latent factor models, especially matrix factorization (MF), which are widely used techniques towards this problem, suffer from the severe data sparsity problem when the number of items in the platform becomes extremely large.

In order to address the sparsity issues, researchers have devoted extensive efforts to leverage various types of side information. Among which, the users' reviews attract lots of research interests. A recent trend is applying deep learning techniques to rating prediction [1,2,4–6]. These approaches differ mainly in how they utilize textual information.

In this paper, we present a novel deep neural framework which is *topic-oriented* and *can selectively focus on informative reviews*. To this end, we first apply the simplest LDA topic model to each review to get the review's initial topic distribution. We then design an attention network which can assign reviews' topical importance to users/items by automatically learning the reviews' attention weights. We finally present a neural prediction layer to include both the latent factors from ratings and textual information from reviews.

© Springer Nature Switzerland AG 2019
G. Li et al. (Eds.): DASFAA 2019, LNCS 11448, pp. 249–253, 2019.
https://doi.org/10.1007/978-3-030-18590-9_22

2   Our Proposed Model

The architecture of proposed ARTAN (**A**lign **R**eviews with **T**opics in **A**ttention **N**etwork) model is shown in Fig. 1.

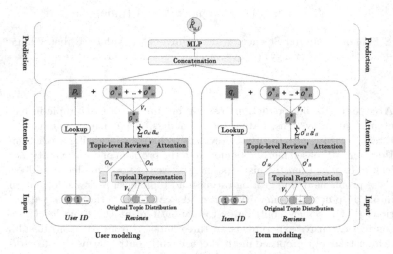

Fig. 1. The architecture of ARTAN

Input Layer. *Initial Latent Factor Representation of User and Item.* We set up a lookup table to transform the one-hot representations of userID and itemID into low-dimensional dense vectors as $\mathbf{p_u} \in \mathbb{R}^K$, $\mathbf{q_i} \in \mathbb{R}^K$.

Topical Representation of Review. We input each review j into the LDA model to obtain its topical distribution θ_j with two hyper-parameters $\alpha = 0.25$ and $\beta = 0.1$. We introduce a matrix $\mathbf{V} \in \mathbb{R}^{T \times K}$ to associate the embeddings to different topics, where T is the number of topics and K is the dimensionality. The topical representation of review j of user u can be calculated by aggregating the weighted embeddings over topics:

$$\mathbf{o_{u,j}} = \sum_{t=1}^{T} \theta_{j,t} \cdot \mathbf{v_t} \tag{1}$$

Attention Layer. The input of the attention network consists of the topical representation of user u's j-th review ($\mathbf{o_{u,j}}$) and a randomly initialized vector ($\mathbf{a'_u}$). The user u's attention weight on a review j is defined as:

$$a^*_{u,j} = \mathbf{h^T} \cdot ReLU(\mathbf{W_o}\mathbf{o_{u,j}} + \mathbf{W_u}\mathbf{a'_u} + \mathbf{b}), \tag{2}$$

where $\mathbf{h} \in \mathbb{R}^K$, $\mathbf{W_o} \in \mathbb{R}^{K \times K}$, $\mathbf{W_u} \in \mathbb{R}^{K \times K}$ and $\mathbf{b} \in \mathbb{R}^K$, are parameters that project the input into the hidden layer. The final attention $a_{u,j}$ is obtained by applying the softmax function to the original $a_{u,j}^*$ as normalization. After that, the feature vector of user u is calculated as the weighted sum of all reviews:

$$a_{u,j} = \frac{exp(a_{u,j}^*)}{\sum_{j=1}^{l} exp(a_{u,j}^*)}, \qquad \mathbf{o_u^*} = \sum_{j=1}^{l} a_{u,j} \cdot \mathbf{o_{u,j}} \tag{3}$$

To further strengthen the impacts of topics, we let the above user feature vector $\mathbf{o_u^*}$ interact with each topic embedding $\mathbf{v_t}$. In this way, the topical representation of user u can be highlighted and the user's interests would be associated with each topic once again.

$$\mathbf{o_u} = \sum_{t=1}^{T} \mathbf{o_u^*} \odot \mathbf{v_t}, \tag{4}$$

where $\odot$ denotes the element-wise product.

Prediction Layer. Given a user u, an item i and their reviews, we now take the latent factor representations on ratings and topical representations on reviews to model the interaction between user u and item i as follows.

$$\mathbf{x^*} = (\mathbf{p_u} + \mathbf{o_u}) \oplus (\mathbf{q_i} + \mathbf{o_i}), \tag{5}$$

where $\oplus$ denotes the concatenation operation. We present a multi-layer perceptron (MLP) based structure on $\mathbf{x^*}$ for modeling the complicated interactions and getting the real-valued rating $\hat{R}_{u,i}$.

Network Training. Since our target is rating prediction, we treat it as a regression task and adopt the square loss as the objective function:

$$\mathcal{L} = \sum_{u,i \in \mathcal{T}} (\hat{R}_{u,i} - R_{u,i})^2, \tag{6}$$

where $\mathcal{T}$ denotes the set of instances for training, i.e., $\mathcal{T} = \{(u, i, r_{u,i}, d_{u,i})\}$, $R_{u,i}$ is ground truth rating in training set $\mathcal{T}$, and $\hat{R}_{u,i}$ is the predicted rating.

3   Experiments

Datasets and Experimental Settings. We use three publicly available datasets from Amazon[1] to evaluate our model, including Patio Lawn and Garden, Automotive, and Grocery and Gourmet Food. We take the 5-core version for experiments [1,2,6]. We randomly split the datasets into training/validation/test sets with a 80/10/10 split and use MSE as the evaluation metric. The statistics of these datasets are shown in Table 1.

---

[1] http://jmcauley.ucsd.edu/data/amazon/links.html.

Table 1. Statistics of the evaluation datasets

Datasets	#users	#items	#ratings	#sparsity
Garden	1686	962	13272	0.9918
Automotive	2928	1834	20473	0.9962
Grocery	14679	8711	151254	0.9988

We choose six state-of-the-art methods as our baselines, including *SentiRec* [4], *TARMF* [5], *MPCN* [6], *NARRE* [1], A^3NCF [2] and *ALFM* [3]. For the baselines, we use the same hyper-parameter settings if they are reported by authors. For our ARTAN model, we set the number of topics $T = 25$, the number of latent factors K is 25, and the learning rate is 0.001.

Performance Evaluation. The performance of ARTAN and the baselines on all datasets are reported in Table 2. From the results, we have the following important observations.

Firstly, our ARTAN model significantly outperforms the state-of-the-art baselines on all the datasets.

Secondly, among four word-based methods (SentiRec, TARMF, MPCN, NARRE), NARRE is the best in consideration of the different weights of reviews. While the word-level textual features are discentralized, its performance is inferior to our model. Not utilizing users' and items' textual representations at testing stage in TARMF, and only selecting the most representative reviews in MPCN lead to the poor result. The poor performance of SentiRec can be due to it not differentiating reviews at all.

Thirdly, two topic-based methods (A^3NCF, ALFM) are not as good as our model since they neglect the difference among various reviews when building topic model. In contrast, our ARTAN is not only topic oriented, but also assigns

Table 2. Performance comparison on four datasets. The best performance among all is in bold while the best one among baselines is marked with an underline. * denotes the index of the baselines.

Method	Index	Results (MSE)			Improvements (%) – (7) vs. *		
		Garden	Automotive	Grocery	Garden	Automotive	Grocery
SentiRec	(1)	1.0671	0.8245	1.0140	9.79	7.50	2.56
TARMF	(2)	1.1025	0.8686	1.0733	12.69	12.19	7.95
MPCN	(3)	1.1664	0.8154	1.0941	17.47	6.46	9.70
NARRE	(4)	0.9904	0.7815	<u>0.9972</u>	2.81	2.41	0.92
A^3NCF	(5)	1.0349	0.8228	1.0199	6.99	7.30	3.13
ALFM	(6)	<u>0.9835</u>	<u>0.7718</u>	1.0018	2.13	1.18	1.38
ARTAN	(7)	**0.9626**	**0.7627**	**0.9880**	–	–	–

varying weights for different reviews with attention mechanism, and thus achieves the best performance.

4 Conclusion

In this paper, we propose a novel model for rating prediction. Our model adopts the deep neural attention network as architecture to extract representations from reviews in addition to latent factors from ratings. Different from previous topic level studies in utilizing reviews, our model can automatically identify the important reviews for each user/item with the attention mechanism. We conduct extensive experiments on three real world datasets. Results demonstrate that our model significantly outperforms the state-of-the-art baselines.

Acknowledgment. The work described in this paper has been supported in part by the NSFC project (61572376).

References

1. Chen, C., Zhang, M., Liu, Y., Ma, S.: Neural attentional rating regression with review-level explanations. In: WWW, pp. 1583–1592 (2018)
2. Cheng, Z., Ding, Y., He, X., Zhu, L., Song, X., Kankanhalli, M.: A^{3}NCF: an adaptive aspect attention model for rating prediction. In: IJCAI, pp. 3748–3754 (2018)
3. Cheng, Z., Ding, Y., Zhu, L., Kankanhalli, M.: Aspect-aware latent factor model: rating prediction with ratings and reviews. In: WWW, pp. 639–648 (2018)
4. Hyun, D., Park, C., Yang, M.C., Song, I., Lee, J.T., Yu, H.: Review sentiment-guided scalable deep recommender system. In: SIGIR, pp. 965–968 (2018)
5. Lu, Y., Dong, R., Smyth, B.: Coevolutionary recommendation model: mutual learning between ratings and reviews. In: WWW, pp. 773–782 (2018)
6. Tay, Y., Luu, A.T., Hui, S.C.: Multi-pointer co-attention networks for recommendation. In: KDD, pp. 2309–2318 (2018)

PSMSP: A Parallelized Sampling-Based Approach for Mining Top-k Sequential Patterns in Database Graphs

Mingtao Lei[1], Xi Zhang[1(✉)], Jincui Yang[1], and Binxing Fang[1,2]

[1] Key Laboratory of Trustworthy Distributed Computing and Service (BUPT), Ministry of Education, Beijing University of Posts and Telecommunications, Beijing, China
{leimingtao,zhangx,jincuiyang,fangbx}@bupt.edu.cn
[2] Institute of Electronic and Information Engineering of UESTC in Guangdong, Dongguan, China

Abstract. We study to improve the efficiency of finding top-k sequential patterns in database graphs, where each edge (or vertex) is associated with multiple transactions and a transaction consists of a set of items. This task is to discover the subsequences of transaction sequences that frequently appear in many paths. We propose **PSMSP**, a Parallelized Sampling-based Approach For Mining Top-k Sequential Patterns, which involves: (a) a parallelized unbiased sequence sampling approach, and (b) a novel PSP-Tree structure to efficiently mine the patterns based on the anti-monotonicity properties. We validate our approach via extensive experiments with real-world datasets.

Keywords: Database graph · Sequential pattern mining · Parallelized sampling

1 Introduction

The problem of sequential pattern mining, which discovers frequent subsequences from a sequence database, has been extensively studied in [3–6]. Motivated by the rich knowledge in the database graph, a related but different problem is proposed [2], which is to perform sequential pattern mining in database graphs, aiming to discover the subsequences that frequently appear in many paths of the graph. It is natural to ask whether we can extend the existing sequential pattern mining methods for this new task. Unfortunately, there is no straightforward way as most of existing methods don't account for a database graph as input. Moreover, the number of transaction sequences induced by all possible

This work has been supported in part by the National Key Research and Development Program of China (No. 2017YFB0803301), the Natural Science Foundation of China (No. U1836215), and DongGuan Innovative Research Team Program (No. 201636000100038).

G. Li et al. (Eds.): DASFAA 2019, LNCS 11448, pp. 254–258, 2019.
https://doi.org/10.1007/978-3-030-18590-9_23

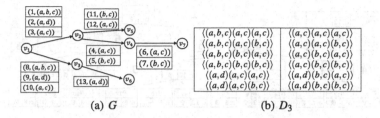

(a) *G*   (b) *D₃*

Fig. 1. An example of a graph G, where each edge is associated with a transaction database. D_3 represents the set of length-3 transaction sequences of the graph G. The first sequence of D_3 is the sequence of transactions with $tid = 1, 4, 6$.

random walks of a database graph is extremely huge, demanding more efficient approaches. Despite the fact that a recent study [2] has proposed to approximate the sequential patterns in a database graph, it still falls short in efficiency.

To fill this gap, we propose to improve the efficiency of finding top-k sequential patterns in database graphs through two-stage optimizations. First, we develop a parallelized sampling method as well as a weighting mechanism for the sampled transaction sequences. Second, we develop a novel Tree of Trees (ToT) structure, PSP-Tree, to mine the approximate top-k patterns by exploring the anti-monotonicity properties.

2   Problem Definition

Let α be an item, I be a set of *items* and an *itemset* X be a subset of I. We assume that items follow an arbitrary fixed order in each itemset. A *transaction* is a tuple $\tau = (tid, X)$, where tid is a unique transaction-id and X is the corresponding itemset.

A *database graph* $G = (V, E, \mathbb{T})$ is directed, where V is a finite set of vertices, $E \subseteq V \times V$ is a set of directed edges and $\mathbb{T} = \{T_e | e \in E\}$ is a set of transaction databases that each transaction database T_e is associated with the edge e.

A *(directed) path* p in G is a sequence of edges $p = \langle e_1, \ldots, e_h \rangle$, where e_i and e_{i+1} ($1 \leq i < h$) are incident, and the *length* of the path is $len(p) = h$. A *transaction sequence* supported by p, denoted by $ts = \langle \tau_1, \ldots, \tau_h \rangle$, is a sequence of transactions such that there exists a path $p = \langle e_1, \ldots, e_h \rangle$ and $\tau_i \in T_{e_i}$ for $1 \leq i \leq h$. The length of ts is $len(ts) = h$. A *sequential pattern* $s = \langle X_1, \ldots, X_h \rangle$ is a series of itemsets, and it is said to be contained by a transaction sequence $ts = \langle \tau_1, \ldots, \tau_h \rangle$ if $X_i \subseteq \tau_i$ for $1 \leq i \leq h$; i.e., X_i is an itemset of the transaction τ_i, and it is also called the i-th itemset of ts.

Denote D_l as the set of all length-l transaction sequences supported by paths in G. The *support* and *frequency* of a pattern s in D_l, denoted by $sup(s)$ and $f(s)$ respectively, are the number and the proportion of unique transaction sequences in D_l that contain s.

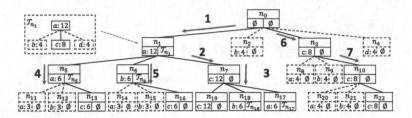

Fig. 2. The PSP-Tree for top-6 length-3 frequent patterns in Fig. 1. $\mathcal{T}_{n_1}$ is a FP-Tree of the node n_1. The numbers in red indicate the mining order and the nodes with dashed lines are pruned during the mining process. (Color figure online)

In this paper, given a database graph G, an integer $k > 0$ and a fixed length $l > 0$, the problem of finding the top-k sequential patterns is to find the top-k patterns of length l that have the largest frequencies in D_l of G.

For example, given G in Fig. 1, the top-6 length-3 patterns when $k = 6$ and $l = 3$ are $\langle(a)(c)(c)\rangle$, $\langle(a,c)(c)(c)\rangle$, $\langle(a)(a)(c)\rangle$, $\langle(a)(b)(c)\rangle$, $\langle(a)(c)(b)\rangle$, and $\langle(a)(c)(a)\rangle$.

3 The Method

In this section, we introduce our PSMSP method, consisting of two stages, the parallelized sampling stage and the PSP-Tree mining stage.

The Parallelized Sampling Stage. We assume that the database graph G has been partitioned into m disjoint groups $\{E_1, \ldots, E_m\}$. Given a partition $\mathcal{G}_i(V_i, E_i)$, which is induced by E_i, and the path length l, denote E^+ as the set of additional edges beyond $\mathcal{G}_i$ that can be reached from an edge $e \in E_i$ through a path whose length is at most l. P_l^+ is the set of all the length-l paths with the starting edges in E_i and the end edges in E^+. $\mathcal{G}_i^+(V_{\mathcal{G}_i^+}, E_{\mathcal{G}_i^+})$ is defined as the l-enhanced partition of $\mathcal{G}_i$, where $V_{\mathcal{G}_i^+} = V_i \cup \{v | v \in p' \wedge p' \in P_l^+\}$ and $E_{\mathcal{G}_i^+} = E_i \cup \{e | e \in p' \wedge p' \in P_l^+\}$. For p and ts supported by p, we can define their weights as $\varphi(p) = |P_l^i|$ and $\varphi(ts) = \varphi(p)Z$, where $Z = \prod_{q=1}^{l} |T_{e_q}|$ is the number of all transaction sequences supported by p.

The parallelized sampling process is shown as follows. Given the enhanced partition set $\mathcal{G}^+ = \{\mathcal{G}_1^+, \ldots, \mathcal{G}_m^+\}$, we draw μ samples from each partition and the total sample size is thus $m\mu$. For each partition $\mathcal{G}_i^+$, in its h-th iteration, we first sample a path p_h uniformly and calculate its weight $\varphi(p_h)$. We then sample a transaction sequence ts_h from p_h and calculate its weight $\varphi(ts_h)$. ts_h and $\varphi(ts_h)$ are then added into the set S_l. At last, we output S_l, which consists of all the sampled transaction sequences and the corresponding weights.

The PSP-Tree Mining Stage. To mine the top-k sequential patterns efficiently, we propose a novel Tree of Trees (ToT) model, PSP-Tree, denoted by $\mathcal{T}$,

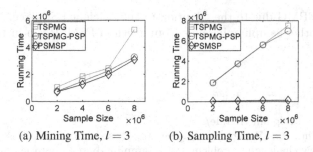

(a) Mining Time, $l = 3$ (b) Sampling Time, $l = 3$

Fig. 3. The mining time and sampling time (in milliseconds). The partition number is 20.

where each node of the main (pattern) tree itself may contain another (itemset-specific) tree. The i-th node of $\mathcal{T}$ is denoted by n_i, which contains an item β_i with its occurrence, and a FP-Tree [4] $\mathcal{T}_{n_i}$. $\mathcal{T}_{n_i}$ is used to represent the itemsets containing β_i with their occurrences, and its root is β_i. The *occurrence* of the q-th itemset X_q is the frequency of transaction sequences containing X_q. With the pre-computed occurrence of β_i, we can determine whether to build n_i in the main tree by comparing this occurrence to the k-th largest frequency of patterns. For clarity, $\mathcal{T}$ is called a main tree and $\mathcal{T}_{n_i}$ is called a itemset-specific tree. When we generate the PSP-Tree, we first generate all qualified nodes for the main tree and then build corresponding itemset-specific trees. This enables us prune unqualified patterns earlier.

The itemsets of each node n_i can be easily extracted from $\mathcal{T}_{n_i}$. For the example in Fig. 2, the main tree has 12 nodes shown in the solid lines. n_1 contains an item a and $\mathcal{T}_{n_1}$, and the occurrence of a is 12. The itemsets that are derived from n_1 are $\{(a), (a, c)\}$. n_5 is a child of n_1 and n_{13} is a child of n_5. The patterns can be obtained by traversing the PSP-Tree in a top-down manner. For example, traversing n_1, n_5 and n_{13} sequentially can obtain the patterns $\langle (a)(a)(c) \rangle$ and $\langle (a, c)(a)(c) \rangle$.

However, the number of remaining patterns in the main tree may be still larger than k. Thus, for all the remaining patterns in the tree, we compute their supports and select top-k patterns that have the largest supports.

4   Experiments

We evaluate the performance of **PSMSP** (using parallelized sampling and PSP-Tree), **TSPMG** [2] (using serial sampling and PrefixSpan) and **TSPMG-PSP** (using serial sampling and PSP-Tree) on the collaboration network [1]. We set $l = 3$ and $k = 100$.

Figure 3(a) shows TSPMG-PSP and PSMSP have similar mining time, as they all adopt the PSP-Tree in the mining stage. In addition, they all run much faster than TSPMG as TSPMG uses the slow PrefixSpan rather than PSP-Tree.

Figure 3(b) shows the sampling time cost increases when the sample size increases due to the larger time cost in sampling processing. TSPMG-PSP is faster than TSPMG due to the adoption of the efficient PSP-Tree. In summary, compared to other sampling baselines, our method PSMSP can further improve the efficiency.

References

1. DBLP. http://dblp.uni-trier.de/
2. Lei, M., Chu, L., Wang, Z.: Mining top-k sequential patterns in database graphs: a new challenging problem and a sampling-based approach. arXiv preprint arXiv:1805.03320 (2018)
3. Li, H., Yi, W., Dong, Z., Ming, Z., Edward, Y.C.: PFP: parallel FP-growth for query recommendation. In: RecSys, pp. 107–114 (2008)
4. Pei, J., et al.: PrefixSpan: mining sequential patterns by prefix-projected growth. In: ICDE, pp. 215–224 (2001)
5. Riondato, M., Upfal, E.: Efficient discovery of association rules and frequent itemsets through sampling with tight performance guarantees. In: ECML PKDD, pp. 25–41 (2012)
6. Riondato, M., Upfal, E.: Mining frequent itemsets through progressive sampling with rademacher averages. In: KDD, pp. 1005–1014 (2015)

Value-Oriented Ranking of Online Reviews Based on Reviewer-Influenced Graph

Yiming Cao, Lizhen Cui$^{(\boxtimes)}$, and Wei He

Shandong University, Jinan, China
sdu_cym@163.com, {clz,hewei}@sdu.edu.cn

Abstract. To mitigate the uncertainty of online purchases, people rely on reviews written by customers who already bought the product to make their decisions. The key challenge in this situation is how to identify the most helpful reviews among a large number of candidate reviews with different quality. Existing work normally employs diversified text and sentiment analysis algorithms to analyze the helpfulness of reviews. Voting on reviews is another popular valuation way adopted by many websites, which also has difficulties to reflect the real helpfulness of the reviews due to the problem of data sparseness. In this paper, a reviewer-influenced graph model is constructed based on the reviewers' historical reviews and voting information to measure the influence of reviewers' quality on the helpfulness of reviews. Experimental results with actual review data from Amazon.com demonstrate the effectiveness of our approach.

Keywords: Reviewer-influenced graph model · Reviewer quality · Reviewer-influenced helpfulness

1 Introduction

Reviews of the items from previous customers are valuable for potential consumers to make the purchase decision. However, facing massive reviews, consumers often need to browse most or even all of the reviews before they can find information with practical reference value. Therefore, it's a huge challenge for consumers to find highly helpful reviews from a large number of online reviews. Highly helpful reviews usually refers to the reviews that truly reflect the experience of the users and have reference value for other users to buy the items.

In the e-commerce platform like Amazon.com, a reviewer always plays two roles: one is the author of reviews, that is, the reviewer can publish reviews on the items purchased. The other is the voter, that is, the reviewer can vote for reviews published by other reviewers. Companies such as Google, Microsoft and Amazon and some prior work like [2] and [4] have implemented complicated algorithms to measure user expertise. However, the works focus on finding user

© Springer Nature Switzerland AG 2019
G. Li et al. (Eds.): DASFAA 2019, LNCS 11448, pp. 259–263, 2019.
https://doi.org/10.1007/978-3-030-18590-9_24

preferences to recommend products for users. As the roles of reviewers mentioned above, *reviewer quality* includes the quality of reviews published by the reviewer (defined as *author quality*), and the quality of other reviews that the reviewer voted as helpful (defined as *voter quality*). High quality authors often write highly helpful reviews and high quality voters tend to vote for highly helpful reviews. At the same time, *the helpfulness of the review* also affects the reviewer quality. In this paper, the helpfulness of reviews is measured by reviewer-influenced helpfulness. The reviewer-influenced graph model is constructed for reviewer-influenced helpfulness, which is used to demonstrate the reviewer's influence on the helpfulness of reviews. We have two nodes in the model: reviewers and reviews. We measure the influence of reviewers on the helpfulness of reviews by introducing three concepts, namely, author quality, voter quality and reviewer-influenced helpfulness. The basic principle is that if a reviewer quality is high and then the reviewer-influenced helpfulness of the reviews is high.

Our contributions are summarized as follows:

1. The reviewer-influenced graph model is constructed to measure the influence of reviewer quality on the helpfulness of review.
2. We compare the experimental results with the top rated method of Amazon.com and the result of crowd using the data of mobile power review, books review and cloth review in Amazon.com to illustrate the accuracy of our method.

The rest of the paper is organized as follows. Reviewer-influenced graph mode is proposed to quantify the reviewer's influence on the helpful of reviews in Sect. 2. We carry out experiments to validate our method in Sect. 3.

2   Reviewer-Influenced Graph Model

We consider that reviewer quality includes the author quality and the voter quality. Based on the relationship between reviewers and reviews, we believe that the helpfulness of reviews will be affected by the reviewer quality and put forward the reviewer-influenced graph model.

The reviewer-influenced graph model is described in the form of graph $G<V, E>$, where $V = V_U \bigcup V_R$ is composed of reviewer vertex sets V_U and review vertex sets V_R. For each two tuple (u, r) of the data set, there is a set of corresponding edges $e(v_u, v_r)$ in the graph $G<V, E>$. The reviewers nodes can be divided into authors and voters. Each review node establishes an edge between with its author node or its voter node, which demonstrates the review came from which author or which voters voted for it. As shown in Fig. 1, the degree of the review node is the sum of the number of author and voters. The degree of the author node is the number of historical reviews published by the author. The degree of the voter node is the number of historical voted reviews of voters. For each review, its helpfulness is quantified by the reviewer quality. The reviewer quality is calculated through the helpfulness of the review to be evaluated and helpfulness of the reviewer's historical reviews or voted reviews.

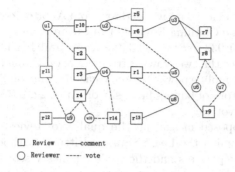

Fig. 1. Reviewer-influenced graph model between reviewers and reviews

The reviewer-influenced graph model $G{<}V, E{>}$ can be formalized into a special adjacency matrix $A_{[i,j]}$. The rows of $A_{[i,j]}$ represent reviews, the columns of $A_{[i,j]}$ represent reviewers. The adjacency matrix $A_{[i,j]}$ of $G{<}V, E{>}$ is defined as the following form.

$$A_{[i,j]} = \begin{cases} 0, \text{if } (u_i, r_j) \notin E(G) \\ 1, \text{if } u_i \text{ is the author of } r_j \\ 2, \text{if } u_i \text{ is the voter of } r_j \end{cases}$$

To quantify the helpfulness of review, author quality and voter quality, we formalize the three conceptions to the as follows:

$$Q(u) = \frac{1}{1 + e^{-(H(r_u) + H(r))}}, \tag{1}$$

where $Q(u)$ denote the reviewer quality. $H(r_u)$ is the average of helpfulness of the historical reviews that the reviewers wrote or voted, $H(r)$ refers to helpfulness of the reviews to be evaluated. To calculate the reviewer quality, we need to first obtain the helpfulness of the review:

$$H(u) = \frac{2}{1 + e^{-((Q(u_w) + \sum_{i=0}^{n} Q(u_v)/n) - 1)}} - 1 \tag{2}$$

To calculate the helpfulness of review, we need to first know the reviewer quality. Therefore, we use the formula mentioned above to conduct iterative computation. When the calculation converges, we can obtain the final quantitative results of three concepts.

3   Experiment

In this section, we experimentally find out helpful reviews using and analyze the experimental results by comparing with other method of review ranking.

The experimental review data we used are from Amazon.com and Amazon.cn. The first data set is the Clothing Reviews Data from Amazon Review of SNAP[1] [1], which contains 581,933 reviews, 39,349 authors and 48,243 voters from Amazon. The other data set that we crawled from Amazon China are 36,407 reviews, including the "PISEN mobile power" from Amazon China on September 26, 2017 and the "PA brief history of human beings"[2] on April 22, 2018, which come from 597 authors and 449 voters.

Paper [3] has proposed measuring the quality of reviews using the review honesty score. We applied the Least Squares Fitting Method to the results of reviewer-influenced helpfulness and the review honesty score, which are shown in Fig. 2(a). From the Fig. 2(a), we can see that the trend of two lines is basically the same in mobile power reviews data and book reviews data. We sorted the reviews in descending order by the values of helpfulness of reviews as the ranking of our method. The "top rated" is the default ranking method of Amazon, which has a strong reference value. We also asked human to sort the helpfulness of each review as the ranking of crowd. 25 graduate students major in computer science were invited as our artificial analysts who have a lot of online shopping experience. The results are as shown in Fig. 2. From the Fig. 2, we can see that the ranking of our method is more close to the ranking of crowd.

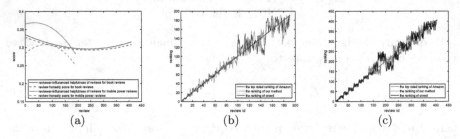

(a)   (b)   (c)

Fig. 2. (a) The scores of two method (b) Mobile power reviews' ranking (c) Books reviews' ranking

4   Conclusion

Analyzing online helpfulness of reviews is an important direction of online helpfulness of reviews research, which helps consumers find helpful review in a short time. In our opinion, helpfulness of reviews can be measured by the reviewer quality. Therefore, influence graph model is used between reviewers and reviews to demonstrate the relationship between reviewer and helpfulness of reviews. The experimental results show that the method in this paper can help consumers to identify helpful reviews so that consumers can understand the information of the product objectively.

---

[1] http://snap.stanford.edu/data/web-Amazon.html.
[2] https://www.amazon.cn.

Acknowledgements. The research work was supported by the NSFC (91846205, 61572295), National Key R&D Program (2017YFB1400102, 2016YFB1000602), SDNSFC (No. ZR2017ZB0420, No. ZR2018MF014, No. ZR2017MF065), and Shandong Major scientific and technological innovation projects (2018YFJH0506).

References

1. McAuley, J., Leskovec, J.: Hidden factors and hidden topics: understanding rating dimensions with review text. In: Proceedings of the 7th ACM Conference on Recommender Systems, RecSys 2013, pp. 165–172. ACM (2013)
2. Mcauley, J.J., Leskovec, J.: From amateurs to connoisseurs: modeling the evolution of user expertise through online reviews. In: International Conference on World Wide Web, pp. 897–908 (2013)
3. Wang, G., Xie, S., Liu, B., Yu, P.S.: Identify online store review spammers via social review graph. ACM Trans. Intell. Syst. Technol. **3**(4), 61 (2012)
4. Yang, X., Steck, H., Liu, Y.: Circle-based recommendation in online social networks. In: ACM SIGKDD International Conference on Knowledge Discovery and Data Mining, pp. 1267–1275 (2012)

Ancient Chinese Landscape Painting Composition Classification by Using Semantic Variational Autoencoder

Bo Yao[✉], Qianzheng Ji, Xiangdong Zhou, Yue Pang, Manliang Cao, Yixuan Wu, and Zijing Tan

School of Computer Science, Fudan University, Shanghai, China
{16110240008,qzji17,xdzhou,ypang15,mlcao17,zjtan}@fudan.edu.cn

Abstract. In the theory of art, composition is based on the placement or arrangement of visual elements or ingredients in a painting to express the thoughts of the artist. Inspired by that, we propose a novel approach called Semantic Variational Autoencoder (SemanticVAE) to deal with the problem of ancient Chinese landscape painting composition classification. Extensive experiments are conducted on a real ancient Chinese landscape painting image dataset collected from museums. The experimental results show that, in contrast to the state-of-the-art deep CNNs, our method significantly improves the performance of ancient Chinese landscape painting composition classification.

Keywords: Ancient Chinese landscape painting composition classification · Semantic feature extraction · Semantic Variational Autoencoder

1 Introduction

With the fast development and breakthrough of deep learning on image and scene classification and understanding, the mission of artistic resource classification seems no longer just for human experts. However, the composition classification of ancient Chinese landscape paintings is a complicated and subtle issue, which is still regarded as a problem suitable for professional experts to deal with. To the best of our knowledge, this is the first attempt to automatically classify ancient Chinese landscape paintings (precisely the "Shan Shui" paintings) by their composition type. "Shan Shui" painting refers to a style of Chinese painting that involves the painting of scenery or natural landscapes with brush and ink. The name literally translates to "mountain-water-painting" (landscape painting).

For the art research aspect, "Shan Shui" painting involves a complicated and rigorous set of requirements for balance, composition and form [5]; and the composition type is usually divided into three types [1]: (1) "Lofty and remote" ("Gao-Yuan"): standing at the foot of the mountain and looking up at the top,

© Springer Nature Switzerland AG 2019
G. Li et al. (Eds.): DASFAA 2019, LNCS 11448, pp. 264–267, 2019.
https://doi.org/10.1007/978-3-030-18590-9_25

(a) "Lofty and remote"   (b) "Deep and remote"   (c) "Wide and remote"

Fig. 1. Ancient Chinese "Shan Shui" paintings with three types of composition. The left one is the painting with a composition label and the right one is the same painting with scene element labels. Best viewed in electronic form.

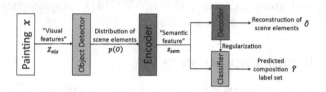

Fig. 2. The framework of SemanticVAE. SemanticVAE uses proper features to deal with proper problems: while handling the "concrete semantic" problem of learning the distribution of scene elements of a painting, it uses "visual features" Z_{vis}; in contrast, while handling the "abstract semantic" problem of painting composition classification, it turns to using a high-level "semantic feature" z_{sem}. Best viewed in electronic form.

as the red arrows shown in Fig. 1(a). (2) "Deep and remote" ("Shen-Yuan"): standing in front of the mountain and peeping the behind, as the green arrows shown in Fig. 1(b). (3) "Wide and remote" ("Ping-Yuan"): standing at the top of the mountain and looking into the distance, as the blue arrows shown in Fig. 1(c).

Actually, most classification problems, including the notable image recognition tasks in ImageNet Large Scale Visual Recognition Challenge [3], are "concrete semantic problem". In contrast, the problem of ancient Chinese "Shan Shui" painting composition classification is an "abstract semantic problem". The state-of-the-art deep CNNs conduct classification using "visual features" [6]. Extensive experiments and practical applications have validated the effectiveness of this kind of deep "visual feature" while performing "concrete semantic problem". However, to classify ancient "Shan Shui" paintings by "abstract semantic" composition types, merely using "visual features" seems insufficient.

There exists inherent relationship between the composition types and the distribution of scene elements of an ancient "Shan Shui" painting. As shown in Fig. 1(a) (composition type "lofty and remote"), the trees and the peaks are far apart. It means the trees are at the foot of the mountains, and gives a hint that the painter is standing at the foot of the mountain and looking up at the top. In Fig. 1(b) (composition type "deep and remote"), the building is surrounded by the peaks and the trees. It means the building is in the mountains and gives a hint that the painter is standing in front of the mountain and peeping the behind. Also, in Fig. 1(c) (composition type "wide and remote"), the trees and the peaks are close and "parallel", which implies that the painter is standing at the top of the mountain and looking into the distance. Inspired by that, we present a novel approach named Semantic Variational Autoencoder (Semantic-VAE) to handle the issue of ancient Chinese "Shan Shui" painting composition classification. Extensive experiments are conducted on a real ancient Chinese "Shan Shui" painting image dataset provided by museums. The experimental results show that, in contrast to the state-of-the-art deep CNNs, our method significantly improves the performance of ancient Chinese "Shan Shui" painting composition classification: increases the Subset Accuracy [7] about 8% compared to AlexNet [2] and about 11% compared to VGG16 [4] (improvements on more evaluation metrics see the experimental section).

2   Methods and Experiments

The framework of SemanticVAE is presented in Fig. 2. Given an ancient Chinese landscape painting x in pixel space, SemanticVAE first maps it to a set of "visual features" Z_{vis}. Then, using these "visual features" Z_{vis}, the object detector of SemanticVAE learns the distribution of scene elements $p(O)$ of the painting x. We use $O = \{o_1, o_2, \cdots\}$ to denote the ground truth scene elements in the painting, and $\tilde{O} = \{\tilde{o}_1, \tilde{o}_2, \cdots\}$ to denote the predicted scene elements.

We use $p(O)$ to learn the "semantic feature" z_{sem} in the training process and $p(\tilde{O}|Z_{vis})$ to construct the "semantic feature" z_{sem} in the inference process. In the training process, the encoder of SemanticVAE learns the posterior distribution $p(z_{sem}|Y, O)$ of the latent random variable z_{sem}, and this process is supervised by the decoder and the classifier simultaneously. The supervision of the classifier is to impel z_{sem} to describe the subtle inherent relationship between the composition types Y and the distribution of scene elements $p(O)$ of the painting x. The supervision of the decoder is a kind of regularization, which impels z_{sem} to "remember" more information about $p(O)$ in order to reconstruct O (we use $\hat{O} = \{\hat{o}_1, \hat{o}_2, \cdots\}$ to denote the reconstructed scene elements). Once z_{sem} "remembers" more information about $p(O)$, it can learn the subtle inherent relationship better. In the inference process, the encoder estimates the posterior distribution $p(z_{sem}|\hat{Y}, \tilde{O})$ of the latent random variable z_{sem}. Then, the classifier predicts the composition label set $\hat{Y}$ of the painting x using the samples from $p(z_{sem}|\hat{Y}, \tilde{O})$.

For training the object detector, the loss L_{obj} is a weighted sum of the localization loss L_{loc} and the confidence loss L_{conf}. For training the encoder, the decoder and the classifier simultaneously, we propose a loss L_{sem} which is a weighted sum of the classification loss L_{cla} and the reconstruction loss L_{rec}:

$$L_{sem} = L_{cla} + \lambda L_{rec}. \tag{1}$$

To evaluate the performance of our method for ancient Chinese "Shan Shui" painting composition classification, we use 900 ancient Chinese "Shan Shui" paintings collected in museums. For each painting, the field experts label its composition types (that is, "lofty and remote", "deep and remote" and "wide and remote" [1]); and some non-expert students help labeling the scene elements (that is, the building, the rock, the bridge, the mountain slope, the peak, the tree, the cascading peaks, the inscription and the stamp) in the painting. In our experiments, we use 10-fold cross validation to test the average performance of different methods. As shown in Table 1, to test the performance of different methods in different aspects, we use a variety of evaluation metrics for the problem of multi-label classification [7]: (1) the Subset Accuracy (SA), (2) the Hamming Loss (HL), (3) the Accuracy (A), (4) the Precision (P), (5) the Recall (R) and (6) the F1-score (F1).

The experimental results are shown in Table 1, it demonstrates that our "semantic feature" based method SemanticVAE significantly outperforms those low-level "visual feature" based deep CNNs in different aspects. The experimental results provide good support for our idea: to deal with the "abstract semantic" and subtle issues such as ancient Chinese landscape painting composition classification, the high-level "semantic feature" is more reasonable and appreciated.

Table 1. Comparisons on a real ancient Chinese "Shan Shui" painting image dataset provided by museums. The "(+)" means the larger, the better, and the "(−)" means the smaller, the better. The best performance is bold. Best viewed in electronic form.

Method	SA (+)	HL (−)	A (+)	P (+)	R (+)	F1 (+)
AlexNet	55.3 ± 0.7	55.2 ± 0.7	75.5 ± 0.2	86.3 ± 0.3	83.4 ± 0.3	84.8 ± 0.2
VGG16	52.0 ± 0.8	60.9 ± 0.6	71.6 ± 0.2	84.0 ± 0.3	82.1 ± 0.2	82.7 ± 0.2
SemanticVAE	$\mathbf{63.5 \pm 0.6}$	$\mathbf{44.9 \pm 0.7}$	$\mathbf{80.4 \pm 0.2}$	$\mathbf{90.6 \pm 0.3}$	$\mathbf{88.6 \pm 0.3}$	$\mathbf{89.5 \pm 0.2}$

Acknowledgements. This work was supported by NSFC grant No. 61370157, NSFC grant No. 61572135 and Shanghai Innovation Action Project No. 17DZ1203-600.

References

1. Foong, P.: Guo Xi's intimate landscapes and the case of "Old trees, level distance". Metrop. Mus. J. **35**, 87–115 (2000)
2. Krizhevsky, A., Sutskever, I., Hinton, G.E.: Imagenet classification with deep convolutional neural networks. In: Advances in Neural Information Processing Systems, pp. 1097–1105 (2012)
3. Russakovsky, O., et al.: Imagenet large scale visual recognition challenge. Int. J. Comput. Vis. **115**(3), 211–252 (2015)
4. Simonyan, K., Zisserman, A.: Very deep convolutional networks for large-scale image recognition. In: International Conference on Learning Representations (2015)
5. Wicks, R.: Being in the dry Zen landscape (2004)
6. Zeiler, M.D., Fergus, R.: Visualizing and understanding convolutional networks. In: Fleet, D., Pajdla, T., Schiele, B., Tuytelaars, T. (eds.) ECCV 2014. LNCS, vol. 8689, pp. 818–833. Springer, Cham (2014). https://doi.org/10.1007/978-3-319-10590-1_53
7. Zhang, M.L., Zhou, Z.H.: A review on multi-label learning algorithms. IEEE Trans. Knowl. Data Eng. **26**(8), 1819–1837 (2014)

Learning Time-Aware Distributed Representations of Locations from Spatio-Temporal Trajectories

Huaiyu Wan[✉], Fuchen Li, Shengnan Guo, Zhong Cao, and Youfang Lin

Beijing Key Laboratory of Traffic Data Analysis and Mining,
School of Computer and Information Technology, Beijing Jiaotong University,
Beijing 100044, China
{hywan,lifuchen,guoshn,caozhong,yflin}@bjtu.edu.cn

Abstract. The goal of location representation learning is to learn an embedded feature vector for each location. We propose a Time-Aware Location Embedding (TALE) method to learn distributed representations of locations from users' spatio-temporal trajectories, in which a novel tree structure is designed to incorporate the temporal information in the hierarchical softmax model. We utilize TALE to improve two location-based prediction tasks to verify its effectiveness.

1 Introduction

Learning the embedded feature representations of locations is a fundamental problem in location mining. The key point of location representation is to accurately model the sequential influence of locations. Motivated by the researches on neural network language models, such as word2vec [3,4] which can effectively capture the sequential semantic relationships among words and has been employed to model users' sequential check-ins [2], recently a new location representation model POI2Vec [1] was proposed to incorporate geographical influence into the word2vec model, and it obtained a very good performance in predicting potential visitors for a given Point of Interest (POI).

In fact, the temporal information involved in users' trajectories can also reflect the characteristics of locations. To incorporate the temporal information in learning more exquisite location representations, we propose a novel *Time-Aware Location Embedding* (TALE) method. A novel tree structure is designed to incorporate the temporal information of spatio-temporal trajectory sequences in the hierarchical softmax.

We finally utilize TALE in two location-based prediction tasks and the experimental results demonstrate the effectiveness of our proposed model.

This work is supported by the Natural Science Foundation of China (No. 61603028).

G. Li et al. (Eds.): DASFAA 2019, LNCS 11448, pp. 268–272, 2019.
https://doi.org/10.1007/978-3-030-18590-9_26

2   The TALE Model

We use U to denote a set of users and L a set of locations, where each location l is associated with its longitude and latitude coordinates (l^{Lon}, l^{Lat}). Let H be the set of historical spatio-temporal trajectories of users U within locations L. Each point (u, l, t) in a trajectory h indicates that user u has visited location l at time t. Our goal is to learn a general latent representation for each location from the user-generated spatio-temporal trajectories by exploiting the sequential relationships among locations and the temporal influence that when a user arrive at a certain location.

2.1   Basic Location Representation Model

Given a user u and his/her current location l_i^u, we define $C(l_i^u) = \{l_j^u, |j-i| \leq \tau\}$ as the set of contextual locations of l_i^u, where τ is a hyperparameter to control the size of the context window. The goal of location sequential modeling is to estimate the probability that a user will visit a location given its contextual locations.

We use a vector $\boldsymbol{w}(l) \in \mathbb{R}^d$ in a d-dimensional latent space to represent a location l, and $P(l|C(l))$ is the probability of l given its contexts $C(l)$. We select the Continuous Bag-of-Words (CBOW) framework to calculate the probability as follows:

$$P(l|c(l)) = e^{(\boldsymbol{w}(l)\phi(c(l)))}/Z(C(l)) \tag{1}$$

where $\phi(C(l)) = \sum_{l_j \in C(l)} \boldsymbol{w}(l_j)$ is the sum of the vectors of all the contextual locations, $Z(C(l)) = \sum_{l' \in L} e^{(\boldsymbol{w}(l')\phi(C(l)))}$ is a normalization factor.

2.2   Incorporating Temporal Influence

In our TALE model we present a novel temporal tree structure for the hierarchy softmax. The temporal tree consists of two parts from top to bottom, as shown in Fig. 1. The top part is a two-layer multi-branch tree, in which the first layer contains only a root node v_0, and the second layer contains T nodes from v_1

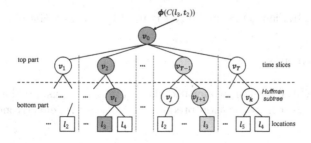

Fig. 1. The temporal tree structure of the TALE model

to v_T. We divide the time of a day into T equal time slices and let each time slice correspond to one leaf node v_t $(1 \leq t \leq T)$. The bottom part of the tree is generated from location sequences of users trajectories. We construct a Huffman subtree for each time slice based on the frequencies of locations appearing within the time slice, taking each leaf node v_t in the top part as the root of the subtree.

2.3   Probability Estimation

After the temporal information of trajectory points is involved into the tree structure, we can compute the probability that a user will visit a location l at time t given the contextual locations $C(l,t)$, i.e., $P(l,t|C(l,t))$. The hierarchical softmax method performs a softmax classification by calculating the probability of a path from the root to the leaf node.

The path from the root v_0 to leaf node l at time t can be defined as a sequence of nodes $path = (v_0^t, v_1^l, v_2^l, \ldots, v_n^l)$. We divide path $path$ into two segments according to the structure of the temporal tree, i.e., $path = path_1 + path_2$, where $path_1 = (v_0^t)$ only contains the root node, and $path_2 = (v_1^l, v_2^l, \ldots, v_n^l)$ belongs to a binary Huffman subtree. Then the probability of observing l at time t along the path $path$ can be estimated by:

$$P(l,t|C(l,t))^{path} = P(v_0^t|\phi(C(l,t))) \prod_{v_i^l \in path_2} P(v_i^l|\phi(C(l,t)))$$

$$= P(t|C(l,t))^{path_1} * P(l|C(l,t),t)^{path_2} \tag{2}$$

The root node v_0 has a latent matrix $M \in \mathbb{R}^{T \times d}$, which can be treated as the parameters of the multi-class classifier. So the first item in Eq. (2) can be calculated as:

$$P(t|C(l,t))^{path_1} = e^{(M^{(v_0^t)}\phi(C(l,t)))}/Z(C(l,t)) \tag{3}$$

The second item in Eq. (2) can be calculated as:

$$P(l|C(l,t))^{path_2} = \prod_{v_i^l \in path_2} \sigma(\boldsymbol{\Psi}(v_i^l) \cdot \phi(C(l,t))) \tag{4}$$

Stochastic Gradient Descent (SGD) is used to learn all the parameters.

3   Experiments

We use our TALE model in two location-based mining tasks, i.e., location crowd prediction and user next location prediction, to verify its effectiveness.

3.1 Dataset

We collected a trajectory dataset from the mobile phone signaling data in Beijing. The data record the switching events between telecommunication base stations of mobile users in 5 consecutive workdays. We treat each base station as a location, and take the base stations accessed by a user in the 5 days as a trajectory. The dataset contains 0.2 million users' trajectories, 14,762 locations, 9.7 million trajectory points and the average length of trajectories is 48.5.

3.2 Experimental Results

Location Crowd Flow Prediction. In this task, we choose LSTM as the host of our TALE model and compare LSTM+TALE with four baselines: ARIMA, LSTM, LSTM+W2V and LSTM+POI2V. We use two metrics, Mean Square Error (MSE) and Mean Absolute Error (MAE), to evaluate the performance of different methods. The experimental results are shown in Fig. 2.

User Next Location Prediction. In this task, we compare the LSTM+TALE method with five baselines: MF, MC, LSTM, LSTM+W2V and LSTM+POI2V. We use the $Pre@N$ matrix to evaluate the performance of different methods. The experimental results are shown in Fig. 3.

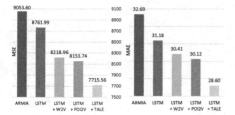
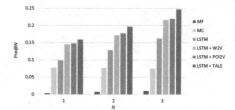

Fig. 2. MSE and MAE of flow prediction

Fig. 3. Pre@N of location prediction

From Figs. 2 and 3 we can find that our TALE method achieves the best results among all the methods. This demonstrates the effectiveness of our TALE model on integrating the temporal information into location representation learning, which makes the presentations embed more "semantic" information of locations.

4 Conclusions

In this paper we propose a novel time-aware location embedding model TALE, which is able to incorporate the temporal information in visitors' mobility trajectories in learning the latent distributed representations of locations.

References

1. Feng, S., Cong, G., An, B., Chee, Y.M.: POI2Vec: geographical latent representation for predicting future visitors. In: AAAI, pp. 102–108 (2017)
2. Liu, X., Liu, Y., Li, X.: Exploring the context of locations for personalized location recommendations. In: IJCAI, pp. 1188–1194 (2016)
3. Mikolov, T., Chen, K., Corrado, G., Dean, J.: Efficient estimation of word representations in vector space. In: ICLR (2013)
4. Mikolov, T., Sutskever, I., Chen, K., Corrado, G., Dean, J.: Distributed representations of words and phrases and their compositionality. In: NIPS, pp. 3111–3119 (2013)

Hyper2vec: Biased Random Walk for Hyper-network Embedding

Jie Huang[1,2], Chuan Chen[1,2(✉)], Fanghua Ye[1,2], Jiajing Wu[1,2(✉)],
Zibin Zheng[1,2], and Guohui Ling[3]

[1] School of Data and Computer Science, Sun Yat-sen University, Guangzhou, China
{chenchuan,wujiajing}@mail.sysu.edu.cn
[2] National Engineering Research Center of Digital Life, Sun Yat-sen University,
Guangzhou, China
[3] Tencent Technology, Shenzhen, China

Abstract. Network embedding aims to obtain a low-dimensional representation of vertices in a network, meanwhile preserving structural and inherent properties of the network. Recently, there has been growing interest in this topic while most of the existing network embedding models mainly focus on normal networks in which there are only pairwise relationships between the vertices. However, in many realistic situations, the relationships between the objects are not pairwise and can be better modeled by a hyper-network in which each edge can join an uncertain number of vertices. In this paper, we propose a deep model called Hyper2vec to learn the embeddings of hyper-networks. Our model applies a biased 2nd order random walk strategy to hyper-networks in the framework of Skip-gram, which can be flexibly applied to various types of hyper-networks.

1 Introduction

In the real world, there are a variety of systems that are composed of objects connected in a particular way, which can be represented as networks. Recently, a series of network embedding models [3,7,9] have been proposed to obtain a low-dimensional representation of vertices in the networks. However, the relationships between the objects may be not pairwise. For instance, giving a set of articles, the number of authors in each article is not fixed at 2, so it is not suitable to model the relationships between the authors using a normal network. A hyper-network in which each edge can connect an uncertain number of vertices [1] can better depict the relationships between the authors instead.

However, most embedding algorithms cannot be applied to hyper-networks directly. Till now, only a few embedding models for hyper-networks have been proposed and these models are not particularly applicable for general cases. A conventional approach to deal with hyper-networks is to convert them into normal networks, which will lose some important structural information of the original hyper-networks.

© Springer Nature Switzerland AG 2019
G. Li et al. (Eds.): DASFAA 2019, LNCS 11448, pp. 273–277, 2019.
https://doi.org/10.1007/978-3-030-18590-9_27

Based on the above considerations, we propose a general hyper-network embedding model called **Hyper2vec**, which is based on biased 2nd order random walks in hyper-networks. In this model, a 1st order random walker moves to the next vertex based on the current vertex only, and a 2nd order random walker moves to the next vertex based on both the current and the previous vertices. Furthermore, we utilize a degree biased random walk to correct the negative bias or introduce a positive bias for random walks.

2   Notations and Definitions

A **hyper-network** is defined as a hypergraph $G(V, E)$ with the vertex set V and the hyperedge set E. A **hyperedge** $e \in E$ indicates the relationship of an uncertain number of vertices, which is a subset of V. A weighted hyper-network defined as $G(V, E, w)$ is a hypergraph that has a weight $w(e)$ associated with each hyperedge e. A hyperedge e is incident with a vertex v if $v \in e$, and if $v \in e$, then $h(v, e) = 1$, otherwise $h(v, e) = 0$.

For a vertex $v \in V$, and a hyperedge $e \in E$, the degree of v is defined as $d(v) = \sum_{e \in E} w(e)h(v, e)$, and the degree of e is defined as $\delta(e) = |e| = \sum_{v \in V} h(v, e)$.

3   Hyper2vec: The Proposed Method

In order to preserve the high-order proximity of hyper-networks, we propose an efficient and scalable biased 2nd order random walk model in the framework of Skip-gram.

The 1st order random walk strategy decides the probability of moving to the next vertex based on the current vertex. To generalize random walks in hyper-networks, we adopt the transition rule described in [12]. Given the current vertex v, we first randomly select a hyperedge e incident with v based on the weight of e, and then pick the next vertex $x \in e$ uniformly at random. The unnormalized transition probability of each walk of the 1st order model can be given as follows:

$$\pi_1(x|v) = \sum_{e \in E} w(e) \frac{h(v, e)}{d(v)} \frac{h(x, e)}{\delta(e)}. \tag{1}$$

To shuttle between homophily and structural equivalence when doing prediction tasks, it's feasible to introduce a 2nd order random walk model [3]. We divide the neighbors of vertex v into three categories, the previous vertex u, the neighbors of the previous vertex u, and the others. The 2nd order search bias is defined as follows:

$$\alpha(x|v, u) = \begin{cases} \frac{1}{p}, & \text{if } x = u \\ 1, & \text{if } \exists e \in E, x \in e, u \in e \\ \frac{1}{q}, & \text{others} \end{cases} \tag{2}$$

The unnormalized transition probability of the 2nd order random walk model is defined as $\pi_2(x|v, u) = \alpha(x|v, u)\pi_1(x|v)$. The relationship between parameters p,

q, and 1 leads to the tendency of **Depth-First Search (DFS)** which encourages outward exploration and **Breadth-First Search (BFS)** which obtains a local view of the start vertex.

Previous studies have shown that DFS and BFS search strategy will introduce some negative bias. For instance, it is observed that BFS introduces a bias towards big degree vertices [6]. On one hand, we need to correct the negative bias [4]. On the other hand, we need to introduce some positive bias to capture some special characteristics of networks or balance the update opportunities for vertices [2,11].

Based on the above considerations, we propose a degree biased random walk model, which has two sampling strategies, **Big-Degree Biased Search (BDBS):** the walker tends to choose a neighbor vertex with bigger degree, and **Small-Degree Biased Search (SDBS):** the walker tends to choose a neighbor vertex with smaller degree.

We define a biased 2nd order random walk by introducing a bias coefficient $\beta(x)$ based on the degree of x with a parameter r. To make the mechanism more flexible, parameter r should not only guide the strategy of BDBS or SDBS but also control the degree of influence. To achieve the above aim, we define degree bias $\beta(x) = d(x) + r$ $(r > 0)$ for BDBS. Obviously, $\beta(x)$ grows linearly with $d(x)$, and as r increases, the degree of BDBS tendency declines. With a similar principle of BDBS, we combine SDBS $(r < 0)$ to produce the final formula as follows:

$$\beta(x) = \begin{cases} d(x) + r, & r > 0 \\ \frac{1}{d(x) - r}, & r < 0 \\ 1, & r = 0 \end{cases} \tag{3}$$

Combined with the 2nd order random walk model, the biased 2nd order transition probability $p_2'(x|v, u)$ is calculated as follows:

$$p_2'(x|v, u) = \frac{\alpha(x|v, u) \cdot \beta(x) \cdot \sum_{e \in E} w(e) \frac{h(v,e)}{d(v)} \frac{h(x,e)}{\delta(e)}}{Z}, \tag{4}$$

where Z is a normalizing factor. Based on the transition probability $p_2'(x|v, u)$, starting from each vertex in the hyper-network, we can generate a set of random walks. We produce the final embedding results via Skip-gram model. Skip-gram [5] is a language model that maximizes the co-occurrence probability among the words that appear within a window in a sentence, which is replaced by a random walk in our model.

4   Experiments

In this section, we conduct classification experiments on DBLP and IMDb datasets. DBLP[1] is a computer science bibliography. We abstract each author as a vertex and each article as a hyperedge. We select some conferences from

---

[1] http://dblp.uni-trier.de/xml/.

276 J. Huang et al.

three different research fields: KDD, WWW from data mining, NIPS, ICML from machine learning, and CVPR, ICCV from computer vision. We use the most frequent fields as the author's label. Authors with degree less than 3 are filtered out, and finally, the hyper-network contains 8329 vertices and 14677 hyperedges. IMDb[2] dataset contains the information about movies and actors. We use a subset of IMDb dataset in our experiments. We abstract each actor as a vertex, and choose the top three actors in each movie to form a hyperedge. We use the top three frequent genres as the actor's labels. Actors with degree less than 3 are filtered out, and finally, the hyper-network contains 4396 vertices and 3875 hyperedges.

Table 1. Results of classification on DBLP and IMDb

Embedding algorithm	DBLP		IMDb	
	Micro-F1	Macro-F1	Micro-F1	Macro-F1
DeepWalk	0.8023	0.7875	0.4798	0.2844
Node2vec	0.8074	0.7929	0.4783	0.2887
LINE	0.7450	0.7242	0.4624	0.2411
HHE	0.4444	0.2081	0.4381	0.0902
HGE	0.5411	0.4764	0.4410	0.2128
Hyper2vec	**0.8180**	**0.8039**	**0.4953**	**0.3101**

We compare the performance of Hyper2vec against normal network embedding algorithms, including DeepWalk [7], Node2vec [3] and LINE [9] and hyper-network embedding algorithms, including HHE [13] and HGE [10]. For normal network embedding algorithms, we use clique expansion [8] to transform a hyper-network into a normal network. Logistic regression is used as the outer classifier and the Micro-F1 and Macro-F1 scores of 5-fold cross-validation are shown in Table 1. From the results, we can draw a conclusion that Hyper2vec performs better than the baselines in classification tasks.

Acknowledgment. The paper was supported by the National Key Research and Development Program (2016YFB1000101), the National Natural Science Foundation of China (11801595, 61503420), the Natural Science Foundation of Guangdong (2018A030310076), the Program for Guangdong Introducing Innovative and Entrepreneurial Teams (2016ZT 06D211) and the CCF Opening Project of Information System.

---

[2] https://www.imdb.com/.

References

1. Berge, C.: Hypergraphs: Combinatorics of Finite Sets, vol. 45. Elsevier, Amsterdam (1984)
2. Feng, R., Yang, Y., Hu, W., Wu, F., Zhuang, Y.: Representation learning for scale-free networks. arXiv preprint arXiv:1711.10755 (2017)
3. Grover, A., Leskovec, J.: node2vec: scalable feature learning for networks. In: KDD, pp. 855–864. ACM (2016). https://doi.org/10.1145/2939672.2939754
4. Kurant, M., Markopoulou, A., Thiran, P.: On the bias of BFS. arXiv preprint arXiv:1004.1729 (2010)
5. Mikolov, T., Chen, K., Corrado, G., Dean, J.: Efficient estimation of word representations in vector space. arXiv preprint arXiv:1301.3781 (2013)
6. Najork, M., Wiener, J.L.: Breadth-first crawling yields high-quality pages. In: WWW, pp. 114–118. ACM (2001). https://doi.org/10.1145/371920.371965
7. Perozzi, B., Al-Rfou, R., Skiena, S.: DeepWalk: online learning of social representations. In: KDD, pp. 701–710. ACM (2014). https://doi.org/10.1145/2623330.2623732
8. Sun, L., Ji, S., Ye, J.: Hypergraph spectral learning for multi-label classification. In: KDD, pp. 668–676. ACM (2008). https://doi.org/10.1145/1401890.1401971
9. Tang, J., Qu, M., Wang, M., Zhang, M., Yan, J., Mei, Q.: Line: large-scale information network embedding. In: WWW, pp. 1067–1077. International World Wide Web Conferences Steering Committee (2015). https://doi.org/10.1145/2736277.2741093
10. Yu, C.A., Tai, C.L., Chan, T.S., Yang, Y.H.: Modeling multi-way relations with hypergraph embedding. In: CIKM, pp. 1707–1710. ACM (2018)
11. Zeng, Z., Liu, X., Song, Y.: Biased random walk based social regularization for word embeddings. In: IJCAI, pp. 4560–4566 (2018)
12. Zhou, D., Huang, J., Schölkopf, B.: Learning with hypergraphs: clustering, classification, and embedding. In: NIPS, pp. 1601–1608 (2007)
13. Zhu, Y., Guan, Z., Tan, S., Liu, H., Cai, D., He, X.: Heterogeneous hypergraph embedding for document recommendation. Neurocomputing **216**, 150–162 (2016). https://doi.org/10.1016/j.neucom.2016.07.030

Privacy-Preserving and Dynamic Spatial Range Aggregation Query Processing in Wireless Sensor Networks

Lisong Wang[⊠], Zhenhai Hu, and Liang Liu

College of Computer Science and Technology,
Nanjing University of Aeronautics and Astronautics, Nanjing 210016, China
wangls@nuaa.edu.cn

Abstract. The existing privacy-preserving aggregation query processing methods in sensor networks rely on pre-established network topology and require all nodes in the network to participate in query processing. Maintaining the topology results in a large amount of energy overhead, and in many cases, the user is interested only in the aggregated query results of some areas in the network, and thus, the participation of the entire network node is not necessary. Aiming to solve this problem, this paper proposes a spatial range aggregation query algorithm for a dynamic sensor network with privacy protection (E^2PDA – Energy-efficient Privacy-preserving Data Aggregation). The algorithm does not rely on the pre-established topology but considers only the query area that the user is interested in, abandoning all nodes to participate in distributing the query messages while gathering the sensory data in the query range. To protect node data privacy, Shamir's secret sharing technology is used to prevent internal attackers from stealing the sensitive data of the surrounding nodes. The analysis and experimental results show that the proposed algorithm outperforms the existing algorithms in terms of energy and privacy protection.

Keywords: Wireless sensor network · Query processing · Privacy protection · Spatial range aggregation query

1 Introduction

In a wireless sensor network, users often submit a spatial range aggregation query to obtain statistical information of an area in the network, such as the average temperature and the maximum humidity of an area in a forest. Sensor nodes are battery powered, have limited energy, and in many cases, sensor nodes are easy to be captured and attacked. Therefore, it is necessary to study the privacy-preserving and energy-efficient spatial range aggregation query processing technique to solve these two problems.

IWQE is an itinerary-based spatial window aggregation query processing algorithm [1], but it does not take privacy-preserving of nodes into consideration. The existing privacy-preserving spatial range aggregation query algorithms rely on pre-configured network topologies, which can be divided into two categories according to their type: tree-based [2–5] algorithms and cluster-based [2, 6–8] algorithms. The network topology frequently changes. Maintaining a pre-established topology consumes

© Springer Nature Switzerland AG 2019
G. Li et al. (Eds.): DASFAA 2019, LNCS 11448, pp. 278–281, 2019.
https://doi.org/10.1007/978-3-030-18590-9_28

considerable energy. Additionally, the existing privacy-preserving aggregation query algorithm, SMART [2], and its variants PECDA [3] and ASSDA [5], hide original and sensitive data by slicing the data. Although these algorithms have certain privacy protection capabilities, they all assume that the query area is the entire monitoring area, and all nodes in the network are required to participate in query processing. Obviously, they require considerable unnecessary energy overhead, and it is not necessary for all nodes to participate in query processing.

2 Privacy-Preserving and Dynamic Spatial Range Aggregation Query Processing in Wireless Sensor Networks-E²PDA

To avoid relying on a pre-constructed topology and to reduce the energy consumption of the nodes in query processing, this paper proposes a privacy-preserving and dynamic spatial range aggregation query algorithm, named E²PDA, in a wireless sensor network.

Figure 1 shows the execution procedure of the E²PDA algorithm, assuming that the average temperature of Q_a is calculated using E²PDA. The nodes in the query region are represented by a, b, and c, and the sensing data of the corresponding nodes is represented by D_a, D_b and D_c. The algorithm divides the query region into several subregions, and the size of the subregions meets the condition that the distance between the adjacent subregion nodes does not exceed the communication radius between the nodes.

The E²PDA algorithm is divided into three stages: (1) the query message's geographic routing stage, which sends the query message to a node in the query region; (2) the query distribution and sensory data aggregation stage, the query message is sent to all nodes in the query region, and the sensing data is collected and aggregated to calculate the query results; (3) the query results generated in the second phase are returned to the query initiation node by geographic routing protocol.

Without loss of generality, the query area is described as a rectangle. The sink node sends the query message to node a in the query region Q_a using the geographic routing protocol [9], while the sink sends a random number r to node a to protect the data of node a. When the query message and random number arrive at node a, the random number is added to D_a to obtain D'_a. Then, node a sends the query message and D'_a to node b. After receiving the query message and D'_a, node b adds D_b and D'_a and sends it to the next node. The final query result in the query area is aggregated until E²PDA reaches the last node y in the query area Node y sends the final aggregated result back to the sink node via a geographic routing protocol. The sink node subtracts the random number r from the returned aggregation results to obtain the real aggregation results in the query region.

In the SMART and PECDA, when establishing a secure connection, node S_i and node S_j can establish a secure connection because S_i and S_j have the same secret key value d_{ij}. The privacy leakage of node data mainly includes two aspects: (1) The third-party node has the same secret key; (2) because of the secret key preallocation stage, each node randomly selects k secret keys from the secret key pool with K secret keys and then establishes a connection if the adjacent nodes have the same secret keys. Therefore, SMART and PECDA have the possibility of privacy leakage during data aggregation. E^2PDA adopts a route-based aggregation method to avoid the problem of sensing data leakage caused by secret key leakage in SMART and PECDA.

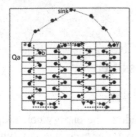

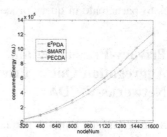

Fig. 1. The execution procedure of the E^2PDA algorithm.

Fig. 2. Influence of the number of network nodes on energy loss

3 Analysis of Experimental Results

In this section, the energy consumption of the E^2PDA is analyzed through experiments. Using the simulator in [10], the E^2PDA, SMART and PECDA in the existing algorithm are implemented. The experimental parameters were selected as follows. According to [11], the energy consumption formula of wireless communication transmission and reception of 1 byte is: $E_t = \alpha + \gamma * d^n$, $E_r = \beta$. The other parameters are shown in Table 1 below.

Table 1. Experimental parameters.

Parameter name	Parameter value
Network coverage area	450 m * 450 m
Number of nodes	600
Perceived data size	30
Query data size	24

Figure 2 shows the effect of the number of network nodes on energy loss. When the number of network nodes is fixed, PECDA and SMART consume significantly more energy than E^2PDA. There are a certain number of nodes. Under the same query area, the SMART algorithm needs to distribute the slicing data to the neighboring nodes in the data slicing stage, and the energy consumed in this process accounts for a large proportion of the total energy consumption. The PECDA algorithm conducts slicing transmission only of leaf node perception data, while the SMART algorithm requires data slicing transmission of all nodes in the network.

4 Conclusion

We proposed a privacy-preserving and dynamic spatial range aggregation query processing algorithm in wireless sensor networks. It does not depend on the pre-constructed topology structure and collects the sensing data of the specified area while ensuring the data privacy of the sensor nodes. The analysis of the algorithm and experimental results show that our algorithm outperforms some existing algorithms such as SMART and PECDA in terms of energy consumption and privacy protection.

Acknowledgments. The research work was supported by the Aeronautical Science Foundation of China (Grant No. 20165515001), the National Natural Science Foundation of China under (Grant No. 61402225), State Key Laboratory for smart grid protection and operation control Foundation, Science and Technology Funds from National State Grid Ltd.

References

1. Xu, Y., Lee, W.-C., Xu, J., Mitchell, G.: Processing window queries in wireless sensor networks. In: Proceedings of the 22nd International Conference on data Engineering, pp. 70–80. IEEE Computer Society (2006)
2. He, W., Liu, X., Nguyen, H., et al.: PDA: privacy-preserving data aggregation in wireless sensor networks. In: INFOCOM 2007, 26th IEEE International Conference on Computer Communications, pp. 2045–2053. IEEE (2007)
3. Wang, T., Qin, X., Ding, Y., et al.: Privacy-preserving and energy-efficient continuous data aggregation algorithm in wireless sensor networks. Wirel. Pers. Commun. **98**(1), 665–684 (2018)
4. Akila, V., Sheela, T.: Preserving data and key privacy in Data Aggregation for Wireless Sensor Networks. In: International Conference on Computing and Communications Technologies, pp. 282–287. IEEE (2017)
5. Hua, P., Liu, X., Yu, J., et al.: Energy-efficient adaptive slice-based secure data aggregation scheme in WSN. Procedia Comput. Sci. **129**, 188–193 (2018)
6. Ozdemir, S., Xiao, Y.: Integrity protecting hierarchical concealed data aggregation for wireless sensor networks. Comput. Netw. **55**(8), 1735–1746 (2011)
7. Elhoseny, M., Yuan, X., El-Minir, H.K., et al.: An energy efficient encryption method for secure dynamic WSN. Secur. Commun. Netw. **9**(13), 2024–2031 (2016)
8. Fang, W., Wen, X.Z., Xu, J., et al.: CSDA: a novel cluster-based secure data aggregation scheme for WSNs. Clust. Comput. **4**, 1–12 (2017)
9. Karp, B., Kung, H.T.: GPSR: greedy perimeter stateless routing for wireless networks. In: Proceedings of the 6th Annual International Conference on Mobile Computing and Networking, New York, pp. 243–254 (2000)
10. Coman, A., Sander, J., Nascimento, M.A.: Adaptive Processing of Historical Spatial Range Queries in Peer-to-Peer Sensor Networks. Kluwer Academic Publishers (2007)
11. Rappaport, T.: Wireless Communications: Principles and Practice. Prentice-Hall Press, New Jersey (1996)

Adversarial Discriminative Denoising for Distant Supervision Relation Extraction

Bing Liu[1], Huan Gao[1], Guilin Qi[1,2(✉)], Shangfu Duan[1], Tianxing Wu[3], and Meng Wang[1]

[1] School of Computer Science and Engineering, Southeast University,
Nanjing 211111, China
{liubing_cs,hg,gqi,sf_duan,meng.wang}@seu.edu.cn
[2] Key Laboratory of Computer Network and Information Integration,
Ministry of Education, Southeast University, Nanjing 211111, China
[3] School of Computer Science and Engineering, Nanyang Technological University,
Singapore, Singapore
wutianxing@ntu.edu.sg

Abstract. Distant supervision has been widely used to generate labeled data automatically for relation extraction by aligning knowledge base with text. However, it introduces much noise, which can severely impact the performance of relation extraction. Recent studies have attempted to remove the noise explicitly from the generated data but they suffer from (1) the lack of an effective way of introducing explicit supervision to the denoising process and (2) the difficulty of optimization caused by the sampling action in denoising result evaluation. To solve these issues, we propose an adversarial discriminative denoising framework, which provides an effective way of introducing human supervision and exploiting it along with the potentially useful information underlying the noisy data in a unified framework. Besides, we employ a continuous approximation of sampling action to guarantee the holistic denoising framework to be differentiable. Experimental results show that very little human supervision is sufficient for our approach to outperform the state-of-the-art methods significantly.

Keywords: Distant supervision · Relation extraction ·
Noise reduction · Adversarial discriminative model

1 Introduction

Distant supervision (DS) is a promising approach to relation extraction (RE). It can generate training data automatically by aligning knowledge base (KB) with text [3,6]. However, it suffers from the introduced noise, which can severely effect the performance of RE. Previous researches have focused on building DS models with noise adaptability [2,7]. However, they did not remove the noise explicitly. To enable explicit noise reduction, a few challenges need to be addressed.

© Springer Nature Switzerland AG 2019
G. Li et al. (Eds.): DASFAA 2019, LNCS 11448, pp. 282–286, 2019.
https://doi.org/10.1007/978-3-030-18590-9_29

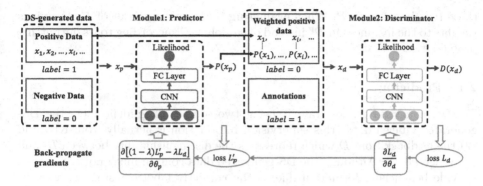

Fig. 1. Overview of the adversarial discriminative denoising framework.

The first challenge comes from the lack of an effective way of introducing explicit supervision to the denoising process. Without introducting human supervision, the unsupervised methods can only make a coarse-grained distinction between the true positive instances and the noise. Another challenge relates to the evaluation of the denoising result. The existing approaches [1,4,5] performed the evaluation by sampling the noisy data according to the noise recognizer and then assessing the resulting subset. The problem is that the sampling action can lead to non-differentiability, which hinders the use of the evaluators that back-propagate gradients to guide the optimization of the noise recognizer in a holistic manner.

To solve the above challenges, we propose an adversarial discriminative denoising framework, which can not only acquire the denoising ability by exploiting the beneficial information underlying DS-generated data but also further get boosted via introducing very few human annotations efficiently. To guarantee the model to be differentiable, we employ a continuous approximation of sampling action when evaluating the denoising result, which helps fast convergence and gains a better solution.

2 Adversarial Discriminative Denoising Framework

The overview of our approach is shown in Fig. 1. The DS-generated dataset to be cleansed can be partitioned into positive data $\mathcal{D}^p$ and negative data $\mathcal{D}^n$ concerning a specific relation r. The adversarial learning process is carried out on a handful of manually labeled data $\mathcal{D}^l$ and a mass of DS-generated data. The framework consists of two core modules: (1) a true positive instance predictor P, which acts as a noise recognizer and outputs the probability that an instance is true positive, and (2) a data source discriminator D, which functions as a critic of the distinguishability between $\mathcal{D}^l$ and $\mathcal{D}^p$ weighted by P. P not only tries to satisfy the supervision from DS but also attempts to assign $\mathcal{D}^p$ with proper weights to make it indistinguishable from $\mathcal{D}^l$. D attempts to improve its ability to distinguish $\mathcal{D}^l$ from the weighted $\mathcal{D}^p$ and back feed the similarity of these two data sets to P. When evaluating the denoising result of P, D assigns weights to the instances in $\mathcal{D}^p$ instead of sampling $\mathcal{D}^p$ according to its outputting probabilities.

2.1 Predictor

To obtain a valid P, we train it using two sources of guidance: (1) the DS-generated labeled data (this supervision information is actually from KB) and (2) the feedback from D which represents the distinguishability between $\mathcal{D}^l$ and the weighted $\mathcal{D}^p$. Although the DS-generated data contain much noise, they can provide beneficial information due to the correctly labeled instances. To take advantage of this information, we use P to classify $\mathcal{D}^p$ and $\mathcal{D}^n$ and take the classification loss as part of its loss function. Another goal of P is to reduce the distinguishability between $\mathcal{D}^l$ and the weighted $\mathcal{D}^p$. Thus, we treat this distinguishability, which is measured by D, as another part of the loss of P.

2.2 Continuous Approximation of Sampling Action

We approximate the sampling action by assigning each instance x in $\mathcal{D}^p$ with $P(x)$ as the weight and let the weights play a role in measuring the similarity between $\mathcal{D}^p$ and $\mathcal{D}^l$. In this continuous approximation setting, the instances with higher weights have more effect on the measurement, which is equivalent to more frequent participation in sampling setting. Therefore, this similarity is controlled by the weights.

2.3 Discriminator

D aims to detect if an instance is from $\mathcal{D}^l$ or the weighted $\mathcal{D}^p$. Essentially, D is the metrics of the weights, and P adjusts itself according to the feedback about the weights. In $\mathcal{D}^p$, the true positive instances are more difficult to be correctly recognized by D than the false positive ones. In order to puzzle D and cause more losses to it, P will assign higher weights on the true positive ones while lower weights on the noisy ones. As the adversary of P, D has to pay more attention to the instances with high weights so as to avoid major losses. This will drive P to avert mistaking noisy data as the correct ones.

2.4 Cleaning Noisy Dataset with Predictor

After the adversarial learning process, we obtain a valid P concerning relation r. Then, we apply P to $\mathcal{D}^p$ and filter out the instances whose scores are below a certain threshold thr. The cleansed positive data will be used as the positive data of relation r in the training stage of RE models.

3 Experiments

Our experiments are carried out on the widely used NYT dataset[1] [6]. Following the previous work [3], we evaluated our method on NYT using the held-out evaluation. We chose a CNN based model with attention mechanism (CNN+ATT) proposed by [2] as the DS RE model and used this CNN+ATT model trained on the original NYT dataset as the baseline. Then, we chose two state-of-the-art denoising methods based on RL [5] (CNN+ATT+RL) and GAN [4] (CNN+ATT+DSGAN) for comparison. To show the effect of human annotations, we trained our approach in two ways, including using none annotation (CNN+ATT+AD+0) and 1500 annotations (CNN+ATT+AD+1500). Figure 2 shows the PR curves of held-out evaluation. It demonstrates that our approach can effectively reinforce the existing models and outperforms state-of-the-art methods.

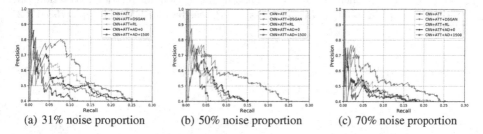

(a) 31% noise proportion   (b) 50% noise proportion   (c) 70% noise proportion

Fig. 2. Aggregate PR curves of CNN+ATT based model upon the held-out evaluation. Our approach (CNN+ATT+AD+1500) outperforms state-of-the-art methods significantly.

4 Conclusion

In this paper, we proposed an adversarial discriminative denoising framework to remove the false positive noise explicitly from DS-generated data. This framework provides an effective way of introducing human supervision and exploiting it along with the potentially useful information underlying the original data in a unified framework.

Acknowledgement. This work was supported by National Key R&D Program of China (2018YFC0830200) and National Natural Science Foundation of China Key Project (U1736204).

---

[1] http://iesl.cs.umass.edu/riedel/ecml/.

References

1. Feng, J., Huang, M., Zhao, L., Yang, Y., Zhu, X.: Reinforcement learning for relation classification from noisy data. In: Proceedings of AAAI, pp. 5779–5786 (2018)
2. Lin, Y., Shen, S., Liu, Z., Luan, H., Sun, M.: Neural relation extraction with selective attention over instances. In: Proceedings of ACL, pp. 2124–2133 (2016)
3. Mintz, M., Bills, S., Snow, R., Jurafsky, D.: Distant supervision for relation extraction without labeled data. In: Proceedings of ACL, pp. 1003–1011 (2009)
4. Qin, P., Xu, W., Wang, W.Y.: DSGAN: generative adversarial training for distant supervision relation extraction. In: Proceedings of ACL, pp. 496–505 (2018)
5. Qin, P., Xu, W., Wang, W.Y.: Robust distant supervision relation extraction via deep reinforcement learning. In: Proceedings of ACL, pp. 2137–2147 (2018)
6. Riedel, S., Yao, L., McCallum, A.: Modeling relations and their mentions without labeled text. In: Balcázar, J.L., Bonchi, F., Gionis, A., Sebag, M. (eds.) ECML PKDD 2010. LNCS (LNAI), vol. 6323, pp. 148–163. Springer, Heidelberg (2010). https://doi.org/10.1007/978-3-642-15939-8_10
7. Zeng, D., Liu, K., Chen, Y., Zhao, J.: Distant supervision for relation extraction via piecewise convolutional neural networks. In: Proceedings of EMNLP, pp. 1753–1762 (2015)

Nonnegative Spectral Clustering for Large-Scale Semi-supervised Learning

Weibo Hu[1,2], Chuan Chen[1,2](✉), Fanghua Ye[1,2], Zibin Zheng[1,2],
and Guohui Ling[3]

[1] School of Data and Computer Science,
Sun Yat-sen University, Guangzhou, China
chenchuan@mail.sysu.edu.cn
[2] National Engineering Research Center of Digital Life,
Sun Yat-sen University, Guangzhou 510006, China
[3] Data Center of Wechat Group,
Tencent Technology, Shenzhen, China

Abstract. This paper proposes a novel clustering approach called *Scalable Nonnegative Spectral Clustering* (SNSC). Specifically, SNSC preserves the original nonnegative characteristic of the indicator matrix, which leads to a more tractable optimization problem with an accurate solution. Due to the nonnegativity, SNSC offers high interpretability to the indicator matrix, that is, the final cluster labels can be directly obtained without post-processing. SNSC also scales linearly with the data size, thus it can be easily applied to large-scale problems. In addition, limited label information can be naturally incorporated into SNSC for improving clustering performance. Extensive experiments demonstrate the superiority of SNSC as compared to the state-of-the-art methods.

Keywords: Spectral clustering · Nonnegativity · Semi-supervised

1 Introduction

Spectral clustering (SC), a widely used graph-based clustering approach, has drawn much attention in recent years [9]. However, there are two main limitations. First, SC involves costly steps including the affinity graph construction and eigen-decomposition. Thus their applicability to large-scale problems is highly restricted [10]. Second, most SC methods relax the entries of the indicator matrix to be arbitrary real values, which may cause severe deviation of the solution from the real one, due to the loss of the nonnegative characteristic [4,6].

To address the aforementioned problems, we propose a novel spectral clustering method, namely *Scalable Nonnegative Spectral Clustering* (SNSC), which deals with the solution deviation issue and scalability issue simultaneously. Thus SNSC can achieve better performance and has the ability to handle large-scale problems. To take full advantage of the valuable prior knowledge, we naturally extend SNSC to make it suitable for semi-supervised clustering.

© Springer Nature Switzerland AG 2019
G. Li et al. (Eds.): DASFAA 2019, LNCS 11448, pp. 287–291, 2019.
https://doi.org/10.1007/978-3-030-18590-9_30

2   The Proposed Approach

Given n samples $X = \{x_1, ..., x_l, x_{l+1}, ..., x_n\}$, the first l samples x_i ($i = 1, ..., l$) are labeled as $y_i \in \{1, ..., k\}$ from k distinct classes, and the remaining samples are unlabeled. We can construct an affinity matrix $W \in \mathbb{R}^{n \times n}$ and a ground truth matrix $Y \in \mathbb{R}^{n \times k}$. Let $F \in \mathbb{R}^{n \times k}$ be the cluster indicator matrix, in which $f_{ik} = 1$ indicates that x_i is assigned to the k-th cluster. We preserve the nonnegative characteristic by adding a nonnegative constraint to spectral clustering as follows:

$$\min_{F} Tr(F^T LF) \quad s.t. \ F^T F = I, F \geq 0. \tag{1}$$

Since directly optimizing Eq. (1) is time consuming, similar to [3], we apply some transformation and introduce an auxiliary matrix $H \in \mathbb{R}^{n \times k}$ to it. Then we further incorporate the label information to Eq. (1). Thus, the final model called SNSC is proposed as follows:

$$\min_{F,H} \|W - FH^T\|_F^2 + \alpha\|F - H\|_F^2 + \beta\|H - Y\|_F^2 \ s.t. \ F^T F = I, H \geq 0, \tag{2}$$

where $\alpha > 0$ and $\beta > 0$ are two trade-off parameters to balance three terms. If β is set to 0, SNSC boils down to the unsupervised version. The first item naturally captures the cluster structure due to the combination of the orthogonal and nonnegative constraint. The second item is a regularization term which forces F and H as close as possible. The third item is a label regularizer to force the predicted labels closer to the ground truth. Compared with Eqs. (1), (2) is a more tractable optimization problem. Due to the nonnegativity, H can be seen as the *label matrix* that gives the final cluster labels. Concretely, the entries of H can be viewed as the propensity that the data points belong to a certain cluster, then the final cluster labels can be obtained from the column index of the largest entry in each row of H.

For the optimization of the objective function, we divide it into two subproblems and then optimize them iteratively. Concretely, we first update H with F fixed and then update F with H fixed, iteratively updating H and F until convergence. To accelerate the process of graph construction, a fast approximation approach is utilized to compute the affinity matrix W [5]. In summary, SNSC consists of three stages: (1) constructing affinity graph, (2) initializing F, and (3) updating H, F iteratively. Each stage can be completed in linear time. Therefore, our proposed approach scales linearly with the data size.

3   Experiments

Datasets and Comparison Schemes. We conduct experiments on four mid-sized and large real-world datasets: COIL-100, USPS, PenDigits and Seismic. We compare SNSC to five relevant clustering methods: SC (Spectral Clustering), CKmeans [1], SCPC [8], APJCSC [7], SCSSR [2]. Furthermore, we implement

two versions of SNSC, in which SNSC-R selects the anchor points by random
sampling and SNSC-K adopts k-means, where anchor points will be used for
graph construction. For all the parameters in compared methods, they will be
set following the suggestion in their original papers or source codes.

Table 1. Clustering ACC (%) comparison on all datasets.

Datasets	Label ratio	SC	CKmeans	SCPC	APJCSC	SCSSR	SNSC-R	SNSC-K
COIL-100	5%	55.73	72.23	59.91	57.38	67.33	74.17	**81.92**
	10%	55.73	74.99	66.39	60.79	73.08	80.87	**88.10**
	15%	55.73	78.21	67.65	64.10	75.71	84.28	**90.15**
USPS	5%	67.71	76.09	43.19	39.43	76.86	89.62	**93.73**
	10%	67.71	79.77	61.48	52.31	82.93	90.47	**94.54**
	15%	67.71	81.42	77.10	67.68	85.10	91.75	**94.96**
PenDigits	5%	64.92	78.32	76.75	72.20	70.88	92.74	**95.93**
	10%	64.92	79.13	87.22	78.85	79.36	93.64	**96.48**
	15%	64.92	80.96	88.01	81.03	83.36	94.05	**96.91**
Seismic	5%	-	69.23	55.07	-	-	71.92	**72.35**
	10%	-	71.02	59.68	-	-	73.62	**73.98**
	15%	-	72.71	64.06	-	-	75.12	**75.36**

Table 2. Computational time (s) with 5% label ratio.

Datasets	SC	SCPC	APJCSC	SCSSR	SNSC-R	SNSC-K
COIL-100	31.42	**5.37**	153.3	1395.7	10.83	24.11
USPS	28.72	2.24	256.35	2216.3	**1.13**	6.3
PenDigits	3.42	2.46	397.13	2986.25	**0.95**	2.29
Seismic	-	1389.18	-	-	**23.14**	46.51

Experimental Results. These clustering results are measured by clustering
accuracy (ACC). For the semi-supervised clustering settings, we randomly choose
5%, 10% and 15% labeled samples from each dataset such that they contain at
least one sample from each class. By repeating experiments 10 times, the average
results are reported in Table 1, where the best results have been bolded. Since
SC, APJCSC and SCSSR are infeasible on Seismic, they are not run on this
dataset. As can be seen, SNSC outperforms all the baselines on different datasets
with different label ratio. For most case, SNSC is significantly better than the
compared methods, gaining at least 10% performance improvement. It is noted
that SNSC consistently produces better performance with the increment of label
information, which indicates that the label information is efficiently utilized by
SNSC to guide the clustering process. Compared with other semi-supervised
spectral clustering methods, the superiority of SNSC verifies that SNSC is able
to obtain a more accurate solution by preserving the nonnegative characteristic.

Turn to SNSC-R and SNSC-K, SNSC-K performs much better on all datasets, which reveals that k-means is more efficacious for anchor points selection.

In Table 2, we report the computational time of all methods on different datasets with 5% label ratio. Note that CKmeans is not included because it is not based on spectral clustering. It is observed that SNSC is quite efficient on large dataset. Concretely, SNSC-R is at least 60 times faster than the accelerated method SCPC, which demonstrates the efficiency of our improvements on spectral clustering.

4 Conclusion

In this paper, we propose a novel and fast clustering method named SNSC, which solves the solution deviation issue and scalability issue simultaneously. By preserving the nonnegativity of the indicator matrix and further incorporating the label information, the solution is much closer to the ideal one and the clustering performance is improved efficiently. Besides, SNSC scales linearly with the data size, thus it can be easily applied to large-scale problems. Extensive experiments are performed to show the effectiveness and superiority of our proposed method.

Acknowledgments. The paper is supported by the National Key R&D Program (2016YFB1000101), the National Natural Science Foundation of China (11801595, U1811462), the Natural Science Foundation of Guangdong (2018A030310076), the Program for Guangdong Introducing Innovative and Entrepreneurial Teams (2016ZT06D211) and the CCF Opening Project of Information System.

References

1. Basu, S., Banerjee, A., Mooney, R.: Semi-supervised clustering by seeding. In: ICML 2002, Citeseer (2002)
2. Jia, Y., Kwong, S., Hou, J.: Semi-supervised spectral clustering with structured sparsity regularization. IEEE Sig. Process. Lett. **25**(3), 403–407 (2018). https://doi.org/10.1109/LSP.2018.2791606
3. Kuang, D., Ding, C., Park, H.: Symmetric nonnegative matrix factorization for graph clustering. In: SDM 2012, pp. 106–117. SIAM (2012). https://doi.org/10.1137/1.9781611972825.10
4. Liu, J., Wang, C., Danilevsky, M., Han, J.: Large-scale spectral clustering on graphs. In: AAAI 2013, pp. 1486–1492. AAAI Press (2013)
5. Liu, W., He, J., Chang, S.F.: Large graph construction for scalable semi-supervised learning. In: ICML 2010, pp. 679–686 (2010)
6. Musco, C., Musco, C.: Recursive sampling for the nystrom method. In: NeurIPS 2017, pp. 3833–3845 (2017)
7. Qian, P., et al.: Affinity and penalty jointly constrained spectral clustering with all-compatibility, flexibility, and robustness. IEEE Trans. Neural Netw. Learn. Syst. **28**(5), 1123–1138 (2017). https://doi.org/10.1109/TNNLS.2015.2511179
8. Semertzidis, T., Rafailidis, D., Strintzis, M.G., Daras, P.: Large-scale spectral clustering based on pairwise constraints. Inf. Process. Manag. **51**(5), 616–624 (2015). https://doi.org/10.1016/j.ipm.2015.05.007

9. Von Luxburg, U.: A tutorial on spectral clustering. Stat. Comput. **17**(4), 395–416 (2007). https://doi.org/10.1007/s11222-007-9033-z
10. Yang, Y., Shen, F., Huang, Z., Shen, H.T., Li, X.: Discrete nonnegative spectral clustering. IEEE Trans. Knowl. Data Eng. **29**(9), 1834–1845 (2017). https://doi.org/10.1109/TKDE.2017.2701825

Distributed PARAFAC Decomposition Method Based on In-memory Big Data System

Hye-Kyung Yang$^{(\boxtimes)}$ and Hwan-Seung Yong

Department of Computer Science and Engineering,
Ewha Womans University, Seoul, Korea
yang88710@ewhain.net, hsyong@ewha.ac.kr

Abstract. We propose IM-PARAFAC, a PARAFAC tensor decomposition method that enables rapid processing of large scalable tensors in Apache Spark for distributed in-memory big data management systems. We consider the memory overflow that occurs when processing large amounts of data because of running on in-memory. Therefore, the proposed method, IM-PARAFAC, is capable of dividing and decomposing large input tensors. It can handle large tensors even in small, distributed environments. The experimental results indicate that the proposed IM-PARAFAC enables handling of large tensors and reduces the execution time.

Keywords: Distributed PARAFAC decomposition · Apache Spark

1 Introduction

Tensors are high-dimensional matrices [1]. Recently, there are increasingly large numbers of studies on data analysis that use tensor data. The reasons are that high-dimensional data have latency factors and patterns. However, it is difficult to analyze the whole tensor because of its sparseness and complexity as compared to a matrix. As such, tensor analysis methods that aid in the identification of potential patterns and elements using tensor decomposition have been developed.

Since the existing tensor tools run on single machines and are unable to handle large amounts of data, a recent studies have focused on large-scalable tensors. In particular, there are several Hadoop-based tensor decomposition studies, such as GigaTensor [2], BigTensor [3] and so on. Although these studies have proved effective in handling large tensors, the slow computational process remains a challenge. This can be attributed to the high disk I/O cost of Hadoop when performing iterative operations. Therefore, we propose a novel tensor decomposition method based on Apache Spark. The fundamental data structure of Apache Spark comprises resilient distributed datasets (RDD) [4]. RDD facilitates various computation and iterative operations on large amounts of data. Therefore, Spark-based tensor decomposition can be realized at a greater

© Springer Nature Switzerland AG 2019
G. Li et al. (Eds.): DASFAA 2019, LNCS 11448, pp. 292–295, 2019.
https://doi.org/10.1007/978-3-030-18590-9_31

speed than Hadoop-based decomposition, given the large number of repetitive operations involved in tensor decomposition. However, it is difficult to handle a large-scalable tensor in a small distributed environment. Therefore, we propose IM-PARAFAC, to use memory more efficiently and to reduce execution time. This study focuses on PARAFAC decomposition algorithm. The proposed method can load and compute non-zero elements in sparse or dense tensors, and it can divide and decompose large input tensors because the memory space is limited. In this paper, we evaluate the performance of IM-PARAFAC, and compare its execution time against state-of-the-art tensor decomposition algorithms using experiments with various tensor datasets. The experimental results indicate that the proposed IM-PARAFAC can handle large tensors while reducing the execution time.

2   Tensor Decomposition Method Based on Apache Spark

PARAFAC decomposition algorithm is factorized into a sum of rank-one tensors [1–4]. In the case of a three-way tensor X, it is decomposed into R components to yield three factor matrices, A, B and C. The alternating least square (ALS) method is used for obtaining PARAFAC decomposition. The PARAFAC-ALS approach fixes two factor matrices to solve for another factor matrix. This process is repeated until either the maximum number of iterations is reached, or convergence is realized [1–4]. The $X_{(1)}(C \odot B)(C^\mathsf{T} C * B^\mathsf{T} B)^{\dagger 1}$ is used for obtaining factor matrix A in PARAFAC-ALS. Here, the product $X_{(1)}(C \odot B)$ leads to intermediate data explosion. The Khatri-rao($\odot$) product of two matrices $C \in \mathbb{R}^{K \times R}$ and $B \in \mathbb{R}^{J \times R}$ is the result of matrices of size $(KJ) \times R$. Especially, distributed in-memory big data systems like Spark system could experience memory overflow errors, because of limited memory spaces. Therefore, this study develops a PARAFAC decomposition algorithm for efficient processing in distributed in-memory environments.

First, when we load the input tensors, only non-zero elements are converted into the RDD format. We then unfold the tensor using *CoordinateMatrix*. Apache Spark supports the *CoordinateMatrix* format, which is a distributed matrix constructed by RDD of MatrixEntry entries. The *CoordinateMatrix* is used when the size of the matrix is large and sparse data [4]. Subsequently, we can set a sub-tensor size when dividing the tensor. For example, if we set the size of the sub-tensors $I \times J \times N$, we obtain K/N sub-tensors; moreover, the factor matrix is also divided by K/N. As mentioned earlier, three-way PARAFAC-ALS is obtained by fixing two matrices and updating one matrix. We modified Khatri-rao product to avoid the intermediate data explosion problem. The Khatri-Rao product is used for factor matrix A using Eq. (1) as follows.

---

[1] $X_{(1)}$ is unfolded by mode I of the tensor $X \in \mathbb{R}^{I \times J \times K}$. The symbol $*$ is the Hadamard product and the symbol $\dagger$ is the pseudo-inverse of the matrix. The Khatri-Rao product is denoted by $\odot$.

$$X_{(1)}(C \odot B)(j,r) = \sum_{s=0}^{S} \sum_{y=0}^{JN-1} X_{(1)_s}(j,y) \, C_s(\lceil \tfrac{y}{I} \rceil, r) \, B(y\%I, r) \tag{1}$$

$$(C^{\mathsf{T}} C * B^{\mathsf{T}} B)^{\dagger} = (\sum_{s=0}^{S} (C_s^{\mathsf{T}} C_s) * B^{\mathsf{T}} B)^{\dagger} \tag{2}$$

Here, S is number of sub-tensors and N indicates the size of the sub-tensor. $X_{(1)_s}$ is unfolding sub-tensors. We calculate $X_{(1)}(C \odot B)$ using the sub-tensors and sub-matrices C_s and factor matrix B. These formulae are computed using the *Map* and *ReduceByKey* operators provided in Apache Spark. In this process, the *HashPartition* functions supported by Apache Spark are used to handle the same hash key on the same node, which reduces the network traffic [4]. Then, $C^{\mathsf{T}} C * B^{\mathsf{T}} B$ can be calculated, as shown in equation (2). We can obtain the factor matrix B in a similar method. However, matrix C is solved differently that matrices A and B. Moreover, given that matrix C is already divided, each C_s matrix should be solved separately.

$$C_s(k,r) = \sum_{y=0}^{N-1} X_{(3)_s}(k,y) \, B(\lceil \tfrac{y}{I} \rceil, r) \, A(y\%I, r)(B^{\mathsf{T}} B * A^{\mathsf{T}} A)^{\dagger} \tag{3}$$

In Eq. (3), by fixing A and B, each C_s matrix is solved. Subsequently, factor matrix C can be sequentially added to the rows of each C_s matrix.

3   Experiments

The experimental environments used Hadoop v2.6 and Apache Spark v1.6.1 and were conducted using a 6-node worker, with a total of 48 GB memory. In Table 1, the tensor datasets used in the experiments are listed and the execution time was compared against existing tensor decomposition tools. The tensor datasets consist of synthesized and real tensors[2]. Table 1 shows the comparison between the execution times of BigTensor[3], S-PARAFAC[4] and the proposed method. IM-PARAFAC set the optimal divided slice size for the each tensor dataset. For example, we can set optimal size of 100 in Dense500, because Fig. 1 shows that slices of size 100 rapidly reduced the execution time. As a result, it can be seen that the proposed method outperforms BigTensor in all aspects. Moreover, BigTensor and S-PARAFAC were unable to process 1998DARPA data in small distributed environments, whereas IM-PARAFAC was able to process data even when divided. Following from this result, we found that IM-PARAFAC processes faster than S-PARAFAC in large dense dataset. We also compared relative error to other tools using synthetic datasets. In Table 2, we observe a similar relative error to the other tools. Furthermore, we know that sparse datasets has low accuracy because reconstruction of the tensor fills missing values.

---

[2] Real datasets are supported by BigTensor. https://datalab.snu.ac.kr/bigtensor/dataset.

[3] BigTensor is Hadoop-based tensor decomposition tool.

[4] S-PARAFAC is Spark-based PARAFAC decomposition tool.

Table 1. Tensor dataset and comparison of execution time (min)

Data Name	Tensor size	Non-zero	BigTensor	S-PARAFAC	IM-PARAFAC
Sparse100	100 × 100 × 100	312,969	1.5	0.33	0.46
Dense100	100 × 100 × 100	10,000,000	3.714	0.76	1.08
Sparse500	500 × 500 × 500	15,000	5.97	0.35	0.43
Dense500	500 × 500 × 500	125,000,000	N/A	40.2	36.46
YELP [3]	70K × 15K × 108	334,166	7.74	1.007	1.4
MovieLens [3]	71K × 10K × 157	10,000,054	75.48	8.48	10.7
1998DARPA [3]	22K × 22K × 23M	28,436,033	N/A	N/A	2304.6

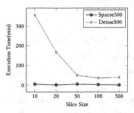

Fig. 1. The comparison of execution times between Sparse500 and Dense500 datasets for varying slice sizes

Table 2. The comparison of relative error on the synthetic datasets.

Data name	Big tensor	S-PARAFAC	IM-PARAFAC
Sparse100	0.43	0.42	0.43
Dense100	0.36	0.35	0.34
Sparse500	0.98	0.98	0.97
Dense500	N/A	0.31	0.31

4   Conclusions

IM-PARAFAC can handle large tensors even in small distributed environments, because it divides and decompose large input tensors. The proposed method outperforms state-of-the-art Hadoop-based methods, in terms of both execution time and the ability to process large tensors. Furthermore, compared to S-PARAFAC running on Spark, IM-PARAFAC can handle a large tensor dataset.

Acknowledgement. This research was supported by Basic Science Research Program through the National Research Foundation of Korea (NRF) funded by the Ministry of Education (NRF-2016R1D1A1B03931529).

References

1. Kolda, T.G., Bader, B.W.: Tensor decompositions and applications. SIAM Rev. **51**(3), 455–500 (2009)
2. Kang, U., Papalexakis, E.E., Harpale, A., Faloutsos, C.: GigaTensor: scaling tensor analysis up by 100 times - algorithms and discoveries. In: Proceedings of the 18th ACM SIGKDD International Conference on Knowledge Discovery and Data Mining, KDD 2012, Beijing, China, 12–16 August 2012, pp. 316–324. ACM (2012)
3. Park, N., Jeon, B., Lee, J., Kang, U.: BIGtensor: mining billion-scale tensor made easy. In: Proceedings of the 25th ACM International on Conference on Information and Knowledge Management (CIKM 2016), pp. 2457–2460. ACM (2016)
4. Yang, H.K., Yong, H.S.: S-PARAFAC: distributed tensor decomposition using Apache Spark. J. Korean Inst. Inf. Sci. Eng. (KIISE) **45**(3), 280–287 (2018)

GPU-Accelerated Dynamic Graph Coloring

Ying Yang, Yu Gu$^{(\boxtimes)}$, Chuanwen Li, Changyi Wan, and Ge Yu

School of Computer Science and Engineering, Northeastern University,
Shenyang 110819, Liaoning, China
guyu@mail.neu.edu.cn

Abstract. The graph coloring is a classic problem in the graph theory, which can be leveraged to mark two objects with a certain relationship with different colors. Existing graph coloring solutions mainly focus on efficiently calculating high-quality coloring of static graphs. However, many graphs in the real world are highly dynamic and the coloring result changes when the graph is updated. Repeated adoption of static graph coloring schemes will incur prohibitive costs. Although some CPU-based incremental graph coloring methods have been proposed recently, they become inefficient when facing dense graphs and large batch updates. In this paper, we explore the dynamic graph coloring solution by utilizing the powerful parallel processing capabilities of GPU and propose a CPU-GPU heterogeneous method. We conduct extensive experiments comparing our algorithm with the existing methods. The results confirm that our algorithm is superior to others in many aspects such as coloring efficiency.

Keywords: Graph coloring · Dynamic graph · GPU ·
Parallel computing

1 Introduction

Graphs can be used to model complex entities and relationships in the real world, and thus graph processing has been becoming a key workload in various data analysis tasks, including business analysis, social networking, engineering simulation, etc. [1]. Therefore, exploring the coloring problem of graphs is of great significance for solving various practical applications.

The exact evaluation of graph coloring is a NP-complete problem [4]. [2,3,6] have proposed GPU-based static graph coloring algorithms, which are better than the existing CPU algorithms in efficiency. However, the real-world graphs are constantly updating which leads to the frequent change of the coloring results. Iterative usage of static graph coloring algorithms can incur considerable processing latency. As the up-to-date solution, an effective dynamic graph coloring method is proposed in [7], which transforms undirected graphs of the real world into a directed graph to satisfy the consistency of the coloring result,

G. Li et al. (Eds.): DASFAA 2019, LNCS 11448, pp. 296–299, 2019.
https://doi.org/10.1007/978-3-030-18590-9_32

and further improves the coloring efficiency by constructing indexes, pruning and other operations. However, for dense graphs and large batch updates, the cost of this method becomes prohibitive.

To crack the nut of efficient and effective graph coloring in the dynamic scenario, we propose a novel GPU-accelerated solution in this paper. Specifically, We propose an algorithm to refresh dynamic updating graphs in the batch mode, which makes full utilization of the parallel processing power of the GPU. We design the *hybrid block* data structure, which can be processed on both CPU and GPU. It is efficient in both intro-block parallel and inter-block serial calculations on GPU. Furthermore, We propose a CPU-GPU-based heterogeneous dynamic coloring method, applying the global idea to ensure that the coloring results are independent of the coloring order during the dynamic graph coloring calculation.

2 Proposed Method

2.1 Overall Process

The key idea of our approach is to solve the problem of coloring updates of dynamic graphs by a GPU based graph coloring algorithm. We first update the preliminary pending vertices in the batch mode. Then, the sub-units suitable for GPU parallel processing are divided by our hybrid partitioning method. Finally, we perform the heterogeneous coloring calculation. Figure 1 depicts the overall work flow of our method, in which step 1, step 3, and step 4 work on CPU, and step 2 works on GPU.

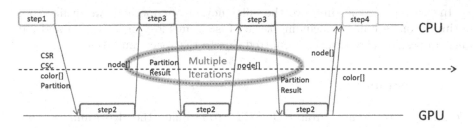

Fig. 1. Overall design process.

2.2 Detailed Introduction

A. Prepossessing, Batch Update and Data Organization. First, we detail the first half of step 1 in Fig. 1. The process is divided into three sub-steps: graph preprocessing, batch updating, and data organization. The preprocessing refers to the directionalization of the original undirected graph G into a directed graph G^*. We adopt the idea proposed in [7]. By transforming the undirected graph into a directed graph, it can be ensured that the coloring result is independent of the insertion order of edges. To fully utilize the computing power of GPU, we propose to adopt the batch processing. Batch update refers to the process of coloring update of a graph after merging a series of delete and insert operations. The most efficient data structure on GPU is the form of arrays. Therefore we adopt the CSR format to store the vertex information.

B. Hybrid Partition Design. Furthermore, we introduce the second part of step 1 and step 3 in Fig. 1, for determining the re-coloring vertex queue q. When we color a vertex, it is necessary to access the neighbors of the vertex to determine the colors which will cause severe access conflicts in parallel processing. To ensure the correctness of the results, we design a greedy hybrid partitioning method. First, we define N partitions, and sort the vertices in q according to the declining order of degree. And then we place the first vertex in the first partition, traverse vertices in turn, and add the vertices which are not adjacent to the vertices in q to the partition. We repeat the above step until the first $(N-1)$ partitions are finished. Finally, we put the remaining vertices in q into the N_{th} partition. The method can guarantee that the vertices in each partition can be parallel on GPU and no data access conflict occurs.

C. Heterogeneous Dynamic Graph Coloring Design. Finally, we describe the step 2 processed in Fig. 1. After receiving the $(n-1)$ partition results from CPU, GPU will implement the intra-partition parallel and inter-partition serial high-concurrency calculation. Each partition P_i creates $|P_i|$ threads, where each thread processes one vertex, implements vertex re-coloring, and returns the results to CPU. Recoloring any vertex u in q is conducted by both CPU and GPU.

In order to further improve the efficiency of our method, we utilize some optimization techniques including kernel fusion, memory organization and parallelism design. The details are omitted due to the space limitation.

3   Experiments

We select 10 real graphs from KONECT[1] to evaluate the algorithm. We compare our method, denoted by RC_Hybrid, with two existing methods, denoted by RC_Local [5] and RC_Orient [7]. We simulate the dynamic graph by inserting/deleting 10% edges randomly into the original graph for each update. All experiments are conducted on a machine with an Intel Core i7 3.2 GHz CPU (8 cores) and an GTX1060 GPU with 6 GB memory.

Figure 2 illustrates the average processing time for each update of the three algorithms and RC_Hybrid uses 8 partitions by default. We can find that, RC-hybrid is superior to RC-Local and RC-Orient. This is because the RC-hybrid algorithm is a GPU accelerated heterogeneous coloring method. When the update size of the graph is sufficiently large, the parallel capability of the GPU can be fully utilized, and high coloring efficiency is achieved.

---

[1] http://konect.uni-koblenz.de/networks.

We illustrate the effect of different number of partitions on the coloring efficiency of the RC-hybrid algorithm in Fig. 3 by illustrating the rendering efficiency in different partitions over the Amazon data set. It can be found that when the number of partitions reaches a certain range, coloring efficiency becomes stable.

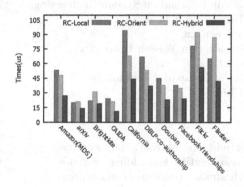

Fig. 2. Coloring efficiency.

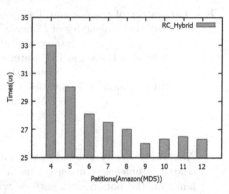

Fig. 3. The number of partitions.

Acknowledgements. This work is supported by the National Key R&D Program of China (2018YFB1003404), the National Nature Science Foundation of China (61872070, U1435216, 61872071 and 61602103) and the Fundamental Research Funds for the Central Universities (N171605001).

References

1. Blum, A.: New approximation algorithms for graph coloring. J. ACM **41**(3), 470–516 (1994)
2. Che, S., Rodgers, G., Beckmann, B., Reinhardt, S.: Graph coloring on the GPU and some techniques to improve load imbalance. In: Parallel and Distributed Processing Symposium Workshop, pp. 610–617 (2015)
3. Chen, X., Li, P., Yang, C.: Efficient and high-quality sparse graph coloring on the GPU. CoRR abs/1606.06025 (2016)
4. Li, Z., Zhu, E., Shao, Z., Xu, J.: Np-completeness of local colorings of graphs. Inf. Process. Lett. **130**, 25–29 (2018)
5. Preuveneers, D., Berbers, Y.: ACODYGRA: an agent algorithm for coloring dynamic graphs. SYNASC **6**, 381–390 (2004)
6. Shi, X., et al.: Frog: asynchronous graph processing on GPU with hybrid coloring model. IEEE Trans. Knowl. Data Eng. **30**(1), 29–42 (2018)
7. Yuan, L., Qin, L., Lin, X., Chang, L., Zhang, W.: Effective and efficient dynamic graph coloring. PVLDB **11**(3), 338–351 (2017)

Relevance-Based Entity Embedding

Weixin Zeng[1], Xiang Zhao[1,3](✉), Jiuyang Tang[1,3], Jinzhi Liao[1],
and Chang-Dong Wang[2]

[1] Key Laboratory of Science and Technology on Information System Engineering,
National University of Defense Technology, Changsha, China
xiangzhao@nudt.edu.cn
[2] School of Data and Computer Science, Sun Yat-sen University,
Guangzhou, China
[3] Collaborative Innovation Center of Geospatial Technology,
Wuhan, China

Abstract. Entity embedding plays an indispensable role in many entity-related problems. Currently, mainstream entity embedding methods build on the notion that entities with similar contexts or close proximity should be placed adjacently in the embedding space. Nonetheless, this goal fails to meet the objectives of many downstream tasks, where the relevance among entities is more significant. To fill this gap, in this paper, a novel relevance-based entity embedding approach, Lead, is proposed, where the relevance is captured via query-document information. The experimental results verify the superiority of our proposal.

Keywords: Entity embedding · Relevance · Knowledge graph

1 Introduction

Over recent years, Knowledge Graph (KG), along with its applications, has proliferated in many fields. As the unique identifier of object, entity is the basic component of KG and also the pivot connecting irregular free text and structured KG. To alleviate the sparse representation and facilitate the computation of similarity, vectorial representation of entity, entity embedding, is proposed, which can be found useful in a variety of tasks. Currently, entity embeddings are mainly generated via traditional word2vec model [1], with words replaced with entities. The target of this approach is to model the proximity of similar terms by predicting adjacent entities given specific entity.

Nonetheless, the objective of aforementioned embedding solution does not necessarily meet the requirement of many downstream tasks, particularly those within the scope of information retrieval from various databases, where the primary objective is to capture relevance, instead of proximity, among entities [2]. Concretely, the notion of proximity is more restricted in comparison to that of relevance. While the former requires that two items should have many features, e.g., contexts, types and description information, in common, the latter merely

© Springer Nature Switzerland AG 2019
G. Li et al. (Eds.): DASFAA 2019, LNCS 11448, pp. 300–304, 2019.
https://doi.org/10.1007/978-3-030-18590-9_33

needs the items to be related in some sense. Consequently, introducing the notion of relevance into entity embedding can broaden the extension of relatedness, and capture a wider range of signals of connections among entities.

As thus, we are motivated to investigate the integration of relevance into entity embedding. Specifically, a novel relevance based entity embedding, Lead, is put forward to explicitly model the relevance among entities by optimizing the objective that given specific information need, i.e., user query, the entities in relevant documents should be predicted. To achieve this goal, we design a standard feedforward neural network with a single hidden layer, which takes query entities as inputs, and the training target is to optimize the network parameters by classifying an entity in document as relevant or irrelevant to query. The neural model is then trained over millions of queries and Wikipedia documents. The effectiveness of Lead is examined against entity embedding based on word2vec [3] via diversified entity recommendation (DER) [4] task.

2 Relevance-Based Entity Embedding

The neural architecture for training relevance-based entity embedding consists of input layer, hidden layer and output layer. It takes query entities as inputs and outputs a vector recording the connections between all the entities in vocabulary V and query entities. Given a query q comprising m entities, $e_1, ..., e_m$, the entities are firstly represented as one-hot sparse vectors, $\mathbf{e}_1, ..., \mathbf{e}_m$, with length N. N denotes the total number of entities in V. Then the input entity vectors are transformed to d-dimensional hidden vectors via the weight matrix $W_{N \times d}$. The m hidden vectors are averaged so as to obtain the final query hidden vector $\mathbf{h}_q = \frac{1}{m} \sum_{i=1}^{m} \mathbf{e}_i \times W_{N \times d}$, where d represents the dimension of the hidden layer. Then the query hidden vector $\mathbf{h}_q$ is forwarded through a fully connected layer so as to obtain the output $\mathbf{y} = \mathbf{h}_q \times W'_{d \times N}$, where $W'_{d \times N}$ is a $d \times N$ matrix that maps the d-dimensional query hidden vector to the output vector with length N. Since $\mathbf{y}$ records the association between all the entities in V and query q, whether a random entity $c_r (0 < r \leq N)$ is related to the query can be measured by $\hat{p}(R = 1|e_r, q; \theta) = \frac{1}{1+\exp(-y_r)}$, where y_r is the r-th value of output vector $\mathbf{y}$, and R is a boolean variable, with 1 representing that entity e is related to query q. θ consists of the parameters in the neural network, i.e., $W_{N \times d}$ and $W'_{d \times N}$. The objective function can be formally defined as:

$$\arg\max \sum_{i=1}^{n} \sum_{e_j \in \mathbb{E}_i^+} [\log \hat{p}(R = 1|e_j, q_i; \theta) + \sum_{e_k \in \mathbb{E}_{i,j}^-} \log \hat{p}(R = 0|e_k, q_i; \theta)], \quad (1)$$

where $\mathbb{E}_i^+$ denotes the positive instances for query q_i, and $\mathbb{E}_{i,j}^-$ represents negative instances for each positive instance e_j.

To obtain the training set from query-document information, firstly given query q_i, the entity distribution $\mathbf{R}_i$ is generated via pseudo-relevance feedback model [5], Then we choose the top-η^+ entities with the highest probabilities to form the positive instances $\mathbb{E}_i^+$ as they can best represent all the relevant entities. And η^- negative instances $\mathbb{E}_{i,j}^-$ are sampled for each positive instance e_j [1].

3 Experiment

Preprocessing and Training. The corpus for training Lead was derived from publicly available AOL query log, where DBpedia spotlight[1] was harnessed to detect entities in text. A total of 3.55 million queries were used for final training after cleaning. Regarding document collection, to avoid the complicated text annotation process, we utilized Wikipedia articles[2], in which anchor texts had already been annotated with Wikipedia identifiers. Lucene[3] was utilized for indexing and retrieval, where top 10 results were considered as relevant documents, in which merely entities were retained.

The model was implemented and trained with Pytorch. Particularity, the learning rate was chosen from $[0.001, 0.01, 0.02, 0.1, 0.2, 1]$, the batch size was selected among $[64, 128, 256, 512, 1024]$, the number of positive instances, η^+ was swept among $[10, 20, 50, 100]$, and η^- was chosen from $[5, 10, 20]$. Note that all the hyper-parameters were fine-tuned on the training set, with the ones minimizing training loss value chosen as the optimized setting. The embedding size and iteration times were set to 300 and 20, respectively. As for the baseline, W2V, it is also trained on Wikipedia documents. The learning rate was set to 0.025, context window to 10, negative samples to 5, and epoch to 10.

Evaluation via Diversified Entity Recommendation. Diversified entity recommendation (DER) [4] aims to recommend a set of relevant and varied entities to a given entity. We used the cosine similarity between the embeddings of a target entity e and other entities for entity ranking and kept those with highest similarities for recommendation. Two datasets were harnessed for evaluation. The first one, **Ori**, replaces the word in query set [4] with its corresponding entity. The second dataset, **Seal**, is originated from SEAL dataset [6], which consists of 13 conceptual lists of entities, and we randomly selected one entity from each list to form the final test set.

We utilized the simplified quality score [4] as evaluation metric. Concretely, given an entity e, denote the top-h entities ranked by the cosine similarity of entity embeddings as $\mathbb{N} = \{n_1, ..., n_h\}$. Let $S(n_i, n_j)$ represent the relatedness score of an entity pair (n_i, n_j). Denote $r(n_p)$ as a special case of $S(n_i, n_j)$ and $r(n_p) = S(e, n_p)$, where $0 < p \leq h$. The quality score of retrieved entities $\mathbb{N}$ for entity e, $Q(\mathbb{N}) = \frac{1}{\binom{2}{h}} \sum_{(n_i, n_j) \in \mathbb{N}} r(n_i) \times r(n_j) \times \exp(-S(n_i, n_j))$, where $r(n_i)$ and $r(n_j)$ represent how relevant entities n_i and n_j are to the given entity e, while $\exp(-S(n_i, n_j))$ captures the diversity between the related entities. Intuitively, to fulfil the task of DER, it is more preferable to retrieve entities that are closely related to the query but less related among themselves. Consequently, higher Q is more desirable. Note that $S(n_i, n_j)$, which represents relatedness score between an entity pair (n_i, n_j), is generated via Milne and Witten [7].

---

[1] https://www.dbpedia-spotlight.org/.
[2] We used Wikipedia dump on 20-Mar-2018.
[3] http://lucene.apache.org/.

Results. We set h value at 5, 10 and 20 and reported the corresponding results in Table 1. It can be easily observed that Lead attains superior results regardless of datasets or h value, and the gap widens as h gets larger. Additionally, with the increase of h, the *quality* of entities retrieved by W2V drops quickly, indicating that most of the top-ranked entities conform to several certain topics and lack diversity. On the contrary, the *quality* of Lead is comparatively consistent, verifying its ability to recommend both relevant and varied entities. Figure 1 visualizes the recommendation results by Lead and W2V. Different colors of nodes represent different types of entities, while the thickness of lines denote how related the entity at the other end is to *Michael Jordan*. It can be easily observed that the related entities recommended by Lead are of various types and the relationships can be explained via diverse aspects.

Table 1. Quality score Q on DER.

Dataset	Method	Top 5		Top 10		Top 20	
Ori	W2V	0.2705	–	0.2606	–	0.2391	–
	Lead	**0.2805**	+3.70%	**0.2699**	+3.57%	**0.2695**	+12.71%
Seal	W2V	0.2671	–	0.2505	–	0.2305	–
	Lead	**0.2730**	+2.21%	**0.2687**	+7.27%	**0.2700**	+17.14%

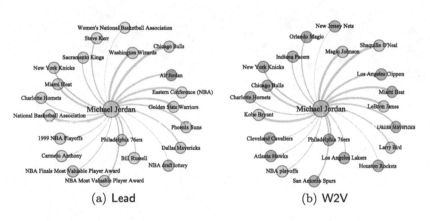

(a) Lead   (b) W2V

Fig. 1. Recommended entities to *Michael Jordan* by Lead and W2V

Acknowledgements. This work was partially supported by NSFC under grants Nos. 61872446, 61876193 and 71690233, and NSF of Hunan province under grant No. 2019JJ20024.

References

1. Mikolov, T., Sutskever, I., Chen, K., Corrado, G.S., Dean, J.: Distributed representations of words and phrases and their compositionality. In: NIPS 2013, pp. 3111–3119 (2013)
2. Zamani, H., Croft, W.B.: Relevance-based word embedding. In: SIGIR 2017, pp. 505–514 (2017)
3. Yamada, I., Shindo, H., Takeda, H., Takefuji, Y.: Joint learning of the embedding of words and entities for named entity disambiguation. In: CONLL 2016, pp. 250–259 (2016)
4. Krishnan, A., Padmanabhan, D., Ranu, S., Mehta, S.: Select, link and rank: diversified query expansion and entity ranking using Wikipedia. In: WISE 2016, pp. 157–173 (2016)
5. Lavrenko, V., Croft, W.B.: Relevance-based language models. In: SIGIR 2001, pp. 120–127 (2001)
6. Wang, R.C., Cohen, W.W.: Language-independent set expansion of named entities using the web. In: ICDM 2007, pp. 342–350 (2007)
7. Witten, I.H., Milne, D.N.: An effective, low-cost measure of semantic relatedness obtained from Wikipedia links. In: AAAI 2008, pp. 25–30 (2008)

An Iterative Map-Trajectory Co-optimisation Framework Based on Map-Matching and Map Update

Pingfu Chao[✉], Wen Hua, and Xiaofang Zhou

School of ITEE, The University of Queensland, Brisbane, Australia
{p.chao,w.hua}@uq.edu.au, zxf@itee.uq.edu.au

Abstract. The digital map has long been suffering from low data quality issues caused by lengthy update period. Recent research on map inference/update shows the possibility of updating the map using vehicle trajectories. However, since trajectories are intrinsically inaccurate and sparse, the existing map correction methods are still inaccurate and incomplete. In this work, we propose an iterative map-trajectory co-optimisation framework that takes raw trajectories and the map as input and improves the quality of both datasets simultaneously. The map and map-matching qualities are quantified by our proposed measures. We also propose two scores to measure the credibility and influence of new road updates. Overall, our framework supports most of the existing map inference/update methods and can directly improve the quality of their updated map. We conduct extensive experiments on real-world datasets to demonstrate the effectiveness of our solution over other candidates.

Keywords: Map update · Map-matching ·
Map-trajectory co-optimisation

1 Introduction

The map quality issue has long been a persistent problem mainly caused by the low update frequency compared with the rapid map change [4]. In the recent years, the pervasive use of GPS-enabled devices produces massive users' trajectories, which can be map-matched [3] to existing maps to support various map-based applications or used to automatically construct or update maps [1,4].

However, the intrinsic inaccuracy of both maps and trajectories poses great challenges to the applications: The inaccurate map causes wrong map-matching results, while the uncertain trajectories lead to a poorly constructed map. To address those challenges, in this paper, we propose an iterative map-trajectory co-optimisation algorithm to improve the quality of both maps and trajectories. Overall, the main contributions of this work can be summarised as follows:

- We propose an iterative framework to combine map-matching, map update and our proposed co-optimisation model. The framework is generalised to support existing map-matching and map inference/update methods and achieves better overall quality through the iterative process.

© Springer Nature Switzerland AG 2019
G. Li et al. (Eds.): DASFAA 2019, LNCS 11448, pp. 305–309, 2019.
https://doi.org/10.1007/978-3-030-18590-9_34

- To the best of our knowledge, this is the first work to propose quality measures for both map-matching and map update. The goal of our model is to maximise both quality measures simultaneously through the update of map.
- We propose confidence score and influence score to measure the correctness of each road update. To this end, we propose a top-k map-matching algorithm and the concept of matching certainty to better evaluate the road influence. Experimental results show that the scores help better identify correct road updates over outliers, and meanwhile detect one-way roads.

2   Related Work

In general, map-matching, map inference and map update are all related to our work. The map-matching process aligns GPS trajectories to the road network. Among both use cases (online or offline), the probability-based method [3] is currently the state-of-the-art solution. The map inference attempts to generate digital maps directly from GPS trajectories. Among other alternatives, the clustering-based methods, such as k-means and Kernel Density Estimation (KDE), are more popular in various map inference and map update work [2,4] due to more balanced performance between efficiency and effectiveness. Unlike map inference, the map update aims to modify a given map using trajectories provided. It usually consists of three steps: (1) Extract unmatched trajectories via map-matching; (2) Conduct map inference to generate new roads; (3) Merge roads to the existing map. Most of the recent work still follows the same idea as in the pioneer work CrowdAtlas [4] with slight changes to the map influence and map merge solution. However, the existing map inference and update approaches are still far from being practically useful as they cannot handle GPS errors and trajectory disparity, making the result accuracy only around 40% (F-Score).

3   Co-optimisation Framework and Model

3.1   Framework Overview

The objective of our framework is to improve the map quality and map-matching quality through the iterative co-optimisation process. The system requires two types of input: the trajectory set R and road network G. As shown in Fig. 1, the framework mainly consists of the following procedures:

Data Preprocessing: Besides the data filtering on R, an initial map-matching process is performed to generate the input of the first iteration, including the unmatched trajectory set R_*^0 and the matching result $M(R, G^0)$.

After the preprocessing, each iteration i consists of three main steps:

Map Update: The map update takes R_*^{i-1} as input and infers new map elements ΔG^i, which are further merged with G^{i-1} to form a temporary map $G^{i'}$. The *confidence score* ω_e for each new edge is generated during map inference.

Map-Matching: We match each trajectory $tr \in R$ to the temporary road network $G^{i'}$ and generate a temporary map-matching result $M(tr, G^{i'})$ and a

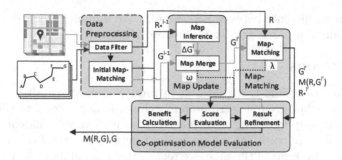

Fig. 1. Framework overview

temporary unmatched trajectory set $R_*^{i'}$. In addition, we compare the map-matching result $M(tr, G^{i'})$ with the result from last iteration $M(tr, G^{i-1})$ to generate the *influence score*, denoted by λ_e, for each new road e.

Co-optimisation Model: First, we identify the correctness of each new edge $e \in \Delta G^i$ according to the scores λ_e and ω_e generated in aforementioned steps. Then, edges that are identified as wrong roads are removed from $G^{i'}$ and an overall benefit is calculated to evaluate the quality improvement during the current iteration. The iteration terminates if the benefit is negligible. Eventually, the final G^i is determined after removing wrong elements from $G^{i'}$, and all previous temporary results, namely $M(R, G^{i'})$ and $R_*^{i'}$ are refined according to the new map G^i and finalised.

3.2   Co-optimisation Model

As mentioned in the framework, the main tasks for co-optimisation model include: (1) Road correctness identification based on influence and confidence scores. (2) Benefit calculation based on the map and matching quality measures.

For each new road e, the confidence score is defined as the number of trajectories contributing to inferring e, which can be easily generated in most map inference/update algorithms. Meanwhile, the influence score denotes the gross probability improvement of all trajectory map-matching results after e is inserted to the map. Following the idea that a matching result is more reliable if its probability outshines all other alternatives, we further propose a top-k map-matching strategy and the concept of matching certainty, which is based on the information entropy, to accurately measure the influence score. Eventually, the correctness of each new road e is determined based on the combination of two scores, and roads with low scores are removed from temporary map $G^{i'}$.

The idea of map quality and map-matching quality is based on the road correctness. For each iteration i, the newly generated map ΔG^i is comprised by a set of correct edges/nodes and a set of wrong edges/nodes, denoted as G_+ and G_- respectively. Hence, the map quality is defined as the total confidence score of G_+ subtracts G_-. Likewise, the map-matching quality is determined by the gross influence difference between G_+ and G_-. We terminate the iteration if the quality measures of G_- is greater than those of G_+.

4   Evaluations

Due to the lack of ground-truth data, we introduce a two-step evaluation process: we first use a public ground-truth map-matching dataset *Global* to validate the accuracy of our map-matching algorithm, then we use our map-matching method to generate "ground-truth" on a large-scale trajectory dataset *Beijing* for evaluation purpose. In terms of the map, we regard the original map as ground-truth and generate our input map by randomly remove a set of road edges. We use precision/recall/F-measure to evaluate both map and map-matching qualities.

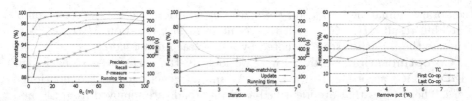

(a) Parameter tuning for map-matching evaluation
(b) Map improvement over iterations
(c) Map quality by varying removal percentage

Fig. 2. Result for effectiveness and efficiency test

Figure 2a proves that our map-matching algorithm [3] achieves a decent 98% F-measure on *Global* dataset when the HMM candidate range (θ_c) is set to 60 meters. In terms of the experiments on *Beijing* dataset, Fig. 2b demonstrates the change of map-matching and map quality over the iterations. It can be observed that both qualities improve gradually as iteration continues, which shows the effectiveness of our iterative process.

We also compare our algorithm with the most outstanding work in map update [4]. Figure 2c shows the quality of map-matching and map update when varying road removal percentage. Compared with the baseline method (denoted by TC), it is clear to see that the final iteration of our co-optimisation algorithm outperforms the baseline by more than 10% F-measure constantly.

5   Conclusion

In this paper, we propose an iterative map-trajectory co-optimisation framework to improve the quality of map and map-matching results simultaneously. We also introduce a co-optimisation model with several concepts, including influence and confidence score, map and map-matching quality measures, top-k map-matching strategy and map-matching certainty for better road correctness identification. The experiment results on real dataset show the superior performance compared with the state-of-the-art map update solutions and the effectiveness of the iterative framework in improving the map and map-matching quality.

References

1. Ahmed, M., Karagiorgou, S., Pfoser, D., Wenk, C.: Map construction algorithms. Map Construction Algorithms, pp. 1–14. Springer, Cham (2015). https://doi.org/10.1007/978-3-319-25166-0_1
2. Biagioni, J., Eriksson, J.: Map inference in the face of noise and disparity. In: Proceedings of the 20th International Conference on Advances in Geographic Information Systems, pp. 79–88. ACM (2012)
3. Newson, P., Krumm, J.: Hidden Markov map matching through noise and sparseness. In: Proceedings of the 17th ACM SIGSPATIAL International Conference on Advances in Geographic Information Systems, pp. 336–343. ACM (2009)
4. Wang, Y., Liu, X., Wei, H., Forman, G., Chen, C., Zhu, Y.: CrowdAtlas: self-updating maps for cloud and personal use. In: Proceeding of the 11th Annual International Conference on Mobile Systems, Applications, and Services, pp. 27–40. ACM (2013)

Exploring Regularity in Traditional Chinese Medicine Clinical Data Using Heterogeneous Weighted Networks Embedding

Chunyang Ruan[1,3], Ye Wang[2], Yanchun Zhang[1,2,3(✉)], and Yun Yang[4]

[1] School of Computer Science, Fudan University, Shanghai, China
cyruan16@fudan.edu.cn
[2] College of Engineering and Science, Victoria University, Melbourne, Australia
Yanchun.Zhang@vu.edu.au
[3] Cyberspace Institute of Advanced Technology, Guangzhou University,
Guangzhou, China
[4] Department of Oncology and Longhua Hospital, Shanghai, China

Abstract. Regularities of prescriptions are important for both clinical practice and novel healthcare development in clinical traditional Chinese medicine (TCM). To address this issue and meet clinical demand for determining treatments, we propose an unsupervised analysis model termed *AMNE* to determine effective herbs for diverse symptoms in prescriptions. Results confirmed by human physicians demonstrate *AMNE* can outperform several previous TCM regularity discovery models in prescriptions.

1 Introduction

The goal of exploring regularity in prescriptions is to investigate the complex interrelation between herbs composition and corresponding symptoms. Determining herbs regularity from clinical prescriptions is an essential topic for clinical treatment and novel prescription development [5]. Previous works developed many methods to investigate regularity in TCM prescriptions [2–5], but they suffer many limitations due to the high unstructured and high dimension of TCM prescriptions. In addition, TCM prescriptions are not precisely like the general sentences, which have their own way of organising the herbs and symptoms in the weakly ordered pattern.

In this paper, We explore how to automatically detect the specific herbs for symptoms in TCM prescriptions. We propose a novel deep autoencoder model termed *AMNE* based on heterogeneous information network (HIN) [1] to address this issue. We maintain the information of TCM prescriptions via HINs and

The work is supported by National Natural Science Foundation of China (No. 61672161), Youth Research Fund of Shanghai municipal health and Family Planning Commission (No. 2015Y0195).

encode the TCM-HINs in a low-dimension vector space. Then, We utilize clustering and link prediction to detect the herb-symptom relations, which can discover the herbs composition and corresponding symptoms. The proposed method can investigate the indications of the effective herbs and assist TCM physicians in prescribing. Out contributions are as follows:

– We construct the TCM-HIN for prescriptions and develop a novel propagation to preserve the structure and semantics of TCM-HIN.
– We leverage autoencoder to learn stable and robust TCM-HIN embedding.
– We utilize the clustering and the link-prediction to explore regularity of prescriptions.

Table 1. Mathematical notations and explanations

G	Graphical representation of TCM-HINs
V	The node set of TCM-HINs
E	The edge set of TCM-HINs
k	Number of layers
$T=\{t_i\}_{i=1}^n$	The adjacency matrix of TCM-HINs
$\mathbf{X}$	The initial TCM data
$\mathbf{Y}$	The low-dimensional representations of TCM data
$\mathbf{W}^{(k)}, \bar{\mathbf{W}}^{(k)}$	The k-th layer weight matrix
$\mathbf{b}^{(k)}, \bar{\mathbf{b}}^{(k)}$	The k-th layer biases
ω	The set of weighted values of all node

2 AMNE

2.1 AMNE Overview

To obtain the whole non-linear structure and rich semantics of TCM-HINs, we develop a deep architecture consisting of multiple non-linear mapping functions to encode the input data to a low-dimensional latent space. In addition, in order to address the structure-preserving and semantic-preserving, we study a deep autoencoder to exploit both the meta-path based proximity and second-order proximity. As a real-world network data, TCM-HINs has multi-type nodes, and two nodes may be connected via multiple typed paths (a.k.a *edges*). Conceptually, each path represents a specific direct or composite semantic relationship between nodes. So, we use weight meta-path based proximity to obtain pairwise similarities of nodes. Meanwhile, we attempt to capture and reconstruct its neighborhood information for each node to preserve the second-order proximity. Accordingly, to optimize the proximity in the process of model learning, our model can preserve the highly-nonlinear local and global network structure well and is robust to other HINs. Finally, we first obtain clusters including herbs and symptoms for specific disease conditions or particular patient groups based on the low-dimensional representation. Then, in each cluster, we use link prediction to detect the strong relations between herbs and symptoms. Given the notation presented in Table 1, the main algorithm is shown in Algorithm 1.

Algorithm 1. AMNE

Input:
  The network G=(V, E, ω) with adjacency matrix T;
  The length threshold l;
  The hyper-parameters α, λ;
Output:
  The Low-dimensional representation Y for TCM-HINs; The updated Parameters:
    $\theta=\{\mathbf{W}^{(k)}, \bar{\mathbf{W}}^{(k)}, \mathbf{b}^{(k)}, \bar{\mathbf{b}}^{(k)}\}_{k=1}^n$;
1: Initial parameters θ via random process;
2: Set X=T;
3: **repeat**
4:    X, θ, Eq.3 $\mapsto$ $\bar{\mathbf{X}}$, $\mathbf{Y}^{(k)}$;
5:    Sample negative nodes randomly via *PRM*;
6:    Based on Eq.8 and *ASGD*, adopt $\partial L/\partial\theta$ to
7:    back-propagate through the entire network
      to get updated parameters θ;
8: **until** converge
9: **return** TCM-HINs representations Y=$\mathbf{Y}^{(k)}$;

---

3 Experiments

3.1 Baseline

We conduct the experiments to compare different models (i.e. CPM [4], PaReCat [2]) for the effective herbs detecting in TCM prescriptions based on a real-world dataset collected from clinical prescriptions for lung tumor.

Table 2. The difference and intersection (maked in bold) herbs prescribed by *AMNE* and TCM physicians according to clinical symptoms of lung tumor.

Symptoms	Herbs	
cough, profuse phlegm, loose stools, fever	*AMNE*	*Physician1* & *Physician2*
	medicinal changium root, indian buead, liquorice, parched white atractylodes rhizome, pinellia tuber, mangnolia officinalis	medicinal changium root, indian buead, parched white atractylodes rhizome, semen dolichoris, dried tangerine peel, pinellia tuber

3.2   Case Study

For the clinical purpose of the effective herbs determination, we compare herbs detecting for specific symptoms via *AMNE* to herbs prescribed by two experienced TCM physicians for some symptoms of lung cancer. Table 2 gives the comparisons of herbs recommendation. We note that *AMNE* prescribed eleven herbs in common with the herbs prescribed by the physicians, which are marked by bold in this table. It is also noticeable that AMNE succeed recommending seven herbs missed by physicians which have been verified to be associated with lung tumor in clinical trials. In addition, we quantificationally evaluate relations learned from over 5,000 lung cancer records. The Micro-F1 and Macro-F1 are employed as the evaluation metrics. Figure 1 gives the comparisons of the results of herbs recommendation.

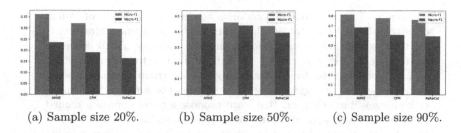

(a) Sample size 20%.   (b) Sample size 50%.   (c) Sample size 90%.

Fig. 1. Performance evaluation of the effectiveness herbs recommendation according to symptoms on clinical dataset with sample size variation.

4   Conclusion

This paper proposes a novel HINs embedding model *AMNE* to investigate the regularity of prescriptions including the complex interrelation of herbs, symptoms of diseases. Empirical evaluations in TCM application confirm that *AMNE* is helpful for clinical practice and research.

References

1. Chen, Y., Wang, C.: HINE: heterogeneous information network embedding. In: DAS-FAA, pp. 180–195 (2017)
2. Huang, E.W., Wang, S., Liu, B., Liu, B., Zhou, X., Zhai, C.X.: PaReCat: patient record subcategorization for precision traditional Chinese medicine. In: ACM International Conference on Bioinformatics, Computational Biology, and Health Informatics, pp. 443–452 (2016)
3. Ruan, C., et al.: THCluster: herb supplements categorization for precision traditional Chinese medicine. In: BIBM, pp. 417–424 (2017)
4. Wang, S., et al.: A conditional probabilistic model for joint analysis of symptoms, diseases, and herbs in traditional Chinese medicine patient records. In: BIBM, pp. 411–418 (2017)
5. Yao, L., Zhang, Y., Wei, B., Zhang, W., Jin, Z.: A topic modeling approach for traditional Chinese medicine prescriptions. IEEE Trans. Knowl. Data Eng. **1**(99), PP (2018)

AGREE: Attention-Based Tour Group Recommendation with Multi-modal Data

Fang Hu, Xiuqi Huang, Xiaofeng Gao$^{(\boxtimes)}$, and Guihai Chen

Shanghai Key Laboratory of Scalable Computing and Systems,
Department of Computer Science and Engineering,
Shanghai Jiao Tong University, Shanghai, China
hu-fang@sjtu.edu.cn, xiuqihwang@gmail.com,
{gao-xf,gchen}@cs.sjtu.edu.cn

Abstract. Tour recommendation aims to design a sequence of Points of Interest (POIs) for a tourist that suits his/her preference. Most existing tour recommenders mainly focus on recommending a POI sequence to a single tourist but cannot be applied to the tour group, which is a common way to travel. Designing a tour group recommender is more challenging in aggregating group preference and tracking influence changes during a tour. Hence we propose a novel approach named AGREE (Attention-based Tour Group Recommendation), which leverages the attention mechanism, to adjust members' influence dynamically. Specifically, our model aggregates group's preference based on members' history data in different modalities, utilizing attention sub-networks to focus on influential ones in each modality across a POI sequence. Then we adopt a bi-directional recurrent unit (Bi-GRU) to generate the POI sequence. Experimental results show that the proposed scheme outperforms benchmark methods on a real-world dataset.

Keywords: Group recommendation · Attention mechanism · Recurrent neural network

1 Introduction

Travel route recommendation system intends to design a sightseeing itinerary route to make tourists experience as gratifying as possible. The majority of existing researches focus on individuals [8], neglecting a common fact that people tend to travel together. Apart from challenges encountered in individual tour planning, two other significant issues must be handled when dealing with tour groups. One is the balance of various tastes, which means that the route needs to

This work was supported by the National Key R&D Program of China [2018YFB1004703]; the National Natural Science Foundation of China [61872238, 61672353]; the Shanghai Science and Technology Fund [17510740200]; the Huawei Innovation Research Program [HO2018085286]; and the State Key Laboratory of Air Traffic Management System and Technology [SKLATM20180X].

© Springer Nature Switzerland AG 2019
G. Li et al. (Eds.): DASFAA 2019, LNCS 11448, pp. 314–318, 2019.
https://doi.org/10.1007/978-3-030-18590-9_36

cater for members' different or even contradictory tastes. The other is influence variation in one trip. For example, in a family tour, after visiting parents' favorite attraction which cannot satisfy their child's taste, then parents are likely to let the child determine the next destination.

To tackle the above issues, we introduce the Attention-based Tour Group Recommendation (AGREE) model. It leverages attention mechanism to balance members' diverse tastes and takes their changeable influence into account. The key contributions of this paper are:

- The proposed model takes a novel deep learning approach to tackle the tour group recommendation problem. To the best of our knowledge, AGREE is the first to exploit the attention mechanism technique with Bi-GRU for tour group recommendation.
- Our model can dynamically adjust the weight for each tourist across groups and recommendation stages.

Recommendation over multi-modal data has been explored in other fields, such as [3,9,10] in video recommendation. Only a limited amount of work has been done in group tour recommendation. [1] formulates the problem and design three methods aimed at distinct objective functions without taking personal preference into account and [5] classifies the tourists into groups first, followed by an optimized algorithm of a variant of the Orienteering problem.

2   Attention-Based Tour Group Recommendation

Our model consists of three major modules: personal preference extraction, group preference aggregation and route generation, as is illustrated in Fig. 1.

Personal Preference Extraction: This module profiles each member's preference from multi-modal data. In this work, we consider members' reviews and photos as two modalities. We use PV-DBOW model [4] to obtain the representation of each review and deep neural network architecture to learn that of each photo. Next, we merge vector representations of all reviews and photos from a member separately by taking the average. In this way, we extract each member's preference from raw data to a vector representation in each modality.

Group Preference Aggregation: After obtaining members' preferences, we aggregate them to form a group preference. This module leverages the attention sub-network to focus on few experts' or influential members' preference in each view. A view refers to a specific modality of data, referring to the principle of multi-view machine learning [7]. In each view, the input of attention sub-network consists of all members' preference representations of the corresponding modality, coming from the output of the first module. To be specific, let V_j be the j-th view (modality j) and $e_{u_i}^j$ be the vector representation of

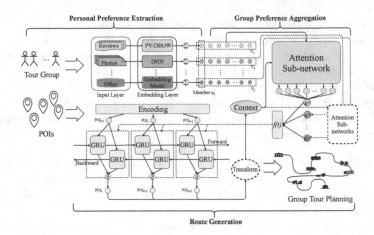

Fig. 1. Overview of the Attention-based Tour Group Recommendation model.

member u_i in the j-th view. In this work, we take V_1 for reviews, V_2 for photos. Then we employ influence weight $\alpha(i,j,t)$ obtained form attention sub-network, which represents the influence of member u_i in j-th view at recommendation stage t. Then the aggregated group preference is calculated by $g_t = f(e^1, e^2, \cdots, e^j) = f(\sum_i \alpha(i,1,t)\mathbf{e}^1_{u_i}, \sum_i \alpha(i,2,t)\mathbf{e}^2_{u_i}, \cdots, \sum_i \alpha(i,j,t)\mathbf{e}^j_{u_i})$, where $e^j = \sum_i \alpha(i,j,t)\mathbf{e}^j_{u_i}$. In this work we use vector concatenation operation ($\oplus$) as the merge function $f(\cdot)$, which concatenates $e^1, e^2, \cdots, e^j$ into one vector.

Route Generation: Finally, we adopt the Bi-GRU to generate a sequence of POIs that satisfies the aggregated group preference. The output of POI vector at the previous stage serves as the input of the next stage. The aggregated group preference g_t is the context of the Bi-GRU, which is fed to stage t. The one-hot representation of the POI predicted at time step t is used as a part of the context of the attention sub-network.

3   Experiments and Analysis

Datasets and Evaluation Metrics: The tourists pool is created based on a set of individual trajectories retrieved from [6] in three European cities. We evaluate our model's performance using two metrics: average score (AS) and diversity at different precision levels, respectively. The average score is the mean value of history scores of all members for POIs in one recommendation route. The diversity is defined by the categorical similarity, which is defined in [2].

Model Performances: Our model are compared against a set of baselines raised by [1] and a simplified version of AGREE without attention sub-network

(Basic Tour Group Recommendation, BTGR). We evaluate our model at precision level $K = 5, 10$, which means that the score is the average of top K predicted routes. As shown in Fig. 2, our model outperforms all baselines in varying sizes (from 2 to 16 members). Besides, the impact of dynamic influence is compared between AGREE and BTGR. We observe that the lack influence variation leads to BTGR at most 29.2% less in diversity (Group size = 16, $K = 10$) and 27.9% less in average score (Group size = 16, $K = 5$).

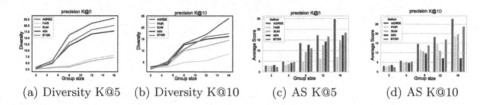

(a) Diversity K@5   (b) Diversity K@10   (c) AS K@5   (d) AS K@10

Fig. 2. Experimental results: (a), (b) shows the diversity against group size and (c), (d) shows the average score against group size at different precision.

4   Conclusions

Aiming to solve the tour group recommendation problem through a deep learning approach, we propose the Attention-based Tour Group Recommendation (AGREE). The model not only utilizes multi-modal information resources to profile members' preference, but also dynamically learns different impact weights of each member across recommendation stages. AGREE is the first model to combine attention mechanism and Bi-GRU in tour group recommendation. Conducting extensive experiments on a real-world dataset, we show the competitive performance of AGREE as compared to baselines.

References

1. Anagnostopoulos, A., Atassi, R., Becchetti, L., Fazzone, A., Silvestri, F.: Tour recommendation for groups. DMKD **31**(5), 1157–1188 (2017)
2. Baral, R., Iyengar, S., Li, T., Balakrishnan, N.: CLoSe: contextualized location sequence recommender. In: RecSys, pp. 470–474 (2018)
3. Huang, Y., Cui, B., Jiang, J., Hong, K., Zhang, W., Xie, Y.: Real-time video recommendation exploration. In: SIGMOD, pp. 35–46 (2016)
4. Le, Q., Mikolov, T.: Distributed representations of sentences and documents. In: ICML, pp. 1188–1196 (2014)
5. Lim, K.H., Chan, J., Leckie, C., Karunasekera, S.: Towards next generation touring: personalized group tours. In: ICAPS, pp. 412–420 (2016)
6. Muntean, C.I., Nardini, F.M., Silvestri, F., Baraglia, R.: On learning prediction models for tourists paths. TIST **8**–**7**, 1–20 (2015)
7. Sun, S.: A survey of multi-view machine learning. Neural Comput. Appl. **23**(7), 2031–2038 (2013)

8. Xia, B., Li, Y., Li, Q., Li, T.: Attention-based recurrent neural network for location recommendation. In: ISKE, pp. 1–6 (2017)
9. Zhou, X., Chen, L., Zhang, Y., Cao, L., Huang, G., Wang, C.: Online video recommendation in sharing community. In: SIGMOD, pp. 1645–1656 (2015)
10. Zhou, X., et al.: Enhancing online video recommendation using social user interactions. VLDB **26-5**, 637–656 (2017)

Random Decision DAG: An Entropy Based Compression Approach for Random Forest

Xin Liu[1], Xiao Liu[1], Yongxuan Lai[2], Fan Yang[1,2(✉)], and Yifeng Zeng[3]

[1] Department of Automation, Xiamen University, Xiamen, China
xinliu@stu.xmu.edu.cn, xiaoliu95@outlook.com, yang@xmu.edu.cn
[2] Shenzhen Research Institute/Software School, Xiamen University, Xiamen, China
laiyx@xmu.edu.cn
[3] School of Computing, Teesside University, Middlesbrough, UK
yifeng.zeng.uk@gmail.com

Abstract. Tree ensembles, such as Random Forest (RF), are popular methods in machine learning because of their efficiency and superior performance. However, they always grow big trees and large forests, which limits their use in many memory constrained applications. In this paper, we propose Random decision Directed Acyclic Graph (RDAG), which employs an entropy-based pre-pruning and node merging strategy to reduce the number of nodes in random forest. Empirical results show that the resulting model, which is a DAG, dramatically reduces the model size while achieving competitive classification performance when compared to RF.

Keywords: Random Forest · Pre-pruning · Directed Acyclic Graph

1 Introduction

Tree ensembles, such as Random Forest (RF) [3], are very popular methods in machine learning in the merits of their computational efficiency and overall good performances. However, in order to deal with high-dimensional data, RFs and other tree ensembles often grow big decision trees and large forests, which may not only deteriorate the generalization performances but limit their use in memory-constrained devices, such as mobile phones [2,6]. In this case, post-pruning methods [4], which need to build and store the whole forest in memory first, is not applicable. To build a lightweight model, pre-pruning methods begin to arouse the interest of researchers. For example, Globally Induced Forests (GIFs) [2] iteratively and greedily deepens multiple trees by optimizing a global function and pre-pruning the undesirable nodes. However, GIF uses

Supported by the Natural Science Foundation of China (61672441, 61673324), the Natural Science Foundation of Fujian (2018J01097), the Shenzhen Basic Research Program (JCYJ20170818141325209).

G. Li et al. (Eds.): DASFAA 2019, LNCS 11448, pp. 319–323, 2019.
https://doi.org/10.1007/978-3-030-18590-9_37

extremely randomized trees to generate nodes, and cannot be directly applied to standard random forests. In this paper, we propose Random decision Directed Acyclic Graph (RDAG), which has the following properties:

- It can be viewed as a pre-pruning method. Similar to [2], it avoids generating a whole forest first, therefore, requiring much fewer memories.
- It is fast with a linear time complexity in the number of training instances.
- It results in a DAG with multiple roots while RF results in many redundant trees.

2   RDAG: A Compression Approach for RF

The motivation of RDAG is that the overfitting risk of the tree increases with the node splitting. Pre-pruning can reduce the risk by preventing node splitting, while node merging can also reduce the risk by reducing the number of nodes. The implementation of node splitting and merging is guided by an Adaptive Information Criterion $AdIC$. In a standard RF, the trees are fully grown and independent of each other, while in RDAG the node splitting is guided by $AdIC$ and node merging is performed across different branches in every iteration. Hence we obtain a directed acyclic graph rather than an ensemble of trees. The framework is shown in Algorithm 1. Given a training set with D observations, the algorithm will generate T root nodes with Bagging (Step 2) and then the

---

Algorithm 1. The Framework of RDAG

Input: $\{x_n, y_n\}_{n=1}^{N}$: training data
   T: tree number
   λ: parameter of regularization term
   n_iter: max depth
Output: F: RDAG model
 1: $gNodes \leftarrow \{\}$: growing nodes
 2: Grow T root nodes and distribute sampled training data to them (Bagging)
 3: $F \leftarrow \{root_1, root_2, \cdots, root_T\}$
 4: **for** $i\_iter = 1$ to n_iter **do**
 5:    update $gNodes$
 6:    **if** $gNodes$ is empty **then**
 7:       **break**
 8:    **end if**
 9:    split and merge nodes
10:    update F
11: **end for**

---

depth of the graph will be increased by iteration. In every iteration, the node set $gNodes$ including those nodes which should be modified is updated (Step 5), and splitting or merging trail is executed on every node in $gNodes$ (Step 9), then the RDAG model F will be updated (Step 10). Once the node set $gNodes$ is empty, the iteration will be terminated before the graph achieves the maximum depth n_iter (Step 6–7). The overall time complexity of RDAG framework is $O(N \times p \times d)$, which is the same as that of RF, with N, p, d denoting the number of instances, features, and nodes in RDAG respectively.

3   Experimental Results

We investigate RDAG on 20 UCI datasets [1], and a summary of these datasets are listed in Table 1. The baseline method, RF, is performed with version 0.19.1 of Scikit-Learn [5]. The reported results are averaged over 10 times of 10-fold CV. In each fold, an internal 10-fold CV is run on the training partition for parameter tuning.

Table 1. Details of datasets used in experiments

Name	Case	Feature	Category	Name	Case	Feature	Category
hayes	132	4	3	ccc	30000	23	2
car	1728	6	4	dermatology	366	34	6
wholesale	440	7	2	hearts	268	44	3
pima	768	8	2	lung-cancer	32	56	3
breast-cancer	699	9	2	bankrupt1	7027	64	2
cmc	1473	9	3	urban	675	147	9
yeast	1485	9	10	musk2	6598	166	2
heart	303	13	5	arrhythmia	452	261	13
crx	684	15	2	colon	62	2000	2
hepatitis	155	19	2	leukemia	72	5147	2

For RF, the number of trees is set to 10 (denoted as RF_{10}), 100 (denoted as RF_{100}) and 1000 (denoted as RF_{1000}) respectively. For node splitting, the parameter $mtry$ which denoted the number of selected features per splitting is set to $\sqrt{p}$ with p denoting the number of features [3]. For RDAG, it takes the same manner and parameter setting of node splitting as RF. We try $T \in [1, 100]$. Other hyperparameters are tuned by cross-validation.

Table 2. Comparison of test accuracy (in percentage) and the average number of nodes of RF and RDAG: the ordering of the accuracy is marked by different gray values for a clearer comparison. The data is displayed in the format of "average accuracy (standard deviation)—number of nodes".

dataset	RF_{10}	RF_{100}	RF_{1000}	RDAG
hayes	81.21 (1.06)—511	80.97 (1.47)—5030	80.91 (1.30)—49921	81.15 (1.91)—166
car	97.21 (0.20)—3204	98.36 (0.16)—32496	98.64 (0.14)—322563	98.48 (0.13)—2679
wholesale	91.20 (0.48)—644	91.77 (0.32)—6390	91.57 (0.36)—64082	91.66 (0.55)—258
pima	73.56 (0.56)—2444	75.99 (0.54)—23794	76.76 (0.44)—236792	76.78 (1.09)—3126
breast-cancer	95.85 (0.38)—564	96.77 (0.21)—5651	96.92 (0.13)—56292	96.81 (0.14)—632
cmc	51.37 (1.12)—9839	51.94 (0.55)—98342	52.17 (0.61)—983361	53.41 (0.45)—3611
heart	56.59 (1.66)—1569	57.28 (0.80)—15678	57.18 (0.61)—156644	57.89 (1.22)—152
crx	86.39 (0.59)—1812	88.08 (0.38)—17884	88.16 (0.37)—178410	87.80 (0.67)—2307
hepatitis	84.82 (2.40)—373	84.58 (1.33)—3748	85.10 (1.51)—37514	85.68 (1.15)—191
ccc	80.55 (0.78)—78219	81.56 (0.50)—782345	81.63 (0.51)—7821372	81.78 (0.80)—158
dermatology	96.80 (0.68)—624	97.26 (0.34)—6244	97.51 (0.19)—62455	97.40 (0.53)—276
hearts	79.68 (1.36)—504	81.16 (1.18)—5015	81.30 (0.68)—50473	81.68 (1.59)—601
lung-cancer	43.50 (6.79)—184	46.67 (3.78)—1850	44.67 (4.09)—18632	47.67 (4.16)—114
bankrupt1	96.98 (0.61)—3431	97.64 (0.37)—32947	97.67 (0.40)—332848	97.74 (0.35)—1462
urban	82.99 (1.02)—1399	86.05 (0.40)—13902	86.19 (0.31)—138402	85.05 (0.67)—1407
musk2	97.47 (0.49)—3238	97.91 (0.49)—32488	98.00 (0.43)—325225	98.12 (0.47)—4521
arrhythmia	69.60 (1.32)—1584	73.90 (0.94)—15855	74.01 (0.50)—158601	74.16 (1.02)—1772
colon	75.24 (4.05)—109	82.00 (2.89)—1071	81.79 (2.22)—10722	80.88 (2.75)—136
leukemia	91.80 (1.69)—77	96.21 (1.42)—777	97.62 (0.92)—7785	95.84 (0.86)—144

The test accuracy and model size are shown in Table 2. As the number of trees in RF increases, the overall performances are getting better, while the node size also increases dramatically. RDAG achieves similar accuracy with RF_{1000}, while it has much fewer nodes. Even compared to RF_{10}, the number of nodes in RDAG is always smaller. Figure 1 shows a comparison on node distribution on *hearts*. It shows the node distribution at different depths so that we can see the effect of compression. The nodes of RF increase exponentially with the depth unless the instances are all allocated to leaf nodes, and a large number of leaf nodes are generated at a relatively shallow depth, while the sample sizes of these nodes are very sparse. In contrast, RDAG limits the growing width of the model and necessarily seeks compensation in depth to obtain sufficient learning ability.

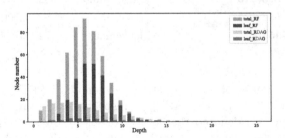

Fig. 1. Comparison on node distribution of RF_{10} and RDAG on *hearts*.

4 Conclusions

In this paper, we propose an entropy-based compression approach for random forests which generates a lightweight classification model. Experimental results show that the resulted RDAG model is compact and accurate.

References

1. Bache, K., Lichman, M.: UCI machine learning repository (2013). http://archive.ics.uci.edu/ml
2. Begon, J.M., Joly, A., Geurts, P.: Globally induced forest: a prepruning compression scheme. In: International Conference on Machine Learning, pp. 420–428 (2017)
3. Breiman, L.: Random forests. Mach. Learn. **45**(1), 5–32 (2001)
4. Elisha, O., Dekel, S.: Wavelet decompositions of random forests: smoothness analysis, sparse approximation and applications. J. Mach. Learn. Res. **17**(1), 6952–6989 (2016)
5. Pedregosa, F., et al.: Scikit-learn: machine learning in Python. J. Mach. Learn. Res. **12**(Oct), 2825–2830 (2011)
6. Shotton, J., Sharp, T., Kohli, P., Nowozin, S., Winn, J., Criminisi, A.: Decision jungles: compact and rich models for classification. In: Advances in Neural Information Processing Systems, pp. 234–242 (2013)

Generating Behavior Features for Cold-Start Spam Review Detection

Xiaoya Tang, Tieyun Qian$^{(\boxtimes)}$, and Zhenni You

School of Computer Science, Wuhan University, Wuhan, Hubei, China
{xiaoyatang,qty,znyou}@whu.edu.cn

Abstract. Existing studies on spam detection show that behavior features are effective in distinguishing spam and legitimate reviews. However, it usually takes a long time to collect such features and is hard to apply them to cold-start spam review detection tasks. In this paper, we exploit the generative adversarial network for addressing this problem. The key idea is to generate *synthetic behavior features* (SBFs) for new users from their *easily accessible features* (EAFs). We conduct extensive experiments on two Yelp datasets. Experimental results demonstrate that our proposed framework significantly outperforms the state-of-the-art methods.

Keywords: Spam review detection · Cold-start problem ·
Generative adversarial network

1 Introduction

The cold-start spam review detection problem, i.e., *a review is just posted by a new reviewer*, is critical in preventing the damage of spams in their early stage. Two embedding models [6,7] show much better performance than the traditional methods. Nevertheless, the inherent problem, i.e., *lack of effective behavior features for new users who post just one review*, remains unsolved.

To solve this problem, we generate the synthetic behavior features (SBFs) for new users who actually do not have such features. Specifically, we first extract six real behavior features (RBFs) for regular users and three types of easily accessible features (EAFs) which exist for both regular users and new user. Secondly, taking these EAFs as the input, we generate SBFs in the generator of the GAN and use the discriminator of the GAN to train the generator. The trained GAN is finally applied to new users to get SBFs which are actually not yet observed for these new users. We conduct extensive evaluations on two real world Yelp datasets. Results demonstrate that our model significantly outperforms the state-of-the-art baseline methods.

2 Real Behavior and Easily Accessible Features

Real Behavior Features (RBFs). We choose six types of real behavior features including activity window (AW) [3,4], maximum number of reviews

© Springer Nature Switzerland AG 2019
G. Li et al. (Eds.): DASFAA 2019, LNCS 11448, pp. 324–328, 2019.
https://doi.org/10.1007/978-3-030-18590-9_38

(MNR) [4,5], percentage of positive reviews (PR) [4,5], review count (RC) [3,4], reviewer deviation (RD) [1,4], and maximum content similarity (MCS) [1,5].

Easily Accessible Features (EAFs). We choose three types of features which can easily accessible for both regular and new reviewers including text features (TF), rating features (RF), and attribute features (AF). All these EAFs are converted into 100-dimension embeddings. We using the convolutional neural network (CNN) like those in [6,7] to get TF from the review text. Moreover, we discretize the deviation of rating score of this review from the average ratings on the same product and the deviation of registered timestamp from posting timestamp to get RF and AF. If the value falls into an interval, the corresponding dimension is set to 1 and other dimensions are set to 0.

3   Our Proposed Model

We aim to exploit generative adversarial network (GAN) [2] to generate the behavior features which are effective in spam detection but new users lack such information in the cold-start scenario. The architecture of our behavior feature generating (bfGAN) model is shown in Fig. 1. The left part in Fig. 1 is the generator which is used to generate SBFs from the input EAFs. The right part is the discriminator which will make a discrimination between SBFs and RBFs using a classifier to guide the training.

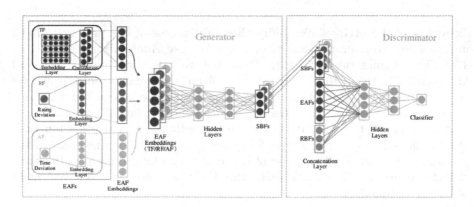

Fig. 1. Architecture of our bfGAN model

Generator. The generator contains six layers. The first three layers are used to do normalization and get EAFs. We then use three non-linear hidden layers to transform EAFs into SBFs.

The generator includes a task loss $\mathcal{L}_t$ to misjudge the discriminator and a closeness loss $\mathcal{L}_c$ to encourage the generator to generate SBFs that have similar distribution as RBFs. We use the cross entropy loss to define the overall loss for the generator as:

$$\mathcal{L}_G = \mathcal{L}_t + \mathcal{L}_c = \min_{\Theta_G} \ J(D(EAF_+ \oplus SBF), 1) + J(RBF, SBF), \quad (1)$$

where D is the discriminative function activated by $tanh$, EAF_+ is the positive EAF matching RBF or SBF and $\oplus$ denotes the concatenation of two embeddings.

Discriminator. The discriminator should judge the (EAF+, RBF) pairs from the realistic training data as real and the (EAF+, SBF) pairs from the generator as fake. Hence we define two loss functions $J(D(EAF_+ \oplus RBF), 1)$ and $J(D(EAF_+ \oplus SBF), 0)$ for this purpose.

Another source of error in the discriminator may come from the unrealistic behavior features. To separate two sources of error, we add a third type of input consisting of RBFs with mismatched EAFs, which the discriminator should learn to score as fake. We formally define a loss function $J(D(RBF \oplus EAF_-), 0)$ to achieve this. The overall function loss $\mathcal{L}_D$ for the discriminator is as follows.

$$\mathcal{L}_D = \min_{\Theta_D} \ J(D(EAF_+ \oplus RBF), 1) + \frac{1}{2} J(D(EAF_+ \oplus SBF), 0) \\ + \frac{1}{2} J(D(RBF \oplus EAF_-), 0), \quad (2)$$

4   Experiments

Experimental Settings. We verify the effectiveness of our proposed model on Hotel and Restaurant datasets of Yelp. We use Adam algorithm to train GAN. The learning rate is 0.00001. The number of iterations for TF, RF, and AF is 300, 21, 19 on Hotel, and 13, 67, 13 on Restaurant, respectively. We stop iteration when the network becomes stable or the loss in the generator becomes the lowest. In the generator network, the number of neurons in two hidden layers is set to 64 and 32, and tanh is adopted as activation function. The last layer is the mapping layer for generating SBFs with 6 neurons. In the discriminator network, the number of neurons in two hidden layers is the same with the generator's. The final classification layer has 1 neuron and uses sigmoid as activation function.

Comparison with Baselines. The comparison results between our model and nine state-of-the-art methods are shown in Table 1.

It is clear that our proposed bfGAN achieves the best performance in terms of F1 and Accuracy on both Hotel and Restaurant datasets. Our bfGAN model combines the traditional approaches in finding effective real behavior features and the recent advances in deep learning to generate synthetic behavior features to simulate the real ones.

Compared to the state-of-the-art method AEDA, our model uses less information and does not integrate domain adaption into the framework, but it reaches

Table 1. Comparison with baselines.

Features	Hotel				Restaurant			
	P	R	F1	Acc	P	R	F1	Acc
LF [3]	54.5	71.1	61.7	55.9	53.8	80.8	64.6	55.8
Supervised-CNN [6]	61.2	51.7	56.1	59.5	56.9	58.8	57.8	57.1
LF+BF [3]	63.4	52.6	57.5	61.1	58.1	61.2	59.6	58.5
BF_EditSim+LF [6]	55.3	69.7	61.6	56.6	53.9	82.2	65.1	56.0
BF_W2Vsim+W2V [6]	58.4	65.9	61.9	59.5	56.3	73.4	63.7	58.2
RE* [6]	62.1	68.3	65.1	63.3	58.4	75.1	65.7	60.8
RE+RRE+PRE* [6]	63.6	71.2	67.2	65.3	59.0	78.8	67.5	62.0
AE [7]	76.7	74.2	75.4	75.8	80.3	66.2	72.6	75.0
AEDA [7]	83.9	74.2	78.7	80.0	82.4	65.1	72.8	75.6
bfGAN(ours)	81.2	85.7	**83.4**	**83.0**	76.7	73.4	**75.1**	**75.7**

*Denotes that the model is trained on the labeled data and a large number of unlabeled data.

an improvement of 6.0% and 3.2% of F1 over AEDA on Hotel and Restaurant, respectively. This clearly demonstrates that our bfGAN model can generate highly effective SBFs whose distribution is close to that of RBFs, and hence it achieves significantly better performance than baseline methods.

5   Conclusion

In this paper, we propose a novel bfGAN model for cold-start spam review detection. To address the problem of lacking effective real behavior features (RBFs) for new users in cold-start scenario, we design a GAN framework to generate synthetic behavior features (SBFs) using EAFs and further present a new implementation of GAN by incorporating an extra loss to explicitly guide SBFs to be close to RBFs. We conduct extensive experiments on two Yelp datasets. Results demonstrate that our bfGAN model significantly outperforms the state-of-the-art baselines.

Acknowledgments. The work described in this paper has been supported in part by the NSFC projects (61572376).

References

1. Fei, G., Mukherjee, A., Liu, B., Hsu, M., Castellanos, M., Ghosh, R.: Exploiting burstiness in reviews for review spammer detection. In: ICWSM, pp. 175–184 (2013)
2. Goodfellow, I., et al.: Generative adversarial nets. In: NIPS, pp. 2672–2680 (2014)
3. Mukherjee, A., Venkataraman, V., Liu, B., Glance, N.: Fake review detection: classification and analysis of real and pseudo reviews. Technical Report UIC-CS-2013-03, University of Illinois at Chicago, Technical report (2013)

4. Mukherjee, A., Venkataraman, V., Liu, B., Glance, N.S.: What yelp fake review filter might be doing? In: ICWSM (2013)
5. Rayana, S., Akoglu, L.: Collective opinion spam detection: bridging review networks and metadata. In: KDD, pp. 985–994 (2015)
6. Wang, X., Liu, K., Zhao, J.: Handling cold-start problem in review spam detection by jointly embedding texts and behaviors. In: ACL, pp. 366–376 (2017)
7. You, Z., Qian, T., Liu, B.: An attribute enhanced domain adaptive model for cold-start spam review detection. In: COLING, pp. 1884–1895 (2018)

TCL: Tensor-CNN-LSTM for Travel Time Prediction with Sparse Trajectory Data

Yibin Shen[1], Jiaxun Hua[2], Cheqing Jin[1], and Dingjiang Huang[1(✉)]

[1] School of Data Science and Engineering,
East China Normal University, Shanghai, China
ybshen@stu.ecnu.edu.cn, {cqjin,djhuang}@dase.ecnu.edu.cn
[2] School of Computer Science and Software Engineering,
East China Normal University, Shanghai, China
vichua@stu.ecnu.edu.cn

Abstract. Predicting the travel time of a given path plays an indispensable role in intelligent transportation systems. Although many prior researches have struggled for accurate prediction results, most of them achieve inferior performance due to insufficient extraction of travel speed features from the *sparse trajectory data*, which confirms the challenges involved in this topic. To overcome those issues, we propose a deep learning framework named Tensor-CNN-LSTM (*TCL*) in this paper, which can extract travel speed effectively from historical sparse trajectory data and predict travel time with better accuracy. Empirical results over two real-world large-scale datasets show that our proposed *TCL* can achieve significantly better performance and remarkable robustness.

1 Introduction and Related Work

Thanks to the popularity of GPS-embedded devices, much more trajectory data has been generated, by analyzing which municipal authorities may identify and optimize the traffic congestion in advance. However, predicting an accurate travel time is still very challenging as the travel time is affected by many dynamic factors, such as dynamic departure time, dynamic traffic conditions and dynamic driver behavior. All these 'dynamics' make it intractable to predict future pattern of traffic with statistic model [6].

With rapid evolution of deep learning, some studies adopt embedding technologies to solve the challenge of dynamics [3,7]. They transform departure time, drivers, weather and locations into low-dimensional learnable real vectors, and construct a deep neural network to predict the travel time. Nevertheless, most of them don't extract travel speed features adequately from *sparse trajectory data* because trajectory data isn't necessarily generated on all road segments at every moment[1], which results in poor performance.

---

[1] In our experiments, there are only 1.00% and 1.56% roads in *Beijing* and *Shanghai* can satisfy this condition, respectively.

© Springer Nature Switzerland AG 2019
G. Li et al. (Eds.): DASFAA 2019, LNCS 11448, pp. 329–333, 2019.
https://doi.org/10.1007/978-3-030-18590-9_39

Meanwhile, tensor/matrix decomposition algorithms have been adopted to solve the data sparsity [5,8]. However, these decomposition methods often take several minutes to restore the travel time/speed on a road, which is almost infeasible in reality. Even worse, tensor/matrix decomposition algorithms can only estimate the previous travel time/speed of a road because there's no future data in the tensor/matrix. Consequently, it cannot be directly applied to the problem of travel time prediction[2].

With the aim of solving the aforementioned challenges, we propose a novel deep learning framework named Tensor-CNN-LSTM (*TCL*) for travel time prediction, which can extract travel speed features effectively from historical sparse trajectory data and predict the travel time of a given path with better accuracy.

2   Model Architecture

In this section, we introduce the framework of our proposed *TCL* model, as is shown in Fig. 1. *TCL* is comprised of three major components: non-negative tensor decomposition, long-short-term speed CNN and LSTM prediction model.

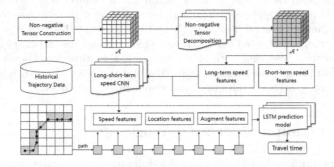

Fig. 1. The framework of *TCL* model.

Non-negative Tensor Decomposition. In the module of non-negative tensor decomposition, we partition an hour into M time slots, and construct three homogenous matrices $A_h, A_m, A_r \in \mathbb{R}^{N^2 \times M}$, where $A_r(i,j) = a$ denotes the i-th grid with travel speed a in time slot j, and A_h is constructed based on historical trajectories over a long period of time (e.g. a week). A_m is a mixed matrix to combine A_r and A_h. After constructing these matrices, we concatenate them together to construct a 3D non-negative tensor $\mathcal{A} \in \mathbb{R}^{N^2 \times M \times 3}$. We fill the missing value in $\mathcal{A}$ by using a fast non-negative CP decomposition algorithm [2].

---

[2] Once future information is added, such as the real travel speed, the problem will no longer be travel time prediction [1].

Long-short-term Speed CNN. In the module of long-short-term speed CNN, we extract the long/short-term speed features from a given path, based on $\mathcal{A}^*$, where the long-term speed features are the travel speed values of the target grid in the past 7 days, and the short-term speed features are the speed distributions in that grid and relevant grids in the previous hour[3]. Afterwards, we construct a CNN to obtain the whole speed features, as is shown in Fig. 2.

LSTM Prediction Model. The LSTM prediction model consists of two parts: feature extraction layer and prediction layer, as is shown in Fig. 3. The former extracts useful features from the path, such as augment features (the driver ID, the departure time, the day of the week, the travel distance and holiday indicator) and location features (the latitude, the longitude and the grid ID). The prediction layer predicts travel time of the path, which consists of a 2-layer LSTM and a multi-layer perceptron (MLP). The loss function of our model is *Mean Absolute Percentage Error*(MAPE), and the optimizer is Adam.

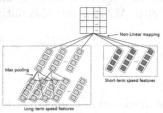

Fig. 2. Architecture of CNN model.   **Fig. 3.** Architecture of LSTM model.

3   Experiments

Datasets: We evaluate the performance of our model on two real-world trajectory datasets, namely *Beijing* and *Shanghai*. The Beijing dataset contains 3,384,847 trajectories of 10,039 drivers from Oct. 1st to Oct. 31st in 2013. The Shanghai dataset contains 9,727,798 trajectories of 13,622 drivers from Apr. 1st to Apr. 30th in 2015. For each dataset, we split the trajectories generated in the last 7 days as the test set and the rest as the training set[4].

Results: We select *TEMP* [4], *XGBoost*, *DeepTTE* [3] as baseline methods for comparison with our model. For *DeepTTE* and *TCL*, we train these models for 50 epochs and repeat each experiment 3 times. The mean and the standard deviation are calculated, and the results are shown in Table 1.

---

[3] The top-k most relevant grids are calculated by time-shifting KL-divergence.

[4] The sampling rates on two datasets are different, *Beijing* has low sampling rates (sampling interval is 60 s), and *Shanghai* owns higher sampling rates (10 s).

Table 1. Performance comparison of travel time prediction.

Dataset	Beijing			Shanghai		
Metrics	MAPE (%)	MAE (sec)	RMSE (sec)	MAPE (%)	MAE (sec)	RMSE (sec)
TEMP	23.92	247.20	395.08	23.05	142.25	228.06
XGBoost	18.53	179.41	274.73	21.80	124.84	192.16
DeepTTE	13.47 ± 0.15	137.32 ± 1.03	221.03 ± 1.22	15.11 ± 0.08	99.77 ± 1.55	164.47 ± 3.83
TCL	**12.40 ± 0.10**	**124.58 ± 0.99**	**200.06 ± 0.38**	**13.08 ± 0.05**	**85.96 ± 0.23**	**142.75 ± 1.07**

As we can see, *TEMP* only considers the information of starting points and destinations, resulting in the worst performance. *XGBoost* performs better than *TEMP* on both datasets because the feature selection of *XGBoost* is consistent with our model. However, *XGBoost* can not handle sequence data, so fixing the length of a path will lose some information. *DeepTTE* consider various factors which may affect the travel time, the performance of *DeepTTE* are much better than aforementioned methods. Our proposed *TCL* captures the travel speed accurately, which is the most important factor affecting travel time. *TCL* scores 12.40% and 13.08% (on two datasets respectively) in MAPE, and also outperforms other models in other metrics.

4 Conclusion

In this paper, we propose a novel deep learning framework, namely *TCL*, to predict the travel time of a given path. Specifically, *TCL* can extract travel speed effectively from historical sparse trajectory data and predict travel time with better accuracy. *TCL* achieves satisfying performance on two real-world datasets, which means that we have conquered the challenges of dynamics and sparsity in the trajectory data.

Acknowledgment. This work is partially supported by the National Natural Science Foundation of China (U1711262, U1811264, 11501204).

References

1. Achar, A., Sarangan, V., Regikumar, R., Sivasubramaniam, A.: Predicting vehicular travel times by modeling heterogeneous influences between arterial roads. In: AAAI, pp. 2063–2070 (2018)
2. Kim, J., Park, H.: Fast nonnegative tensor factorization with an active-set-like method. In: Berry, M., et al. (eds.) High-Performance Scientific Computing - Algorithms and Applications, pp. 311–326. Springer, London (2012). https://doi.org/10.1007/978-1-4471-2437-5_16
3. Wang, D., Zhang, J., Cao, W., Li, J., Zheng, Y.: When will you arrive? Estimating travel time based on deep neural networks. In: AAAI, pp. 2500–2507 (2018)
4. Wang, H., Kuo, Y., Kifer, D., Li, Z.: A simple baseline for travel time estimation using large-scale trip data. In: ACM SIGSPATIAL, pp. 61:1 61:4 (2016)

5. Wang, Y., Zheng, Y., Xue, Y.: Travel time estimation of a path using sparse trajectories. In: SIGKDD, pp. 25–34 (2014)
6. Wang, Z., Fu, K., Ye, J.: Learning to estimate the travel time. In: SIGKDD, pp. 858–866 (2018)
7. Zhang, H., Wu, H., Sun, W., Zheng, B.: Deeptravel: a neural network based travel time estimation model with auxiliary supervision. In: IJCAI, pp. 3655–3661 (2018)
8. Zhou, X., Luo, Q., Zhang, D., Ni, L.M.: Detecting taxi speeding from sparse and low-sampled trajectory data. In: Cai, Y., Ishikawa, Y., Xu, J. (eds.) APWeb-WAIM 2018. LNCS, vol. 10988, pp. 214–222. Springer, Cham (2018). https://doi.org/10.1007/978-3-319-96893-3_16

A Semi-supervised Classification Approach for Multiple Time-Varying Networks with Total Variation

Yuzheng Li[1,2], Chuan Chen[1,2(✉)], Fanghua Ye[1,2], Zibin Zheng[1,2], and Guohui Ling[3]

[1] School of Data and Computer Science, Sun Yat-sen University, Guangzhou, China
chenchuan@mail.sysu.edu.cn
[2] National Engineering Research Center of Digital Life, Sun Yat-sen University, Guangzhou 510006, China
[3] Data Center of Wechat Group, Tencent Technology, Shenzhen, China

Abstract. In recent years, we have seen a surge of research on semi-supervised learning for improving classification performance due to the extreme imbalance between labeled and unlabeled data. In this paper, we innovatively propose a semi-supervised classification model for multiple time-varying networks, i.e., Multiple time-varying Networks Classification with Total variation (*MNCT*), which can integrate the multiple time-varying networks and select relevant ones. From a numerical point of view, the optimization is decomposed into two sub-problems, which can be solved efficiently under the alternating direction method of multipliers (ADMM) framework. Experimental results on both synthetic and real-world datasets empirically demonstrate the advantages of *MNCT* over state-of-the-art methods.

Keywords: Semi-supervised learning · Time-varying · Total variation

1 Introduction

With the development of social networks, graph-based learning is playing an important role in machine learning and data mining. Semi-supervised learning (SSL) for classification is one of the crucial research fields, which integrates labeled and unlabeled data within a unified model [5,7,8,10]. Considering the time dimension, the multiple time-varying networks are obtained by observing an underlying relationship at different epochs.

Intuitively, inspired by multi-view learning [9], we propose a novel graph-based SSL approach to handle the time dimensional data, i.e., Multiple time-varying Networks Classification with Total variation (*MNCT*). In order to utilize the temporal information, we introduce total variation constraint [6], which encourages the weight of each network to be locally consistent, and accords with the fact that gaps between any adjacent networks within a consecutive period are generally small.

G. Li et al. (Eds.): DASFAA 2019, LNCS 11448, pp. 334–337, 2019.
https://doi.org/10.1007/978-3-030-18590-9_40

2   Method

For simplicity of illustration, we focus on the binary classification problem. In multiple time-varying networks settings, we have T networks with affinity matrices $\{W^{(t)}\}_{t=1}^{T}$, and $\{L^{(t)}\}_{t=1}^{T}$ as the set of the corresponding graph Laplacian matrices. The similarity between x_i and x_j in network t is expressed as $W_{ij}^{(t)}$. The semi-supervised vector is defined as $y = (y_1, y_2, \ldots y_n)^\intercal$. Without loss of generality, we suppose that the first ℓ instances are labeled by $y_i \in \{\pm 1\}$, where $0 \leq i \leq \ell \ll n$. Then the remaining are unlabeled ($y_i = 0$), which are in fact our prediction target. Illuminated by label propagation algorithm, we innovatively propose the *MNCT* model:

$$\min_{f,w} \sum_{t=1}^{T} w^{(t)} f^\intercal L^{(t)} f + \lambda_1 \|f - y\|_2^2 + \lambda_2 \|Dw\|_1, \quad s.t. \ w^\intercal 1 = 1, w \geq 0, \quad (1)$$

where vector w represents the network weights, λ_1 and λ_2 are positive parameter. The third term is the total variation constraint, in which $D \in \mathbb{R}^{(T-1) \times T}$ is called first-order differential matrix where $\|Dw\|_1 = \sum_{i=1}^{T-1} |w_i - w_{i+1}|$.

The total variation constraint expresses the time correlation between networks, which makes the network weights distribution tend to be piecewise constant. This fits well with the fact that information gaps between adjacent time-varying networks are generally small. Through the proper parameter settings, the model could obtain sparse but smooth weights for time-varying networks, which results in an optimal combination of networks information. In the meantime, label propagation procedure estimates optimal label prediction f under this combination to achieve a higher classification performance.

To solve the problem given by Eq. (1), we alternately minimize the objective function with respect to f and w. Noteworthy, since each sub-problem is convex, the existence of global optimal in each iteration is guaranteed. For the f sub-problem, we can obtain an analytical solution on account of no constraint. For the w sub-problem, we apply ADMM [2] framework to solve a quadratic programming, where an auxiliary variable and soft-shrinkage operator [1] are there to deal with ℓ_1-norm.

3   Experimental Result

In this section, we mainly compare our algorithm with the four methods: Single-view Semi-supervised Classification (*SSC*) [10], Optimized Multiple Graph-based Semi-Supervised Learning (*OMG-SSL*) [7], Sparse Multiple Graph Integration (*SMGI*) [5], and Semi-supervised Time Series Classification (*STSC*) [8]. In the following experiments, the parameters of each method are tuned for optimal performance. Considering some methods can only process one network per runtime, we evaluate the average performance over all networks for them. We perform each method on three kinds of datasets with different label rates over 10 runs and report the average performance.

Table 1. Average accuracy (std.) on synthetic datasets over 10 runs.

Dataset	Label rate	SMGI	OMG-SSL	SSC	**MNCT**
Grow	1%	0.5495(0.0300)	0.5580(0.0315)	0.5188(0.0099)	**0.5708(0.0351)**
	3%	0.5603(0.0171)	0.5638(0.0206)	0.5621(0.0131)	**0.5935(0.0232)**
	5%	0.5908(0.0247)	0.6065(0.0318)	0.5984(0.0164)	**0.6195(0.0353)**
Merge	1%	0.5210(0.0172)	0.5205(0.0123)	0.5124(0.0081)	**0.5523(0.0214)**
	3%	0.5397(0.0175)	0.5445(0.0193)	0.5435(0.0177)	**0.6040(0.0277)**
	5%	0.5775(0.0317)	0.5880(0.0324)	0.5794(0.0296)	**0.6580(0.0370)**
Mixed	1%	0.4096(0.0266)	0.4143(0.0251)	0.4190(0.0258)	**0.4439(0.0315)**
	3%	0.4594(0.0170)	0.4688(0.0185)	0.4843(0.0156)	**0.5336(0.0232)**
	5%	0.4961(0.0142)	0.5039(0.0149)	0.5262(0.0146)	**0.5828(0.0211)**

Synthetic Dataset Experiment. We firstly perform the evaluation on synthetic graph datasets, which are generated by an approach for simulating simple dynamic networks [4]. The nodes in those datasets will move across all clusters following several behaviour patterns, which causes the labels changing at difference time epochs, together with some noise and misleading information. We report the results in Table 1, which indicate that *MNCT* could fit the changing of time-varying networks well, then achieve the highest predictive performance among all methods on synthetic datasets.

Daily and Sports Activities Dataset Experiment. In this section, we employ a real-world dataset named Daily and Sports Activities Dataset [3]. We first split 19 categories into three groups, each contains 125 time-varying networks and 1200 samples per network, and the edges are weighted by vector Euclidean distances between pairs of samples. With similar settings, the evaluation results are reported in Table 2, which demonstrates effectiveness of *MNCT*.

Table 2. Average accuracy (std.) on daily and sports activities datasets over 10 runs.

Dataset	Label rate	SMGI	OMG-SSL	SSC	**MNCT**
1–6	1%	0.8349(0.0834)	0.8180(0.0959)	0.5079(0.0039)	**0.8373(0.0819)**
	3%	0.8421(0.0603)	0.7925(0.0965)	0.5416(0.0251)	**0.8428(0.0614)**
	5%	0.8539(0.0643)	0.8229(0.0941)	0.5700(0.0327)	**0.8543(0.0659)**
7–12	1%	0.1753(0.0005)	0.1750(0.0000)	0.2065(0.0042)	**0.2689(0.0261)**
	3%	0.1918(0.0002)	0.1917(0.0000)	0.2033(0.0030)	**0.2727(0.0165)**
	5%	0.2083(0.0000)	0.2083(0.0000)	0.2135(0.0014)	**0.2739(0.0179)**
13–19	1%	0.6529(0.0471)	0.5311(0.0633)	0.4026(0.0080)	**0.6560(0.0399)**
	3%	**0.6981(0.0197)**	0.4966(0.0623)	0.4229(0.0138)	0.6961(0.0230)
	5%	0.7113(0.0161)	0.5133(0.0382)	0.4388(0.0109)	**0.7144(0.0150)**

4   Conclusion

Exploiting multiple time-varying networks for data learning remains a challenging problem, and a robust and flexible approach is highly valued. In this paper, we propose a new method *MNCT* to deal with classification task under multiple time-varying networks. We innovatively introduce total variation constraint to fit the continuity between time-varying networks. The experimental results verify that *MNCT* could outperform other state-of-the-art SSL methods under multiple time-varying networks and low label rates.

Acknowledgements. The paper is supported by the National Key Research and Development Plan (2018YFB1003803), the National Natural Science Foundation of China (11801595, U1811462), the Natural Science Foundation of Guangdong (2018A-030310076) and the CCF Opening Project of Information System.

References

1. Bioucas-Dias, J., Figueiredo, M.: A new TwIST: two-step iterative shrinkage/thresholding algorithms for image restoration. IEEE Trans. Image Process. **16**(12), 2992–3004 (2007). https://doi.org/10.1109/tip.2007.909319
2. Boyd, S., Parikh, N., Chu, E., Peleato, B., Eckstein, J., et al.: Distributed optimization and statistical learning via the alternating direction method of multipliers. Found. Trends® Mach. Learn. **3**(1), 1–122 (2011). https://doi.org/10.1561/2200000016
3. Dheeru, D., Karra Taniskidou, E.: UCI machine learning repository (2017). http://archive.ics.uci.edu/ml
4. Granell, C., Darst, R.K., Arenas, A., Fortunato, S., Gómez, S.: Benchmark model to assess community structure in evolving networks. Phys. Rev. E **92**(1), 012805 (2015). https://doi.org/10.1103/physreve.92.012805
5. Karasuyama, M., Mamitsuka, H.: Multiple graph label propagation by sparse integration. IEEE Trans. Neural Netw. Learn. Syst. **24**(12), 1999–2012 (2013). https://doi.org/10.1109/tnnls.2013.2271327
6. Rudin, L.I., Osher, S., Fatemi, E.: Nonlinear total variation based noise removal algorithms. Phys. D Nonlinear Phenom. **60**(1–4), 259–268 (1992). https://doi.org/10.1016/0167-2789(92)90242-f
7. Wang, M., Hua, X.S., Hong, R., Tang, J., Qi, G.J., Song, Y.: Unified video annotation via multigraph learning. IEEE Trans. Circ. Syst. Video Technol. **19**(5), 733–746 (2009). https://doi.org/10.1109/tcsvt.2009.2017400
8. Wei, L., Keogh, E.: Semi-supervised time series classification. In: SIGKDD-06, pp. 748–753. ACM Press (2006). https://doi.org/10.1145/1150402.1150498
9. Xu, C., Tao, D., Xu, C.: A survey on multi-view learning, vol. 37, pp. 2531–2544, arXiv preprint arXiv:1304.5634 (2013). https://doi.org/10.1109/tpami.2015.2417578
10. Zhou, D., Bousquet, O., Lal, T.N., Weston, J., Schölkopf, B.: Learning with local and global consistency. In: NIPS 2004, pp. 321–328 (2004)

Multidimensional Skylines
over Streaming Data

Karim Alami$^{(\boxtimes)}$ and Sofian Maabout$^{(\boxtimes)}$

LaBRI, University of Bordeaux, Bordeaux, France
{karim.alami,sofian.maabout}@u-bordeaux.fr

Abstract. We consider a stream where each record is described by a set of dimensions D. The records have a validity time interval of size ω. The queries we consider consist in retrieving the valid skyline records with respect to subsets D' (subspace) of D. Answering multidimensional skyline queries over streaming data is a hard task because of the data velocity and even index structures that optimize these queries need to be continuously updated. To overcome this difficulty, we propose a framework that handles the streaming data in a micro-batch mode together with an incrementally maintainable index structure.

1 Introduction

The skyline operator [1] is relevant for retrieving the *best* elements w.r.t. user preferences. In this work, we investigate the problem of answering multidimensional skyline queries over streaming data. As a motivating scenario, consider a data analytics agency which receives a stream of statistical data about tweets. Each record represents a tweet statistics (UserId, TweetId, #retweets, #likes, #comments, #shares_on_other_social_nets, retweet_depth, #followers). The agency is interested by the skyline tweets w.r.t. several subsets of attributes in a 24 h sliding window. Considering the last 6 attributes representing statistics, there are 63 distinct skyline queries ($2^6 - 1$) that can be issued on this stream.

We propose a framework MSSD (Multi Skyline Streaming Data) that handles (i) a buffer B where records are first collected during k units of time, (ii) a main dataset R that stores records arrived in a *window* of size ω and (iii) an index structure NSCt which consists in storing for each record in R a lossless summary of the subspaces where it is **not** in the skyline. This idea has been presented for a static data context in [2], and its efficiency w.r.t. time and space complexity has been proven. We adopt a *micro-batch* processing approach, i.e., the stream source emits at most one record every θ units of time. Our framework collects the records into a buffer during k units of time. Thereafter, the buffered records are inserted into R and the outdated ones are removed from R. Simultaneously, the maintenance of the index structure NSCt is triggered. When a subspace skyline query is issued, NSCt is used in order to compute the skyline. Continuing with the analytics agency example, suppose that it is interested in querying a 24 h window, i.e., ω, and sets the batch interval, i.e., k, to 15 min with a processing

© Springer Nature Switzerland AG 2019
G. Li et al. (Eds.): DASFAA 2019, LNCS 11448, pp. 338–342, 2019.
https://doi.org/10.1007/978-3-030-18590-9_41

at {HH:00, HH:15, HH:30, HH:45}, then at 13:40, R covers the window (13 : 30($-1\ day$), 13 : 30]. [5] has been first to tackle the problem of computing a *single* skyline in a streaming context. In [4], authors have noted the importance of multidimensional skyline queries: each user may prefer a subset of attributes with respect to which the skyline is computed. However, none of the previous approaches are suited for handling multidimensional skyline over streaming data.

2   Preliminaries

We consider a stream where each record is described, among others, by a set of dimensions $D = \{D_1, \ldots, D_d\}$. A subset X of D is called a subspace. Let r and r' two records, wlog we consider that $r[D_i]$ is preferred to $r'[D_i]$ iff $r[D_i] < r'[D_i]$. Every record r has a timestamp corresponding to its arrival time denoted $arr(r)$. To simplify, we consider that $arr(r) \in \mathbb{N}$. We also consider that all records share the same validity period ω. By convention, the current timestamp, denoted t_c corresponds to the timestamp of the most recent record in a stream S. That is, $t_c = \underset{r \in S}{\mathrm{argmax}}\ arr(r)$. Given two records r_1 and r_2, we say that r_1 dominates r_2 w.r.t. $X \subseteq D$, noted $r_1 \prec_X r_2$, iff (i) $\forall D_i \in X : r[D_i] \leq r'[D_i]$ and (ii) $\exists D_j \in X$ s.t $r[D_j] < r'[D_j]$. Then, the subspace skyline of S w.r.t. X over an interval $[a, b]$, denoted $Sky_{[a,b]}(X, S)$, is the set $\{r \in S[a, b]|\ \nexists r' \in S[a, b] \text{ s.t } r' \prec_X r\}$ such that $S[a, b]$ represents those records arrived within the interval $[a, b]$. We simplify the notation $Sky_{[a,b]}(X, S)$ by $Sky(X)$ when $[a, b]$ represents the current valid interval $[t_c - \omega, t_c]$.

3   MSSD Framework

MSSD consists of three data structures, (i) a buffer B, (ii) a main dataset R and (iii) an index structure NSCt. It integrates a *micro-batch* processing: (i) during a time interval of size k, records are first inserted into the buffer B, afterwards (ii) the content δ^+ of B is inserted into the dataset R, (iii) the outdated records δ^- are deleted from R, and finally (iv) the update of NSCt is triggered. The framework is clocked by the parameter θ which determines the delay between two timestamps t_i and t_{i+1}, i.e., the delay between two successive records.

The NSCt index structure consists of storing for every record *subspaces* where it is dominated in a form of pairs of subspaces. Given r and r' two valid records, we compute the pair $compare(r, r') = \langle X|Y \rangle$ such that X represents subspaces where r' is strictly better than r and Y represents subspaces where they are equal.

Example 1. Let $T(A, B, C)$ a stream, and $r = (1, 0, 4)$ and $r' = (0, 0, 3)$ two records in T. then $compare(r, r') = \langle AC|B \rangle$.

Hence, from the set of pairs of a given record r, we deduce the subspaces where this record is dominated by computing the *coverage* of its pairs:

$cover(\langle X|Y\rangle) = \{Z \subseteq D| Z \subseteq XY \text{ and } Z \cap X \neq \emptyset\}$. Hence, given Z a subspace, a record $r \notin Sky(Z)$ if it has a pair that covers Z.

In the following, we present our approaches to (i) compute the pairs of a newly inserted record into R along with the organization of the pairs and (ii) update the set of pairs of an existing record.

Handling a New Record. Let r be a newly inserted record into R, we compute its pairs w.r.t. records in R and organize them as follows: we allocate to r a **set of buckets** that we call $Pairs(r)$ where each bucket $Pairs(r).Buck_i$ contains pairs computed w.r.t. a transaction in R. Since there exists $\frac{\omega}{k}$ transactions in R, then each record will have $\frac{\omega}{k}$ buckets. Due to lack of space we just illustrate the process in Example 2.

Example 2. We report in Tables 1 and 2 the dataset R at two successive timestamps and in Table 3 the pairs of newly inserted records r_6 and r_7 into R at timestamp 7. The first bucket is computed w.r.t. the oldest transaction in R, i.e., $\{r_2, r_3\}$.

Table 1. Dataset R at timestamp 6

Id	A	B	C	arr. time
r_0	5	4	1	0
r_1	3	4	2	1
r_2	5	1	3	2
r_3	1	1	3	3
r_4	1	0	4	4
r_5	0	1	5	5

Table 2. Dataset R at timestamp 7

Id	A	B	C	arr. time
r_2	5	1	3	2
r_3	1	1	3	3
r_4	1	0	4	4
r_5	0	1	5	5
r_6	2	0	3	6
r_7	2	1	1	7

Table 3. Pairs of r_6 and r_7

Id	1	2	3
r_6	$\langle A|C\rangle, \langle \emptyset|C\rangle$	$\langle A|\emptyset\rangle, \langle A|B\rangle$	$\langle C|A\rangle$
r_7	$\langle A|B\rangle, \langle \emptyset|B\rangle$	$\langle A|B\rangle, \langle AB|\emptyset\rangle$	$\langle B|A\rangle$

Indeed, the number of pairs in $Pairs(r)$ can be reduced without loss of *covering* information by computing an **equivalent** subset, i.e. P and Q are equivalent, $P \equiv Q$ iff $cover(P) = cover(Q)$.

Updating Pairs of an Existing Record. Let r be a record in R having a set of buckets $\{Buck_1, \ldots, Buck_m\}$. At the maintenance time, on one hand, the *oldest* bucket, i.e., bucket computed with respect to the oldest transaction in R, is discarded from $Pairs(r)$, here $Buck_1$. On the other hand, a new bucket is computed w.r.t. records in δ^+ and is appended to $Pairs(r)$.

Example 3. At timestamp 6, r_5 had the set of pairs $[\{\}, \{\langle C|A\rangle\}, \{\langle A|\emptyset\rangle\}]$. At timestamp 7, a new *bucket* containing pairs related to r_6 and $r7$ is added and the last *bucket* is discarded. The set of pairs become $[\{\langle C|A\rangle\}, \{\langle A|\emptyset\rangle\}, \{\langle CB|\emptyset\rangle\}]$.

Query Answering. Given a subspace Z, to evaluate $Sky(Z)$, through NSCt we look for records having no pairs covering Z. To optimize this search, we construct a subspace index that consists of a map which for every subspace XY, it associates the set of pairs $\langle r|Y\rangle$ such that $\langle X|Y\rangle \in Pairs(r)$. Now, to evaluate $Sky(Z)$, we look **only** for subspaces XY such that $Z \subseteq XY$. Let XY be such a subspace, then from every pair $\langle r|Y\rangle$ associated to XY, if $Z \neq Y$ then we deduce that $r \notin Sky(Z)$.

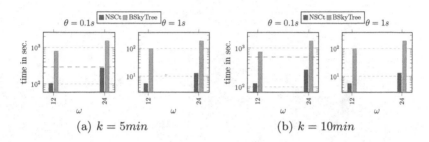

(a) $k = 5min$ (b) $k = 10min$

Fig. 1. Query answering with INDE data (Color figure online)

4   Experiments

We evaluate the ability of MSSD to answer all possible skyline queries during a batch interval and we compare it to a baseline approach that computes the skyline using state of the art algorithm BSkyTree [3]. The goal is to show that despite its maintenance process, NSCt is much more efficient.

We consider an independent synthetic dataset of 12 dimensions as input and we vary the values of MSSD parameters as follows: θ in $\{0.1\,\text{s},\ 1\,\text{s}\}$, ω in $\{12\,\text{h},\ 24\,\text{h}\}$, k in $\{5\,\text{mn},\ 10\,\text{mn}\}$.

All algorithms are implemented in C++. Source code is available on GitHub[1].

Figure 1 depict the results. Red dashed lines represent the value of k. When it is exceeded, it means that the approach cannot answer all issued queries during the batch interval. Recall that BSkyTree does not require any maintenance so query evaluation can start as soon as a new transaction is inserted into R, while for NSCt we include the maintenance time. We point out two observations from this experiment: (i) NSCt is faster with more than one order of magnitude in all cases despite the fact that its maintenance time is also included, and (ii) BSkyTree is unable to answer all the issued queries for several scenarios, for e.g. in Fig. 1(a), BSkyTree takes more than 5 min to answer all queries.

---

[1] https://github.com/karimalami7/MSSD.

References

1. Börzsönyi, S., et al.: The skyline operator. In: Proceedings of ICDE Conference, pp. 421–430 (2001)
2. Hanusse, N., et al.: Computing and summarizing the negative skycube. In: Proceedings of CIKM Conference, pp. 1733–1742 (2016)
3. Lee, J., et al.: BSkyTree: scalable skyline computation using a balanced pivot selection. In: Proceedings of EDBT Conference (2010)
4. Pei, J., et al.: Towards multidimensional subspace skyline analysis. ACM TODS **31**(4), 1335–1381 (2006)
5. Lin, X., et al.: Stabbing the sky: efficient skyline computation over sliding windows. In: Proceedings of ICDE Conference 2005, pp. 502–513. IEEE (2005)

A Domain Adaptation Approach for Multistream Classification

Yue Xie[1(✉)], Jingjing Li[1], Mengmeng Jing[1], Ke Lu[1], and Zi Huang[2]

[1] University of Electronic Science and Technology of China, Chengdu 610054, China
yuexie0806@gmail.com, lijin117@yeah.net
[2] University of Queensland, Brisbane, Australia
huang@itee.uq.edu.au

Abstract. In this paper, we formulate cross-domain multistream classification as a domain adaptation problem. Then we propose a novel algorithm that utilizes low-rank representation and graph embedding to preserve data structures, which benefits in dealing with concept drifts and concept revolution. In addition, we deploy MMD metric to minimize the distribution discrepancy between the source data stream and the target data stream. Experiment results on Office+Caltech dataset with DeCAF$_6$ features verified the effectiveness of our algorithm.

Keywords: Cross-domain multistream classification ·
Domain adaptation · Low-rank representation

1 Introduction

Datastream classification is one of the most challenging tasks in data mining [1,2], which generally needs label information of the streamed data. Manually labeling data costs too much time and efforts. To solve this problem, researchers proposed to train a classifier with labeled data samples generated from a non-stational process, then to classify the unlabeled data generated from a different non-stational process. The labeled data derive from the source data stream, while unlabeled data derive from the target data stream. Such a learning paradigm is well-known as multistream classification [3,4], in which concept drifts and concept revolution are two main issues [3,4].

In this paper, we propose a novel algorithm, which addresses the multistream classification where the source and target datastreams have distribution gap. Firstly, we learn a shared subspace with low-rank constraint in which the target samples can be reconstructed by source samples. At the same time, we deploy both conditional and marginal MMD to minimize the distribution gap. At last, we use graph embedding to preserve the manifold structure of data. Notably, this approach can efficiently filter out noisy information and avoid the negative influence. By minimizing the distribution gap between the source and target data streams, concept drifts and concept revolution will be automatically handled. To summarize, the main contributions of this paper are: (1) We propose

© Springer Nature Switzerland AG 2019
G. Li et al. (Eds.): DASFAA 2019, LNCS 11448, pp. 343–347, 2019.
https://doi.org/10.1007/978-3-030-18590-9_42

a novel algorithm to address the cross-domain multistream classification. Our algorithm not only minimizes the distribution discrepancy between the different datastreams but also takes care of concept drift and concept revolution. (2) We formulate the multistream classification as a domain adaptation problem. The discrepancies of both marginal and conditional probability distributions are minimized to align data streams. A low-rank subspace guarantees that data common structures are preserved.

2 Related Work

In datastream classification, traditional techniques assumed that stream data comes from a single non-stationary process [5,6]. Recently, several researchers proposed multistream classification [2–4], which have source and target data streams that generating labeled data and unlabeled data, respectively.

Domain adaptation [7–9] leverages the knowledge from a labeled source domain to facilitate the learning of an unlabeled target domain. Currently, the existing domain adaptation methods can be roughly divided into two categories. The first group specifically defines a metric to measure the distribution difference [10,11]. The other category discovers the common factors shared by the source and target domains [12].

3 The Proposed Method

To align the distributions of the two domains and preserve the common structure of different domains, we learn a shared subspace spanned by P, in which we reconstruct the target data by Z with low-rank constraint. Since the rank minimization problem is non-convex, we use the nuclear norm of Z to replace low-rank constraint. Furthermore, we introduce sparse matrix E to formulate the noisy information. Then, we have:

$$\min_{P,Z,E} \|Z\|_* + \alpha\|Z\|_1 + \beta\|E\|_1 + \lambda\|P\|_F^2 \qquad s.t. \ P^T X_t = P^T X_s Z + E, \qquad (1)$$

where $\alpha > 0, \beta > 0$ and $\lambda > 0$ are balanced parameters. The $\|Z\|_1$ term guarantees each target sample can be reconstructed by only related source samples instead of all source data. The $\|P\|_F^2$ term controls the complexity of P.

In this paper, we further introduce MMD [10], an explicit metric to measure the distance of different distributions. Specifically, we jointly minimize the marginal and conditional MMD. In addition, we utilize graph embedding to preserve the geometric information of data. We add the two terms:

$$\eta \sum_{c=0}^{c} tr(P^T X M_c X^T P) + \gamma tr(P^T X L_w X^T P), \qquad (2)$$

where $\eta > 0$, $\gamma > 0$ are tradeoff parameters. M_c and L_w indicate the MMD and graph laplacian matrixs. Morever, to utilize the discriminative ability of the source data in the common space and expand the margins between different

---

Algorithm 1. A Domain Adaptation Approach for Multistream Classification

---

Input: $X_s, X_t, Y, B, M_c, L_w, \alpha, \beta, \gamma, \lambda$ and η

Initlize: $N = 1$; $Z = Z_1 = Z_2 = 0$; $E = 0$; $Y_1 = Y_2 = Y_3 = 0$; $\mu_{max} = 10^7$;
$\mu = 0.1$; $\rho = 1.01$; $\epsilon = 10^{-7}$

While not converged **do**

1. Fix other variables and update P by solving $P^* = (X_s X_s^T + \mu G_2 G_2^T + 2\lambda I + 2\eta X M_c X^T + 2\gamma X L_w X^T)^{-1}(X_s G_1^T + \mu G_2 G_3^T)$ where $G_1 = Y + B \odot N$, $G_2 = X_t - X_s Z$, and $G_3 = E - \frac{Y_1}{\mu}$.

2. Fix other variables and update Z by solving $Z^* = (\mu X_s^T P P^T X_s + 2\mu I)^{-1}(G_5 + G_6 - X_s^T P G_4)$ where $G_4 = P^T X_t - E + \frac{Y_1}{\mu}$, $G_5 = Z_1 - \frac{Y_2}{\mu}$, and $G_6 = Z_2 - \frac{Y_3}{\mu}$.

3. Fix other variables and update Z_1 by solving $Z_1^* = \vartheta_{1/\mu}(Z + \frac{Y_2}{\mu})$.

4. Fix other variables and update Z_2 by solving $Z_2^* = \mathrm{shrink}(Z + \frac{Y_3}{\mu}, \frac{\alpha}{\mu})$.

5. Fix other variables and update E by solving $E^* = \mathrm{shrink}(P^T X_t - P^T X_s Z - E + \frac{Y_1}{\mu}, \frac{\beta}{\mu})$.

6. Fix other variables and update N by solving $N = max((P^T X_s - Y) \circ B, 0)$.

7. Update the multipliers and parameters by $Y_1 = Y_1 + \mu(P^T X_t - P^T X_s Z - E)$, $Y_2 = Y_2 + \mu(Z - Z_1)$, $Y_3 = Y_3 + \mu(Z - Z_2)$ and $\mu = \min(\rho\mu, \mu_{max})$.

8. Check the convergence conditions
$\|P^T X_t - P^T X_s Z - E\|_\infty < \epsilon$, $\|Z - Z_1\|_\infty < \epsilon$, $\|Z - Z_2\|_\infty < \epsilon$

End while

Output: P,Z,E

---

classes, we follow [13] to add a discriminative learning function. Thus, the object function is:

$$\min_{P,Z,E} \|Z\|_* + \alpha\|Z\|_1 + \beta\|E\|_1 + \lambda\|P\|_F^2 + \eta \sum_{c=0}^{c} tr(P^T X M_c X^T P)$$
$$+ \gamma tr(P^T X L_w X^T P) + \|P^T X_s - (Y + B \odot N)\|_F^2 \qquad (3)$$
$$s.t. \ P^T X_t = P^T X_s Z + E, \quad N \geq 0,$$

where Y is a source data label matrix, B is a constant matrix, and N is a non-negative label relaxation matrix. To optimize Eq. (3), we get the augmented lagrange function by deploying the Augmented Lagrange Multiplier [14] as:

$$\mathcal{L} = \frac{1}{2}\|P^T X_s - (Y + B \odot N)\|_F^2 + \|Z_1\|_* + \alpha\|Z_2\|_1 + \beta\|E\|_1$$
$$+ \lambda\|P\|_F^2 + \eta \sum_{c=0}^{C} tr(P^T X M_c X^T P) + \gamma tr(P^T X L_w X^T P)$$
$$+ \langle Y_1, P^T X_t - P^T X_s Z - E \rangle + \langle Y_2, Z - Z_1 \rangle + \langle Y_3, Z - Z_2 \rangle \qquad (4)$$
$$+ \frac{\mu}{2}\|P^T X_t - P^T X_s Z - E\|_F^2 + \frac{\mu}{2}(\|Z - Z_1\|_F^2 + \|Z - Z_2\|_F^2),$$

where Y_1, Y_2 and Y_3 are Lagrange multipliers and μ is a positive scalar. $\langle \cdot \rangle$ represents the trace of matrix. Z_1 and Z_2 are two relaxed variables. Algorithm 1 lists every step in which variables can be updated by solving their corresponding equations. Specifically, the shrink implies the shrinkage operator.

Table 1. Multistream classification accuracies (%) on the Office+Caltech dataset.

Source	Target	NN	GFK	TCA	JDA	TJM	JGSA	DMM	Ours
C	A	87.3	88.2	89.8	89.6	88.8	91.4	92.4	**93.0**
	W	72.5	77.6	78.3	85.1	81.4	86.8	87.5	**92.2**
	D	79.6	86.6	85.4	89.8	84.7	**92.6**	90.4	88.5
A	C	71.7	79.2	82.6	83.6	84.3	84.9	84.8	**87.0**
	W	68.1	70.9	74.2	78.3	71.9	81.0	84.7	**89.8**
	D	74.5	82.2	81.5	80.3	76.4	88.5	**92.4**	89.8
W	C	55.3	69.8	80.4	84.8	83.0	85.0	81.7	**88.2**
	A	62.6	76.8	84.1	90.3	87.6	**90.7**	86.5	**90.7**
	D	98.1	**100.0**	**100.0**	**100.0**	**100.0**	**100.0**	98.7	**100.0**
D	C	42.1	71.4	82.3	85.5	83.8	86.2	83.3	**87.9**
	A	50.0	76.3	89.1	91.7	90.3	92.0	90.7	**92.5**
	W	91.5	99.3	99.7	99.7	99.3	99.7	99.3	**100.0**
Avg.		71.1	81.5	85.6	88.2	86.0	90.0	89.4	**91.6**

4   Experiments

The Office+Caltech dataset with DeCAF$_6$ features [15,16] contains 10 common classes from A (amazon), W (webcam), D (dslr), and C (caltech) domains. We randomly choose two different domains, in turn, as the source and the target domains respectively. The results on this dataset compared with different algorithms are shown in Table 1.

It can be seen that our algorithm outperforms others in 10 out of 12 tasks from the experiment results. Compared with JGSA [17], currently the state-of-the-art method, our algorithm gains a 1.6% improvement. What's more, half of 12 pairs experiments achieve over 90% accuracies, which varifies the effectiveness of our algorithm.

5   Conclusion

In this paper, we propose a novel algorithm to challenge the multistream classification problem. The proposed algorithm takes advantage of low-rank representation, graph embedding and distribution alignment into a generalized framework. The algorithm formulates multistream classification as a domain adaptation problem. In our future work, we are going to investigate online datastream classification tasks, which needs almost real-time processing.

References

1. Wagde, J., Deshkar, P.A.: A review on method of stream data classification through tree based approach. In: 2016 World Conference on Futuristic Trends in Research and Innovation for Social Welfare (Startup Conclave), pp. 1–4 (2016)
2. Haque, A., Khan, L., Baron, M., Thuraisingham, B., Aggarwal, C.: Efficient handling of concept drift and concept evolution over stream data. In: ICDE, pp. 481–492 (2016)
3. Chandra, S., Haque, A., Khan, L., Aggarwal, C.: An adaptive framework for multistream classification. In: IKM, pp. 1181–1190 (2016). https://doi.org/10.1145/2983323.2983842
4. Haque, A., Chandra, S., Khan, L., Hamlen, K., Aggarwal, C.: Efficient multistream classification using direct density ratio estimation. In: ICDE, pp. 155–158 (2017)
5. Last, M.: Online classification of nonstationary data streams. Intell. Data Anal. **6** 129–147
6. Aggarwal, C.C., Han, J., Wang, J., Yu, P.S.: On demand classification of data streams. In: KDD, pp. 503–508 (2004). https://doi.org/10.1145/1014052.1014110
7. Li, J., Wu, Y., Zhao, J., Lu, K.: Low-rank discriminant embedding for multiview learning. IEEE Trans. Cybern. **47**(11), 3516–3529 (2017)
8. Li, J., Lu, K., Huang, Z., Zhu, L., Shen, H.T.: Heterogeneous domain adaptation through progressive alignment. IEEE TNNLS. https://doi.org/10.1109/TNNLS.2018.2868854
9. Li, J., Lu, K., Huang, Z., Zhu, L., Shen, H.T.: Transfer independently together: a generalized framework for domain adaptation. IEEE TCYB. https://doi.org/10.1109/TCYB.2018.2820174
10. Long, M., Wang, J., Ding, G., Sun, J., Yu, P.S.: Transfer feature learning with joint distribution adaptation. In: ICCV. https://doi.org/10.1109/ICCV.2013.274
11. Zellinger, W., Lughofer, E., Saminger-Platz, S., Grubinger, T., Natschläger, T.: Central moment discrepancy (CMD) for domain-invariant representation learning. Stat 1050, 4 (2017)
12. Baktashmotlagh, M., Harandi, M.T., Lovell, B.C., Salzmann, M.: Unsupervised domain adaptation by domain invariant projection. In: ICCV, pp. 769–776 (2013)
13. Xu, Y., Fang, X., Wu, J., Li, X., Zhang, D.: Discriminative transfer subspace learning via low-rank and sparse representation. TIP **25**, 850–863 (2016)
14. Lin, Z., Chen, M., Ma, Y.: The augmented lagrange multiplier method for exact recovery of corrupted low-rank matrices. J. Struct. Biol. **181**(2), 116–127 (2013)
15. Saenko, K., Kulis, B., Fritz, M., Darrell, T.: Adapting visual category models to new domains. In: Daniilidis, K., Maragos, P., Paragios, N. (eds.) ECCV 2010. LNCS, vol. 6314, pp. 213–226. Springer, Heidelberg (2010). https://doi.org/10.1007/978-3-642-15561-1_16
16. Griffin, G., Holub, A., Perona, P.: Caltech-256 object category dataset
17. Zhang, J., Li, W., Ogunbona, P.: Joint geometrical and statistical alignment for visual domain adaptation. In: CVPR, pp. 5150–5158 (2017)

Gradient Boosting Censored Regression for Winning Price Prediction in Real-Time Bidding

Piyush Paliwal(✉) and Oleksii Renov

LoopMe Ltd., London, UK
{piyush.paliwal,alexey}@loopme.com

Abstract. The demand-side platform (DSP) is a technological ingredient that fits into the larger real-time-bidding (RTB) ecosystem. DSPs enable advertisers to purchase ad impressions from a wide range of ad slots, generally via a second-price auction mechanism. In this aspect, predicting the auction winning price notably enhances the decision for placing the right bid value to win the auction and helps with the advertiser's campaign planning and traffic reallocation between campaigns. This is a difficult task because the observed winning price distribution is biased due to censorship; the DSP only observes the win price in case of winning the auction. For losing bids, the win price remains censored. Erstwhile, there has been little work that utilizes censored information in the learning process. In this article, we generalize the winning price model to incorporate a gradient boosting framework adapted to learn from both observed and censored data. Experiments show that our approach yields the hypothesized boost in predictive performance in comparison to classic linear censored regression.

Keywords: Demand-side platform · Real-time bidding
Display advertising · Censored regression · Gradient boosting

1 Introduction

With growing popularity and usage [5], real-time bidding (RTB) has monetized the ad tech industry to a new scale. In the RTB auction process, buying and selling of online ad impressions are facilitated through an ad exchange. Whenever a visitor lands on a publisher's webpage or app, an ad request is initiated for a particular ad slot. An instant auction is invoked at the ad exchange wherein participating advertisers place their bids. The demand-side platform (DSP) assists advertisers to gainfully set a bid value on an impression, based on relevance of the user towards the potential ad. Several advertisers are bidding with the help of DSPs. The advertiser with the highest bid wins the auction and pays the second highest bid. The winner's ad is then displayed on the publisher's webpage or app. The entire auction process finishes within 100 ms.

© Springer Nature Switzerland AG 2019
G. Li et al. (Eds.): DASFAA 2019, LNCS 11448, pp. 348–352, 2019.
https://doi.org/10.1007/978-3-030-18590-9_43

For DSP, the winning price is the highest price set by its competitors. None of the DSPs has information about what their competitors are bidding. This is where predicting the winning price becomes necessary. However, data censorship makes this difficult. DSPs only observe the winning price of the bids which they win in auction. The full winning price distribution consists of the winning prices of all win bids plus the winning prices of all lose bids. Since the latter is missing in the training data, it becomes substantially challenging to model the winning price. However, in case of losing the auction, the DSP at least knows a lower bound of the winning price, i.e., its own bid price. This is a right censored case [2,8]. Few prior works have sought to deal with censorship for winning price prediction. [8] implements a linear censored model and [7] generalizes the winning price to use a deep censored learning framework. In this paper, we generalize the winning price model to incorporate a gradient boosting framework [3] adapted to learn from both observed and censored data. Section 3 shows how gradient boosting censored regression (tree-based) outperforms the results of [8].

2 Methodology

A detailed technical report about our work can be found on[1]. Following linear censored regression (LCR) [8], $y \sim N(\theta^T x, \sigma^2)$ as true win price, w observed win price, b bid price and x feature vector, the negative log-likelihood function is:

$$\sum_{i \in \mathcal{W}} -\log \phi \left(\frac{w_i - \theta^T x_i}{\sigma} \right) + \sum_{i \in \mathcal{L}} -\log \Phi \left(-\left(\frac{b_i - \theta^T x_i}{\sigma} \right) \right) \quad (1)$$

Where $\mathcal{W}$ represents the set of all winning bids and $\mathcal{L}$ represents the set of all losing bids. Inspired by LCR, we propose a new winning price model that uses xgboost [1], an optimized distributed gradient boosting library, adapted to censored regression. Inspired by **X**gboost and the idea of **C**ensored **R**egression, for brevity we call our winning price algorithm XCR. In the boosting mechanism, each base learner sequentially compensates shortcomings made by previous learners, to improve the model, $F(x)$. For simplicity we define:

$$z_i = \begin{cases} \frac{w_i - F(x_i)}{\sigma}, & i \in \mathcal{W} \\[2mm] \frac{b_i - F(x_i)}{\sigma}, & i \in \mathcal{L} \end{cases} \quad (2)$$

For XCR, true win price $y \sim N(F(x), \sigma^2)$. By replacing $\theta^T x_i$ with $F(x_i)$ in Eq. (1) and by using Eq. (2), we achieve a loss function for XCR as in (3), whose first and second order derivatives for ith observation (x_i, y_i) are in (4) and (5).

$$L(y, F(x)) = \sum_{i \in \mathcal{W}} -\log \phi(z_i) + \sum_{i \in \mathcal{L}} -\log \Phi(-z_i) \quad (3)$$

---

[1] https://github.com/paliwal90/xcr_win_price_pred/blob/master/xcr_wpp.pdf.

$$\frac{\partial L(y_i, F(x_i))}{\partial F(x_i)} = -\frac{z_i}{\sigma}\mathbb{1}_{\{i \in \mathcal{W}\}} - \frac{1}{\sigma}\frac{\phi(z_i)}{\Phi(-z_i)}\mathbb{1}_{\{i \in \mathcal{L}\}} \tag{4}$$

$$\frac{\partial^2 L(y_i, F(x_i))}{\partial^2 F(x_i)} = \frac{1}{\sigma^2}\mathbb{1}_{\{i \in \mathcal{W}\}} + \frac{1}{\sigma}\frac{\partial L(y_i, F(x_i))}{\partial F(x_i)}\left(z_i + \sigma\frac{\partial L(y_i, F(x_i))}{\partial F(x_i)}\right)\mathbb{1}_{\{i \in \mathcal{L}\}}$$
$$\tag{5}$$

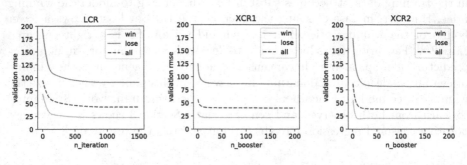

Fig. 1. Validation RMSE w.r.t iterations on win, lose and all bids of day 2013-06-10 for LCR, XCR1 and XCR2.

3   Experiments

We conduct and evaluate our algorithms on iPinYou RTB dataset [9]. For LCR, we choose the Adam optimizer [4]. Two variants of XCR: XCR1 (with linear base learners) and XCR2 (with tree base learners) are evaluated. We use the data of Season 2 and Season 3, apply the data preparation trick and features processing as proposed in [8]. We use Bayesian optimisation [6] for parameters tuning using validation set. Figure 1 shows a smooth convergence for all three models.

The reported RMSE in Table 1 shows that, for most days, XCR2 outperforms LCR and XCR1. This holds true for the win bids, the lose bids and overall. The tree based XCR2 is capable of learning the interactions between features, which is not possible for the linear models, LCR and XCR1. Although XCR1 also uses a gradient boosting mechanism, but due to the underlying linear base learners, the end results are still linear. Hence XCR1 and LCR generally perform similar, if converged well. For all three models, the performance on season 2 is better than on season 3 because the amount of data in season 2 is significantly higher than in season 3 [8,9]. The high win rates introduces a skewed distribution biased more towards the win bids. Hence, for all three models, the error on lose bids is significantly higher than the error on win bids.

Table 1. Root mean squared error on testing set of simulated dataset of iPinYou Season 2 (2013-06-06 till 2013-06-12) and Season 3 (2013-10-19 till 2013-10-27).

Day	RMSE all			RMSE win			RMSE lose		
	LCR	XCR1	XCR2	LCR	XCR1	XCR2	LCR	XCR1	XCR2
2013-06-06	45.59	46.44	**43.91**	25.63	25.59	**24.43**	95.57	98.02	**92.36**
2013-06-07	45.12	45.66	**43.19**	26.32	25.17	**23.58**	96.31	99.52	**94.45**
2013-06-08	47.06	47.84	**46.14**	24.53	24.47	**23.78**	99.35	101.53	**97.73**
2013-06-09	45.51	45.74	**43.56**	26.08	24.66	**23.53**	93.42	95.88	**91.25**
2013-06-10	41.31	40.41	**38.10**	22.03	21.91	**20.51**	88.86	86.43	**81.70**
2013-06-11	45.11	41.76	**41.20**	23.39	22.44	**20.65**	98.11	**89.83**	90.45
2013-06-12	41.19	42.06	**39.80**	24.27	22.61	**21.36**	85.22	89.95	**85.19**
2013-10-19	59.61	60.09	**57.46**	33.96	34.34	**32.56**	103.29	104.04	**99.74**
2013-10-20	57.42	56.14	**55.34**	37.65	34.34	**33.98**	94.31	95.02	**93.52**
2013-10-21	57.48	55.97	**54.83**	40.15	35.36	**33.80**	**89.33**	91.30	90.33
2013-10-22	61.33	58.29	**57.27**	41.35	35.44	**34.22**	97.56	96.98	**95.90**
2013-10-23	59.36	59.07	**57.64**	38.88	37.24	**36.20**	98.23	99.45	**97.19**
2013-10-24	**54.46**	56.13	55.22	34.50	34.83	**34.05**	**101.44**	105.97	104.13
2013-10-25	56.42	57.89	**55.67**	34.90	35.95	**34.12**	98.42	100.81	**97.47**
2013-10-26	62.37	59.02	**56.56**	36.33	34.49	**33.84**	106.41	100.57	**95.60**
2013-10-27	61.61	58.24	**56.63**	34.88	32.40	**31.75**	110.54	105.09	**101.93**

4 Conclusions

This paper generalized the winning price model to incorporate a gradient boosting framework adapted to censored regression in RTB display advertising. We evaluated two variants: XCR1 which uses linear base learners and XCR2 which uses tree base learners, against classic linear censored regression (LCR). Our proposed XCR2 demonstrates superiority over LCR.

References

1. Chen, T., Guestrin, C.: XGBoost: a scalable tree boosting system. In: Proceedings of the 22Nd ACM SIGKDD International Conference on Knowledge Discovery and Data Mining, KDD 2016, pp. 785–794. ACM, New York (2016). https://doi.org/10.1145/2939672.2939785
2. Elandt-Johnson, R.C., Johnson, N.L.: Survival Models and Data Analysis. Wiley, New York (1980)
3. Freund, Y., Schapire, R.E.: A decision-theoretic generalization of on-line learning and an application to boosting. J. Comput. Syst. Sci. **55**(1), 119–139 (1997). https://doi.org/10.1006/jcss.1997.1504
4. Kingma, D.P., Ba, J.: Adam: a method for stochastic optimization. CoRR abs/1412.6980 (2014). http://arxiv.org/abs/1412.6980

5. Muthukrishnan, S.: Ad exchanges: research issues. In: Leonardi, S. (ed.) WINE 2009. LNCS, vol. 5929, pp. 1–12. Springer, Heidelberg (2009). https://doi.org/10.1007/978-3-642-10841-9_1
6. Snoek, J., Larochelle, H., Adams, R.P.: Practical Bayesian optimization of machine learning algorithms. In: Proceedings of the 25th International Conference on Neural Information Processing Systems, NIPS 2012, vol. 2, pp. 2951–2959. Curran Associates Inc., USA (2012). http://dl.acm.org/citation.cfm?id=2999325.2999464
7. Wu, W., Yeh, M.Y., Chen, M.S.: Deep censored learning of the winning price in the real time bidding. In: Proceedings of the 24th ACM SIGKDD International Conference on Knowledge Discovery & Data Mining, KDD 2018, pp. 2526–2535. ACM, New York (2018). https://doi.org/10.1145/3219819.3220066
8. Wu, W.C.H., Yeh, M.Y., Chen, M.S.: Predicting winning price in real time bidding with censored data. In: Proceedings of the 21st ACM SIGKDD International Conference on Knowledge Discovery and Data Mining, KDD 2015, pp. 1305–1314. ACM, New York (2015). https://doi.org/10.1145/2783258.2783276
9. Zhang, W., Yuan, S., Wang, J.: Real-time bidding benchmarking with iPinYou dataset. CoRR abs/1407.7073 (2014). http://arxiv.org/abs/1407.7073

Deep Sequential Multi-task Modeling for Next Check-in Time and Location Prediction

Wenwei Liang, Wei Zhang$^{(\boxtimes)}$, and Xiaoling Wang

East China Normal University, Shanghai, China
51174500033@stu.ecnu.edu.cn, zhangwei.thu2011@gmail.com,
xlwang@sei.ecnu.edu.cn

Abstract. In this paper, we address the problem of next check-in time and location prediction, and propose a deep sequential multi-task model, named Personalized Recurrent Point Process with Attention (PRPPA), which seamlessly integrates user static representation learning, dynamic recent check-in behavior modeling, and temporal point process into a unified architecture. An attention mechanism is further included in the intensity function of point process to enhance the capability of explicitly capturing the effect of past check-in events. Through the experiments, we verify the proposed model is effective in location and time prediction.

Keywords: Multi-task learning · Check-in prediction ·
Deep recurrent modeling · Temporal point process

1 Introduction

The proliferation of location-based social networks has triggered amount of studies on user spatial behavior modeling [7]. In essence, **Where** and **When** are the two most important factors that determine the user behavior choices. However, most pioneering studies [3] heavily focus on the spatial aspect of user mobility, while ignore to explicitly consider the temporal aspect, let alone model both aspects jointly. In this paper we study the problem of next check-in time and location prediction, which is defined as follows:

Problem Formulation: Suppose each check-in record c is a triplet $(u, l, t) \in U \times L \times T$, where U, L and T represent a set of unique users, locations and timestamps, respectively. We arrange all the check-ins belonging to the same user in chronological order as user check-in trajectory sequence $C_u = \{c_1, c_2, \ldots, c_n\}$. We further divide C_u into several non-overlapping trajectories $S_{u,1:k}$ with respect to a maximum time interval w and a maximum trajectory length d. Given the user u and current trajectory $S_{u,k}$, for a next check-in c_{n+1} occurs, the goal is to predict the location l_{n+1} and time t_{n+1}.

This work was supported in part by Shanghai Sailing Program (17YF1404500), SHMEC (16CG24), and NSFC (61702190, U1609220).

© Springer Nature Switzerland AG 2019
G. Li et al. (Eds.): DASFAA 2019, LNCS 11448, pp. 353–357, 2019.
https://doi.org/10.1007/978-3-030-18590-9_44

We propose a deep sequential multi-task model, named Personalized Recurrent Point Process with Attention (PRPPA), to address the above problem. The key idea of PRPPA is to seamlessly integrate user static representation learning, dynamic recent check-in behavior modeling, and temporal point process into a unified architecture, enabling the simultaneous training of next check-in time prediction and location prediction benefit each other for better performance. An attention mechanism is included in the intensity function of point process to enhance the capability of explicitly capturing the effect of past check-in events. We have conducted the experiments on real world datasets, indicating that the superiority of the proposed model, both in location and time prediction tasks.

2   Model and Learning

Our model PRPPA is an end-to-end learning framework which takes location ID, user ID, time interval (hours) and timestamp (hours in a week) as inputs and outputs location and time interval prediction. In practice, we feed numbered features including location ID, user ID and timestamp into the corresponding embedding layer. Afterwards, we concatenate these dense vectors as the inputs to the subsequent recurrent layer. GRU is selected as the building block for its simplicity. We can obtain the hidden state sequence $\{h_1, \ldots, h_o, \ldots, h_j\}$ after recurrent computation for each time step. To simplify description, we take time step k as an example.

Check-in Time Prediction: Previous studies [1,5] propose to leverage the most recent hidden state h_k of GRU to characterize the intensity function of point process, which does not fully consider the influence of historical check-ins in the trajectory. In this work, we devise a simple but effective attention based mechanism to automatically quantify the effect of past check-in events on future event in the conditional intensity function. We get the integrated representation $\tilde{h}_k$ to be: $\tilde{h}_k = \sum_{o=1}^{k} \text{Softmax}(h_k \cdot h_o / \sqrt{d_{h_k}})h_o$, where $\sqrt{d_{h_k}}$ is the scaling factor. Upon this, we can formulate the conditional intensity function for the next check-in time prediction by:

$$\lambda^*(t) = exp(v^{t^T} \cdot \tilde{h}_k + w^u \cdot v_u + w^t \cdot g_k + b_t), \tag{1}$$

where v^{t^T}, w^u, w^t, b_t are the parameters to be learned, $g_k = t - t_k$ is the time interval since the last check-in, v_u is the user static representation. Finally, the conditional density function of proposed deep point process is specified as: $f^*(t) = \lambda^*(t)exp(-\int_{t_k}^{t} \lambda^*(\epsilon)d\epsilon)$. The expected return time of next check-in $\hat{t}_{k+1}$ can be computed as: $\hat{t}_{k+1} = \int_{t_k}^{\infty} t \cdot f^*(t)dt$. In general, we can apply commonly used numerical integration techniques to approximates the predictions.

Check-in Location Prediction: For predicting the next check-in location, we apply a fully connected layer with Softmax as activation function: $y_z^* = \text{Softmax}(W_z[h_k; v_u] + b_z)$, where $[h_k; v_u]$ is the concatenation of user recent dynamic preference and user static preference, W_z, b_z are the parameters and have the hidden size with number of unique location. We could select the maximum element as the most possible prediction l_{k+1}.

Multi-task Training: Having two tasks at hand and the inputs, we consider solving them jointly. The final time loss is: $L_{time}(\tilde{h}_k, v_u, g_k) = -log f^*(t)$. The loss of the location prediction task is the cross entropy loss: $L_{loc}(h_k, v_u) = -\sum_{z=1}^{Z} y_z log(y_z^*)$, where y_z is the one-hot encoding of the target location. The final objective function is defined as follows:

$$L_{obj} = \sum_{u \in U} \sum_{S_u \in C_u} \sum_{k=1}^{j} L_{loc}(h_k, v_u) + \beta L_{time}(\tilde{h}_k, v_u, g_k) + \gamma \|\Theta\|_2^2, \quad (2)$$

where β and γ are hyperparameters used for tuning relative influence from the time prediction loss and regularization loss about model parameters. Θ is the set of model parameters. The whole framework can be trained using backpropagation in an end-to-end fashion.

3 Experiments

We conduct our experiments on two public datasets **Gowalla**[1] and **Foursquare**[2]. Each check-in is a tuple of user ID, location ID, GPS coordinates, and timestamp. We restrict the geographical area to New York City according to its coordinate and apply appropriate preprocessing procedures to both datasets. We set maximum time interval $w = 72$ h and maximum trajectory length $d = 10$. At last, Gowalla consists 1063 users and 57338 check-ins while Foursquare contains 1079 users and 139878 check-ins. For each user, we select the earliest 80% trajectories as training data, the remaining 20% as testing data. To evaluate the performance of each method on location prediction task, two metrics are used: Precision (Prec@1) and Recall (Rec@10). We adopt Mean Square Error (MSE) for the time interval prediction task. For the dimension of location embedding and hidden state size, we set to 10 for all methods. The user embedding size and hyperparameter β is carefully selected within a certain range due to its performance.

We compare our methods with several state-of-the-art methods for next check-in location prediction task, i.e., Markov model, RMTPP [1], DeepMove [2], CARA [4]. As results are shown in Table 1. Without the help of user modeling, Markov model and RMTPP are not competitive with the other models. Deep-Move and CARA are newly emerged representative methods for user mobility data modeling. Compared with them, our proposed PRPPA considers the spatial and temporal factors from the multi-task learning perspective and achieves the best performance. This phenomenon also makes sense as more supervised signals from the next check-in time prediction task is injected into the model optimization process through back-propagation.

---

[1] https://snap.stanford.edu/data/loc-gowalla.html.

[2] https://sites.google.com/site/yangdingqi/home/foursquare-dataset.

Table 1. Method comparison on location prediction task.

		Markov	RMTPP	DeepMove	CARA	PRPPA
Gowalla	Rec@10	0.15365	0.18306	0.25701	0.27282	0.29513
	Prec@1	0.07200	0.06592	0.11359	0.10700	0.12424
Foursquare	Rec@10	0.29841	0.30441	0.48442	0.48274	0.50384
	Prec@1	0.12392	0.10954	0.17592	0.16515	0.18313

For time prediction task, we compare PRPPA with the following methods: Average popular time (Avg pop), Feedforward deep Neural Network (NN), RMTPP and RSTPP [6]. Experimental results are shown in Table 2. Generally speaking, our approach has considerable improvements on both datasets over all the baselines, due to the multi-task learning and the incorporation of attention mechanism in conditional intensity function.

Table 2. MSE performance on time prediction task. Unit: 4 h.

	Avg pop	NN	RMTPP	RSTPP	PRPPA
Gowalla	35.22	24.54	28.08	24.47	22.67
Foursquare	34.42	23.91	27.26	24.93	21.29

4 Conclusion

We have addressed the next check-in location and time prediction from a multi-task learning perspective, and proposed PRPPA, a natural extension of RMTPP by: (1) introducing user static representation to denote user long-term preference and combine it with user recent dynamic preference; (2) proposing an attention based method to explicitly capture the effect of past check-in events for the generation of next check-in event. Comprehensive experiments on two real datasets have shown PRPPA is superior among all the adopted methods for both tasks.

References

1. Du, N., Dai, H., Trivedi, R., Upadhyay, U., Gomez-Rodriguez, M., Song, L.: Recurrent marked temporal point processes: embedding event history to vector. In: SIGKDD, pp. 1555–1564. ACM (2016)
2. Feng, J., et al.: DeepMove: predicting human mobility with attentional recurrent networks. In: WWW, pp. 1459–1468 (2018)
3. Liang, Y., Ke, S., Zhang, J., Yi, X., Zheng, Y.: GeoMAN: multi-level attention networks for geo-sensory time series prediction. In: IJCAI, pp. 3428–3434 (2018)
4. Manotumruksa, J., Macdonald, C., Ounis, I.: A contextual attention recurrent architecture for context-aware venue recommendation. In: SIGIR, pp. 555–564 (2018)

5. Qiao, Z., Zhao, S., Xiao, C., Li, X., Qin, Y., Wang, F.: Pairwise-ranking based collaborative recurrent neural networks for clinical event prediction. In: IJCAI, pp. 3520–3526 (2018)
6. Yang, G., Cai, Y., Reddy, C.K.: Recurrent spatio-temporal point process for check-in time prediction. In: CIKM, pp. 2203–2211 (2018)
7. Zhang, W., Wang, J.: Location and time aware social collaborative retrieval for new successive point-of-interest recommendation. In: CIKM, pp. 1221–1230 (2015)

SemiSync: Semi-supervised Clustering by Synchronization

Zhong Zhang, Didi Kang, Chongming Gao, and Junming Shao[(✉)]

School of Computer Science and Engineering,
University of Electronic Science and Technology of China, Chengdu 611731, China
{zhongzhang,ddkang,chongming.gao}@std.uestc.edu.cn
junmshao@uestc.edu.cn

Abstract. In this paper, we consider the semi-supervised clustering problem, where the prior knowledge is formalized as the Cannot-Link (CL) and Must-Link (ML) pairwise constraints. We propose an algorithm called SEMISYNC that tackles this problem from a novel perspective: synchronization. The basic idea is to regard the data points as a set of (constrained) phase oscillators, and simulate their dynamics to form clusters automatically. SEMISYNC allows dynamically propagating the constraints to unlabelled data points driven by their local data distributions, which effectively boosts the clustering performance even if little prior knowledge is available. We experimentally demonstrate the effectiveness of the proposed method.

Keywords: Semi-supervised clustering · Synchronization

1 Introduction

Clustering with a priori knowledge is referred to as constrained clustering or semi-supervised clustering. Extensive studies have shown that once a priori knowledge (commonly formalized as instance-level Cannot-Link (CL) and Must-Link (ML) pairwise constraints) is incorporated, the clustering performance can be greatly improved. In this paper, we are going to tackle the semi-supervised clustering problem from a different perspective: synchronization.

SYNC [4] along with its variants [7–9] are novel clustering models, which are derived from an interesting physical phenomenon synchronization. It basically defines a discrete dynamic system. The idea is to regard each object as a phase oscillator and simulate the *local* interaction behavior with its neighborhood over time under a specified interaction model. As time evolves, similar objects will synchronize together and form distinct clusters. Inspired by SYNC, we develop a novel semi-supervised clustering method called SEMISYNC. It incorporates CL and ML constraints in an intuitive way by introducing an additional *global* interaction paradigm. Thanks to the dynamic property, once two or more objects have synchronized together over time (i.e., they have the same position), they can be

ⓒ Springer Nature Switzerland AG 2019
G. Li et al. (Eds.): DASFAA 2019, LNCS 11448, pp. 358–362, 2019.
https://doi.org/10.1007/978-3-030-18590-9_45

merged into a prototype. This provides a natural way to propagate the constraints within the synchronized objects. Therefore SEMISYNC supports to find high-quality clusterings even if only limited prior knowledge is available.

2 The Proposed Method

We first introduce some notations. $\boldsymbol{x}_i(t)$ denotes the i-th data point at the t-th time stamp. For brevity, we omit the time stamp in the following statement. Let $\boldsymbol{p}_i$ denote the i-th prototype and w_i denote the weight of $\boldsymbol{p}_i$. w_i is the number of data points that $\boldsymbol{p}_i$ represents, we will explain them later. $\mathcal{C} = \{(\boldsymbol{x}_i, \boldsymbol{x}_j) | l_i \neq l_j\}$ and $\mathcal{M} = \{(\boldsymbol{x}_i, \boldsymbol{x}_j) | l_i = l_j\}$ denote the CL and ML constraint sets, respectively, where l_i is the label of the i-th data points. For convenience, we use $\mathcal{C}(\boldsymbol{x}_i) = \{\boldsymbol{x}_j | (\boldsymbol{x}_i, \boldsymbol{x}_j) \in \mathcal{C}\}$ and $\mathcal{M}(\boldsymbol{x}_i) = \{\boldsymbol{x}_j | (\boldsymbol{x}_i, \boldsymbol{x}_j) \in \mathcal{M}\}$ to denote a set of data points that cannot link and must link to $\boldsymbol{x}_i$, respectively. $\mathcal{N}_\epsilon^c(\boldsymbol{x}_i) = \{\boldsymbol{x}_j | dist(\boldsymbol{x}_i, \boldsymbol{x}_j) \leq \epsilon, (\boldsymbol{x}_i, \boldsymbol{x}_j) \notin \mathcal{C})\}$ denotes the exclusive ϵ-range neighborhood.

The overall clustering algorithm is a discrete dynamic system, which simply requires a interaction model and a stopping criterion. Different from [4], we define the dynamic system over a set of prototypes rather than original data points. We first present the interaction model and explain it in details.

Definition 1 (Semi-supervised Interaction Model). Given the prototype $\boldsymbol{p}_i \in \mathbb{R}^m$ and its neighbors' weights w_j at the t-th iteration, the constraint sets $\mathcal{C}$ and $\mathcal{M}$. The semi-supervised interaction model is defined as follows.

$$
\boldsymbol{p}_i \longleftarrow \boldsymbol{p}_i + \frac{\alpha \sum_{\boldsymbol{p}_j \in \mathcal{N}_\epsilon^c(\boldsymbol{p}_i)} w_j \sin(\boldsymbol{p}_j - \boldsymbol{p}_i)}{\sum_{\boldsymbol{p}_j \in \mathcal{N}_\epsilon^c(\boldsymbol{p}_i)} w_j} + \frac{(1-\alpha) \sum_{\boldsymbol{p}_j \in \widetilde{\mathcal{M}}(\boldsymbol{p}_i)} w_j \sin(\boldsymbol{p}_j - \boldsymbol{p}_i)}{\sum_{\boldsymbol{p}_j \in \widetilde{\mathcal{M}}(\boldsymbol{p}_i)} w_j}
$$
$$
- \frac{\sum_{\boldsymbol{p}_j \in \mathcal{C}(\boldsymbol{p}_i)} a_{ij} w_j \sin(\boldsymbol{p}_j - \boldsymbol{p}_i)}{\sum_{\boldsymbol{p}_j \in \mathcal{C}(\boldsymbol{p}_i)} w_j},
$$

$$(1)$$

Note that on the right side the second term is the original local synchronization interaction [4]. The last two terms are the global synchronization and desynchronization interaction for the ML and CL constraints, respectively. $\widetilde{\mathcal{M}}(\boldsymbol{p}_i) = \mathcal{M}(\boldsymbol{p}_i) \setminus \mathcal{N}_\epsilon^c(\boldsymbol{p}_i)$ makes no duplicate interaction. $a_{ij} = e^{-\frac{||\boldsymbol{p}_i - \boldsymbol{p}_j||^2}{2\epsilon^2}}$ is a decay weight for desynchronization. Since we only need to desynchronize cannot-link points when they are significantly close and prevent side effect from other distant cannot-link points. $\alpha \in [0, 1.0]$ is a parameter that balances the local and global synchronization. We need to restrict the synchronization coupling strength to 1 to ensure convergence.

During the dynamic interaction process some data points may synchronize together, i.e., they have the identical positions at the end of the t-th time stamp. A prototype is a representative of the synchronized data points. The synchronized data points will have the identical interaction behaviors, thus we can replace the original data points with prototypes with proper weights. The advantages are, the number of prototypes is monotonically decreasing as the iteration proceeds, which significantly alleviates the computation and memory burden. More importantly, it provides a natural way to propagate constraints.

We assume the pairwise constraint links are local consistent. Thereby, a prototype can inherit all the constraints from the merging points. However, it must be careful with the conflict. Suppose we have two synchronized points x_i and x_j, they can merge into a prototype p_i only when the following two conditions are satisfied: (1) $\mathcal{M}_c(x_i) \cap \mathcal{C}(x_j) = \emptyset$; (2) $\mathcal{M}_c(x_j) \cap \mathcal{C}(x_i) = \emptyset$. $\mathcal{M}_c(x_i)$ is a must-link closure of x_i since ML constraints are transitive. Note the constraints are propagated in a local way, which prevents the error constraints from spreading.

Finally, we define an adjusted cluster order parameter r_a to indicate convergence. The algorithm stops when r_a reaches to 1.0 or barely changes.

Definition 2 (Adjusted Cluster Order Parameter). The adjusted cluster order parameter characterizes the degree of synchronization defined as follows.

$$r_a = \frac{1}{N} \sum_{i=1}^{N} \frac{1}{|\mathcal{N}_\epsilon^c(p_i) \cup \mathcal{M}(p_i)|} \sum_{p_j \in \mathcal{N}_\epsilon^c(p_i) \cup \mathcal{M}(p_i)} e^{-||p_j - p_i||}. \tag{2}$$

3   Experiments

We evaluate the semi-supervised clustering performance of the proposed method on eight real-world data sets from the UCI and UCR repositories. For comparison, we select six typical different type of semi-supervised clustering algorithms. It includes MPCK-means [3], LCVQE [5], CECM [2], C1SP [6], CPSNMF [10] and SKMS [1]. We use two unsupervised clustering algorithms SYNC and K-means as the baseline. All data sets are normalized to $[0, \pi]$. For all algorithms, we exhaustively tune their parameters to achieve the best performance.

Figure 1 shows the semi-supervised clustering results on all eight real-world data sets while varying the number of pairwise constraints. Overall, SEMISYNC achieves the best results on all data sets except for the Car data set, but it still yields the second best result. SEMISYNC can work well when only a few constraints are incorporated. The rationale is that SEMISYNC propagates the constraints building upon the local and global interactions, simultaneously.

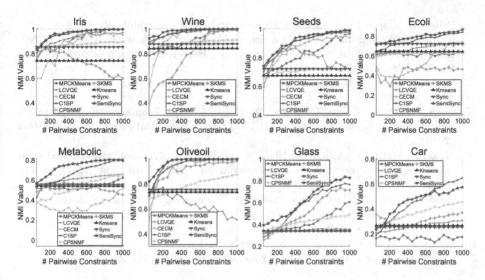

Fig. 1. Semi-supervised clustering results. *Some experiments are aborted due to unsolvable issues: CECM on Plan, Glass, Car.

4 Conclusion

We propose a novel semi-supervised clustering algorithm SEMISYNC from a different perspective: synchronization. SEMISYNC utilizes the local and global interaction paradigms to preserve the intrinsic structure of the data set and incorporate the pairwise constraints. Besides, SEMISYNC supports an intuitive constraint propagation, which helps improve the clustering performance. We experimentally demonstrate the effectiveness of the proposed method.

Acknowledgments. This work is supported by the National Natural Science Foundation of China (61403062, 61433014, 41601025), Science-Technology Foundation for Young Scientist of SiChuan Province (2016JQ0007), Fok Ying-Tong Education Foundation for Young Teachers in the Higher Education Institutions of China (161062) and National key research and development program (2016YFB0502300).

References

1. Anand, S., Mittal, S., Tuzel, O., Meer, P.: Semi-supervised kernel mean shift clustering. IEEE Trans. Pattern Anal. Mach. Intell. **36**(6), 1201–1215 (2014)
2. Antoine, V., Quost, B., Masson, M.H., Denoeux, T.: CECM: constrained evidential C-means algorithm. Comput. Stat. Data Anal. **56**(4), 894–914 (2012)
3. Bilenko, M., Basu, S., Mooney, R.J.: Integrating constraints and metric learning in semi-supervised clustering. In: ICML, p. 11 (2004)
4. Böhm, C., Plant, C., Shao, J., Yang, Q.: Clustering by synchronization. In: KDD, pp. 583–592 (2010)

5. Pelleg, D., Baras, D.: K-means with large and noisy constraint sets. In: Kok, J.N., Koronacki, J., Mantaras, R.L., Matwin, S., Mladenič, D., Skowron, A. (eds.) ECML 2007. LNCS, vol. 4701, pp. 674–682. Springer, Heidelberg (2007). https://doi.org/10.1007/978-3-540-74958-5_67
6. Rangapuram, S.S., Hein, M.: Constrained 1-spectral clustering. In: AISTATS, vol. 30, p. 90 (2012)
7. Shao, J., He, X., Böhm, C., Yang, Q., Plant, C.: Synchronization-inspired partitioning and hierarchical clustering. IEEE Trans. Knowl. Data Eng. $25(4)$, 893–905 (2013)
8. Shao, J., Wang, X., Yang, Q., Plant, C., Böhm, C.: Synchronization-based scalable subspace clustering of high-dimensional data. Knowl. Inf. Syst. $52(1)$, 83–111 (2017)
9. Shao, J., Yang, Q., Dang, H.V., Schmidt, B., Kramer, S.: Scalable clustering by iterative partitioning and point attractor representation. ACM Trans. Knowl. Discov. Data $11(1)$, 5 (2016)
10. Wang, D., Gao, X., Wang, X.: Semi-supervised nonnegative matrix factorization via constraint propagation. IEEE Trans. Cybern. $46(1)$, 233–244 (2016)

Neural Review Rating Prediction with Hierarchical Attentions and Latent Factors

Xianchen Wang[1], Hongtao Liu[1(✉)], Peiyi Wang[1], Fangzhao Wu[2],
Hongyan Xu[1], Wenjun Wang[1], and Xing Xie[2]

[1] College of Intelligence and Computing, Tianjin University, Tianjin, China
{wangxc,htliu,wangpeiyi9979,hongyanxu,wjwang}@tju.edu.cn
[2] Microsoft Research Asia, Beijing, China
wufangzhao@gmail.com, xing.xie@microsoft.com

Abstract. Text reviews can provide rich useful semantic information for modeling users and items, which can benefit rating prediction in recommendation. Different words and reviews may have different informativeness for users or items. Besides, different users and items should be personalized. Most existing works regard all reviews equally or utilize a general attention mechanism. In this paper, we propose a hierarchical attention model fusing latent factor model for rating prediction with reviews, which can focus on important words and informative reviews. Specially, we use the factor vectors of Latent Factor Model to guide the attention network and combine the factor vectors with feature representation learned from reviews to predict the final ratings. Experiments on real-world datasets validate the effectiveness of our approach.

Keywords: Recommendation · Rating prediction · Attention

1 Introduction

Using text reviews to model user preferences and item features for rating prediction in recommendation has been an active research topic in recent years [1–3,6–8]. Kim et al. [4] adopt convolutional neural network to extract semantic features of reviews. Lu et al. [6] introduce attention mechanism to build recommender models. However, existing works ignore different words in a review and different reviews are differentially informative. Besides, they utilize a general attention for all items and users, which may be unreasonable since different users and items should be personalized.

Hence we develop a Hierarchical Attentions model which incorporates Latent Factor model (HALF) for rating prediction. We utilize hierarchical attention mechanism to focus important words and informative reviews. Specially, we use the factor vectors obtained in Latent Factor Model (LFM) [5] as query vectors to guide the review level attention mechanism. Moreover, we combine the feature representation learned from text reviews with the factor vectors in LFM to

© Springer Nature Switzerland AG 2019
G. Li et al. (Eds.): DASFAA 2019, LNCS 11448, pp. 363–367, 2019.
https://doi.org/10.1007/978-3-030-18590-9_46

compute the ratings. Our experimental results on real-world datasets indicate that HALF considerably outperforms previous methods.

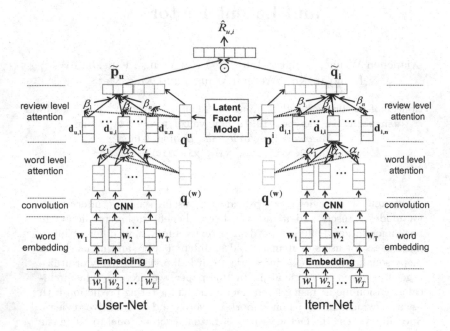

Fig. 1. Overview architecture of our model.

2   Methodology

In this section, we will present our model HALF including User-Net and Item-Net. Since the architectures of User-Net and Item-Net are similar, we will describe the User-Net in detail only. The overview of our model is shown in Fig. 1.

2.1   Review Encoder with Word Level Attention

First, we denote the user set as U, item set as I, the rating matrix as $\mathbf{R} \in \mathcal{R}^{|U| \times |I|}$ and the text review collection as $D \in \mathcal{R}^{|U| \times |I|}$ and each review is a word sequence. For a review $d_{u,i} = \{w_1, \cdots, w_T\}$, we first transform the word sequences into an embedding matrix $\mathbf{M}_{u,i}$ via word embeddings. Then we apply convolutional neural networks to extract feature matrix about the review: $\mathbf{C} \in \mathcal{R}^{K \times T}$, and $\mathbf{C_j} = \sigma(\mathbf{W_j} * \mathbf{M_{u,i}} + \mathbf{b_j}), 1 \leq j \leq K$ where $*$ is the convolution operator, $\mathbf{W_j}$ is the weight matrix of the j-th filter and K is the number of filters.

Hence, each column in $\mathbf{C}$ (denoted as $\mathbf{z_i} \in \mathcal{R}^K$) represent the semantic feature of the i-th word in a review. To highlight the important words, we employ the attention pooling mechanism in word level, denoted as:

$$g_i = \mathbf{q}^{(w)}\mathbf{A}\mathbf{z_i}, \tag{1}$$

$$\alpha_i = \frac{\exp(g_i)}{\sum_{j=1}^{T}\exp(g_j)}, \quad \alpha_i \in (0,1), \tag{2}$$

where $\mathbf{A}$ and $\mathbf{q}^{(w)}$ are the attention parameters. Finally, we obtain the representation of the i-th review of user u via aggregating feature vectors of all words: $\mathbf{d_{u,i}} = \sum_{j=1}^{T}\alpha_j\mathbf{z_j}$.

2.2 Review Level Attention Guided by Latent Factor Model

In this section, we employ an attention mechanism in review level based on Latent Factor Model (LFM) [5] to focus on more informative and personalized reviews. Latent Factor Model predicts the rating $R_{u,i}$ between user u and item i as follows: $R_{u,i} = \mathbf{q_u}^T\mathbf{p_i} + b_u + b_i + \mu$ where b_u, b_i and μ are the user bias, item bias and global rating bias. $\mathbf{q_u}$ and $\mathbf{p_i}$ are the factor vectors of user u and item i respectively.

Given the review set of a user $d_u = \{d_{u,1}, d_{u,2}, \cdots, d_{u,N}\}$, we compute the weight β_j about the j-th review of the user u as follows:

$$e_j = \mathbf{q_u}\mathbf{A_2}\mathbf{d_{u,j}}, \tag{3}$$

$$\beta_j = \frac{\exp(e_j)}{\sum_{k=1}^{N}\exp(e_k)}, \quad \beta_j \in (0,1), \tag{4}$$

where A_2 is the parameter matrix in attention; $\mathbf{q_u}$ is the factor vector specifically for the user u. The user feature vector is denoted as $\mathbf{m_u} = \sum_{i=1}^{N}\beta_j\mathbf{d_{u,i}}$ via aggregating all the review features. Similarly, we can obtain the item features denoted as $\mathbf{m_i}$.

2.3 Prediction Layer: Fusion of Attention and Latent Factor

We calculate the rating between a user u and an item i via fusing of attention model and Latent Factor Model as shown in Fig. 1. First, we combine the factor vectors in LFM and feature vectors learned from text reviews as the final representation of users and items: $\tilde{\mathbf{p_u}} = \mathbf{m_u} \oplus \mathbf{q_u}$ and $\tilde{\mathbf{q_i}} = \mathbf{m_i} \oplus \mathbf{p_i}$ where $\oplus$ is the concatenation operation. Afterwards, we compute the final prediction rating that user u would score item i in a form of LFM:

$$\hat{R}_{u,i} = \texttt{ReLU}(\mathbf{W}(\tilde{\mathbf{p_u}} \odot \tilde{\mathbf{q_i}}) + b_u + b_i + \mu) \tag{5}$$

where $\odot$ is the element-wise inner-product operation and $\mathbf{W}$ is the parameter matrix. In addition, we employ the mean squared error (MSE) as the loss function.

3 Experiments

Dataset. Following the previous work [6], we use five public real-world datasets for evaluation. Yelp 2013 and Yelp 2014 are selected from Yelp Dataset Challenge[1]. Electronics, Video Games and Gourmet Foods are selected from Amazon 5-core[2]. The details of datasets can be found in [6].

Results. Table 1 shows the results of our model and some recent state-of-art methods in terms of MSE. We can conclude that our model HALF can consistently outperform all the baseline methods which indicates the effectiveness of our method.

Table 1. Results of our HALF model and baseline methods in terms of MSE.

	Yelp 2013	Yelp 2014	Electronics	Video Games	Gourmet Foods
JMARS [2]	0.970	0.998	1.244	1.133	1.114
ConvMF+ [3]	0.917	0.954	1.241	1.092	1.084
NARRE [1]	0.879	0.913	1.215	1.112	0.986
TARMF [6]	**0.875**	0.909	1.147	1.043	1.019
HALF	**0.875**	**0.903**	**1.097**	**1.016**	**0.947**

4 Conclusion

In this paper, we propose a neural hierarchical personalized attention model HALF which integrates latent factor model into the attention mechanism for rating prediction in recommendation. Experimental results show that HALF significantly outperforms the state-of-the-art models.

Acknowledgments. This work was supported by the National Social Science Foundation Project (15BTQ056), the National Key R&D Program of China (2018YFC0809800, 2016QY15Z2502-02, 2018YFC0831000), the National Natural Science Foundation of China (91746205, 91746107), the Key R&D Program of Tianjin (18YFZCSF01370).

---

[1] https://www.yelp.com/dataset/challenge.
[2] http://jmcauley.ucsd.edu/data/amazon/.

References

1. Chen, C., Zhang, M., Liu, Y., Ma, S.: Neural attentional rating regression with review-level explanations. In: WWW, pp. 1583–1592 (2018)
2. Diao, Q., Qiu, M., Wu, C.Y., Smola, A.J., Jiang, J., Wang, C.: Jointly modeling aspects, ratings and sentiments for movie recommendation (JMARS). In: SIGKDD, pp. 193–202 (2014)
3. Kim, D., Park, C., Oh, J., Lee, S., Yu, H.: Convolutional matrix factorization for document context-aware recommendation. In: RecSys, pp. 233–240 (2016)
4. Kim, Y.: Convolutional neural networks for sentence classification. In: EMNLP, pp. 1746–1751 (2014)
5. Koren, Y., Bell, R., Volinsky, C.: Matrix factorization techniques for recommender systems. IEEE Comput. 42(8), 30–37 (2009)
6. Lu, Y., Dong, R., Smyth, B.: Coevolutionary recommendation model: mutual learning between ratings and reviews. In: WWW, pp. 773–782 (2018)
7. McAuley, J., Leskovec, J.: Hidden factors and hidden topics: understanding rating dimensions with review text. In: RecSys, pp. 165–172. ACM (2013)
8. Wang, C., Blei, D.M.: Collaborative topic modeling for recommending scientific articles. In: SIGKDD, pp. 448–456. ACM (2011)

MVS-match: An Efficient Subsequence Matching Approach Based on the Series Synopsis

Kefeng Feng, Jiaye Wu, Peng Wang[✉], Ningting Pan, and Wei Wang

School of Computer Science, Fudan University, Shanghai, China
{11110240010,wujy16,pengwang5,ntpan17,weiwang1}@fudan.edu.cn

Abstract. Subsequence matching is a fundamental task in mining time series data. The UCR Suite approach can deal with normalized subsequence matching problem (NSM), but it needs to scan full time series. In this paper, we propose to deal with the subsequence matching problem based on a simple series synopsis, the mean values of the disjoint windows. We propose a novel problem, named constrained normalized subsequence matching problem (cNSM), which adds some constraints to NSM problem. We propose a query processing approach, named MVS-match, to process the cNSM query efficiently. The experimental results verify the effectiveness and efficiency of our approach.

1 Introduction

The subsequence matching task is important for many time series mining problems. Specifically, given a long time series X, for any query series Q and a distance threshold ε, the subsequence matching problem finds all subsequences from X, whose distance with Q falls within the threshold ε.

FRM [1] is the pioneer work of subsequence matching. Many approaches have been proposed to improve the efficiency or to deal with various distance functions. However, *all* these approaches only consider the raw subsequence matching problem (RSM for short). In recent years, researchers realize the importance of the subsequence normalization [2]. UCR Suite [2] is the first work to solve the normalized subsequence matching problem (NSM for short). The RSM approaches build index to process queries. However, they cannot support the normalization. On the other hand, the state-of-the-art NSM approaches need to scan the full time series X, which is prohibitively expensive for long time series.

In this paper, we propose a new type of subsequence matching problem, called *constrained normalized subsequence matching problem* (cNSM for short). Two constraints, one for mean value and the other for standard deviation, are added

The work is supported by the Ministry of Science and Technology of China, National Key Research and Development Program (2016YFB1000700), NSFC (61672163, 61170006), and National Key Basic Research Program of China under No. 2015CB358800.

G. Li et al. (Eds.): DASFAA 2019, LNCS 11448, pp. 368–372, 2019.
https://doi.org/10.1007/978-3-030-18590-9_47

to the traditional NSM problem. With the constraint, the cNSM problem offers us the opportunity to utilize a simple synopsis to process the query. Moreover, it provides a knob to flexibly control the degree of offset shifting (represented by mean value) and amplitude scaling (represented by standard deviation). We propose an approach to utilize the common synopsis, mean values of disjoint windows, to support the cNSM query. Compared to the popular time series index, this synopsis is extremely easy to compute and maintaining it requires a very small space cost. We propose an efficient approach to process the query.

2   Preliminary Knowledge and Theoretical Foundation

A *time series* is a sequence of ordered values, denoted as $T = (t_1, t_2, \cdots, t_n)$, where $n = |T|$ is the length of T. A length-l *subsequence* of T is a shorter time series, denoted as $T_{i,l} = (t_i, t_{i+1}, \cdots, t_{i+l-1})$, where $1 \leq i \leq n - l + 1$. For any subsequence $S = (s_1, s_2, \cdots, s_m)$, μ^S and σ^S are the *mean value* and *standard deviation* of S respectively. Thus the *normalized series* of S, denoted as $\hat{S}$. In this paper, we utilize the most established distance measure, Euclidean distance.

The cNSM problem adds two constraints to the NSM problem. Thresholds α ($\alpha \geq 1$) and β ($\beta \geq 0$) are introduced to constrain the degree of amplitude scaling and offset shifting.

Constrained Normalized Subsequence Matching (cNSM): Given a long time series T, a query sequence Q, a distance threshold ε, and the constraint thresholds α and β, find all subsequences S of length $|Q|$ from T, which satisfy

$$ED(\hat{S}, \hat{Q}) \leq \varepsilon \,, \quad \frac{1}{\alpha} \leq \frac{\sigma^S}{\sigma^Q} \leq \alpha \,, \quad -\beta \leq \mu^S - \mu^Q \leq \beta$$

The larger α and β, the looser the constraint. In this case, we call that S and Q are *in* $(\varepsilon, \alpha, \beta)$-*match*.

Now, we first propose the theoretical foundation of our approach. Specifically, assume query Q and subsequence S are length-m. μ^S and μ^Q are the mean values of S and Q, σ^S and σ^Q are the standard deviations, $\hat{S}$ and $\hat{Q}$ are the normalized S and Q respectively.

Let $Q_{i,w}$ and $S_{i,w}$ be two length-w aligned windows of Q and S respectively. To simplify the notation, later we denote them as Q_i and S_i. Similarly, we denote the mean value and standard deviation of Q_i (or S_i) as μ_i^Q and σ_i^Q (or μ_i^S and σ_i^S). Given Q, we can obtain a range for μ_i^S, denoted as $R_i = [L_i, U_i]$. If S is in ε-match with Q, for any i ($1 \leq i \leq m - w + 1$), μ_i^S must fall within R_i. If any μ_i^S is outside the range, we can filter S safely. Now, we introduce the range for cNSM query.

Lemma 1. *If length-m sequence S and Q are in $(\varepsilon, \alpha, \beta)$-match under ED measure, that is, $ED(\hat{S}, \hat{Q}) \leq \varepsilon$, then μ_i^S ($1 \leq i \leq m - w + 1$) satisfies*

$$\mu_i^S \in \left[v_{\min} + \mu^Q - \beta, v_{\max} + \mu^Q + \beta \right] \tag{1}$$

where

$$v_{\min} = \min\left(\alpha \cdot (\mu_i^Q - \mu^Q - \tfrac{\varepsilon\sigma^Q}{\sqrt{w}}), \tfrac{1}{\alpha} \cdot (\mu_i^Q - \mu^Q - \tfrac{\varepsilon\sigma^Q}{\sqrt{w}})\right),$$
$$v_{\max} = \max\left(\alpha \cdot (\mu_i^Q - \mu^Q + \tfrac{\varepsilon\sigma^Q}{\sqrt{w}}), \tfrac{1}{\alpha} \cdot (\mu_i^Q - \mu^Q + \tfrac{\varepsilon\sigma^Q}{\sqrt{w}})\right).$$

Due to the space limitation, we omit the proof.

3   The Proposed Approach MVS-match

In this paper, we utilize the most common synopsis, mean value of the disjoint windows to process the query. Formally, we split series T into disjoint length-w windows, denoted as $(W_1, W_2, \cdots, W_{\lfloor \frac{n}{w} \rfloor})$, where $W_i = (t_{(i-1)*w+1}, \cdots, t_{i*w})$. For each window W_i ($1 \le i \le \lfloor \frac{n}{w} \rfloor$), we compute the mean value $\mu_{(i-1)*w+1}^T = \frac{\sum_{j=(i-1)*w+1}^{i*w} t_i}{w}$. Later, we denoted it as μ_i to simplify the notation. We maintain the sequence $(\mu_1, \mu_2, \cdots, \mu_{\lfloor \frac{n}{w} \rfloor})$, named MVS, to process the query.

The proposed approach consists of two phases. In the first phase, we visit MVS to filter unqualified subsequences and generate a set of candidates, denoted as CS. In the second phase, all subsequences in CS will be verified by fetching the raw subsequences, normalizing them and computing the distance from $\hat{Q}$.

Now we formally present our approach. Given length-m query Q, we read MVS from disk and scan MVS sequentially to check all possible subsequences of T. It is a two-layer loop process, in which the outer loop controls the subsequence to be compared, and the inner loop controls for which window the range is compared. First, we test $T_{1,m}$ by comparing $\mu_1 \in R_1$, $\mu_2 \in R_{w+1}$ and so on. If all windows fall within the corresponding range, we add $T_{1,m}$ into CS. Then we move to $T_{2,m}$ by checking $\mu_2 \in R_w$, $\mu_3 \in R_{2w-1}$ and so on. In general, for subsequence $T_{i,m}$, the windows surrounded by $T_{i,m}$ are $W_{\lceil \frac{i-1}{w} \rceil+1}, W_{\lceil \frac{i-1}{w} \rceil+2}, \cdots$. In the second phase, we fetch the raw time series of candidates in CS to verify them. All consecutive candidates are read together to reduce the I/O cost.

Considering the locality of the time series, it is often that the mean values of adjacent sliding windows are similar. In consequence, if μ_i falls outside certain range, say R_j, it is very likely that μ_i falls outside R_{j+1} either. Inspired from this observation, we extend the basic algorithm by merging adjacent ranges. The basic idea is that instead of comparing mean values with ranges of adjacent windows one by one, we merge the consecutive ranges of Q into a broader range to reduce the number of comparisons. In this approach, we do not compare subsequences to Q one by one, we compare w number of subsequences with Q in a batch fashion.

4   Experiments

We concatenate the time series in UCR archive [3] to obtain the time series. To generate a time series T, we execute the following steps repeatedly until T is fully generated: (i) randomly choose a dataset from UCR archive and a length l;

(ii) select a length-l subsequence from a time series randomly selected from this dataset. All experiments are conducted on a PC powered by Linux, with two Intel Xeon E5 1.8 GHz CPUs, 64 GB memory, 5 TB HDD storage. The code is implemented in C++. Similar to other subsequence matching approaches, we evaluate the performance under different selectivities, which is the ratio between the number of true results and that of all possible subsequences. For cNSM problem, we collect queries for each $\langle \alpha, \beta \rangle$ pair under particular selectivity by controlling ε. We generate query series by extracting subsequences of T starting from random offsets. To test the performance of processing queries with arbitrary lengths, we generate queries of length $128, 256, \cdots, 8192$. For each length, 100 different query series are generated. All experimental results are averaged over 100 runs. We compare with UCR Suite [2] and FAST [4], which is a recent improvement approach on UCR Suite. The experiment is conducted on length-10^9 real dataset. The results are shown in Table 1.

Table 1. Results of cNSM queries

Selectivity	MVS-match (s)				UCR	FAST
	$\alpha \backslash \beta'$	0.5	1.0	5.0	10.0	Avg. (s)	Avg. (s)
10^{-9}	1.1	4.18	4.34	4.60	4.93	59.84	86.05
	1.5	4.89	5.07	5.45	5.98		
	2.0	5.69	5.84	6.47	7.19		
10^{-8}	1.1	4.35	4.54	5.00	5.48	60.17	86.09
	1.5	5.18	5.37	6.62	7.77		
	2.0	6.07	6.39	8.30	10.63		
10^{-7}	1.1	5.00	5.32	6.91	8.76	65.25	87.79
	1.5	6.46	7.56	13.56	20.10		
	2.0	7.90	9.35	19.17	29.55		

It can be seen that when the selectivity increases, the runtime of MVS-match increases steadily. When the selectivity is fixed, the runtime increases as α and β increase. Because UCR Suite and FAST always scans the whole dataset, the runtime is more stable and dominated by I/O cost. The extra lower-bounds in FAST seem not efficient, due to its overhead of data preparation. In many cases, our approach can achieve the performance improvement of one order of magnitude compared to them.

5 Conclusion and Future Work

In this paper, we propose a novel constrained normalized subsequence matching problem (cNSM), which enables users to filter unqualified subsequences by only checking the raw subsequences. We propose a query processing approach, named MVS-match, to support the query processing based on the mean value synopsis. The experimental results verify the effectiveness and efficiency of our approach.

References

1. Faloutsos, C., Ranganathan, M., Manolopoulos, Y.: Fast subsequence matching in time-series databases. In: SIGMOD, pp. 419–429 (1994)
2. Rakthanmanon, T., Campana, B., et al.: Searching and mining trillions of time series subsequences under dynamic time warping. In: SIGKDD, pp. 262–270 (2012)
3. Chen, Y., et al.: The UCR time series classification archive. www.cs.ucr.edu/~eamonn/time_series_data/
4. Li, Y., Tang, B., Hou, U.L., Yiu, M.L., Gong, Z.: Fast subsequence search on time series data (poster paper). In: EDBT, pp. 514–517 (2017)

Spatial-Temporal Recommendation for On-demand Cinemas

Taofeng Xue[1,2], Beihong Jin[1,2(✉)], Beibei Li[1,2], Kunchi Liu[1,2], Qi Zhang[3], and Sihua Tian[3]

[1] State Key Laboratory of Computer Sciences, Institute of Software, Chinese Academy of Sciences, Beijing, China
Beihong@iscas.ac.cn
[2] University of Chinese Academy of Sciences, Beijing, China
[3] Beijing iQIYI Cinema Management Co., Ltd., Beijing, China

Abstract. The on-demand cinema is an emerging offline entertainment venue, guiding a new mode of watching movies in recent years. As a breakthrough in the development of the post-movie industry, the on-demand cinemas rely on private booths, high-quality hardware and rich movie resources to provide audiences with new and fresh watching experiences. The recommendation system for on-demand cinemas is to recommend to cinemas movies that may be of interest to their potential audiences, and provide an individualized recommendation service for preparing movie storage of on-demand cinemas to meet the audiences' preferences and instant watching needs. The characteristics implied in the audience behaviors of on-demand cinemas make the recommendation method for them different from those for online videos, items in offline stores or a group of users. In this paper, we describe the challenges and build a system for this application scenario, which fuses the historical on-demand records of cinemas, the POI (Point of Interest) information around cinemas and the content descriptions of movies, and explores the temporal dynamics and spatial influences rooted in audience behaviors. A WeChat applet customized for on-demand cinema staffs/hosts, as the client of our system, has been put into in practice.

Keywords: Recommendation system · On-demand cinema · Spatial-temporal effect

1 Introduction

In recent years, due to the impact of consumption upgrades and technological progress, mass culture and entertainment have gradually diversified. On-demand cinemas, which combine advantages of online video sites and traditional offline cinemas, emerge and continue to evolve. The on-demand cinemas are usually

This work was supported by the National Natural Science Foundation of China (No. 61472408) and the joint project with iQIYI (No. LUM18-200032).

newly-built entertainment venues or adapted venues of KTVs for this kind of extension business. The on-demand cinemas, in the form of private booths, are usually open to small groups of audiences (e.g., friends, couples or families, ranged in the number from 1 to 20). Compared with online video websites, on-demand cinemas, as a type of offline venues, can provide a full range of services; compared with traditional cinemas, on-demand cinemas own rich movie resources, flexible movie arrangement and more choices of watching time. The above advantages make the on-demand cinemas into an explosive development.

In general, what on-demand cinemas provide audiences is the movies with ultra-clear HD pictures and HiFi sounds, but existing networks are unable to download movies in real time for supporting instant watching. In order to ensure that audiences can instantly watch the satisfactory movies, the on-demand cinemas need to choose the movies that are of potential interest to audiences from mass online copyrighted videos in advance and then download and store them on local devices. Therefore, it is very necessary to provide an individualized recommendation service for movie storage of on-demand cinemas.

We note that the recommendation system for offline on-demand cinemas differs a lot from that for online videos. Firstly, the latter aims to recommend movies to individuals while the former to cinemas. Since the audiences of on-demand cinemas arrive in an anonymous and random manner and the consumption frequency of single audience is too low, it is not necessary and also impossible to achieve accurate personalized recommendations directly to audiences at present. Secondly, as for temporal distributions of watching behaviors, online users have more choices of watching time, while the accumulation of audience watching behaviors of on-demand cinemas shows the characteristics of temporal dynamics. Specically, the audiences of on-demand cinemas usually go to the movies along with others, generally for the purpose of dating and gathering, and prefer to choose special times, such as evenings or weekends. In addition, the cinemas show the characteristics of preference locality, i.e., the preference of audiences in different cinemas may vary a lot. When the location of a cinema falls within a typical functional region (e.g., a central business district or a residential area), then the cinemas also imply the characteristics of category locality.

Faced with the above challenges, we build a recommendation system for on-demand cinemas. The system fuses the historical on-demand records of cinemas, the POI (Point of Interest) information around cinemas and the content descriptions of movies, and explores the temporal dynamics and spatial influences rooted in audience behaviors.

To the best of our knowledge, the system built is the first recommendation system for on-demand cinemas. Although this is only a preliminary work, it provides a practical way to build recommendation systems for physical stores, which is convenient for stores to prepare goods or items in advance.

2   System Overview

From the perspective of the cinemas, the higher the on-demand number of a movie at a certain time interval is, the higher the corresponding rating on this movie at this time interval is. Thus, the recommendation system first transforms the on-demand numbers of different movies into ratings at different intervals, which are used to measure the importance of the movies to the cinemas at the time interval. Thus, we obtain the time-aware sparse rating matrix of on-demand cinemas, denoted as $\mathcal{R}$. In addition, we utilize the content descriptions of the movies, including the movie name, introduction, type, stars, etc., to construct the movie content matrix Y. We also collect the POI data around cinemas, mainly utilize the types of POI data, such as catering services, shopping services, residential buildings, etc., to construct the POI type matrix X.

Our goal is to explore rating matrix $\mathcal{R}$ transformed from on-demand records, POI matrix X, and movie content matrix Y to complete the time-aware spare rating matrix $\mathcal{R}$.

We propose a spatial-temporal recommendation method, which is a variant of matrix factorization [1,2] and build a recommendation system. Our recommendation system consists of four modules, which are shown in Fig. 1.

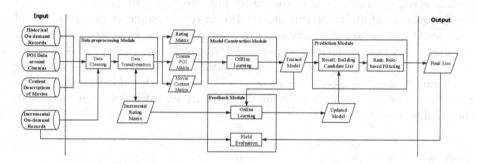

Fig. 1. System architecture

- Data preprocessing module: This module first preprocesses the historical on-demand records of cinemas, the content descriptions of movies, and the POI data around cinemas, where a binning function is employed to perform local smoothing of on-demand numbers at different time intervals to construct the time-aware rating matrix.
- Model construction module: This module implements the proposed recommendation method, conducts offline learning at the initial stage of recommendation system and saves the trained model.
- Prediction module: This module utilizes the trained model to predict the ratings for different movies and generate final recommendation results. Specifically, there are two stages, i.e., the recall stage and the ranking stage. At the recall stage, we employ the trained model to predict the ratings and generate candidate recommendation results. At the ranking stage, we employ the customized rules (e.g., directly add newly-released movies to results or classify results into different lists for different crowds based on the categories of

movies and priori preferences of different crowds) to further filter and sort the candidate results. Currently, we obtain the recommendation lists for couples, friends and families, respectively.

- Feedback module: This module periodically collects new on-demand records of cinemas, incrementally updates the old model and conducts field evaluation. Specifically, we first obtain the ratings of the new on-demand records from the preprocessing module, and then employ online learning to update old model, and finally use the latest model to generate new recommendation results.

3 System Deployment

Our recommendation system for on-demand cinemas provides services via a customized WeChat applet (see in Fig. 2) for staffs/hosts of on-demand cinemas.

Every day, our recommendation system pushes the latest recommendation list to the WeChat applet. The recommendation lists are individualized at the cinema level, i.e., the recommendation for each cinema is different.

The cinema staffs/hosts can browse the latest recommendation results and initiates the downloaded of recommended movies to the local storage device if lacking. While some movies are requested to download, the applet will automatically send a message with the remote address of the movie to the agent for download. When the agent receives it, it will initiate a download request to the remote address of the movie, automatically start downloading, and store it in the local storage device.

     (a) Homepage   (b) Recommendation list   (c) Details

Fig. 2. GUI of WeChat applet

References

1. Mnih, A., Salakhutdinov, R.R.: Probabilistic matrix factorization. In: Advances in Neural Information Processing Systems, pp. 1257–1264 (2008)
2. Koren, Y.: Collaborative filtering with temporal dynamics. In: Proceedings of the 15th ACM SIGKDD International Conference on Knowledge Discovery and Data Mining, 447–456. ACM (2009)

Finding the Key Influences on the House Price by Finite Mixture Model Based on the Real Estate Data in Changchun

Xin Xu[1], Zeyu Huang[1], Jingyi Wu[1], Yanjie Fu[2], Na Luo[1], Weitong Chen[3], Jianan Wang[1], and Minghao Yin[1(✉)]

[1] Northeast Normal University, Changchun, China
{xux894,huangzy149,wujy050,luon110,wangjn,ymh}@nenu.edu.cn
[2] Missouri University of Science and Technology, Rolla, USA
yanjiefoo@gmail.com
[3] University of Queensland, Brisbane, Australia
w.chen9@uq.edu.au

Abstract. Nowadays it's difficult for us to analyze the development law of real estate. What's more, predictable house price and understandable key influences can also build a healthier real estate market. Therefore, we propose a model which can predict the house price, while it can find key influences which are important influences on the house price. Our method is inspired by the finite mixture model (FMM) and information gain ratio (IGO). Specifically, we collect data that includes detail information about houses and communities from Anjuke Inc. which is an online platform for house sales. Then, according to the data, we find the scope of latent groups number by cluster methods to avoid blind searching the number of latent groups. Next, we use IGO to rank the features and weight them and we build a regression model based on the finite mixture model. Finally, the experimental results demonstrate our method performance on predicting house price, and we find key influences on house price.

Keywords: Real estate · Finite mixture model ·
Information gain ratio · Expectation Maximization Algorithm

1 Introduction

Artificial intelligence has achieved success in many domains, such as predicting the illness severity of patients in an ICU [2], recognizing the human intention based on Electroencephalography (EEG) signals [3], measuring liquidity in real estate markets [7] and so on. In the field of real estate, some researches are concerned on the house price using some effective data by training interpretable models [5] or use some features to predict the value of real estate [4]. The key influences of previous work such as violence [1], immigration [8] are not suitable for China, and the previous work has rarely mentioned about the differences in the importance of these influences. Not only do our research predict the house

G. Li et al. (Eds.): DASFAA 2019, LNCS 11448, pp. 378–382, 2019.
https://doi.org/10.1007/978-3-030-18590-9_49

price, but find the key influences of house price of Changchun. In this study, our dataset has more than 18000 real estate transaction records which include community information, house information, and house price.

2   Methodology

2.1   Framework Design

As Fig. 1 shown, we collect the features of the house and the community, then we propose a model based on FMM for alleviating the heterogeneity of Changchun house price. Finally, we rank the features according to IGO and find the key influences affecting house price in Changchun.

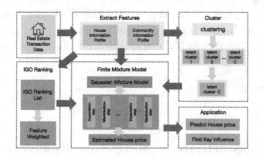

Fig. 1. An overview of framework.

2.2   Parameter Estimation

Objective Function. According to the statistical results of house price in Changchun City, we define the objective function as:

$$l(W) = \arg\max_{W} \sum_{i=1}^{N} \ln \sum_{z=1}^{C} \alpha_z \cdot \frac{\exp \frac{-\left(y_i - \omega_z^T \cdot x_i\right)^2}{2\sigma_z^2}}{\sqrt{2\pi}\sigma_z} \tag{1}$$

y_i is the real house price of the ith house, x_i is the feature vector of the ith house; $W = (\omega_1, \omega_2, ...\omega_C)$, ω_z is the learning matrix of sub-cluster; $\sigma = (\sigma_1, \sigma_2, ...\sigma_C)$, σ_z is common noise variance of zth component of this Finite Mixture Model; α_z is the mixing coefficient of sub-cluster [6].

2.3   Information Gain Ratio and Weighting Features

According to the definition of entropy, conditional entropy, and information gain, we get the definition of information gain ratio. Information gain ratio(IGO) is a ratio of information gain to the intrinsic information.

$$g_r(L, F) = \frac{g(L, F)}{H_F(L)} \tag{2}$$

Table 1. The comparison with baselines.

Method	MSE	MAE	RMSE	SDE
Linear regression	1.7013×10^{22}	1.4680×10^9	1.3043×10^{11}	1.3043×10^{11}
Lasso	4094.4920	35.9385	63.9882	52.9425
SupportVectorRegression	3473.8674	49.8484	58.9395	31.4485
KNeighbors regressor	1384.9328	20.1515	37.2147	31.2866
AdaBoost regressor	676.4514	19.1638	26.0000	**17.5676**
Our method	**619.3120**	**7.2361**	**24.8860**	23.8107

$F = (f_{1i}, f_{2i}, ..., f_{ni}), n = 1, 2, 3, ...,$ f_{ni} is the ith house's n^{th} feature. $L = (l_{1i}, l_{2i}, ..., l_{mi}), m = 1, 2, 3, ...,$ l_{mi} represents the ith house belong to m^{th} category in the label. The higher IGO value dedicates the more important influence in the model. According to the IGO ranking, we weight the features.

3   Experiments

3.1   Baseline

We compare the method in this paper with five baseline algorithms for estimating the house price. The baseline algorithms mentioned here are based on regression.

3.2   Evaluation Metrics

Prediction stability and comprehensiveness are used to evaluate the performance of the proposed model. Mean Square Error (MSE), Mean Absolute Error (MAE), Root Mean Squared Error (RMSE), Standard Deviation of Error (SDE) are the evaluation metrics used in the paper.

3.3   Comparison with Baseline Algorithms

After cluster experimenting, we found the evaluation metrics have smaller values when the latent groups are 2, and our model achieves the optimal effect. Table 1 shows that our approach consistently outperforms other baselines in terms of most evaluation metrics.

3.4   Features Weighted and Information Gain Ratio Ranking

Figure 2 shows the ranking of all 18 features through its IGO value. Therefore, in Changchun, the unit price, area, house type are the most three key influences of the house price. The result in Table 2 demonstrates that the performance of our model with weighted features is better. Therefore the IGO ranking of features is reasonable, and we find the key influences on house price in Changchun.

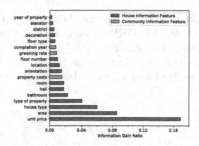

Fig. 2. The ranking of features.

Table 2. The comparison of weighted features and unweighted features.

Feature type	MSE	MAE	RMSE	SDE
Features unweighted	820.2493	7.7493	28.6400	27.5716
Features weighted	**619.3120**	**7.2361**	**24.8860**	**23.8107**

4 Conclusion

The method in this paper took advantages of clustering to avoid blind researching the number of latent groups. Then we weighted features by IGO ranking and estimated the parameters of our model by Expectation Maximization Algorithm. Finally, we used our model to predict house price and to find key influences.

References

1. Besley, T., Mueller, H.: Estimating the peace dividend: the impact of violence on house prices in Northern Ireland. Am. Econ. Rev. **102**(2), 810–33 (2012)
2. Chen, W., Wang, S., Long, G., Yao, L., Sheng, Q.Z., Li, X.: Dynamic illness severity prediction via multi-task RNNs for intensive care unit. In: 2018 IEEE International Conference on Data Mining (ICDM), pp. 917–922. IEEE (2018)
3. Chen, W., et al.: EEG-based motion intention recognition via multi-task RNNs. In: Proceedings of the 2018 SIAM International Conference on Data Mining, pp. 279–287. SIAM (2018)
4. Fu, Y., Xiong, H., Ge, Y., Yao, Z., Zheng, Y., Zhou, Z.-H.: Exploiting geographic dependencies for real estate appraisal: a mutual perspective of ranking and clustering. In: Proceedings of the 20th ACM SIGKDD International Conference on Knowledge Discovery and Data Mining, pp. 1047–1056. ACM (2014)
5. Jang, H., Ahn, K., Kim, D., Song, Y.: Detection and prediction of house price bubbles: evidence from a new city. In: Shi, Y., et al. (eds.) ICCS 2018. LNCS, vol. 10862, pp. 782–795. Springer, Cham (2018). https://doi.org/10.1007/978-3-319-93713-7_76
6. Xu, X., et al.: Dr. right!: embedding-based adaptively-weighted mixture multi-classification model for finding right doctors with healthcare experience data. In: 2018 IEEE International Conference on Data Mining (ICDM), pp. 647–656. IEEE (2018)

7. Zhu, H., et al.: Days on market: measuring liquidity in real estate markets. In: Proceedings of the 22nd ACM SIGKDD International Conference on Knowledge Discovery and Data Mining, pp. 393–402. ACM (2016)
8. Zhu, J., Brown, S., Pryce, G.B.: Immigration and house prices under various regional economic structures in england and wales. Urban Studies (Sage) (2018)

Semi-supervised Clustering with Deep Metric Learning

Xiaocui Li[1], Hongzhi Yin[2(✉)], Ke Zhou[1(✉)], Hongxu Chen[2], Shazia Sadiq[2], and Xiaofang Zhou[2]

[1] Wuhan National Laboratory for Optoelectronics,
Huazhong University of Science and Technology, Wuhan, China
{LXC,k.zhou}@hust.edu.cn
[2] School of Information Technology and Electrical Engineering,
The University of Queensland, Brisbane, QLD, Australia
{h.yin1,hongxu.chen}@uq.edu.au, {shazia,zxf}@itee.uq.edu.au

Abstract. Semi-supervised clustering has attracted lots of reserach interest due to its broad applications, and many methods have been presented. However there is still much space for improvement, (1) How to learn more discriminative feature representations to assist the traditional clustering methods; (2) How to make use of both the labeled and unlabelled data simultaneously and effectively during the process of clustering. To address these issues, we propose a novel semi-supervised clustering based on deep metric learning, namely SSCDML. By leveraging deep metric learning and semi-supervised learning effectively in a novel way, SSCDML dynamically update the unlabelled to labeled data through the limited labeled samples and obtain more meaningful data features, which make the classifier model more robust and the clustering results more accurate. Experimental results on Mnist, YaleB, and 20 Newsgroups databases demonstrate the high effectiveness of our proposed approach.

Keywords: Clustering · Semi-supervised learning ·
Deep metric learning

1 Introduction

Data mining has been a research hotspot for decades and many methods have been proposed [1–4]. However, how to extract useful features and learn an appropriate metric for high-dimensional data without any supervised information is a challenging task. Consequently, some supervised clustering algorithms were proposed to improve the clustering effect, and they indeed achieved limited achievements. These methods have great limitations in real practical applications, as it is almost impossible for all data to have labels, and tagging each data manually is a waste of human resources and time, and is unrealistic. In fact in most of real-world applications, we can only obtain quite limited labeled data. Based on the above problems, semi-supervised based clustering methods [7] emerged.

© Springer Nature Switzerland AG 2019
G. Li et al. (Eds.): DASFAA 2019, LNCS 11448, pp. 383–386, 2019.
https://doi.org/10.1007/978-3-030-18590-9_50

Although the existing semi-supervised clustering algorithms have achieved good results, there are still two important issues that will hinder the performance of clustering. (i) Most of these methods extract features or learn a distance metric through traditional SVM, neural networks or linear mapping, which limits its performance. (ii) They only use the labeled data to guide the process of the clustering, which can not be full used of the data especially unlabeled data. Motivated by the above analysis, we propose a semi-supervised clustering with deep metric learning (SSCDML), which can extract more discriminative features by using deep learning technique with nonlinear transformation, and simultaneously make full use of all data by combining semi-supervised learning.

2   Proposed Method

2.1   Semi-supervised Deep Metric Learning and Classification Network

We design a semi-supervised deep metric learning and classification network. The main training process of the network consists of the following three steps.

Step 1: First, extract discriminable features through CNNs, then use the features to train a classifier. We design a new loss function for semi-supervised deep metric learning and classification network as follows:

$$\min L = L_m + \lambda_1 L_c + \lambda_2 \|W\|_F^2, \tag{1}$$

where, λ_1 and λ_2 are a tunable positive parameter. $\|W\|_F^2$ is a regular term to prevent overfitting. L_m and L_c are metric learning loss and classification loss, respectively. They can be computed as follows:

$$L_m = \frac{1}{N} \sum_{i=1}^{N} (Y_i D(G(X_1), G(X_2)) + (1 - Y_i) max(\mu - D(G(X_1), G(X_2)), 0)), \tag{2}$$

where $D(,)$ is the Euclidean Distance function and $G(.)$ represent the outputs of the feature extracting network. $Y_i \in \{0, 1\}$ is the label of input pair samples.

$$L_c = - \sum_{G(X)} p(G(X)) \log q(G(X)), \tag{3}$$

where $p(.)$ is the expected outputs, and $q(.)$ is the actual outputs of the classification network.

Step 2: Encode the labeled and unlabeled data. Assume that $S_1 = \{(s_{1i}, l_{1i}) | i = 1, 2, \dots, N_1\}$ and $S_2 = \{(s_{2i} | i = 1, 2, \dots, N_2\}$ separately represent the init labeled data and unlabeled data, where N_1 is the number of labeled samples, N_2 is the number of unlabeled samples, and $l_{1i} \in \{1, 2, \dots, C\}$, where C is the number of classes. We use $S_1' = \{s_{1i}' | i = 1, 2, \dots, N_1\}$ and $S_2' = \{s_{2i}' | i = 1, 2, \dots, N_2\}$ represent the outputs of the S_1 and S_2 by CNNs.

Step 3: Tag the unlabeled data according to the classification network. There-fore, S_2 can be denoted as $S_2' = \{(s_{2i}', l_{2i}^1) | i = 1, 2, \ldots, N_2\}$, where l_{2i}^1 is the classification label of the s_{2i}'.

2.2 Semi-supervised Clustering Labeling Propagation Network

To acquire the strong label of the unlabeled data, we design a semi-supervised labeling propagation network. It includes two parts: semi-supervised clustering and labeling propagation.

In the process of the semi-supervised clustering, we applied the traditional k-means clustering algorithm to the data and mark the S_2' according to the clustering results, and record as $S_2' = \{(s_{2i}', l_{2i}^2) | i = 1, 2, \ldots, N_2\}$, where l_{2i}^2 is the clustering label of the s_{2i}'.

When both the classification label and clustering label of the unlabeled data S_1 are obtained, we can implement labeling propagation strategy.

3 Experiments

3.1 Datasets and Compared methods

We implement experiments on three publicly available datasets including: Mnist, YaleB [5] and 20 Newsgroups [6] and compare our approaches with some state-of-the-art related methods including: FSLSC [7], DFCM [8]. To evaluate the performance of our proposed methods and compared methods, we use clustering accuracy (AC):

$$AC = \frac{1}{N} \sum_{i=1}^{K} \max(C_i | L_i), \qquad (4)$$

Table 1. Clutering results of proposed methods and three semi-supervised clustring methods with different percentages of labeled data on Mnist, YaleB and 20 newsgroups datasets.

Datasets	Percentages	FSLSC	DFCM	SSCDML
Mnist	5%	69.0 ± 1.1	87.4 ± 1.6	$\mathbf{89.6 \pm 1.3}$
	10%	75.2 ± 0.9	90.4 ± 1.6	$\mathbf{92.3 \pm 1.3}$
	20%	85.6 ± 1	93.2 ± 0.8	$\mathbf{94.2 \pm 1.1}$
YaleB	5%	52.8 ± 1.3	68.8 ± 1.5	$\mathbf{75.7 \pm 1.5}$
	10%	58.7 ± 1.8	73.8 ± 2.3	$\mathbf{78.3 \pm 1.3}$
	20%	66.4 ± 1	77.9 ± 1.8	$\mathbf{81.2 \pm 0.9}$
20 Newsgroups	5%	$29.1 \pm 0.$	50.3 ± 2.2	$\mathbf{53.5 \pm 1.1}$
	10%	38.6 ± 1.6	52.7 ± 2.2	$\mathbf{57.1 \pm 1.4}$
	20%	46.2 ± 2.3	56.2 ± 1.5	$\mathbf{60.2 \pm 1.2}$

3.2 Results and Analysis

In this subsection, we conduct experiment to evaluate the clustering performance of our proposed SSCDML approach. To evaluate the clustering performance of our proposed approach, we increase the percentage of labeled data from 5% to 20%. Table 1 report the AC results of our proposed method and three semi-supervised clustering methods on three datasets. We can obviously see that our SSCDML approach performs better than compared semi-supervised clustering methods.

4   Conclusion

In this paper, we propose a novel semi-supervised clustering with deep metric learning approach named SSCDML. SSCDML comprises a semi-supervised deep metric learning network and a labeling propagation network. Semi-supervised deep metric learning network can extract more powerful features, and then learn a more discriminative metric. After that, labeling propagation network is used to label new data. Experimental results on Mnist, YaleB and 20 Newsgroups datasets have shown the high performance and effectiveness of our SSCDML approach.

Acknowledgement. This work was supported by ARC Discovery Early Career Researcher Award (DE160100308) and ARC Discovery Project (DP170103954; DP190101985).

References

1. Yin, H. Zou, L. et al.: Joint event-partner recommendation in event-based social networks. In: 34th International Conference on Data Engineering (2018)
2. Yin, H. Wang, Q. et al.: Social influence-based group representation learning for group recommendation. In: 35th ICDE (2019)
3. Chen, H. Yin, H. et al.: PME: projected metric embedding on heterogeneous networks for link prediction. In: The 2018 ACM SIGKDD(2018)
4. Xie, M. Yin, H. et al.: Learning graph-based POI embedding for location-based recommendation. In: The 25th ACM CIKM (2016)
5. Cui, G., Li, X., Dong, Y.: Subspace clustering guided convex nonnegative matrix factorization. Neurocomputing **292**, 38–48 (2018)
6. Chen, G.: Deep learning with nonparametric clustering. arXiv preprint arXiv:1501.03084 (2015)
7. Guan, R., Wang, X. et al.: A feature space learning model based on semi-supervised clustering. In: IEEE International Conference on CSE (2017)
8. Arshad, A., Riaz, S., et al.: Semi-supervised deep fuzzy c-mean clustering for software fault prediction. IEEE Access **6**, 25675–25685 (2018)

Spatial Bottleneck Minimum Task Assignment with Time-Delay

Long Li, Jingzhi Fang, Bowen Du, and Weifeng Lv[✉]

BDBC and SKLSDE Lab, Beihang University, Beijing, China
li_long@buaa.edu.cn, fangjz0707@gmail.com, {dubowen,lwf}@buaa.edu.cn

Abstract. Spatial crowdsourcing (SC) services become very popular. And one basic problem in SC is how to appropriately assign tasks to workers for better user experience. Most of existing researches focus on utilitarian optimization objectives for the benefit of the platform, such as maximizing the number of performed tasks, maximizing the total utility of the assignment, and minimizing the total cost to perform all tasks. However, users (*i.e.*, task-requesters and workers) usually only care about their own cost (*i.e.*, each user hopes his/her cost in the assignment to be small) instead of such those utilitarian optimization objectives. From the perspective of users, we propose an egalitarian version of online task assignment problem in SC, namely Minimizing Bottleneck with Time-Delay in Spatial Crowdsourcing (MBTD-SC). We further devise a heuristic algorithm to solve it. Finally, we validate the effectiveness of the proposed algorithm on both synthetic and real datasets.

Keywords: Spatial Crowdsourcing · Dynamic task assignment ·
Online bipartite matching

1 Introduction and Problem Definitions

With the rapid development of smartphones and mobile Internet, various applications based on spatial crowdsourcing (SC) such as online taxi-calling service (*e.g.*, Uber), food delivery service (*e.g.*, Ele.me) become very popular. In such services, requesters (*i.e.*, passengers or customers) usually release requests (*i.e.*, tasks), which are assigned to workers by the platform, and workers usually physically travel to the locations of tasks to perform them. One fundamental problem in SC is how to assign tasks to suitable workers especially in the online scenario, namely the *online task assignment (OTA)* problem in SC [6], where both tasks and workers dynamically appear on the platform.

There are various existing researches working on OTA, and most of them formulate OTA as a utilitarian optimization problem such as maximizing the number of performed tasks, maximizing the total utility of the assignment, and minimizing the total cost to perform all tasks [5,7,8,10]. These optimization objectives only consider the benefit of the platform. From the perspective of requesters or workers, these optimization objectives are not their concern and

© Springer Nature Switzerland AG 2019
G. Li et al. (Eds.): DASFAA 2019, LNCS 11448, pp. 387–391, 2019.
https://doi.org/10.1007/978-3-030-18590-9_51

their own user experience, namely the cost of their own task-worker matching pairs, is what matters. And every single requester or worker expects the cost of their corresponding matching pair to be small. Accordingly, we treat each task-worker matching pair equally for the benefit of requesters and workers by considering an egalitarian optimization objective, namely *minimizing bottleneck cost*. The egalitarian optimization objective means minimizing the maximum cost of all matching pairs, which seeks to ensure fairness because every awful task-worker matching pair may push the final bottleneck cost to a higher level.

In this paper, we propose a new and egalitarian version of OTA, called the Minimizing Bottleneck with Time-Delay in Spatial Crowdsourcing (MBTD-SC) problem. To the best of our knowledge, no existing work has studied this problem and we summarize 3 major challenges in MBTD-SC: 1. sensitivity of bottleneck cost: even a single task-worker matching pair with unusual large cost can result in unusual large bottleneck cost of the whole assignment; 2. time-delay setting: only reasonable time-delay of tasks and workers can result in ideal bottleneck cost; 3. fully online setting: both tasks and workers arrive at the platform online and the fully online bottleneck matching problem has not been studied before. And we formally define the MBTD-SC problem as follows.

Definition 1: Task/Worker. A spatial task/work $(t =< l_t, a_t >/w =< l_w, a_w >)$, is located at l_t/l_w in a 2D space and appears on the platform at time a_t/a_w.

Definition 2: Matching Triple. A matching triple $(m =< t_m, w_m, a_m >)$, indicates the platform assigns the task t_m to the worker w_m at time a_m.

Definition 3: Cost of A Matching Triple. The cost of a matching triple is defined as $cost(m) = d(l_{t_m}, l_{w_m}) + (a_m - a_t) + (a_m - a_w)$. $d(l_{t_m}, l_{w_m})$ is the travel delay for w_m physically traveling to t_m's location to perform t_m. $(a_m - a_t)$ is the time-delay of task t before getting matched. So is the $(a_m - a_w)$.

Definition 4: MBTD-SC Problem. Given a set of tasks T and a set of workers W, whose spatial information of tasks and workers is unknown before they appear, the MBTD-SC aims to find an assignment M to minimize the maximum cost of all matching triples *i.e.*, $cost(M) = \max_{m \in M} cost(m)$ and meet the following constraints. 1. invariable constraint: once a task is assigned to a worker, it cannot be revoked; 2. capacity constraint: a task can only be assigned to one worker and vice versa; 3. cardinality constraint: the output assignment should have the same cardinality as the offline optimal assignment.

2   Algorithms and Experimental Evaluation

Baseline: Candidate Constrained Greedy (CCG) Algorithm. The greedy strategy is an instinctive strategy to handle OTA, and tasks are greedily assigned to their nearest neighbor workers. To adapt the greedy strategy to MBTD-SC, we do some tony experiments and observe that the time when unmatched candidate neighbors are sufficient is the right time to match. So, we propose the Candidate Constrained Greedy (CCG) algorithm, and define a candidate constraint K and the count of unmatched tasks/worker must reach K befor matching. CCG is not illustrated due to the space limitation.

Algorithm 1. Candidate Constrained Local Optimiation Algorithm

---

1 (i) a task t arrives:
2 $T \leftarrow T \cup t$; //add t to the task set T.
3 (ii) a worker w arrives:
4 $W \leftarrow W \cup w$; //add w to the worker set W.
5 (iii) matching procedure:
6 **if** $|T| > c$ & $|W| > c$ **then**
7 $\quad$ calculate the optimal bottleneck assignment A of available T and W;
8 $\quad$ sort all triples of A in the descending order according to triples' time-delay;
9 $\quad$ //permit the first $|A| - c$ matching triples and make $|A| = c$.
10 $\quad$ **foreach** *matching triple* $m \in A$ **do**
11 $\quad\quad$ **if** m *is one of top* $|A| - c$ *matching triples in* A **then**
12 $\quad\quad\quad$ $M \leftarrow M \cup (t, w)$; //match task t and worker w of m.
13 $\quad\quad\quad$ delete t and w in T and W;

Candidate Constrained Local Optimization (CCLO) Algorithm. The information about tasks and workers is unknown before they appear due to the fully online setting. And if we hold tasks and workers for some time-delay before matching, the arrival time and locations of tasks and workers over the period of time is all the information we can gather and make use of. One of the best strategies to squeeze out all the value of these information is doing local optimal bottleneck assignment and utilizing the assignment to decide how to match current tasks and workers. Similarly with the candidate constraint idea as mentioned in CCG, we introduce the Candidate Constrained Local Optimization (CCLO) algorithm which is illustrated in Algorithm 1.

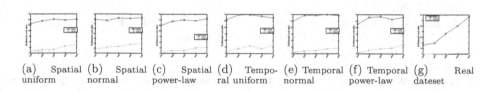

(a) Spatial uniform (b) Spatial normal (c) Spatial power-law (d) Temporal uniform (e) Temporal normal (f) Temporal power-law (g) Real dateset

Fig. 1. Results on different spatial, temporal distributions and real dataset.

Dataset Settings. We conduct experiments on both synthetic and real datasets. We first introduce the settings of these datasets briefly and the detailed settings is omitted due to the space limitation. Note that the Euclidean distance is used as the travel delay function $d(\cdot, \cdot)$. 1. **Synthetic Dataset Settings.** we use a 2000*2000 2D grid space as the working space. Inspired by [11], we consider 3 different spatial and temporal distributions, namely uniform, normal and power-law distributions. 2. **Real Dataset Settings**. We utilize the electric car ridesharing records during June 2017 in Beijing and we sample records as tasks/workers from all 46,261 trip records with corresponding pick-up time and GPS locations.

Metrics. We use the Bottleneck Cost Ratio (BCR, the ratio of our algorithm's bottleneck cost to the optimal bottleneck cost) as the performance measurement. **Summary of Results.** The bottleneck cost of CCLO is much lower than that of CCG, which shows the effectiveness of the local optimization strategy. Note that the horizontal ordinate in Fig. 1 is the number of matching triples which varies from 1000 to 5000.

3   Related Works and Conclusion

Related Works of Task Assignment in Spatial Crowdsourcing. Existing studies usually tend to obtain the assignment of tasks and workers according to different objectives, *e.g.*, maximizing the number of performed tasks [10], maximizing the total utility of the assignment [5,8], minimizing the total cost to perform all tasks [7], maximizing the total expected revenue [9], and minimizing the total latency [12]. Our paper aims to minimize the maximum cost (bottleneck cost) of the assignment, which is different from existing studies.
Related Works of Bottleneck Bipartite Matching. Offline bottleneck bipartite matching and its applications have been studied for decades [3,4]. Only fewer studies focus on one-side online bottleneck bipartite matching [1,2] where only tasks appear online and all workers are known in advance.
Conclusion. In this paper, we first formulate MBTD-SC. We then introduce a heuristic algorithm (CCLO). Finally, extensive experiments on both synthetic and real datasets shows CCLO can output reasonable and satisfying results.

References

1. Anthony, B.M., Chung, C.: Online bottleneck matching. J. Comb. Optim. **27**(1), 100–114 (2014)
2. Anthony, B.M., Chung, C.: Serve or skip: the power of rejection in online bottleneck matching. J. Comb. Optim. **32**(4), 1232–1253 (2016)
3. Gross, O.: The bottleneck assignment problem. Technical report, RAND CORP SANTA MONICA CALIF (1959)
4. Long, C., Wong, R.C.W., Yu, P.S., Jiang, M.: On optimal worst-case matching. In: SIGMOD, pp. 845–856 (2013)
5. Song, T., et al.: Trichromatic online matching in real-time spatial crowdsourcing. In: ICDE, pp. 1009–1020 (2017)
6. Tong, Y., Chen, L., Shahabi, C.: Spatial crowdsourcing: challenges, techniques, and applications. PVLDB **10**(12), 1988–1991 (2017)
7. Tong, Y., She, J., Ding, B., Chen, L., Wo, T., Xu, K.: Online minimum matching in real-time spatial data: experiments and analysis. PVLDB **9**(12), 1053–1064 (2016)
8. Tong, Y., She, J., Ding, B., Wang, L., Chen, L.: Online mobile micro-task allocation in spatial crowdsourcing. In: ICDE, pp. 49–60 (2016)
9. Tong, Y., Wang, L., Zhou, Z., Chen, L., Du, B., Ye, J.: Dynamic pricing in spatial crowdsourcing: a matching-based approach. In: SIGMOD, pp. 773–788 (2018)
10. Tong, Y., et al.: Flexible online task assignment in real-time spatial data. PVLDB **10**(11), 1334–1345 (2017)

11. Tong, Y., Zeng, Y., Zhou, Z., Chen, L., Ye, J., Xu, K.: A unified approach to route planning for shared mobility. PVLDB **11**(11), 1633–1646 (2018)
12. Zeng, Y., Tong, Y., Chen, L., Zhou, Z.: Latency-oriented task completion via spatial crowdsourcing. In: ICDE, pp. 317–328 (2018)

A Mimic Learning Method for Disease Risk Prediction with Incomplete Initial Data

Lin Yue, Haonan Zhao, Yiqin Yang, Dongyuan Tian,
Xiaowei Zhao$^{(\boxtimes)}$, and Minghao Yin

School of Information Science and Technology, Northeast Normal University,
Changchun, China
zhaoxw303@nenu.edu.cn

Abstract. Huge amounts of electronic health records (EHRs) accumulated in recent years have provided a rich foundation for disease risk prediction. However, the challenging problems of incompletion in raw data and interpretability of prediction model are not solved very well so far. In this study, we present a mimic learning approach for disease risk prediction with large ratio of missing values, called SR-DF, as one of the early attempts. Specifically, we adopt spectral regularization for incomplete medical data learning, on which the missingness among raw data can be more accurately measured and imputed. Moreover, by utilizing deep forest, we get an effective method that takes advantages of interpretable and reliable model for disease risk prediction, which requires far fewer parameters and is less sensitive to parameter settings. As we will report in the experiments, the proposed method outperforms the baselines and achieves relatively consistent and stable results.

Keywords: Spectral regularization · Deep forest · Mimic learning

1 Introduction

The benefits of EHRs analysis can be generalized into several aspects: EHRs analysis can help to predict future illness, length of stay, mortality and readmission with the view to improving clinical decisions and optimizing clinical pathways [4, 5].

Incompletion in initial data is one of the challenging problems frequently encountered in EHRs analysis. Main causes of large amounts of missing data include subjective reasons such as subjective error and historical limitation, and objective reasons such as loss to follow-up, experimental instrument failure, or financial burden on ordering all lab tests for each patient. Giving up those missing records will beget difference between the incomplete observational data and the complete observational data. Besides, although deep learning techniques got praise for outstanding performance on a series of EHRs analysis tasks. One of the main criticisms is that almost all

---

This research is supported by Fundamental Research Funds for the Central Universities (Grant No. 2412017QD028), China Postdoctoral Science Foundation (Grant No. 2017M621192), the Scientific and Technological Development Program of Jilin Province (Grant Nos. 20180520022JH and 20190302109GX).

G. Li et al. (Eds.): DASFAA 2019, LNCS 11448, pp. 392–396, 2019.
https://doi.org/10.1007/978-3-030-18590-9_52

deep models are hard to interpret. For this issue, Che [1–3] proposed the powerful knowledge-distillation approach, which used gradient boosting trees to learn interpretable models, at the same time, achieved strong prediction performance as deep learning models. Motivated by the problems and solutions, our proposal improves these shortcomings by introducing missing value imputation mechanism, and utilizing deep forest for disease risk prediction.

2 Proposed Method

The overview of the proposed scheme is shown in Fig. 1. Specifically, we adopt spectral regularization for matrix completion, on which the missingness among raw data can be more accurately measured and imputed. The method is capable of lessening the uncertainty in missing medical data, and easily handling the matrix of large dimension by exploiting the problem structure. It is particularly effective in the cases when the specific domain knowledge is lacking, or the initial data is missing at a high level. Moreover, by integrating deep forest, we get an effective learning scheme that takes advantages of interpretable and reliable model for disease risk prediction, which also makes the scheme requiring far fewer parameters and less sensitive to parameter settings.

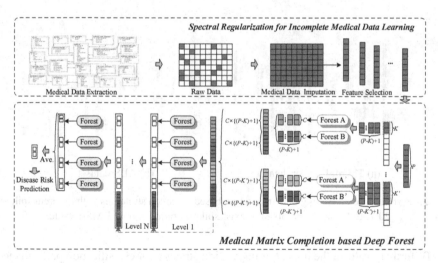

Fig. 1. The illustration of the work flow. To solve the problem of missing in medical data, spectral regularization algorithm (SR) [6] is used for incomplete data learning. Then, the imputed data will be reorganized and restructured into deep forest via feature selection. Deep forest (DF) [7] utilizes layer-by-layer and multi-grained structure to make prediction model interpretable.

3   Experiments

3.1   Benchmark Methods and Datasets

We evaluate the performance of seven missing data imputation strategies based on five different spike-in ratios. Moreover, we test the effectiveness of prediction based on imputed matrix by comparing different classification approaches of RF, KNN, SVM, DNN, AE, and GBTmimic [1–3].

There are two datasets involved in our experiments, i.e., amyotrophic lateral sclerosis (ALS) dataset (https://nctu.partners.org/ProACT/Document/DisplayLatest/9) and thyroid cancer dataset provided by the First Hospital of Jilin University.

3.2   Result Analysis

To evaluate the impact of modelling missingness and the effectiveness of data imputation strategy, we evaluated missing data imputation strategies on two datasets separately. The trends observed in Fig. 2 show the results of different imputation strategies in terms of RMSE on five spike-in ratios. On both thyroid cancer and ALS datasets, SR performs best among other methods, while SVD shares the most similar effect to SR. The deep model of AE outperforms Median and NMF imputation methods steadily. When the spike-in ratio increases to a high level, SR can still get relatively consistent performance without major fluctuations.

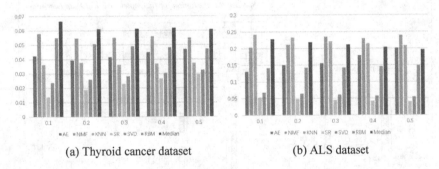

(a) Thyroid cancer dataset   (b) ALS dataset

Fig. 2. Performance of missing data imputation based on seven strategies with different spike-in ratios (10%, 20%, 30%, 40% and 50%). x-axis: spike-in ratio; y-axis: RMSE value.

To further evaluate the impact of imputation strategy for classification performance, we respectively tested the classification methods on raw data and imputed data. As illustrated in Table 1, the imputation strategy will improve the data quality, resulting in better performance for all classification methods, in terms of Accuracy.

On this basis, we further narrowed down the features of thyroid cancer data from 216 to 26, and ALS data from 59 to 44, respectively, in which the most important 13 features ranked by random forest regressor are involved. These risk factors have also been evaluated by physicians.

Table 1. Comparisons of classification performance on raw data and imputed data.

Thyroid cancer			ALS		
Method	Raw data	Imputed data	Method	Raw data	Imputed data
DF	0.8167	0.8667	DF	0.4360	0.5274
RF	0.7667	0.8167	RF	0.4499	0.4792
KNN	0.7667	0.7667	KNN	0.4185	0.4492
SVM	0.7833	0.8333	SVM	0.4813	0.5025
DNN	0.7167	0.8000	DNN	0.4769	0.4989
AE	0.7500	0.8167	AE	0.4704	0.4901
GBTmimic	0.8163	0.8667	GBTmimic	0.2743	0.3115

Table 2. Comparisons of prediction performance on thyroid cancer dataset.

Method	Spike-in ratio				
	10%	20%	30%	40%	50%
SR-DF	**0.9000**	**0.9333**	**0.9000**	**0.8833**	**0.9167**
SR-RF	0.8192	0.7833	0.8075	0.7950	0.8000
SR-KNN	0.7667	0.7667	0.7833	0.8333	0.8167
SR-SVM	0.8167	0.8333	0.8667	0.8500	0.8333
SR-DNN	0.8167	0.7500	0.8167	0.8333	0.7833
SR-AE	0.7667	0.8500	0.8167	0.7833	0.8333
SR-GBTmimic	0.8167	0.7833	0.9167	0.8500	0.7667

Table 3. Comparisons of prediction performance on ALS dataset.

Method	Spike-in ratio				
	10%	20%	30%	40%	50%
SR-DF	**0.5164**	**0.5274**	**0.5186**	**0.5135**	**0.5332**
SR-RF	0.4850	0.4660	0.4953	0.4748	0.4960
SR-KNN	0.4485	0.4390	0.4536	0.4646	0.4631
SR-SVM	0.5164	0.5201	0.5113	0.5128	0.5215
SR-DNN	0.5069	0.5150	0.5018	0.5113	0.4872
SR-AE	0.5157	0.5113	0.5033	0.5077	0.4799
SR-GBTmimic	0.3107	0.3115	0.3095	0.3105	0.3100

In general, DNNs often requires large amounts of training data, while DF performs well with only small training data size. To prove this property of the model, we performed experiments on different scales of data. From the observation in Table 2, the SR-DF runs well across all classification methods in terms of Accuracy among 10%–50% ratios of spike-in missing data. Moreover, when the spike-in ratio increases to a high level, i.e., from 10% to 50%, the proposed method can get relatively stable performance without large fluctuations. The similar trend can also be seen vividly in Table 3. However, the overall Accuracy rate is not as high as shown in thyroid cancer dataset. We further analyzed the reasons for this result, and believed that it is mainly due to the difference of the certainty of etiology between the two diseases. That is to say, the cause of ALS is still unknown, both genetic defects and some environmental factors may cause motor neuron damage. Future work should collect more indicators on the basis of existing ALS database such as genetic data to explore more relevant features.

4   Conclusion

In this paper, we present a mimic learning method to address the challenging problems, i.e., the incompleteness of medical data and interpretability of prediction model. With spectral regularization for incomplete data learning, the classification methods can get stable performance without large fluctuations, even when the spike-in ratio increases to a high level. As one of interpretable mimic learning methods, deep forest uses cascaded structure and multi-grained scanning to make the model interpretable, at the same time, achieves strong disease risk prediction performance as deep learning models.

References

1. Che, Z., Purushotham, S., Khemani, R., Liu, Y.: Distilling knowledge from deep networks with applications to healthcare domain. arXiv preprint arXiv:1512.03542 (2015)
2. Che, Z., Purushotham, S., Khemani, R., Liu, Y.: Interpretable deep models for ICU outcome prediction. In: AMIA Annual Symposium Proceedings 2016, p. 371. American Medical Informatics Association
3. Che, Z., Purushotham, S., Liu, Y.: Distilling knowledge from deep networks with applications to computational phenotyping. In: NSF Workshop on Data Science, Learning and Applications to Biomedical and Health Sciences (DSLA-BHS), pp. 1–6 (2016)
4. Chen, W., Wang, S., Long, G., Yao, L., Sheng, Q.Z., Li, X.: Dynamic illness severity prediction via multi-task RNNs for intensive care unit. In: 2018 IEEE International Conference on Data Mining (ICDM) 2018, pp. 917–922. IEEE
5. Chen, W., et al.: EEG-based motion intention recognition via multi-task RNNs. In: Proceedings of the 2018 SIAM International Conference on Data Mining 2018, pp. 279–287. SIAM (2018)
6. Mazumder, R., Hastie, T., Tibshirani, R.: Spectral regularization algorithms for learning large incomplete matrices. J. Mach. Learn. Res. **11**, 2287–2322 (2010)
7. Zhou, Z.-H., Feng, J.: Deep forest: towards an alternative to deep neural networks. In: Proceedings of the 26th International Joint Conference on Artificial Intelligence 2017, pp. 3553–3559. AAAI Press (2017)

Hospitalization Behavior Prediction Based on Attention and Time Adjustment Factors in Bidirectional LSTM

Lin Cheng[1], Yongjian Ren[1], Kun Zhang[1,2], Li Pan[1], and Yuliang Shi[1,2(✉)]

[1] School of Software, Shandong University, Jinan, China
chenglin123_sdu@163.com, ryjsdu@outlook.com, kunzhangcs@126.com,
{panli,shiyuliang}@sdu.edu.cn
[2] Dareway Software Co., Ltd., Jinan, China

Abstract. Predicting the future medical treatment behaviors of patients from historical health insurance data is an important research hotspot. The most important challenge of this issue is how to correctly model such temporal and high dimensional data to significantly improve the prediction performance. In this paper, we propose an Attention and Time adjustment factors based Bidirectional LSTM hospitalization behavior prediction model (ATB-LSTM). The model uses a hidden layer to preserve the impact state of medical visit sequences at different time on future prediction, and introduces the attention mechanism and the time adjustment factor to jointly determine the strength of the hidden state at different moments, which significantly improves the predictive performance of the model.

Keywords: Health insurance · Medical visit sequences ·
Attention mechanism · Time adjustment factor ·
Hospitalization behavior prediction

1 Introduction

In recent years, with the rapid expansion of the population of China, the number of people who have been hospitalized has increased. Due to the differences in levels of healthcare between hospitals, more patients are willing to go to the hospital with higher levels of healthcare, resulting in the waste of hospital resources and unreasonable distribution of health insurance funds. How to accurately predict the future medical treatment behavior of patients has become an important research issue in health insurance, and the prediction of the hospitalization behavior is particularly important.

Existing work mainly solves this problem by Recurrent Neural Networks (RNNs) [2,3]. However, the predictive power of RNNs drops significantly when the length of the patient visit sequences is large. In addition, the method ignores the influence of the time interval within visit sequences on the modeling.

© Springer Nature Switzerland AG 2019
G. Li et al. (Eds.): DASFAA 2019, LNCS 11448, pp. 397–401, 2019.
https://doi.org/10.1007/978-3-030-18590-9_53

In response to the above issues, we introduce an Attention and Time adjustment factors based Bidirectional LSTM hospitalization behavior prediction model (ATB-LSTM). The key contributions are as follows:

1. Considering the impact of time information within the visit sequences, we propose a hospitalization behavior prediction model in a bidirectional LSTM framework to predict which hospital patients will go to in the future.
2. On the basis of bidirectional LSTM framework, we introduce an attention mechanism and time adjustment factor to improve the prediction performance of the model.

2   ATB-LSTM Model

The goal of the proposed model is to predict the $(t+1)^{th}$ visit hospital. Figure 1 shows the overview of the proposed model. Given the visit information from time 1 to t, the ith visit x_i can be embedding into a vector v_i. The vector v_i is fed into the Bidirectional LSTM, which outputs a hidden state H_i. Along with the set of hidden state H_i, we can compute the degree of correlation a_i between the hidden state and medical treatment behavior at each moment by attention operation. Finally, from the visit sequence v_i and hidden state H_i at all time, we can get the final prediction through the softmax function.

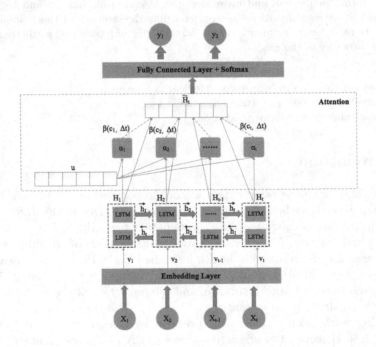

Fig. 1. The proposed ATB-LSTM Model

Implemented in the Attention mechanism as follows:

$$u_t = tanh(WH_t + b) \tag{1}$$

$$a_t = \frac{exp(u_t^T u)}{\sum_{t=1}^{T} exp(u_t^T u)} \tag{2}$$

$$\delta = \sum_t a_t H_t \tag{3}$$

Considering the effect of time interval on medical treatment behavior, we introduce a time adjustment factor Δt to construct a time adjustment function, which is obtained by Eq. (5).

$$\beta(c_n, \Delta t) = \frac{1}{1 + e^{-(\theta_{c_n} - u_{c_n} \Delta t)}} \tag{4}$$

$$\Delta t = \tau_i - \tau_j \tag{5}$$

where θ_{c_n} and u_{c_n} are learnable parameters. Δt is the time interval between visit τ_j and the visit we need to do prediction for, τ_i. Then, we use a sigmoid function to transform $\theta_{c_n} - u_{c_n} \Delta t$ into a probability between 0 and 1.

The output of attention $\widetilde{H}$ is:

$$\widetilde{H} = \sum_{n=1}^{N} \beta(c_n, \Delta t)\delta \tag{6}$$

3   Experiments

3.1   Experimental Baselines

We evaluate the predictive performance of our proposed model on real datasets. Among them, we use Hospital-A, Hospital-B and other forms to represent the name of the hospitals. The Baselines include: RNN, BLSTM, Dipole [4], Timeline [1].

Table 1. The accuracy of prediction methods

Method	Tumor dataset	Coronary heart disease dataset	Diabetes dataset	Pneumonia dataset
RNN	0.8127	0.8087	0.7998	0.8213
BLSTM	0.8429	0.8318	0.8279	0.8410
Dipole	0.8630	0.8475	0.8515	0.8670
Timeline	0.8693	0.8496	0.8542	0.8790
ATB-LSTM	**0.8864**	**0.8674**	**0.8618**	**0.8893**

3.2   Performance Comparison and the Effect of Time Interval

Table 1 shows the accuracy of the proposed model and methods on real datasets for the hospitalization behavior prediction task. We use the visit sequences of two patients as an example to verify the impact of the time interval on the prediction of the hospitalization behavior. Figure 2(a) and (b) show that we change time interval t and show how prediction probabilities change with t.

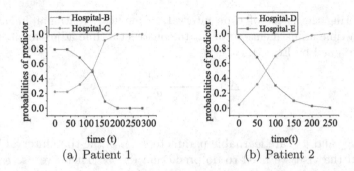

(a) Patient 1   (b) Patient 2

Fig. 2. We change time interval t and show how prediction probabilities change with t

4   Conclusion

How to correctly construct a prediction model based on the patients historical health insurance data has become an important research hotspot in health insurance. In this paper, we propose a novel model, named ATB-LSTM.

Acknowledgment. This work was supported by the National Key Research and Development Plan of China (No. 2018YFC0114709), the Natural Science Foundation of Shandong Province of China for Major Basic Research Projects (No. ZR2017ZB0419), the TaiShan Industrial Experts Program of Shandong Province of China (No. tscy2015 0305) and the Key Research & Development Program of Shandong Province of China (No. 2016ZDJS01A09).

References

1. Bai, T., Zhang, S., Egleston, B.L., Vucetic, S.: Interpretable representation learning for healthcare via capturing disease progression through time. In: Proceedings of the 24th ACM SIGKDD International Conference on Knowledge Discovery & Data Mining, KDD 2018, London, UK, 19–23 August 2018, pp. 43–51 (2018)
2. Che, C., Xiao, C., Liang, J., Jin, B., Zho, J., Wang, F.: An RNN architecture with dynamic temporal matching for personalized predictions of Parkinson's disease. In: Proceedings of the 2017 SIAM International Conference on Data Mining, Houston, Texas, USA, 27–29 April 2017, pp. 198–206 (2017)

3. Choi, E., Bahadori, M.T., Sun, J., Kulas, J., Schuetz, A., Stewart, W.: RETAIN: an interpretable predictive model for healthcare using reverse time attention mechanism. In: Lee, D.D., Sugiyama, M., Luxburg, U.V., Guyon, I., Garnett, R. (eds.) Advances in Neural Information Processing Systems, vol. 29, pp. 3504–3512. Curran Associates, Inc. (2016)
4. Ma, F., Chitta, R., Zhou, J., You, Q., Sun, T., Gao, J.: Dipole: diagnosis prediction in healthcare via attention-based bidirectional recurrent neural networks. In: Proceedings of the 23rd ACM SIGKDD International Conference on Knowledge Discovery and Data Mining, Halifax, NS, Canada, 13–17 August 2017, pp. 1903–1911 (2017)

Modeling Item Categories for Effective Recommendation

Bo Song[1], Yi Cao[1], Weike Pan[2], and Congfu Xu[1($\boxtimes$)]

[1] College of Computer Science and Technology, Zhejiang University,
Hangzhou, China
{bosong16,cao_yi,xucongfu}@zju.edu.cn
[2] College of Computer Science and Software Engineering, Shenzhen University,
Shenzhen, China
panweike@szu.edu.cn

Abstract. *One-class collaborative filtering* and item *cold-start* are two of the most important and challenging problems in recommender systems. In this paper, we focus on addressing these two issues by taking item category information into account. Item categories embody rich information about product attributes, which are available in most E-commerce websites. However, existing methods usually ignore such information or utilize them at a shallow level. For example, the category information is used to regularize model parameters or to extract hand-crafted features. As a response, we propose to model users' different preference spaces over different item domains. Specifically, we design a unified method called CatRec in order to model the complex interactions among a user, an item and the item's category information. Empirically, our method consistently outperforms the state-of-the-art methods on two real-world datasets.

Keywords: Top-N recommendation · Item category · Item cold start

1 Introduction

Modern E-commerce services usually adopt recommender systems to help retailers gain a better understanding of users' purchase behaviors. *One-class collaborative filtering (OCCF)* and *cold-start* are two critical issues for the task of recommendation. *OCCF* suffers from the ambiguity of implicit feedback (e.g., browsing history and purchase history). Many existing methods resort to auxiliary information (e.g., topics and labels) to mitigate the *cold-start* problem. Our work is in line with these works, that is, we exploit the item category information in this paper.

Specifically, we propose a new method called category-aware recommendation (CatRec). Our CatRec represents each user and item as points in a latent 'transition space'. Moreover, each item is associated with an additional 'translation vector' to capture the semantic relationship between users and the item.

© Springer Nature Switzerland AG 2019
G. Li et al. (Eds.): DASFAA 2019, LNCS 11448, pp. 402–405, 2019.
https://doi.org/10.1007/978-3-030-18590-9_54

After that, the item-specific translation vector is projected to the corresponding category-specific preference space. Finally, the interactions between a user and an item are captured by some personalized translation operations.

2   Related Work

Item taxonomy has been shown to be effective in improving the recommendation accuracy. For example, [5] proposed ReMF to learn hierarchical feature influence for recommendation by recursive regularization along the feature hierarchy. The key idea of ReMF is to automatically discover co-influence of hierarchical features by recursively weighting each feature influence in the feature tree. The learned feature influence is then used to regularize user or item latent vectors. Recently, [4] proposed HieVH to model item taxonomy in both vertical and horizontal dimensions. HieVH can model feature path in a vertical dimension and feature relationships in a horizontal dimension. However, it relies on the predefined features extracted from the data, which may hinder its application.

3   Our Solution: CatRec

We use $\mathcal{U}$ to denote a set of users with m = $|\mathcal{U}|$, and $\mathcal{I}$ to denote a set of items with n = $|\mathcal{I}|$. In collaborative filtering with implicit feedback, we have a user-item interaction matrix $\mathbf{R} \in \{0,1\}^{m \times n}$, where $R_{i,j}$ is the (i,j)-th entry of $\mathbf{R}$, indicating whether user i has interacted with item j or not. We also assume that an item taxonomy is available. In our CatRec, users and items are connected through some relations. And we link each item with an additional dense vector which is taken as the representation of the relation. Specifically, given a user-item pair (i,j), we first convert it to a triplet (i, r_j, j), where r_j is the relation. The raw relation is then mapped into a user's preference space with operation $\mathbf{M}_p$ as follows,

$$r_{pj} = \mathbf{M}_p r_j, \tag{1}$$

where $\mathbf{M}_p$ is a path-specific projection matrix constructed from the item category tree, r_j denotes the vector representation of r_j, and r_{pj} is the projected relation vector of r_j. At last, the projected relations are represented as translations in the embedding space which connect user embedding and item embedding. In particular, if (i,j) is observed, then the embedding of the item j plus some vector that relies on the relation r_j should be close to the embedding of user i,

$$\hat{R}_{i,j} = -d(u_i, v_j + r_{pj}), \tag{2}$$

where $d(\cdot)$ is some distance measurement. In the above equation, r_{pj} can be regarded as modeling item's category-specific attributes, and v_j captures item's general attributes. r_{pj} and v_j together represent item's full attributes. The compatibility between a user and an item is measured by the distance.

Hierarchical Preference Encoders. The key idea of hierarchical preference encoders is that each node in the category tree corresponds to a user preference space, which is modeled by some projection matrices. And all the projection matrices along a path of the category tree are blended to form a final projection matrix. We explore weighted addition composition for the formation of the final projection matrix. Weighted addition composition sums up projection matrices in different levels with different weights to represent the final projection matrix as follows,

$$\mathbf{M}_p = \sum_{i=1}^{L} \alpha_i \mathbf{M}_i = \alpha_1 \mathbf{M}_1 + \cdots + \alpha_L \mathbf{M}_L, \tag{3}$$

where L is the number of layers in the hierarchy, $\mathbf{M}_i$ is the projection matrix of i-th layer, and α_i is a hyperparameter.

Model Learning. Given a user i, the associated interacted items $\mathcal{I}_i^+$ and the corresponding categories C, our model learns vector embeddings of the users, items and categories. The goal is to rank an observed item $j \in \mathcal{I}_i^+$ higher than all non-observed items $(j' \in \mathcal{I}/ \mathcal{I}_i^+)$. The basic idea is to make $v_j + r_{pj} \approx u_i$ when (i,j) is observed (u_i should be a nearest neighbor of $v_j + r_{pj}$), while $v_j + r_{pj}$ should be far away from u_i otherwise. Therefore, it is a natural choice to optimize the pairwise ranking between j and j'. To this end, we optimize the following margin-based pairwise ranking objective function,

$$\min_{\Theta} \sum_{(i,j,j')\in E} \max(0, \gamma - \hat{R}_{i,j} + \hat{R}_{i,j'}) + \Omega(\Theta), \tag{4}$$

where γ is a margin separating j and j', and $\Omega(\Theta)$ is the ℓ_2 regularization term. In this paper, we use ℓ_2 distance. Thus, $\hat{R}_{i,j}$ is defined as follows,

$$\hat{R}_{i,j} = -\|r_{pj} + v_j - u_i\|_2^2. \tag{5}$$

4   Experiments

Datasets and Evaluation Metrics. We put the statistics of the two datasets in Table 1. Notice that there is a category tree in each dataset shown in the last column of Table 1. We use AUC (area under the curve) and the leave-one-out protocol used by [1] for performance evaluation.

Baselines. We compare against the following recommendation methods, including (i) BPR-MF [3]; (ii) ReMF [5]; (iii) HieVH [4]; and (iv) FM-BPR [2]. For a fair comparison, we modify the loss functions of ReMF and HieVH to the pairwise ranking loss, which are denoted as ReRank and HieRank, respectively.

Results. We report the results in Table 2. We can see that the taxonomy-aware methods, i.e., ReMF, HieVH and FM-BPR, achieve better performance than the non-taxonomy-aware methods in all cases. Among the baselines, FM-BPR is able to outperform all the other methods by a large margin. Nevertheless, CatRec achieves the best performance in all cases. In *All* setting, CatRec is able to outperform the strongest baseline FM-BPR by a relative improvement of 2.14% ∼ 4.93% on the two datasets. In *ColdItem* setting, CatRec achieves even larger improvements of 9.88% ∼ 10.03%.

Table 1. Statistics of the datasets used in the experiments.

Dataset	# Users	# Items	# Feedback	Category tree
Beauty	9,726	59,760	166,552	1: 6: 29
Clothing	3,571	18,054	62,918	1: 1: 33

Table 2. Recommendation performance of our CatRec and the baseline methods, where the best performance are marked in bold. We use *ColdItem* to denote the setting where only items associated with fewer than 5 users are taken into consideration in performance evaluation. Notice that '%Improv.' indicates the relative improvements that our CatRec achieves over the corresponding best baseline in each case.

Dataset	Setting	BPR-MF	ReRank	HieRank	FM-BPR	CatRec	%Improv.
Beauty	*ColdItem*	0.4576	0.4823	0.4901	0.5907	**0.6502**	10.03%
	All	0.6783	0.6850	0.7010	0.7547	**0.7919**	4.93%
Clothing	*ColdItem*	0.4656	0.4811	0.4860	0.6185	**0.6796**	9.88%
	All	0.5851	0.6815	0.6930	0.7245	**0.7400**	2.14%

5 Conclusions and Future Work

In this paper, we propose a novel category-aware method for collaborative filtering with implicit feedback. Experimental results on two real-world purchase datasets show that our method outperforms the state-of-the-art baselines. For future work, we would like to explore how to incorporate sequential signals into the category-aware recommendation.

Acknowledgement. This work is supported by the National Natural Science Foundation of China (NSFC) No. 61672449 and No. 61872249.

References

1. He, R., Lin, C., Wang, J., McAuley, J.: Sherlock: Sparse hierarchical embeddings for visually-aware one-class collaborative filtering. In: Proceedings of the 25th International Joint Conference on Artificial Intelligence, IJCAI 2016, pp. 3740–3746 (2016)
2. Rendle, S.: Factorization machines. In: Proceedings of the 10th IEEE International Conference on Data Mining, ICDM 2010, pp. 995–1000 (2010)
3. Rendle, S., Freudenthaler, C., Gantner, Z., Schmidt-Thieme, L.: BPR: Bayesian personalized ranking from implicit feedback. In: Proceedings of the 25th Conference on Uncertainty in Artificial Intelligence, UAI 2009, pp. 452–461 (2009)
4. Sun, Z., Yang, J., Zhang, J., Bozzon, A.: Exploiting both vertical and horizontal dimensions of feature hierarchy for effective recommendation. In: Proceedings of the 21st AAAI Conference on Artificial Intelligence, AAAI 2017, pp. 189–195 (2017)
5. Yang, J., Sun, Z., Bozzon, A., Zhang, J.: Learning hierarchical feature influence for recommendation by recursive regularization. In: Proceedings of the 10th ACM Conference on Recommender Systems, RecSys 2016, pp. 51–58 (2016)

Distributed Reachability Queries
on Massive Graphs

Tianming Zhang[1], Yunjun Gao[1(✉)], Congzheng Li[2], Congcong Ge[1], Wei Guo[1],
and Qiang Zhou[2]

[1] College of Computer Science and Technology, Zhejiang University,
Hangzhou, China
{tianmingzhang,gaoyj,gcc,weiguo}@zju.edu.cn
[2] College of Data Science and Software Engineering, Qingdao University,
Qingdao, China
{congzlee,zhouqiang}@qdu.edu.cn

Abstract. Reachability querying is one of the most fundamental graph
problems, and it has many applications in graph analytics and processing. Although a substantial number of algorithms are proposed, most
of them are centralized and cannot scale to massive graphs that are distributed across multiple data centers. In this paper, we study the problem
of distributed reachability queries, and present an efficient distributed
reachability index, called *Parallel Vertex Labeling* (*PVL*), together with
two lemmas to accelerate the query. Using real datasets, we demonstrate
that our solutions achieve better indexing performance and significantly
outperform the state-of-the-art distributed techniques.

Keywords: Graph · Reachability query · Distributed index ·
Algorithm

1 Introduction

Existing studies on reachability queries mostly aim at designing centralized algorithms. However, the real-world graphs grow rapidly, and are usually stored dis-
tributedly in multiple data sites. In view of this, we devise distributed algorithms
for reachability queries. To handle distributed reachability queries, a straightforward approach is to perform parallel breadth-first search (*BFS*) over graphs
directly. This method requires no index, but needs $O(|V| + |E|)$ time for every
query, where $|V|$ and $|E|$ are the number of vertices and edges, respectively. It is
unacceptable for large graphs. Another naïve approach is to collect all distributed
subgraphs into one machine and then call a centralized algorithm [1,5,6], which
is limited to the main memory of a single machine. Obviously, this solution has
scalability issues as graphs become increasingly large. Nowadays, we are aware
of two state-of-the-art distributed methods, namely, *disReach* [2] and *DSR* [3].
Both of them have to builds graph-based index structures, and then performs
DFS when a query is issued, which is inefficient.

© Springer Nature Switzerland AG 2019
G. Li et al. (Eds.): DASFAA 2019, LNCS 11448, pp. 406–410, 2019.
https://doi.org/10.1007/978-3-030-18590-9_55

We propose an efficient distributed reachability index, called *Parallel Vertex Labeling* (*PVL*), which is a labeling scheme. We use a parameter k to control *PVL* size, define a strict vertex level l and *minimal labels* to govern *PVL* so that *PVL* only maintains important vertices. To boost query efficiency, two lemmas are also presented. To sum up, the key contributions are summarized as follows:

- We propose an efficient indexes, i.e., *PVL*, for supporting *distributed reachability queries*.
- We present *PVL* based *reachability query* algorithm, and develop two lemmas to improve query performance.
- We conduct experiments to verify that *PVL* achieves better indexing performance, and significantly outperform the state-of-the-art methods [2,3].

The rest of the paper is organized as follows. Section 2 formalizes our studied problem. Section 3 describes *PVL* based method. Section 4 reports experimental results and our findings. Finally, Sect. 5 concludes the paper.

2   Problem Formulation

Let $G = (V, E)$ be a directed acyclic graph (DAG), in which V is a set of vertices, $E \subseteq V \times V$ is a set of directed edges, where $(u, v) \in E$ represents a directed edge from u to v. A directed path p from a vertex s to another vertex t is a consecutive of vertices $\langle s = v_0, v_1, \cdots, v_n = t \rangle$ such that $\forall i \in [0, n)$, $(v_i, v_{i+1}) \in E$.

Definition 1. Reachability Query. *Given a directed acyclic graph $G = (V, E)$ and a pair of vertices s and t, a* **reachability query,** *denoted as $RQ(s, t)$, determines whether source vertex s can reach terminal vertex t. If there exists a directed path p from s to t on G, then s can reach t. Otherwise, s cannot reach t.*

3   *PVL* Based Method

We present an efficient index structure *PVL*, it uses a parameter k to control its size, and picks up *minimal labels* (to be defined later) so that it maintains important vertices. In addition, *PVL* is governed by a strict vertex level l, which assigns a unique level $l(v)$ to every vertex v. In this paper, we simply define $l(v)$ as the order of v after all vertices are sorted in descending order of their degrees (i.e., $|N_i(v)| + |N_o(v)|$), where $N_i(v)$ and $N_o(v)$ are sets of in-neighbors and out-neighbors of v respectively, and $|N_i(v)|$ (resp. $|N_o(v)|$) is the cardinality of $N_i(v)$ (resp. $N_o(v)$). Based on $l(v)$, we define minimal in-labels, minimal out-labels of vertex v, and *PVL* index below.

Definition 2. *Given a DAG $G = (V, E)$, a vertex v, and a directed path p from another vertex u ($\neq v$) to v, v's* **minimal in-label** *l_{in} w.r.t. p is defined in the form of $(minV, level)$, where $minV$ is a vertex in V and $level = l(minV)$. It indicates that there is a vertex $minV$ passed by p (i.e., $minV \in p$) such that (i) $minV \neq v$ and (ii) $\nexists v' \in p$ having $l(v') < level$.*

The minimal out-label of a vertex v is defined similarly, it indicates the existence of a vertex ($\neq v$) with the smallest level value in a directed path from v.

Definition 3. (**PVL** *Index*). *Given a DAG $G = (V, E)$ and an integer k, a* **PVL index** *is a labeling scheme, where each vertex $v \in V$ is associated with two sets, $Lin(v)$ and $Lout(v)$. Minimal in-labels (resp. out-labels) la are recorded by $Lin(v)$ (resp. $Lout(v)$) if la.level is one of the top-k smallest levels among those of minimal in-labels (resp. out-labels) of v.*

To efficiently construct PVL index, we propose PVL construction algorithm. First, each vertex v sends a message $msg = (v, l(v))$ to v's out-neighbor w with $l(v) < l(w)$ or in-neighbor u with $l(v) < l(u)$. Then, the vertices v which receive messages msg update PVL index by Definition 3, and propagate msg to v's out-neighbors or in-neighbors. PVL index is computed until there is no message in transmit.

We now discuss how we use PVL to answer distributed reachability queries. To boost efficiency, we first present two lemmas.

Lemma 1. *Given a PVL index built on a DAG G and a reachability query $RQ(s, t)$, $RQ(s, t)$ returns $s \rightsquigarrow t$ if (i) $\exists$ label $ls \in Lout(s)$ having $ls.minV = t$, or (ii) $\exists$ label $lt \in Lin(t)$ having $lt.minV = s$, or (iii) $\exists$ label $ls \in Lout(s)$ and label $lt \in Lin(t)$ such that $ls.minV = lt.minV$.*

Lemma 2. *Given a PVL index built on a DAG G and a reachability query $RQ(s, t)$, $RQ(s, t)$ returns $s \not\rightsquigarrow t$ if one of the following two conditions holds. (i) Let l_t and l_s be the minimal level value of any label in $Lout(t)$ and $Lout(s)$ respectively, $l_s > l_t$ holds. (ii) Let l'_t and l'_s be the minimal level value of any label in $Lin(t)$ and $Lin(s)$ respectively, $l'_s < l'_t$ holds.*

Based on the aforementioned lemmas, we present a distributed algorithm for answering $RQ(s, t)$. The algorithm first utilizes Lemmas 1 and 2 to examine whether the query answer can be determined. If the above lemmas cannot tell, the algorithm processes the query by checking if any of the descendants of s can reach t.

Table 1. Statistics of the datasets used

Graph	$|V|$	$|E|$	$d(G)$
twitter	41,652,230	1,468,365,182	23
wikilink	12,150,976	378,142,420	10
liveJournal	4,847,571	68,475,391	20

4   Experimental Evaluation

We use 3 real graphs[1] to evaluate the efficiency of *PVL* based method. The statistics are summarized in Table 1, where $d(G)$ is the diameter of a graph G. We compare *PVL* with two state-of-the-art distributed methods (namely, *DSR* [3] and *disReach* [2]) and an optimized bi-directional breadth-first search (*Bi-BFS*) method. *disReach* was implemented on MapReduce, and other algorithms were implemented in Blogel [4] that is deployed with MPICH 3.2.1. All experiments are conducted on a 13-node Dell cluster. Each node has two processors with 12 cores, 128 GB RAM, and 3TB disk. Note that, for *PVL*, we set $k = 3$, and fix the number of workers to 15. For each method, the total query time of 100 queries is reported.

Table 2. Index construction time (in seconds)

Graph	*PVL*	*DSR*	*disReach*
twitter	**5.285**	1230.285	41
wikilink	**2.673**	15.155	32
liveJournal	**0.551**	39.154	32

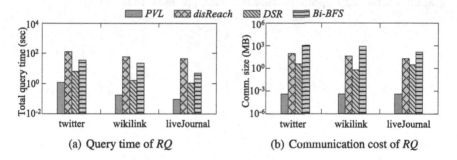

(a) Query time of *RQ*   (b) Communication cost of *RQ*

Fig. 1. Performance costs of *RQ*

Table 2 reports the index construction time. The bold values in the tables represent the best results. It is observed that, the index construction time of *PVL* is 102 times lower than that of *DSR*, and is 25 times lower than that of *disReach* in average. The total query time and communication cost of every method are plotted in Fig. 1. It is observed that, *PVL* significantly outperforms other methods. In particular, *PVL* is 8 times faster than *DSR*, is 73 times faster than *Bi-BFS*, and is two orders of magnitude faster than *disReach* in average.

---

[1] KONECT is available at konect.uni-koblenz.de/.

5   Conclusion

In this paper, we propose a distributed index PVL, and develop two lemmas for efficiently answering reachability queries. Experiments demonstrate that our solutions significantly outperform the state-of-the-art distributed techniques.

Acknowledgments. This work was supported in part by the National Key R&D Program of China under Grant No. 2018YFB1004003, the 973 Program of China under Grant No. 2015CB352502, the NSFC under Grant No. 61522208, the NSFC-Zhejiang Joint Fund under Grant No. U1609217, and the ZJU-Hikvision Joint Project.

References

1. Cheng, J., Huang, S., Wu, H., Fu, A.W.: TF-label: A topological-folding labeling scheme for reachability querying in a large graph. In: SIGMOD, pp. 193–204 (2013)
2. Fan, W., Wang, X., Wu, Y.: Performance guarantees for distributed reachability queries. PVLDB 5(11), 1304–1315 (2012)
3. Gurajada, S., Theobald, M.: Distributed set reachability. In: SIGMOD, pp. 1247–1261 (2016)
4. Yan, D., Cheng, J., Lu, Y., Ng, W.: Blogel: a block-centric framework for distributed computation on real-world graphs. PVLDB 7(14), 1981–1992 (2014)
5. Yildirim, H., Chaoji, V., Zaki, M.J.: GRAIL: a scalable index for reachability queries in very large graphs. VLDB J. 21(4), 509–534 (2012)
6. Zhu, A.D., Lin, W., Wang, S., Xiao, X.: Reachability queries on large dynamic graphs: a total order approach. In: SIGMOD, pp. 1323–1334 (2014)

Edge-Based Shortest Path Caching in Road Networks

Detian Zhang[1], An Liu[1(✉)], Gaoming Jin[1], and Qing Li[2]

[1] Institute of Artificial Intelligence, School of Computer Science and Technology,
Soochow University, Suzhou, China
detian.cs@gmail.com, anliu@suda.edu.cn, gamijin0@gmail.com
[2] Department of Computing, The Hong Kong Polytechnic University,
Kowloon, Hong Kong
csqli@comp.polyu.edu.hk

Abstract. In this paper, we propose an edge-based shortest path cache that can efficiently handle large-scale path queries without needing any road information. We achieve this by designing a totally new edge-based path cache structure, an efficient R-tree-based cache lookup algorithm, and a greedy-based cache construction algorithm. Experiments on a real road network and real POI datasets are conducted, and the results show the efficiency of our proposed caching techniques.

Keywords: Shortest path queries · Path cache · Large-scale

1 Introduction

The shortest path query in road networks becomes more and more popular and important in our daily lives. However, when there are a large number of path queries arrived concurrently or in a short while, a path server has to endure a high workload and then may lead to a long response time to users. To accelerate such large-scale path query processing, we propose an edge-based path cache in this paper, which stores the path results of frequent queries and reuse them to answer the coming queries directly. Different from most of existing path cache systems [1, 3], our proposed path cache system does not need any road information, which is much more practical in the real world.

2 System Architecture and Cache Structure

The system architecture of this paper consists of three entities, i.e., users, a path cache and a path server, as is shown in Fig. 1. Our proposed edge-based cache structure includes the following three indexes:

This work was supported in part by the National Natural Science Foundation of China under Project 61702227, 61572336 and 61632016, and Natural Science Research Project of Jiangsu Higher Education Institution under Project 18KJA520010.

G. Li et al. (Eds.): DASFAA 2019, LNCS 11448, pp. 411–414, 2019.
https://doi.org/10.1007/978-3-030-18590-9_56

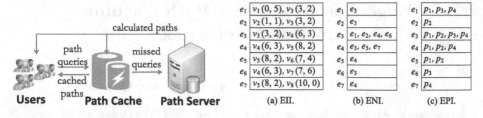

Users — path queries — cached paths — Path Cache — missed queries — Path Server

Fig. 1. System architecture.

(a) EII.

e_1	$v_1(0,5), v_3(3,2)$
e_2	$v_2(1,1), v_3(3,2)$
e_3	$v_3(3,2), v_4(6,3)$
e_4	$v_4(6,3), v_5(8,2)$
e_5	$v_5(8,2), v_6(7,4)$
e_6	$v_4(6,3), v_7(7,6)$
e_7	$v_5(8,2), v_8(10,0)$

(b) ENI.

e_1	e_3
e_2	e_3
e_3	e_1, e_2, e_4, e_6
e_4	e_3, e_5, e_7
e_5	e_4
e_6	e_3
e_7	e_4

(c) EPI.

e_1	p_1, p_3, p_4
e_2	p_2
e_3	p_1, p_2, p_3, p_4
e_4	p_1, p_2, p_4
e_5	p_1, p_2
e_6	p_3
e_7	p_4

Fig. 2. Edge-based cache structure.

p_1	$v_1(0,5,0), v_3(3,2,4.2), v_4(6,3,5), v_5(8,2,2.2), v_6(7,4,2.2)$
p_2	$v_2(1,1,0), v_3(3,2,2.2), v_4(6,3,5), v_5(8,2,2.2), v_6(7,4,2.2)$
p_3	$v_1(0,5,0), v_3(3,2,4.2), v_4(6,3,5), v_7(7,6,3.2)$
p_4	$v_1(0,5,0), v_3(3,2,4.2), v_4(6,3,5), v_5(8,2,2.2), v_8(10,0,2.8)$

(a) By nodes.

p_1	e_1, e_3, e_4, e_5
p_2	e_2, e_3, e_4, e_5
p_3	e_1, e_3, e_6
p_4	e_1, e_3, e_4, e_7

(b) By edges.

Fig. 3. Cached paths represented by nodes and edges.

Edge Information Index (EII). This index is used to store the essential information of each edge in the cache, i.e., the start and end locations of each edge. Figure 2a is an EII example of paths shown in Fig. 3(a). Figure 3(b) is an example of cached paths represented by edges after using EII.

Edge Neighbor Index (ENI). To save cache space, ENI is used to store the neighbors of edges, i.e., each edge has an adjacency list to store its directly linked edges in the cached paths, as shown in Fig. 2b.

Edge Path Index (EPI). This index is used to record the cached path list of edges, i.e., each edge has a path list to store the paths in the cache that contains the edge, as shown in Fig. 2c.

3 Cache Lookup

Edge Locating Based on R-tree. To quickly locate the edges, we employ R-tree to organize the straight-line edges in the cache. More specifically, during building EII for the cache, we can check whether an edge is straight or not based on the path information; then, for all straight-line edges in EII, we index them by using R-tree, as shown in Figs. 4 and 5.

After having built an R-tree for straight-line edges in EII, locating the edge that a point p lies on becomes easy and efficient, which has the following two main steps: *Step 1: find the candidate edges by R-tree.* Starting from the root, we continuously check each node u in the R-tree to see whether the minimum bounding rectangle (MBR) of u covers p (i.e., $p \in MBR(u)$) or not, and reports each edge e_i in u as a candidate edge if u is leaf node and $p \in MBR(e_i)$. *Step 2: validate the candidate edges based on slope.* After having found a candidate

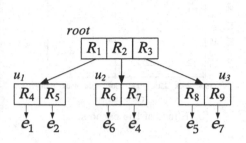

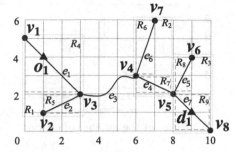

Fig. 4. Straight-line edges by R-tree.   **Fig. 5.** Example visualization.

edge e_i (say its start and end nodes are v_s and v_t, respectively), we propose an efficient method based on slope to validate whether the given point p actually lies on it or not. As e_i is a straight line and $p \in MBR(e_i)$, if the slope of the edge is the same with the slop of the line from v_s to p, based on the fundamental knowledge of geometry, we can have that p lies on e_i.

Cache Lookup Based on Edges. To process a path query $q_i = (o_i, d_i)$, the cache first tries to find the edges in EII that o_i and d_i lie on as stated above. If there is no straight-line edge in EII across o_i or d_i, the cache can not answer the query, i.e., a cache miss happens. Otherwise, i.e., there are straight-line edges e^{o_i} and e^{d_i} across o_i and d_i, respectively, the cache will perform a lookup operation by examining EPIs of e^{o_i} and e^{d_i} as follows: (1) If $EPI(e^{o_i}) \cap EPI(e^{d_i}) = \emptyset$, i.e., the two edges do not belong to the same path in the cache, that is to say there is no cached path across o_i and d_i at the same time, so a cache miss occurs; (2) If $EPI(e^{o_i}) \cap EPI(e^{d_i}) \neq \emptyset$, there is at least one path p_c ($p_c \in EPI(e^{o_i}) \cap EPI(e^{d_i})$) in the cache contains o_i and d_i at the same time, i.e., a cache hit happens. Next, the cache will construct the complete path $p(o_i, d_i)$ for the query. The cache starts from e^{o_i} and visits its neighbor edge e_i whenever the EPI of e_i contains p_c until reach e^{d_i}, to get all of the edges of path p_i; then, it replaces the edges with nodes in EII to retrieve the result path $p(o_i, d_i)$.

4   Cache Construction

Initially, the path cache is totally empty and does not have any knowledge about the underlying road networks. Therefore, we plan to employ historical paths to build a path cache as in [1,2]. Given an empty path cache $\mathcal{C}$, a set of historical path queries $\mathcal{Q}_h = \{q_1, q_2, ..., q_n\}$ and their corresponding result path set $\mathcal{P}_h = \{p_1, p_2, ..., p_n\}$, since the cache capacity is limited, i.e., $|\mathcal{C}| \leq \Psi$, our goal is to build an optimal path cache that it can answer the maximum number of historical queries in $\mathcal{Q}_h$, in expecting that the cache has a good performance in the future.

Based on [2], building such an optimal path cache on a given cache size is NP-complete, so we also utilize a greedy algorithm as [1,2] to populate a path cache. First, we calculate the sharing ability per edge (as the sharing ability per edge in [2]) with respect to $\mathcal{Q}_h$ for each path in $\mathcal{P}_h$. Next, the path p_s with the

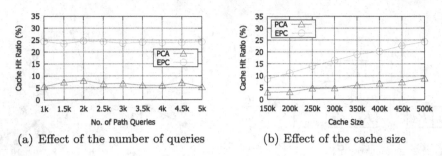

(a) Effect of the number of queries   (b) Effect of the cache size

Fig. 6. Experimental results.

largest sharing ability per edge is selected and placed into the path cache $\mathcal{C}$. Then, we remove q_s and p_s (including each q_i and p_i that $p_i \subset p_s$) from $\mathcal{Q}_h$ and $\mathcal{P}_h$, respectively, and continuously carry out the above two steps until $\mathcal{C}$ is full or $\mathcal{P}_h$ becomes empty.

5 Performance Evaluation

As far as we know, PCA [2] is the only shortest path caching algorithm that can answer path queries when the underlying road network topology is unknown, so we compare our proposed Edge-based Path Caching algorithm (denoted as EPC) with PCA. All of our experiments are implemented based on a real road network (i.e., California road network) with real POI datasets, which contain 21,693 edges, 21,048 nodes and 104,770 POIs. We evaluate the scalability, efficiency and applicability of EPC and PCA with respect to the number path queries and the cache size. The experimental results are shown in Fig. 6.

6 Conclusion

In this paper, we have proposed an effective edge-based path cache to accelerate large-scale path query processing. Unlike the most existing path caches that the underlying road network topology is still needed to process a path query if its querying origin or destination lies on edges, our proposed path cache can effectively handle path queries without needing any road information, which is much more practical in the real world, and also more compact and efficient than an existing path cache system.

References

1. Thomsen, J.R., Yiu, M.L., Jensen, C.S.: Effective caching of shortest paths for location-based services. In: SIGMOD (2012)
2. Zhang, D., Liu, Y., Liu, A., Mao, X., Li, Q.: Efficient path query processing through cloud-based mapping services. IEEE Access 5, 12963–12973 (2017)
3. Zhang, Y., Hsueh, Y.L., Lee, W.C., Jhang, Y.H.: Efficient cache-supported path planning on roads. IEEE TKDE 28(4), 951–964 (2016)

Extracting Definitions and Hypernyms
with a Two-Phase Framework

Yifang Sun[✉], Shifeng Liu, Yufei Wang, and Wei Wang

The University of New South Wales, Sydney, Australia
{yifangs,weiw}@cse.unsw.edu.au, {shifeng.liu,yufei.wang}@unsw.edu.au

Abstract. Extracting definition sentences and hypernyms is the key step in knowledge graph construction as well as many other NLP applications. In this paper, we propose a novel supervised two-phase machine learning framework to solve both tasks simultaneously. Firstly, a joint neural network is trained to predict both definition sentences and hypernyms. Then a refinement model is utilized to further improve the performance of hypernym extraction. Experiment result shows the effectiveness of our proposed framework on a well-known benchmark.

Keywords: Definition extraction · Hypernym extraction

1 Introduction

Both definition extraction and hypernym extraction are fundamental tasks in knowledge graph construction. For example, the first step of Wikipedia BiTaxonomy Project [4], which produced a taxonomized version of Wikipedia, is to extract definitions and hypernyms. They also play important roles in many other NLP tasks such as relation extraction and question answering.

To solve these problems, traditional lexico-syntactic pattern based methods focus on finding hypernym–hyponym pairs in one sentence and take the sentence as definitional [8]. The patterns are sequences of words such as *"is a"* or *"refers to"*, which are either manually crafted or semi-automatically generated. Pattern based methods suffer from both low precision and low recall. On one hand, the patterns are usually noisy, which hurts the precision of the methods. On the other hand, the coverage of the patterns is limited by the highly variable syntactic structures, which affects the recall.

Machine learning technique is another option, as definition extraction can be modeled as a binary classification problem, and hypernym extraction can be modeled as a sequence labeling classification problem. However, there are several drawbacks of the previous machine learning based methods. For example, the two tasks are separately solved as they are modeled as different classification problems. Thus the correlation between them is not well employed.

This research was partially funded by ARC DPs 170103710 and 180103411, and D2DCRC DC25002 and DC25003.

© Springer Nature Switzerland AG 2019
G. Li et al. (Eds.): DASFAA 2019, LNCS 11448, pp. 415–419, 2019.
https://doi.org/10.1007/978-3-030-18590-9_57

In this paper, we propose a novel machine learning framework to extract definitions and hypernyms *simultaneously*. Our framework contains two phases. In phase I, we employ a joint neural network model to predict (a) whether the sentence is definitional, and (b) the best k label sequences for the sentence. In phase II, we train a refinement model to improve the prediction quality of hypernyms in phase I. Unlike most existing machine learning methods, in our framework, the features are effective but easy to obtain.

We demonstrate the effectiveness of the proposed framework by experimenting it on a well-known benchmark of textual definition and hypernym extraction [9]. We show that our proposed framework substantially improves the performance for both tasks, leading to the new state-of-the-art.

2   Proposed Two-Phase Framework

2.1   Problem Definition

Given a sentence, our objectives are (1) classifying the sentence as definitional (labeled as `True`) or non-definitional (labeled as `False`), and (2) labeling each word in the sentence as at the beginning of (e.g., `B-HYP`), inside of (e.g., `I-HYP`), or outside (e.g., `O`) a hypernym. In this paper, we propose to solve the problem with a supervised learning framework. A set of sentences, their labels, and annotated labels for each word in the sentences are given as the training data.

2.2   Phase I: A Joint Neural Network Model

Figure 1 shows the architecture of the neural network in phase I. There are mainly four parts in the neural network.

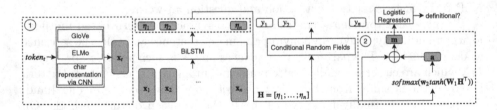

Fig. 1. The architecture of the neural network in phase I

As the first step in the neural network, we take the sequence of tokens from the sentence as input, and for each token, we generate its representation (e.g., the dashed box 1 in Fig. 1). The representation includes the character level representation which is generated by a CNN, the GloVe word embedding, and the ELMo word representation.

The representations for the tokens (e.g., $\mathbf{x}_i$) are then fed into the BiLSTM layer. We use LSTMs instead of Recurrent neural networks (RNNs) to overcome the gradient vanishing/exploding issue and capture long-distance dependencies. We use the bi-directional LSTM (BiLSTM) to access to both the left and right contexts for each token, which leads to a better performance. The output of the BiLSTM layer will be used for both of the following two parts.

As shown in dashed box 2 in Fig. 1, we utilize a self-attention mechanism [7] to encode a variable length sentence into a fixed size embedding (e.g., an embedding for the sentence). The embedding $\mathbf{m}$ is generated by a linear combination of the BiLSTM output vectors. $\mathbf{m}$ will be used as the input of a multilayer perceptron (MLP) to determine whether the sentence is definitional or not.

We use CRF [5] to predict the hypernyms as it is known to be one of the most effective solutions for sequence labeling tasks. The input of the CRF in our framework is the output vector sequence of the BiLSTM layer, and the output of the CRF is a label sequence with length n.

The loss function of the neural network (e.g., $\mathcal{L}_1$) is a combination of the loss for both tasks:

$$\mathcal{L}_1 = \mathcal{L}_{def} + \mathcal{L}_{hyp} = -\sum_i y \log p(y \mid \mathbf{m}) - \sum_i \log(p(\mathbf{y} \mid \eta)).$$

2.3 Phase II: Refinement Model

We observe that in phase I, the performance on labeling hypernyms is relatively low. However, the true hypernyms usually can be labeled correctly in at least one of the k (e.g., $k = 5$) label sequences with highest scores. This motivates us to use another classifier to refine the result of the best k label sequences of the CRF layer.

We use XGBoost [3] to do the multiclass classification, with softmax to calculate the probabilities. Given token x, the probability of all the possible labels (i.e., 0, B-HYP, and I-HYP) are computed using the softmax function and stored in the vector $\mathbf{s}$:

$$\mathbf{s} = softmax(\mathbf{W}^\top f(x)),$$

where $f(x)$ is the feature vector of token x. $\mathbf{W}$ is the weight matrix.

We use cross-entropy with regularization as the loss function:

$$\mathcal{L}_2 = -\sum_i y \log p(y \mid x) + \lambda \|\mathbf{W}\|_2^2.$$

In order to achieve better performance, in addition to best k labels, we also utilize the similarity to hyponym, POS tag, and dependency parsing information as features in phase II.

We present the detail about the proposed framework in the full version of this paper [10].

3   Experiments

We evaluate our model on a public benchmark of definition extraction and hypernym extraction [9]. The benchmark contains 4,619 sentences (1,908 of them are annotated as *definitional*), 1,908 hyponyms and 2,046 hypernyms.

We evaluate the performance of different models by comparing the predicted results on the test set using *Precision* (P), *Recall* (R), and F_1 score. We also report *Accuracy* (Acc) for definition extraction. Following the previous work [8], hypernyms are evaluated in substring level. All the results are averaged over 10-fold cross validation. The ratio of training samples, develop samples and test samples is 8:1:1. All the experiments are performed on Intel Xenon Xeon(R) CPU E5-2640 (v4) with 256 GB main memory and Nvidia 1080Ti GPU.

Table 1. Evaluation results

Method	Definition extraction				Hypernym extraction		
	P	R	F1	Acc	P	R	F1
WCL-3 [8]	**98.8**	60.7	75.2	83.5	78.6	60.7	68.6
Boella and Di Caro [2]	88.0	76.0	81.6	89.6	83.1	68.6	75.2
Li et al. [6]	90.4	92.0	91.2	–	–	–	–
Espinosa-Anke et al. [1]	–	–	–	–	**84.0**	76.1	79.9
Proposed framework	96.8	**96.5**	**96.6**	**97.3**	83.8	**83.4**	**83.5**

Table 1 concludes the results for both tasks. Our framework significantly outperforms the other methods by at least 5.4 in F_1 score for definition extraction task and at least 3.6 F_1 score for hypernym extraction task. For detailed analysis and more experiment results please refer to the full version of this paper [10].

4   Conclusion

In this paper, we propose a two-phase framework to tackle definition extraction and hypernym extraction tasks simultaneously. A joint neural network is used to predict for both tasks, with the performance further enhanced by a refinement model. The experiment shows the effectiveness of our framework.

References

1. Espinosa-Anke, L., Ronzano, F., Saggion, H.: Hypernym extraction: combining machine-learning and dependency grammar. In: Gelbukh, A. (ed.) CICLing 2015. LNCS, vol. 9041, pp. 372–383. Springer, Cham (2015). https://doi.org/10.1007/978-3-319-18111-0_28
2. Boella, G., Di Caro, L.: Extracting definitions and hypernym relations relying on syntactic dependencies and support vector machines. In: ACL, pp. 532–537 (2013)

3. Chen, T., Guestrin, C.: XGBoost: a scalable tree boosting system. In: KDD, pp. 785–794. ACM (2016)
4. Flati, T., Vannella, D., Pasini, T., Navigli, R.: Two is bigger (and better) than one: the Wikipedia bitaxonomy project. In: ACL, vol. 1, pp. 945–955. ACL (2014)
5. Lafferty, J.D., McCallum, A., Pereira, F.C.N.: Conditional random fields: probabilistic models for segmenting and labeling sequence data. In: ICML (2001)
6. Li, S.L., Xu, B., Chung, T.L.: Definition extraction with LSTM recurrent neural networks. In: Sun, M., Huang, X., Lin, H., Liu, Z., Liu, Y. (eds.) CCL/NLP-NABD -2016. LNCS (LNAI), vol. 10035, pp. 177–189. Springer, Cham (2016). https://doi.org/10.1007/978-3-319-47674-2_16
7. Lin, Z., Feng, M., dos Santos, C.N., Yu, M., Xiang, B., Zhou, B., Bengio, Y.: A structured self-attentive sentence embedding. CoRR abs/1703.03130 (2017)
8. Navigli, R., Velardi, P.: Learning word-class lattices for definition and hypernym extraction. In: ACL, pp. 1318–1327. ACL (2010)
9. Navigli, R., Velardi, P., Ruiz-Martínez, J.M.: An annotated dataset for extracting definitions and hypernyms from the web. In: LREC (2010)
10. Sun, Y., Liu, S., Wang, Y., Wang, W.: Extracting definitions and hypernyms with a two-phase framework. arXiv e-prints, January 2019

Tag Recommendation by Word-Level Tag Sequence Modeling

Xuewen Shi[1,2], Heyan Huang[1,2], Shuyang Zhao[1], Ping Jian[1,2(✉)],
and Yi-Kun Tang[1,2]

[1] School of Computer Science and Technology, Beijing Institute of Technology,
Beijing 100081, People's Republic of China
{xwshi,hhy63,zsyprich,pjian,tangyk}@bit.edu.cn
[2] Beijing Engineering Research Center of High Volume Language Information
Processing and Cloud Computing Applications, Beijing, China

Abstract. In this paper, we transform tag recommendation into a word-based text generation problem and introduce a sequence-to-sequence model. The model inherits the advantages of LSTM-based encoder for sequential modeling and attention-based decoder with local positional encodings for learning relations globally. Experimental results on *Zhihu* datasets illustrate the proposed model outperforms other state-of-the-art text classification based methods.

Keywords: Tag recommendation · Tag generation ·
Multi-label classification

1 Introduction

In recent years, online Q&A community and social network platform have become important modes for information transfer, such as Zhihu and Sina Weibo. These corpora contain a form of metadata tags marking its keywords or topics. These tags are useful in many real-world applications. Therefore, automatic tag recommendation has gained a lot of research interests recently and there have been proposed several neural network based tag recommendation approaches. Most of them cast this problem as multi-label classification [3–5], and they typically treat different tags as separate categories, while very few consider the rich relations between tags.

However, the labels tagged on a specific text are often related to each other. This association includes the word-level similarity and the hierarchical semantic relations. With the above novel insights, we propose a new word-based tag recommendation model, *i.e.*, re-formalizing this task as a word-based sequence-to-sequence text generation problem. Formally, given a sequence of input tokens $\mathbf{x} = \{\mathbf{x}_1, .., \mathbf{x}_N\}$, our method seeks to find a tag word sequence $\mathbf{y} = \{y_1, .., y_T\}$ with tag delimiters in it that maximizes the conditional probability of $\mathbf{y}$ given $\mathbf{x}$, *i.e.*, $\arg\max_{\mathbf{y}} p(\mathbf{y}|\mathbf{x})$. Our approach is achieved by a novel encoder-decoder [2] architecture named LSTM-Attention (L2A).

© Springer Nature Switzerland AG 2019
G. Li et al. (Eds.): DASFAA 2019, LNCS 11448, pp. 420–424, 2019.
https://doi.org/10.1007/978-3-030-18590-9_58

The proposed tag recommendation model has several interesting character-istics and advantages:

- It learns the latent rules of grouping words into tags. By the use of richer word-level information, it gains improved generability and applicability.
- An essential LSTM-Attention model is designed for modeling the semantic relations among tags globally and learning the sequential dependencies among the words from the same tags.
- A local positional encoding strategy is integrated into the attention-based decoder to address the complex relations in tag sequences.
- To better leverage the semantics and regularize the generation order of tags, two ordering strategies of tag sequence are explored by considering the top-down and bottom-up relations between tags.

We evaluate our model on *Zhihu* datasets. The experiments demonstrate that our model achieves better performances compared with state-of-the-art methods.

2   Method

Model Architecture. We re-formalize tag recommendation as word-level text generation using a novel sequence-to-sequence model named LSTM-Attention. Tag sequences involve rich semantic relations between tags and sequential depen-dencies between tag words. But there is no obvious sequential dependency between tags. Therefore, in our model, an LSTM encoder is designed to cap-ture the sequential dependencies between input words via its recurrent nature, and an attention-based decoder (Transformer [6]) is applied to learn semantic relations globally (instead of sequentially).

Local Positional Encoding. For tag sequence, there exist strong sequential relations among tag-words and rich semantic (instead of sequential dependen-cies) among tags. Thus the above purely sequential location encoding is limited, as it fails to reflect the independence of tags within tag sequence. To remedy this, we propose here a local positional encoding strategy. The difference with Transformer's positional encoding [6] is that the position index *pos* in Eq. (1) [6] is the inside relative position of a word in each tag instead of the global position in the whole sequence.

$$PE_{(pos,2i)} = sin(pos/10000^{2i/d_m}), \quad PE_{(pos,2i+1)} = cos(pos/10000^{2i+1/d_m}).$$

(1)

Tag Sequence Reordering. In tag recommendation, the orders of the tag sequences are usually given randomly. However, the order of the tag sequence could affect the tag generation result. It can be reasonably assumed that the higher the frequency of the labels is, the more general the meaning they represent. Based on this assumption, we propose to reorder the tag sequence

422   X. Shi et al.

in ascending order of frequency (*Order 1*) and in descending order of frequency (*Order 2*) respectively. **Order 1:** for label generation, given a tag in a specific topic, it is easier to predict following tags in more abstract meanings. So that for *Order 1*, the model is trained to firstly predict more specific tags and then generates high-frequency tags. The ambiguity of this process is relatively small compared to the opposite process. **Order 2:** for *Order 2*, we assume that the tags with lower frequency are derived from high-frequency tags. So the decoder is trained to generate high-frequency tags preferentially and then predict less frequent tags based on previous outputs. In this way, first the model determines the macro topic of the input, and then the previous output will gradually decrease the search scope of the decoder step by step.

N-Best Voting. Inspired by [8], we apply beam search into the decoder. We propose a voting-based label screening method called N-best voting. In this method, we count the total frequency of tags in the N-best list and select the tags with the frequency higher than a threshold.

Table 1. Experimental results on *Zhihu* dataset. *Random*, *Order 1* and *Order 2* means three different tag ordering strategies. See *Tag Sequence Reordering*. in Sect. 2 for more details.

Methods	Dev			Test		
	P ↑	R	F_1 ↑	P ↑	R	F_1 ↑
CNN [3]	32.9	46.8	38.6	27.6	40.2	32.7
LSTM [4]	35.1	50.0	41.3	31.4	45.8	37.3
Topical attention [5]	38.1	**54.3**	44.7	34.8	**47.2**	40.1
L2A-word (*Random*)	43.2	38.7	40.8	37.5	35.3	36.4
L2A-word w/ *Order 1*	**47.0**	42.9	**44.9**	**43.3**	38.7	**40.9**
L2A-word w/ *Order 2*	43.7	43.5	43.6	38.5	41.0	39.7

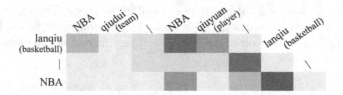

Fig. 1. Can the attention based decoder capture the sematic relations? The pre-generated tags and the new predicted tags lie on the x-axis and the y-axis, respectively. Darker colors represent higher attention weights. It shows that, when predicting "*lanqiu*" and "*NBA*", our model orients attention to those semantically related tags generated previously, which shows that the proposed model is able to capture the rich semantic relations between tags.

3 Experiment

Setups. We implement a word-based model (L2A-word). The model consists of a 2-layer encoder and a 4-layer decoder. The input and target vocabulary consist of 80,000 and 18,966 words respectively. The beam size and the N in N-best voting are set to 48. We compare the proposed model with three baselines: (i) *CNN* [3], (ii) *LSTM* [4], and (iii) *Topical Attention* [5]. And *Zhihu* datasets[1] are used for thoroughly accessing the performance of the proposed approach. The Chinese part for each corpus is segmented by the LTP [1]. All experimental results are evaluated on positional weighted precision (P), recall (R) and F_1-score (F_1).

Performance on Tag Recommendation on *Zhihu*. Experimental results are listed in Table 1. We can find that the proposed sequence-to-sequence based model with *Order 1* outperforms other traditional sentence classification methods. And both *Order 1* and *Order 2* outperform their counterparts with random order, proving the effectiveness of the proposed reordering strategies.

Can the Attention Based Decoder Capture the Sematic Relations? To give an in-depth analysis of the proposed attention-based decoder, we show a visual example of the attentions learned by the decoder in Fig. 1. As we can see from Fig. 1, the newly predicted tag-words, such as *"lanqiu"* and "NBA", are mostly influenced by the semantically related words in the prior predicted tags, such as "NBA", *"qiuyuan"* and *"lanqiu"*.

4 Conclusions

We propose a novel word-based tag recommendation method, which tackles tag recommendation as word-based tag-sequence generation. Our approach is achieved by a carefully designed LSTM-Attention model, which is able to effectively capture the rich semantic relations and sequential dependencies within tag sequences. We extend the Transformer based decoder with a local positional encoding strategy. In addition, two tag ordering methods are proposed for better leveraging the semantic relations. Experimental results show our model outperforms other state-of-the-art multi-class classifier based tag recommendation models. In the future, we would like to introduce Pointer Networks [7] to handle the challenges of generating unseen tags.

Acknowledgments. This work was supported by the National Key Research and Development Program of China (Grant No. 2017YFB1002103) and the National Natural Science Foundation of China (No. 61732005).

---

[1] http://tcci.ccf.org.cn/conference/2018/taskdata.php.

References

1. Che, W., Li, Z., Liu, T.: LTP: A Chinese language technology platform. In: 23rd International Conference on Computational Linguistics, COLING 2010, pp. 13–16 (2010)
2. Cho, K., et al.: Learning phrase representations using RNN encoder-decoder for statistical machine translation. In: Proceedings of the 2014 Conference on Empirical Methods in Natural Language Processing, pp. 1724–1734 (2014)
3. Gong, Y., Zhang, Q.: Hashtag recommendation using attention-based convolutional neural network. In: Proceedings of the Twenty-Fifth International Joint Conference on Artificial Intelligence, IJCAI 2016, pp. 2782–2788 (2016)
4. Li, J., Xu, H., He, X., Deng, J., Sun, X.: Tweet modeling with LSTM recurrent neural networks for hashtag recommendation. In: International Joint Conference on Neural Networks, IJCNN 2016, pp. 1570–1577 (2016)
5. Li, Y., Liu, T., Jiang, J., Zhang, L.: Hashtag recommendation with topical attention-based LSTM. In: Proceedings of COLING 2016, the 26th International Conference on Computational Linguistics: Technical Papers, pp. 3019–3029 (2016)
6. Vaswani, A., et al.: Attention is all you need. In: Advances in Neural Information Processing Systems, pp. 5998–6008 (2017)
7. Vinyals, O., Fortunato, M., Jaitly, N.: Pointer networks. In: Proceedings of the 28th International Conference on Neural Information Processing Systems, pp. 2692–2700 (2015)
8. Wang, J., Yang, Y., Mao, J., Huang, Z., Huang, C., Xu, W.: CNN-RNN: a unified framework for multi-label image classification. In: IEEE Conference on Computer Vision and Pattern Recognition, CVPR 2016, pp. 2285–2294 (2016)

A New Statistics Collecting Method with Adaptive Strategy

Jin-Tao Gao$^{(\boxtimes)}$, Wen-Jie Liu, Zhan-Huai Li, Hong-Tao Du, and Ou-Ya Pei

Northwestern Polytechnical University, Xi'an 710072, China
{gaojintao,peiouya2013}@mail.nwpu.edu.cn,
{liuwenjie,lizhh,duhongtao}@nwpu.edu.cn

Abstract. Collecting statistics is a time- and resource-consuming operation in distributed database systems. It is even more challenging to efficiently collect statistics without affecting system performance, meanwhile keeping correctness in a distributed environment. Traditional strategies usually consider one dimension during collecting statistics, which is lack of generalization. In this paper, we propose a new statistics collecting method with adaptive strategy (APCS), which well leverages collecting efficiency, correctness of statistics and effect to system performance. APCS picks appropriate time to trigger collecting action and filter unnecessary tasks, meanwhile reasonably allocates collecting tasks to appropriate executing locations with right executing model.

Keywords: Statistics collecting · Distributed database ·
Adaptive strategy · Query optimization

1 Introduction

Statistics play an important role at query optimization in database systems. Incorrect statistics will seriously affect the quality of query optimization [1]. Fortunately, there are tons of strategies [1,2,6,7] to keep this right. Under distributed environment, besides considering correctness of statistics, we should not ignore the importance of collecting efficiency and the negative effect to system performance caused by collecting statistics. Some systems use partial collecting global merging strategy to collect statistics [5], which can make use of rich parallel computing resources to improve collect efficiency, but facing the losses of correctness caused by merging operation and adopting immediately executing model without considering system state maybe bringing negative effect to system performance.

Collecting statistics is one special operation in database systems, reflected in its high resources consuming, approximate character and supporting off-line

Supported by Key Research and Development Program of China (2018YFB1003403), National Natural Science Foundation of China (61732014,61672432,61672434) and Natural Science Basic Research Plan in Shaanxi Province of China (No. 2017JM6104).

G. Li et al. (Eds.): DASFAA 2019, LNCS 11448, pp. 425–429, 2019.
https://doi.org/10.1007/978-3-030-18590-9_59

model, which determines that we can flexibly control collecting actions according to system state to improve collecting efficiency meanwhile mitigate negative effect to system performance. Current strategies [3,4,9,10] usually consider one dimension in collecting statistics ignoring multiple interaction factors.

APCS employs three adaptive levels to well leverage collecting efficiency, correctness of statistics and effect to system performance, including adaptive triggering used to more reasonably trigger collecting action by synthetically considering rich information, adaptive scheduling being responsible for allocating collecting tasks to appropriate places with right collecting model, and adaptive executing which can adaptively decide whether to execute collecting tasks according to system state. Relying on the protections provided by these three adaptive levels, our strategy can reasonably handle various situations during collecting statistics.

2   Collecting Statistics with APCS

We split the procedure of collecting statistics into three phases, which are **WHEN** to collect, **HOW** to collect and **WHERE** to store. Compared to general statistics collecting strategies, APCS can adaptively collect statistics by incorporating multi-level adaptive components, including adaptive triggering (1st level), adaptive scheduling (2nd level), and adaptive executing (3rd level), which well leverage system performance, correctness of statistics and collecting efficiency. The procedure of collecting statistics based on APCS is showed at Fig. 1.

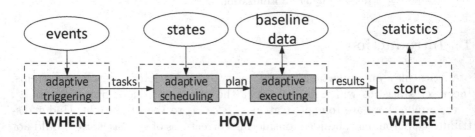

Fig. 1. Procedure of collecting statistics based on APCS.

2.1   Adaptive Triggering

Unlike traditional strategies, adaptive triggering component can synthetically consider rich information to wisely judge triggering time by three thresholds (WT, HT, ST) defined as follows:

$$WT = b + \theta * X, HT = \sigma.\sigma > 0, ST \sum_{i=1}^{M}(\kappa_i.F_{mem} \geq AVG(\{\kappa.F_{mem}\})) \quad (1)$$

WT is the delta data threshold, which is similar to threshold used at PostgreSQL, where b represents base thresh (default is 50), X is number of tuples, and θ is the scale factor (default is 0.1). Frequency of querying one table can represent the importance of this table. We use query heat threshold (HT) to filter tasks with low importance, and set σ as $\dfrac{\sum_{j=1}^{N} h(t_j)}{N}$, which is average of query heat of current tables $(h(t_j))$. For fully using system resources, we introduce state threshold (ST), which represents the number of node (κ) whose state is better than average level. M is the number of total nodes, and F_{mem} is the quantity of free memory.

2.2 Adaptive Scheduling

Facing different collecting tasks and various system states, directly executing tasks may break the balance among collecting efficiency, correctness of statistics and system performance. To resolve this problem, adaptive scheduling component first assigns appropriate executing model for every task, then reasonably allocates tasks to appropriate nodes with assigned executing model. We summary three kinds of executing models from current literatures, including partial executing model (PCM) [5], partial executing with merging model (PeCM) [4] and global executing model (GCM) [8].

- Collecting efficiency (from low to high): $GCM \prec PeCM \prec PCM$
- Correctness of statistics (from low to high): $PCM \prec PeCM \prec GCM$

None of a single model can well leverage the collecting efficiency and correctness of statistics. For adaptively picking appropriate model under different situations, APCS employs the following function to assign executing model to some task (τ) based on the number of nodes occupied by this task $(W(\tau))$.

$$DM(\tau) = \begin{cases} PCM & W(\tau) > \Theta \\ GCM & W(\tau) < 75\% * \Theta \\ \zeta(PCM, GCM) & others \end{cases} \qquad (2)$$

Because PeCM is similar to PCM but with lower efficiency, we only adaptively select other two executing models. Θ is the threshold whose value is $\varepsilon * \log_2 Q/2$. Q is the number of nodes in distributed system and ε is the scale factor (default is 2). The threshold (Θ) is reasonable due to that large task should be executed with high efficiency, while results of small task should be kept high correctness. For reducing the risk caused by Θ, we randomly select the executing model in the interval [75%, 100%].

2.3 Adaptive Executing

When executing tasks, states of executing places may be not allowing to immediately handle these tasks. With adaptive executing component, we can reasonably

428 J.-T. Gao et al.

decide when to execute tasks, which can reduce negative effect to system performance. When detecting the state of current node is lower than some threshold (default is 55%), the task will be executed and store the executing result according to executing model.

3   Related Work

Researchers usually concentrate on how to improve the correctness of cardinality estimation based on statistics [1, 2, 6, 7]. Under distributed environment, the negative effect to system performance leaded by collecting statistics usually is so big that cannot be ignored.

Under distributed environment, relying on centralized controller [3–5] to handle large datasets is a common phenomena. The collecting jobs of Orca [4] are executed in the *memo* structure of main memory in master node. The global statistics are generated by merging local statistics, which are stored at groups. Similarly, Microsoft's PDW QO [5] stores the statistics in the shell database, and its collecting strategy is to execute standard SQL in every node. The global statistics is also derived by every nodes' local statistics.

Besides the above static strategies, one can take collecting statistics as a problem-fixing method by incorporating adaptive thoughts [9, 10]. Oracle 12c employs adaptive statistics [9] to repair the cardinality estimation errors by triggering collecting action when these errors appearing during compiling time or running time. But it adopts appending model only handling small amount of data. In this paper, we propose an adaptive strategy (APCS), which allocates collecting tasks with reasonable executing mode meanwhile minimize negative effect to system performance when handling large scale data set.

References

1. Harmouch, H., Naumann, F.: Cardinality estimation: an experimental survey. Proc. VLDB Endow. **11**(4), 499–512 (2018)
2. Woodruff, D.P., Zhang, Q.: Distributed statistical estimation of matrix products with applications (2018)
3. Chen, J., Jindel, S., Walzer, R., et al.: The MemSQL query optimizer. Proc. VLDB Endow. **9**(13), 1401–1412 (2016)
4. Soliman, M.A., Antova, L., Raghavan, V., et al.: Orca: a modular query optimizer architecture for big data. In: Proceedings of the 2014 ACM SIGMOD International Conference on Management of Data, pp. 337–348. ACM (2014)
5. Shankar, S., Nehme, R., Aguilar-Saborit, J., et al.: Query optimization in Microsoft SQL server PDW. In: Proceedings of the 2012 ACM SIGMOD International Conference on Management of Data, pp. 767–776. ACM (2012)
6. Grohe, M., Schweikardt, N.: First-order query evaluation with cardinality conditions (2017)
7. Müller, M., Moerkotte, G., Kolb, O.: Improved selectivity estimation by combining knowledge from sampling and synopses. Proc. VLDB Endow. **11**(9), 1016–1028 (2018)

8. Chakkappen, S., Cruanes, T., Dageville, B., et al.: Efficient and scalable statistics gathering for large databases in Oracle 11g. In: ACM SIGMOD International Conference on Management of Data, DBLP (2008)
9. Chakkappen, S., Budalakoti, S., Krishnamachari, R., et al.: Adaptive statistics in Oracle 12c. Proc. VLDB Endow. **10**(12), 1813–1824 (2017)
10. Macke, S., Zhang, Y., Huang, S., et al.: Adaptive sampling for rapidly matching histograms. Proc. VLDB Endow. **11**(10), 1262–1275 (2017)

Word Sense Disambiguation with Massive Contextual Texts

Ya-fei Liu and Jinmao Wei$^{(\boxtimes)}$

College of Computer Science, Nankai University, Tianjin, China
liuyf@mail.nankai.edu.cn, weijm@nankai.edu.cn

Abstract. Word sense disambiguation is crucial in natural language processing. Both unsupervised knowledge-based and supervised methodologies try to disambiguate ambiguous words through context. However, they both suffer from data sparsity, a common problem in natural language. Furthermore, the supervised methods are previously limited in the all-word WSD tasks. This paper attempts to collect all publicly available contexts to enrich the ambiguous word's sense representation and apply these contexts to the simplified Lesk and our M-IMS systems. Evaluations performed on the concatenation of several benchmark fine-grained all-word WSD datasets show that the simplified Lesk improves by 9.4% significantly and our M-IMS has shown some improvement as well.

Keywords: WSD · Massive contextual texts · Simplified Lesk · M-IMS

1 Introduction

Word sense disambiguation (WSD) is an open problem in natural language processing, which identifies word sense used in a given context. It's considered as the fundamental cornerstone for machine translation, information extraction and retrieval, parsing, and question answer. What's bad is that all methods on WSD highly depend on knowledge sources like corpora of texts which may be unlabeled or annotated with word sense [1]. Ineluctably these knowledge sources all suffer from data sparsity to varying degrees. Apart from the sparsity, a common agreement is that supervised methods are restricted in the all-word tasks as labeled data for the full lexicon is sparse and difficult to obtain [2], while knowledge-based methods only requiring an external knowledge source are more suitable for the all-word tasks [4]. In summary, this paper is chiefly involved with the data sparsity and the adaptability of supervised algorithms. Accordingly, two main contributions are summarized as follows:

- We relieve the data sparsity by assembling almost all publicly available contextual texts from different corpora.
- We modify It Make Sense (IMS) [7] by embedding a knowledge-based method to ensure the latter starts to work in case the former fails.

© Springer Nature Switzerland AG 2019
G. Li et al. (Eds.): DASFAA 2019, LNCS 11448, pp. 430–433, 2019.
https://doi.org/10.1007/978-3-030-18590-9_60

2 Methodology

2.1 Corpora Sources

The first main point of this paper lies in more corpora with massive instance sentences uniformly annotated by one sense repository. Here are five publicly available corpora annotated with WordNet: WordNet, SemCor [3], OMSTI [6], MASC[1], GMB[2]. WordNet is not only a lexical dictionary as the sense repository here but also a source of example sentences.

2.2 M-IMS

Preprocessing and Feature Extraction. Preprocessing aims to convert various texts from different corpora into formatted instance sentences. In contrast to IMS, we include two additional procedures: Standardization and Sense Mapping. Standardization intends to unify the formats and preserve texts with POS, annotation and lemma. While Sense Mapping deals with the annotation version problem according to the sense key.

Feature Extraction is conducted on the massive contexts (MC) as how IMS does. A small modification to surrounding words feature here is that the surrounding words can be only in the current sentence, not including the adjacent sentences, because we disambiguate ambiguous words on sentence-level.

Classification. Another major contribution of this paper lies in the modification here. The Classification comprises three components: Supervised Classification, Decision Component, and Knowledge-based Classification.

Supervised Classification and Knowledge-Based Classification. The supervised classification part is almost the same as the classifier in the IMS. As for the knowledge-based, we select simplified Lesk as the knowledge-based algorithm to make the disambiguation. The overlapping way of simplified Lesk to calculate the similarity between gloss and context conforms to the characteristic of the MC.

Decision Component. The rhombus with a question mark inside in Fig. 1 represents the decision component. It determines whether or not the knowledge-based methods are introduced into the disambiguation. Here we recommend two boundary conditions for the decision:

- Strict condition: Only if annotations for a word cover all senses of this word with the same part of speech, can the decision output yes/y, otherwise no/n.
- Loose condition: As long as annotations for a word cover at least one sense of this word with the same part of speech, the decision outputs yes.

This paper adopts relatively loose setting: As long as annotations for a word cover at least two senses of this word with the same part of speech, we consider the trained model is helpful in a way.

---

[1] http://www.anc.org/data/masc/.

[2] http://gmb.let.rug.nl/.

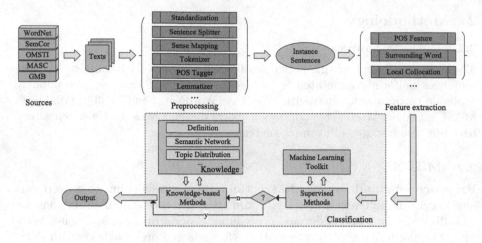

Fig. 1. M-IMS system architecture

3   Experiments and Results

The first experiment aims at showing the ability of massive contextual texts to relieve the data sparsity. The second makes a comparison among M-IMS, IMS, simplified Lesk etc. And we choose the concatenation of the five standardized datasets (Sem-Union) from [5] as the test dataset.

3.1   Results

In Table 1, we have found that contextual texts, like instance sentences, extremely suit for word matching pattern contemporarily. Furthermore, the increment by our MC offers more possibility for previously annotation-lacking senses and relieve the data sparsity to a certain degree.

Table 1. The overlap rates and annotation coverages of several {sources}-context pairs.

Coverage type	{Sources}-context pairs	Rate (%)	Accuracy (%)
Overlap rate	Gloss	16.1	53.7
	WordNet	**82.1**	57.6
	SemCor	66.0	63.9
	MC	72.5	**67.0**
Annotation coverage	SemCor	69.0	–
	OMSTI	71.5	
	MC	**76.7**	

In Table 2, it's remarkable that simplified Lesk with MC obtains a much better performance and pushes the overlap-based algorithms to a new high.

What's more, M-IMS uniformly performs better than IMS both on SemCor and MC, but not with a significant margin implying that the performance of knowledge-based algorithms is required to be promoted in the future.

Table 2. Comparison of IMS, M-IMS and SL with different sources on Sem-Union.

Systems	Sources	Sem-Union (%)
SL	WordNet	57.6
	MC	**67.0**
IMS	SemCor	67.1
	MC	67.5
M-IMS	SemCor	67.4
	MC	**67.7**

4 Conclusion

This paper mainly deals with the data sparsity in WSD with massive contexts and the adaptability of supervised methods. Note that this work is still in progress and we shall release MC in our later research work along with relevant API to enable various applications with detail documentations.

Acknowledgements. This work was partially supported by the National Natural Science Foundation of China (61772288), and the Natural Science Foundation of Tianjin City (18JCZDJC30900).

References

1. Borah, P.P., Talukdar, G., Baruah, A.: Approaches for word sense disambiguation-a survey. IJRTE **3**(1), 35–38 (2014)
2. Chaplot, D.S., Salakhutdinov, R.: Knowledge-based word sense disambiguation using topic models. arXiv preprint arXiv:1801.01900 (2018)
3. Miller, G.A., Leacock, C., Tengi, R., Bunker, R.T.: A semantic concordance. In: Proceedings of the Workshop on Human Language Technology, pp. 303–308. ACL (1993)
4. Miller, T., Biemann, C., Zesch, T., Gurevych, I.: Using distributional similarity for lexical expansion in knowledge-based word sense disambiguation. In: Proceedings of the 24th COLING, pp. 1781–1796 (2012)
5. Raganato, A., Camacho-Collados, J., Navigli, R.: Word sense disambiguation: a unified evaluation framework and empirical comparison. In: Proceedings of the 15th Conference of ECACL, vol. 1, pp. 99–110 (2017)
6. Taghipour, K., Ng, H.T.: One million sense-tagged instances for word sense disambiguation and induction. In: Proceedings of the 19th CoNLL, pp. 338–344 (2015)
7. Zhong, Z., Ng, H.T.: It makes sense: a wide-coverage word sense disambiguation system for free text. In: Proceedings of the ACL 2010 System Demonstrations, pp. 78–83. ACL (2010)

Learning DMEs from Positive and Negative Examples

Yeting Li[1,2], Chunmei Dong[1,2], Xinyu Chu[1,2], and Haiming Chen[1(✉)]

[1] State Key Laboratory of Computer Science, Institute of Software,
Chinese Academy of Sciences, Beijing 100190, China
{liyt,dongcm,chuxy,chm}@ios.ac.cn
[2] University of Chinese Academy of Sciences, Beijing, China

Abstract. The presence of a schema for XML documents has numerous advantages. Unfortunately, many XML documents in practice are not accompanied by a (valid) schema. Therefore, it is essential to devise algorithms to infer schemas from XML documents. The fundamental task in XML schema inference is learning regular expressions. In this paper we consider unordered XML, where the relative order among siblings is ignored, and focus on the subclass called *disjunctive multiplicity expressions* (DMEs) which are proposed for unordered XML. Previous work in this direction lacks inference algorithms that support for learning DME from both positive and negative examples. In this paper, we provide an algorithm to learn DMEs from both positive and negative examples based on genetic algorithms.

Keywords: Unordered XML · Disjunctive multiplicity expressions ·
Positive and negative examples

1 Introduction

EXtensible Markup Language (XML), as a standard format for data representation and exchange, is widely used in various applications on the Web. Its schemas contribute a lot to data processing, automatic data integration, static analysis of transformations and so on [2,5,11]. However, a survey showed that XML documents with corresponding schema definitions on the Web only account for 24.8%, with the proportion of 8.9% for valid ones [8]. Therefore, it is necessary to learn a valid schema for XML documents. Besides, schema inference is also useful in situations where a schema is already available, such as in schema cleaning and dealing with noise [3]. It is known that the essential task in schema inference is inferring regular expressions (REs) from a set of given samples [3]. However, a seminal result by Gold [7] shows that the class of REs could not be identifiable from positive examples only. Consequently, researchers have turned to study subclasses of REs and their corresponding inference algorithms.

Work supported by the National Natural Science Foundation of China under Grant Nos. 61872339 and 61472405.

© Springer Nature Switzerland AG 2019
G. Li et al. (Eds.): DASFAA 2019, LNCS 11448, pp. 434–438, 2019.
https://doi.org/10.1007/978-3-030-18590-9_61

When XML is used for document-centric applications, the relative order among the elements is typically important e.g., the relative order of sections in papers. On the other hand, in case of data-centric XML applications, the order among the elements may be unimportant e.g., the order among student names in a study group [1]. In this paper we focus on the latter use case, in which the relative order among siblings is ignored and the number of occurrences of each sibling can be different. The practical motivations of learning unordered schemas have been explained in [6]. In particular, we focus on the subclass called *disjunctive multiplicity expressions* (DMEs), which define unordered content model only, and are better suited for unordered XML [4]. A DME uses unordered concatenation "∥" to gather all conditions and uses disjunction "|" to specify alternatives. An algorithm for learning DMEs from positive examples has been given [6], in which it uses an approximate algorithm to find maximal clique, a problem known to be NP-complete.

In this paper we consider learning DMEs from positive and negative examples. The motivations of using negative examples have been stated in [6]. However, it has been proved that given an alphabet Σ, learning DMEs from a set of positive and negative examples is NP-complete. It was proved by reduction from *3SAT* which is known as being NP-complete [4,6]. So the inference algorithm in [6] only supports positive examples. We solve this problem by using genetic algorithms, which have been used to solve NP problems. We design an algorithm based on genetic approaches, which can learn DMEs from both positive examples and negative examples. As far as we know, we are the first to solve the problem of learning DMEs from positive and negative examples. Besides, we even support learning with only negative examples.

Related Work. Ciucanu and Staworko proposed two subclasses DME and ME in [4,6]. They support to describe completely unordered data without the sequential order and ME even uses no disjunction operator. The inference algorithm of DME is discussed in [6], which supports learning DMEs from positive examples. There are also some restricted subclasses of regular expressions with interleaving or unordered concatenation[1] [9,10,12,14], but all of them support learning from only positive examples.

2   Learning Algorithm

Due to the limitation of space, we omit the definition of unordered word, multiplicity and DME introduced in [6].

Definition 1 Candidate Region (CR). *We use candidate region to define the skeleton structure of a DME. For a DME $r := D_1^{M_1} \parallel \cdots \parallel D_n^{M_n}$, it belongs to the candidate region $|D_1| \parallel \cdots \parallel |D_n|$.*

For example, DME $r_1 = (a|b) \parallel (c|d)$ and $r_2 = a^+ \parallel b \parallel (c^*|d^+|e^?)$ belong to the candidate region $2 \parallel 2$ and $1 \parallel 1 \parallel 3$, respectively. For a given alphabet

---

[1] Unordered concatenation can be viewed as a weaker form of interleaving.

$|\Sigma| = n$, it is easy to see there are 2^{n-1} candidate regions. Consider the alphabet $\Sigma = \{a, b, c, d\}$ and $|\Sigma| = 4$, we can get 8 candidate regions: 1 ∥ 1 ∥ 1 ∥ 1, 1 ∥ 1 ∥ 2, 1 ∥ 2 ∥ 1, 2 ∥ 1 ∥ 1, 2 ∥ 2, 1 ∥ 3, 3 ∥ 1, 4. Because of the unordered features of ∥ operator, we find candidate regions 1 ∥ 1 ∥ 2, 1 ∥ 2 ∥ 1 and 2 ∥ 1 ∥ 1 are actually equivalent, and we can merge them together. After the merger of some equivalent candidate regions, we get the simplified candidate regions (SCR) and we use the simplified ones in our algorithm.

Our learning algorithm aims to infer accurate and precise DME from positive examples and negative examples, which is mainly based on genetic algorithm (GA). The main procedures of the algorithm are illustrated as follows.

1. Scan positive examples S_+ and negative examples S_- to get the alphabet Σ, then initialize all the simplified candidate regions according to the alphabet size $|\Sigma|$.
2. Select the best DME (with the highest fitness value) for each SCR based on genetic algorithm, and put them in the candidate set C.
3. According to the fitness function, select the best DME from the candidate set C and output it.

Now we give a detailed explanation of second step used in the main algorithm. For a given simplified candidate region (SCR), positive examples S_+ and negative examples S_-, we employ the GA approach to infer the best candidate DME that accepts positive examples and rejects negative examples as many as possible and is most precise. The main procedures are as follows.

1. Initialize the population of candidate character sequences according to the population size.
2. Select a best multiplicity sequence for each character sequence based on genetic approaches with a number of genetic operators, including multiplicity crossover and multiplicity mutation.
3. Decode each pair of character sequences, corresponding with best multiplicity sequences and the given SCR to get the population of candidate DMEs.
4. Calculate a fitness value for each individual DME by the fitness function. The fitness function gives priority to accuracy and then compare the preciseness (CC values introduced in [10]). that is, on the basis of selecting the individual that can accept more positive examples and reject more negative examples, then consider the more precise ones.
5. Use roulette-wheel selection to generate a next generation from the current population of DMEs by comparing the fitness value between individuals. Select some pairs of character sequences to conduct crossover and mutation.
6. Iterate 1–5 steps until the number of generations reaches the given threshold. Finally, we select the best DME from the last generation of DMEs in the given simplified candidate region, which has the maximum value of accuracy rate and the minimum value of CC.

3   Conclusions

In this paper, we provided an algorithm to learn a DME from positive and negative examples based on genetic approaches. To the best of our knowledge, we are the first to solve the problem of learning DMEs from positive and negative examples. Besides, we even support learning with only negative examples.

Actually, expressions in DMEs are not only applied to describing the content models of schemas to generalize unordered XML, but they are also one subclass of *regular bag expressions*, which is the fundamental content of a novel schema formalism for RDF under development by W3C [13]. *Regular bag expressions* allow that the disjunction and the unordered concatenation operators can be nested each other. Clearly, this kind of expressions do not have a restricted form as DME. Then we cannot directly apply our learning algorithm to learn a good result. In the future, we aim to propose a learning framework to learn subclasses of *regular bag expressions* effectively, which will promote the development of schemas for RDF.

References

1. Abiteboul, S., Bourhis, P., Vianu, V.: Highly expressive query languages for unordered data trees. Theory Comput. Syst. **57**(4), 927–966 (2015)
2. Benedikt, M., Fan, W., Geerts, F.: Xpath satisfiability in the presence of DTDs. J. ACM **55**(2), 8:1–8:79 (2008)
3. Bex, G.J., Neven, F., Vansummeren, S.: Inferring XML schema definitions from XML Data. In: Proceedings of the 33rd International Conference on VLDB, pp. 998–1009 (2007)
4. Boneva, I., Ciucanu, R., Staworko, S.: Simple schemas for unordered XML. In: Proceedings of the 16th International Conference on WebDB, pp. 13–18 (2013)
5. Che, D., Aberer, K., Özsu, M.T.: Query optimization in XML structured-document databases. J. VLDB **15**(3), 263–289 (2006)
6. Ciucanu, R., Staworko, S.: Learning schemas for unordered XML. In: Proceedings of the 14th International Conference on DBPL (2013)
7. Gold, E.M.: Language identification in the limit. Inf. Control **10**(5), 447–474 (1967)
8. Grijzenhout, S., Marx, M.: The quality of the XML Web. J. Web Sem **19**, 59–68 (2013)
9. Li, Y., Mou, X., Chen, H.: Learning concise relax NG schemas supporting interleaving from XML documents. In: Gan, G., Li, B., Li, X., Wang, S. (eds.) ADMA 2018. LNCS (LNAI), vol. 11323, pp. 303–317. Springer, Cham (2018). https://doi.org/10.1007/978-3-030-05090-0_26
10. Li, Y., Zhang, X., Xu, H., Mou, X., Chen, H.: Learning restricted regular expressions with interleaving from XML data. In: Trujillo, J.C., et al. (eds.) ER 2018. LNCS, vol. 11157, pp. 586–593. Springer, Cham (2018). https://doi.org/10.1007/978-3-030-00847-5_43
11. Martens, W., Neven, F.: Frontiers of tractability for typechecking simple XML transformations. In: Proceedings of the 23rd International Conference on PODS, pp. 23–34 (2004)
12. Peng, F., Chen, H.: Discovering restricted regular expressions with interleaving. In: Cheng, R., Cui, B., Zhang, Z., Cai, R., Xu, J. (eds.) APWeb 2015. LNCS, vol. 9313, pp. 104–115. Springer, Cham (2015). https://doi.org/10.1007/978-3-319-25255-1_9

13. Staworko, S., Boneva, I., Gayo, J.E.L., Hym, S., Prud'hommeaux, E.G., Solbrig, H.R.: Complexity and expressiveness of shex for RDF. In: Proceedings of the 18th International Conference on ICDT, pp. 195–211 (2015)
14. Zhang, X., Li, Y., Cui, F., Dong, C., Chen, H.: Inference of a concise regular expression considering interleaving from XML documents. In: Phung, D., Tseng, V.S., Webb, G.I., Ho, B., Ganji, M., Rashidi, L. (eds.) PAKDD 2018. LNCS (LNAI), vol. 10938, pp. 389–401. Springer, Cham (2018). https://doi.org/10.1007/978-3-319-93037-4_31

Serial and Parallel Recurrent Convolutional Neural Networks for Biomedical Named Entity Recognition

Qianhui Lu[1,2], Yunlai Xu[1,2], Runqi Yang[1,2], Ning Li[1,2], and Chongjun Wang[1,2(✉)]

[1] National Key Laboratory for Novel Software Technology, Nanjing University, Nanjing, China
cheonhye95@gmail.com, yunlaixu@gmail.com, runqiyang@gmail.com, {ln,chjwang}@nju.edu.cn
[2] Department of Computer Science and Technology, Nanjing University, Nanjing, China

Abstract. Identifying named entities from unstructured biomedical text is an important part of information extraction. The irrelevant words in long biomedical sentences and the complex composition of the entity make LSTM used in the general domain less effective. We find that emphasizing the local connection between words in a biomedical entity can improve performance. Based on the above observation, this paper proposes two novel neural network architectures combining bidirectional LSTM and CNN. In the first architecture S-CLSTM, a CNN structure is built on the top of bidirectional LSTM to keep both long dependencies in a sentence and local connection between words. The second architecture P-CLSTM combines bidirectional LSTM and CNN in parallel with the weighted loss to take advantage of the complementary features of two networks. Experimental results indicate that our architectures achieve significant improvements compared with baselines and other state-of-the-art approaches.

Keywords: Information extraction · Deep neural network · Biomedical text mining · Named entity recognition

1 Introduction

Biomedical named entity recognition (BioNER) is one of the most important stages in biomedical information extraction, aiming to identify biomedical target entities (e.g., genes, proteins) from text (e.g., medical articles, medical records).

This paper is supported by the National Key Research and Development Program of China (Grant No. 2016YFB1001102), the National Natural Science Foundation of China (Grant Nos. 61876080, 61502227), the Fundamental Research Funds for the Central Universities No.020214380040, the Collaborative Innovation Center of Novel Software Technology and Industrialization at Nanjing University.

G. Li et al. (Eds.): DASFAA 2019, LNCS 11448, pp. 439–443, 2019.
https://doi.org/10.1007/978-3-030-18590-9_62

Compared with the general domain, the names of many biomedical entities are typically long and contain more than three words or rarely used words, making the boundary identification much more difficult. By exploring different neural networks and unique features of BioNER tasks, we find that emphasizing the local connection between words in an entity can improve performance.

This paper combines the advantages of LSTM and CNN which utilizes convolutional layers for selecting local features. And two novel neural network architectures of different combinations are proposed: (1) A serial architecture named S-CLSTM using bidirectional LSTM to capture long-distance interactions between input word vectors. Then the output is fed into CNN to learn useful local features. A fully connected layer followed by a CRF layer is used to predict the probabilities for each label of the word. (2) A parallel architecture named P-CLSTM is a combination of bidirectional LSTM and CNN with the weighted loss, which aims to learn a representation of complementary features including contextual and local semantic information. Then the concatenated output is passed to a fully connected layer with a CRF layer. Experimental results on three widely used datasets indicate that our architectures achieve significant improvements compared with baselines and other state-of-the-art approaches.

2 Neural Network Architecture

2.1 S-CLSTM

In the first architecture, LSTM and CNN are connected in series to improve performance. We call this approach S-CLSTM. The recurrent structure captures contextual information as far as possible when learning word representation. The convolution kernel matches a specific keyword or key phrase to learn useful local features. The proposed serial architecture is shown in Fig. 1.

Specifically, the input to the architecture is a sequence of word vectors consisting of the character representation vectors[1] concatenated with 200-dimensional word embeddings[2]. We build bidirectional LSTM (BiLSTM) to process the sequence in two directions, the state size of LSTM is set to 128. The architecture of CNN is similar to GRAM-CNN [1] using 90 filters for three different kernel sizes: 1, 3 and 5. On the top of architecture, a fully connected layer followed by a CRF layer gives the probabilities for each label of the word. We adopt the Viterbi algorithm to infer the best label sequence during prediction.

2.2 P-CLSTM

In the second architecture, we do a combination of LSTM and CNN in parallel. We call this approach P-CLSTM which can learn the contextual information and local semantic information at the same time. The proposed parallel architecture is shown in Fig. 2.

---

[1] The character representation is generated by BiLSTM, with 100 hidden states.
[2] https://github.com/cambridgeltl/BioNLP-2016.

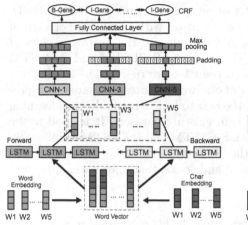
Fig. 1. BioNER S-CLSTM architecture

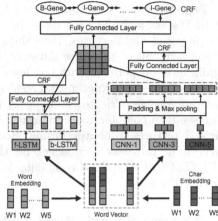
Fig. 2. BioNER P-CLSTM architecture

In the left and the right parts of Fig. 2, the output $\tilde{h}_t$ from BiLSTM and $\tilde{a}_t$ from CNN are forwarded to the fully connected layer, then pass through a CRF layer, respectively. In the middle part, $\tilde{h}_t$ and $\tilde{a}_t$ are concatenated into one vector denoted as z_{concat}, where $z_{concat} = [\tilde{h}_t; \tilde{a}_t]$.

We use z_{concat} as the final word representation to predict the label of each word in a sentence because it contains both local connection and contextual information. In order to train the neural network better and make full use of these three representations, a weighted loss function is introduced as an optimization objective [2]. $\alpha_1, \alpha_2,$ and α_3 are $0.5, 0.3,$ and $0.2,$ that control the contribution of $z_{concat}, \tilde{h}_t,$ and $\tilde{a}_t$. Our networks are optimized by Adam. The learning rate is initialized to $\eta_0 = 0.001$, and is updated on each epoch of training as $\eta_t = \eta_0 \times \rho^t$, with $\rho = 0.9$.

3   Experiment

To demonstrate the advantage of our combined architectures, the first baseline combines bidirectional LSTM with CRF model (BiLSTM-CRF), the second one (CNN-CRF) uses CNN for word representation learning.

Table 1. Comparison with baselines. (Bold: higher scores than baselines, *: best scores)

Approach	BC4CHEMD			NCBI-Disease			JNLPBA		
	Precision	Recall	F1-score	Precision	Recall	F1-Score	Precision	Recall	F1-Score
CNN-CRF	88.14	84.28	86.17	84.44	83.82	84.13	69.11	75.27	72.06
BiLSTM-CRF	89.95	85.50	87.67	83.97	82.05	83.00	70.17*	76.13	73.03
S-CLSTM	89.33	87.52*	88.42*	84.82*	84.90	84.86	69.35	77.88*	73.36
P-CLSTM	90.11*	86.75	88.40	84.74	85.53*	85.13*	69.47	77.86	73.42*

Table 1 presents experimental results of our architectures on BC4CHEMD, NCBI-Disease and JNLPBA datasets in comparison with baselines. For each of our proposed architectures, the values in the table are the average results of 10 runs. For all of the three datasets, the proposed S-CLSTM and P-CLSTM significantly outperform both baselines on the F1-score ($p < 0.01$).

Table 2 presents experimental results of our architectures compared with previous approaches. Crichton et al. [3] are the first to utilize the multi-task learning framework to combine more information in existing datasets. It is good to see our architectures still outperform it. Both S-CSLTM and P-CLSTM significantly outperform previous approaches on the F1-score ($p < 0.01$), indicating that S-CLSTM and P-CLSTM combining local and contextual information are valid for BioNER tasks.

Table 2. Comparison with other approaches. (Bold: higher scores than others, *: best scores)

Approach	BC4CHEMD			NCBI-Disease			JNLPBA		
	Precision	Recall	F1-score	Precision	Recall	F1-Score	Precision	Recall	F1-Score
Crichton [3]	-	-	83.02	-	-	80.37	-	-	70.09
Ma and Hovy [4]	90.83*	83.19	86.84	86.89*	78.75	82.62	70.28*	75.26	72.68
Habibi [5]	88.25	83.17	85.63	86.43	82.92	84.64	-	-	-
Char Attention [6]	-	-	84.53	-	-	-	-	-	72.70
S-CLSTM	89.33	87.52*	88.42*	84.82	84.90	84.86	69.35	77.88*	73.36
P-CLSTM	90.11	86.75	88.40	84.74	85.53*	85.13*	69.47	77.86	73.42*

4 Conclusions

This paper proposes two novel end-to-end architectures named S-CLSTM and P-CLSTM, incorporating the contextual and local representation of a word to recognize biomedical named entities. S-CLSTM connects BiLSTM and CNN in series to learn long-term dependencies and the local connection between words. P-CLSTM applies BiLSTM and CNN in parallel to generate complementary representation with the weighted loss. Experimental results indicate that S-CLSTM and P-CLSTM combining local and contextual information are valid for BioNER tasks.

References

1. Zhu, Q., Li, X., Conesa, A., Pereira, C.: GRAM-CNN: a deep learning approach with local context for named entity recognition in biomedical text. Bioinformatics **34**(9), 1547–1554 (2017)
2. Wu, Z., Dai, X.-Y., Yin, C., Huang, S., Chen, J.: Improving review representations with user attention and product attention for sentiment classification. In: Proceedings of the Thirty-Second AAAI Conference on Artificial Intelligence, New Orleans, Louisiana, USA, 2–7 February 2018

3. Crichton, G., Pyysalo, S., Chiu, B., Korhonen, A.: A neural network multi-task learning approach to biomedical named entity recognition. BMC Bioinform. **18**(1), 368 (2017)
4. Ma, X., Hovy, E.: End-to-end sequence labeling via bi-directional LSTM-CNNS-CRF. In: Proceedings of the 54th Annual Meeting of the Association for Computational Linguistics. ACL, Berlin, Germany, Volume 1: Long Papers, 7–12 August 2016
5. Habibi, M., Weber, L., Neves, M., Wiegandt, D.L., Leser, U.: Deep learning with word embeddings improves biomedical named entity recognition. Bioinformatics **33**(14), i37–i48 (2017)
6. Rei, M., Crichton, G.K.O., Pyysalo, S.: Attending to characters in neural sequence labeling models. In: COLING 2016, 26th International Conference on Computational Linguistics, Proceedings of the Conference: Technical Papers, Osaka, Japan, pp. 309–318, 11–16 December 2016

DRGAN: A GAN-Based Framework for Doctor Recommendation in Chinese On-Line QA Communities

Bing Tian, Yong Zhang$^{(\boxtimes)}$, Xinhuan Chen, Chunxiao Xing, and Chao Li

RIIT, TNList, BNRist, Department of Computer Science and Technology,
Tsinghua University, Beijing, China
{tb17,xh-chen13}@mails.tsinghua.edu.cn
{zhangyong05,xingcx,li-chao}@tsinghua.edu.cn

Abstract. Recently, more and more people choose to seek health-related information in health-related on-line QA communities. Doctor recommendation is very essential for users in these communities since it is difficult for them to find a proper doctor without assistance from medical staffs. In this paper, we develop a Generative Adversarial Nets (GANs)-based doctor recommendation framework utilizing data in Chinese on-line QA communities. We conduct extensive sets of experiments on a real-world dataset. The experimental results show that our framework significantly outperforms the state-of-the-art baselines.

Keywords: Doctor recommendation · Generative Adversarial Nets · QA communities

1 Introduction

Recently, some health related on-line QA communities in China have been built to provide various healthcare resources and services. Many people come to seek health-related information from these communities due to limited off-line medical resources. However, since there are too many doctors and no assistance from medical staffs, it is hard for users in on-line communities to find a proper doctor to diagnose their basic issues. Therefore, offering personalized doctor recommendation service utilizing the existing doctor patient interaction data is very essential for promoting the development of health-care service.

Recently, neural networks have been widely applied in many fields such as health-care [8], text classification [5] and financial analysis [2]. In the field of recommendation, many neural networks based methods have been proposed including content based system, collaborative filtering system and hybrid system [4]. Although these models have achieved promising results, they have been biased towards certain fields such as entertainment industry with rich and large datasets. Generative Adversarial Networks (GANs) were first introduced in [1] and IRGAN [6] was the first paper to propose the GAN-based framework to

© Springer Nature Switzerland AG 2019
G. Li et al. (Eds.): DASFAA 2019, LNCS 11448, pp. 444–447, 2019.
https://doi.org/10.1007/978-3-030-18590-9_63

train recommender systems and obtained promising performance. Motivated by this, in this paper, we propose a Generative Adversarial Networks based Doctor Recommendation (DRGAN). For measuring the relevance between queries (e.g., askers, patients, historical questions) and responses (e.g., doctors, answers, textual documents), we propose a novel scoring function which combines CNNs and multi-layer neural networks. We conduct a comprehensive evaluation on a real world dataset. Experimental results show the effectiveness of our framework.

2   Proposed Framework

As shown in Fig. 1, our framework contains four stages: data collection, data preprocessing, model training and doctor recommendation.

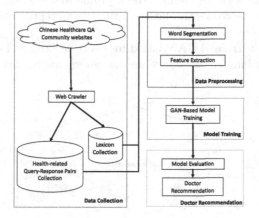

Fig. 1. Framework architecture

We collect and parse the on-line pages in data collection stage. A text parser is applied to extract various information from these pages including:

- **query:** question and asker ID, asker age, question content etc.;
- **response:** answer and doctor ID, doctor's name, professional title, answer content, number of customers and thumbed up, flag of the best answer etc.

Data preprocessing consists of two steps: word segmentation and feature extraction. We utilize a Chinese word segmentation tool, Jieba[1], to remove stop words. And we extract three types of features contained in queries and responses namely numeric features, profile attributes and textual content.

Our method consists of two kinds of models: a generative model and a discriminative model. The former tries to select most relevant response that looks like the ground-truth response to generate query-response tuples (q, r) from the

---

[1] https://pypi.org/project/jieba/.

candidate responses for the given query q and r_{true}. The latter attempts to discriminate well-matched query and response pair (q, r) from ill-matched ones, where the goodness of matching given by $f_\phi(r, q)$ depends on the relevance of r to q. We unify these two different types of models in a minimax schema and the overall objective is as following:

$$O^{g,d} = \min_\theta \max_\phi \sum_{i=1}^{N} \{r_{true}[log(d_\phi(P|q))] + r_{g_\theta(r_i|q, r_{true})}[log(1 - d_\phi(P|q))]\} \quad (1)$$

The Scoring Function. The basic idea of the scoring function is that we learn a distributed vector representation of a given query and its response candidates, and then apply a similarity metric to measure the matching degree. As shown in Fig. 2, in this scoring function, we first obtain the overall representation of queries and responses by combining their profile and textual content features. Then, we calculate the cosine similarity of the overall representation of the query and response as score S_{cos}.

Lastly we use the trained GAN-based model to recommend the top-N doctors for the asker according to the scores among askers' query and responses.

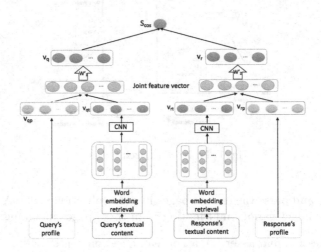

Fig. 2. Scoring function S_{cos} of queries and responses

3   Evaluation

We evaluate the effectiveness of our model on a real-world dataset: xywy.com. And we collect all pages posted between October 2005 and September 2015 from xywy.com. For evaluating the proposed models, we use standard ranking performance measures such as $precision@N$ and Normalised Discounted Cumulative Gain ($NDCG@N$). We provide the overall performance in Table 1. From the table, we can see that our DRGAN-based model outperforms all the baselines by a substantial margin on all the evaluation metrics.

Table 1. Performance against baselines

Method	Precision@1	Precision@3	NCDG@3
QA-CNN [3]	0.1318	0.1578	0.3229
DNS-CNN [7]	0.4374	0.1801	0.4927
IRGAN-ir [6]	0.3744	0.2731	0.6171
DRGAN	**0.4690**	**0.2602**	**0.6522**

4 Conclusions

In this paper, we propose a GAN-based doctor recommendation framework which consists of four stages. We further devise a scoring function for measuring the relevance between doctors and patients by integrating Convolutional Neural Networks (CNNs) and multi-layer neural networks. Experimental results on the real world datasets show that our proposed framework outperforms state-of-the-art methods by a substantial margin.

Acknowledgement. This work was supported by NSFC(91646202), National Key R&D Program of China(SQ2018YFB140235), and the 1000-Talent program.

References

1. Goodfellow, I.J., et al.: Generative adversarial nets. In: NIPS, pp. 2672–2680 (2014)
2. Luo, L., et al.: Beyond polarity: interpretable financial sentiment analysis with hierarchical query-driven attention. In: IJCAI, pp. 4244–4250 (2018)
3. dos Santos, C.N., Tan, M., Xiang, B., Zhou, B.: Attentive pooling networks. CoRR abs/1602.03609 (2016)
4. Singhal, A., Sinha, P., Pant, R.: Use of deep learning in modern recommendation system: a summary of recent works. arXiv preprint arXiv:1712.07525 (2017)
5. Wang, J., Wang, Z., Zhang, D., Yan, J.: Combining knowledge with deep convolutional neural networks for short text classification. In: IJCAI, pp. 2915–2921 (2017)
6. Wang, J., et al.: IRGAN: a minimax game for unifying generative and discriminative information retrieval models. In: SIGIR, pp. 515–524 (2017)
7. Zhang, W., Chen, T., Wang, J., Yu, Y.: Optimizing top-n collaborative filtering via dynamic negative item sampling. In: SIGIR, pp. 785–788 (2013)
8. Zhao, K., et al.: Modeling patient visit using electronic medical records for cost profile estimation. In: Pei, J., Manolopoulos, Y., Sadiq, S., Li, J. (eds.) DASFAA 2018. LNCS, vol. 10828, pp. 20–36. Springer, Cham (2018). https://doi.org/10.1007/978-3-319-91458-9_2

Attention-Based Abnormal-Aware Fusion Network for Radiology Report Generation

Xiancheng Xie[1], Yun Xiong[1,2(✉)], Philip S. Yu[1,2,3], Kangan Li[4], Suhua Zhang[5], and Yangyong Zhu[1,2]

[1] Shanghai Key Laboratory of Data Science, School of Computer Science, Fudan University, Shanghai, China
{17212010043,yunx}@fudan.edu.cn
[2] Shanghai Institute for Advanced Communication and Data Science, Fudan University, Shanghai, China
[3] Computer Science Department, University of Illinois at Chicago, Chicago, IL 60607, USA
[4] Shanghai General Hospital, Shanghai, China
[5] Department of Nephrology, Suzhou Kowloon Hospital, Shanghai Jiaotong University, School of Medicine, Suzhou 2015028, Jiangsu Province, China

Abstract. Radiology report writing is error-prone, time-consuming and tedious for radiologists. Medical reports are usually dominated by a large number of normal findings, and the abnormal findings are few but more important. Current report generation methods often fail to depict these prominent abnormal findings. In this paper, we propose a model named Attention-based Abnormal-Aware Fusion Network (A3FN). We break down sentence generation into abnormal and normal sentence generation through a high level gate module. We also adopt a topic guide attention mechanism for better capturing visual details and develop a context-aware topic vector for model cross-sentence topic coherence. Experiments on real radiology image datasets demonstrate the effectiveness of our proposed method.

Keywords: Radiology report generation · Attention · Abnormal-aware

1 Introduction

In recent years, radiology images are playing a vital role in the auxiliary diagnosis. Most of the existing captioning methods [3,4,6] perform poorly since report usually consists of multiple long, structural and informative sentences. Besides generating short captions, pioneering radiology report generation method [2] adopts a hierarchical LSTM framework combined with a co-attention mechanism to generate paragraph on IU X-Ray dataset [1]. However, their generated reports tend to describe normal findings with some repetitions and are incapable of capturing rare but prominent abnormality. Xue et al. [5] adopt a feedback

© Springer Nature Switzerland AG 2019
G. Li et al. (Eds.): DASFAA 2019, LNCS 11448, pp. 448–452, 2019.
https://doi.org/10.1007/978-3-030-18590-9_64

mechanism to learn long sequence dependency. However, their method is prone to generate fluent but general report without prominent abnormal narratives.

In this paper, we propose an Attention-based Abnormal-Aware Fusion Network (A3FN). Our method can effectively capture these rare but prominent abnormal observations. For more accurate abnormal descriptions, we adopt a topic guide attention mechanism to support abnormal findings with its detailed visual context (e.g., location, size, and severity etc.). We also develop a context-aware topic vector to control topic coherence and descriptive completeness. Our proposed A3FN method achieves the highest detection accuracy of positive abnormality terminologies.

2 The Proposed A3FN Method

This section describes our proposed A3FN in detail. Figure 1 shows the architecture of A3FN. We will go through vital parts of the model in following section.

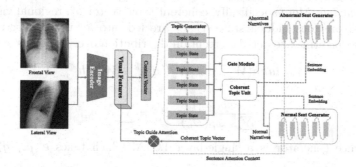

Fig. 1. The architecture of the proposed A3FN method.

Topic Generator and Coherent Topic Unit. Topic generator is a single-layer LSTM which generates a sequence of high-level topic vectors q, one for each sentence. Generally, topic generator can be written as:

$$c_i^t = F_{attn}^t(h^c, h_{i-1}^t), h_i^t = F_{RNN}^t(c_i^t, h_{i-1}^t),$$

$$q_i = tanh(W_t h_i^t + b_t), y_i = softmax(W_{stop} h_i^t + b_{stop}).$$

where F_{attn}^t is the attention mechanism same as [4], and F_{RNN}^t is a single-layer LSTM. The probability y_i controls whether to stop generating next topic vector. After generate all topic states, we enhance current topic state q_i with global topic vector q_g and last generated sentence embedding r_{i-1} for topic coherence as follows:

$$q_g = \sum_{i=1}^{M} \alpha_i q_i, \;\; where \; \alpha_i = \frac{\|q_i\|_2}{\sum_i \|q_i\|_2}.$$

We first transform sentence embedding r_{i-1} to q_{i-1}^h through a 2-layer fully-connected unit. Then we merge q_{i-1}^h and q_i as follows:

$$q_i' = \beta_i q_i + (1 - \beta_i) q_{i-1}^h, \beta_i = sigmoid(W_{c,i} q_i - W_{c,i-1} q_{i-1}^h).$$

The obtained vector q_i' is then coupled with global topic vector q_g using a gating function implemented by a single-layer GRU.

$$q_i^c = F_{GRU}^c (q_i', q_g).$$

Gate Module. Given the i-th topic vector q_i, previous topic vector q_{i-1}, the gate module generate a distribution u_i over $\{abnormal = 0, normal = 1\}$, that is:

$$u_i = softmax(W_u tanh(W_{u,i} q_i + W_{u,i-1} q_{i-1})).$$

Attentional Sentence Generator. Different from [2] that directly feed topic vector into generator, we enhance our sentence generator with a *topic guide attention module*. More specifically, coherent topic vector q_i^c, regional visual feature v and previous hidden state $h_{i,j-1}^s$ are fed into a fully connected layer, followed by a softmax to get the attention distribution over k regions as follows:

$$\mathscr{A} = W_{att} tanh(W_v^{att} v + W_t^{att} q_i^c 1^k + W_s^{att} h_{i,j-1}^s 1^k).$$

$$\alpha_n = \frac{exp(\mathscr{A}_n)}{\sum_n exp(\mathscr{A}_n)}, c_{i,j}^s = \sum_{n=1}^{k} \alpha_n v_n.$$

The sentence generator is a single-layer LSTM which takes $e_{i,j-1}, q_i^c, c_{i,j}^s$ as inputs:

$$h_{i,j}^s = F_{RNN}^s (e_{i,j-1}, [q_i^c, c_{i,j}^s], h_{i,j-1}^s),$$

$$a_{i,j} = softmax(W_s h_{i,j}^s + b_s), s_{i,j} = argmax(a_{i,j}), e_{i,j} = \mathbf{E} 1_{s_{i,j}}.$$

where $\mathbf{E}$ is learnable embedding matrix and $1_{y_{i,j}}$ is one-hot vector.

3   Experiments and Analysis

Settings. We evaluate the proposed A3FN method on public IU X-Ray dataset [1]. There are 2914 reports associated with 5828 images after filtering out reports without two complete image views. We randomly pick 10% reports as testing set. We compare the proposed A3FN with state-of-the-art report generation methods including CNN-RNN [3], Soft-ATT [4], ATT-RK [6], Co-ATT [2] and Multi-Modal [5]. For comparison, we reimplement baseline models [2,5] using Pytorch since the codes of these models are not available. We report the performance of all models on frequently used metrics BLEU-{1,2,3,4}, CIDEr and ROUGE in Table 1. We compute the keywords accuracy (KA) used in [5] and positive abnormality terminology detection accuracy (Acc) as a measurement of abnormal detection. Table 2 shows evaluation results of KA and Acc.

Table 1. Automatic evaluation results.

Dataset	Model	BLEU-1	BLEU-2	BLEU-3	BLEU-4	CIDEr	ROUGE
IU X-Ray	CNN-RNN [3]	0.311	0.218	0.137	0.092	0.124	0.262
	Soft-ATT [4]	0.351	0.237	0.161	0.12	0.278	0.314
	ATT-RK [6]	0.341	0.221	0.153	0.106	0.187	0.302
	Co-ATT [2]	0.421	0.324	0.225	0.174	0.331	0.341
	Multi-Modal [5]	0.434	0.331	0.234	0.177	0.312	0.346
	A3FN	**0.443**	**0.337**	**0.236**	**0.181**	**0.374**	**0.347**

Table 2. Positive abnormality terminology detection accuracy and KA

Dataset	Model	KA (%)	Acc (%)
IU X-Ray	Multi-Modal [5]	58.7	10.1
	Co-ATT [2]	57.3	11.37
	A3FN	**61.0**	**13.25**

Results and Analysis. The proposed A3FN method outperforms all other baselines. Compared with these hierarchical models [2,5], our A3FN method outperforms Co-ATT [2] and Multi-Modal [5] by a large margin, with respectively **11.3%** and **19.9%** relative improvement on CIDEr score. Furthermore, the proposed A3FN model achieves best KA score and best positive abnormality detection accuracy score, which means our generated reports detect more positive abnormality and cover more topics.

4   Conclusion

In this paper, we propose a novel model named Attention-based Abnormal-Aware Attention Network (A3FN) which aims to generate structured, detailed, topic coherent, and abnormal-aware radiology reports. Our model achieves the competitive results compared with all state-of-the-art models.

Acknowledgements. This work is supported in part by the National Natural Science Foundation of China Projects No. U1636207, No. 91546105, the Shanghai Science and Technology Development Fund No. 16JC1400801, No. 17511105502, No. 17511101702, Suzhou Science and Technology Bureau Technology Demonstration Project (SS201712, SS201812).

References

1. Demner-Fushman, D., et al.: Preparing a collection of radiology examinations for distribution and retrieval. JAMIA **23**, 304–310 (2015)
2. Jing, B., Xie, P., Xing, E.P.: On the automatic generation of medical imaging reports. In: Proceedings of ACL (2018)

3. Vinyals, O., Toshev, A., Bengio, S., Erhan, D.: Show and tell: a neural image caption generator. In: Proceedings of CVPR (2015)
4. Xu, K., et al.: Show, attend and tell: neural image caption generation with visual attention. In: Proceedings of ICML (2015)
5. Xue, Y., et al.: Multimodal recurrent model with attention for automated radiology report generation. In: Frangi, A.F., Schnabel, J.A., Davatzikos, C., Alberola-López, C., Fichtinger, G. (eds.) MICCAI 2018. LNCS, vol. 11070, pp. 457–466. Springer, Cham (2018). https://doi.org/10.1007/978-3-030-00928-1_52
6. You, Q., Jin, H., Wang, Z., Fang, C., Luo, J.: Image captioning with semantic attention. In: Proceedings of CVPR (2016)

LearningTour: A Machine Learning Approach for Tour Recommendation Based on Users' Historical Travel Experience

Zhaorui Li, Yuanning Gao, Xiaofeng Gao(✉), and Guihai Chen

Shanghai Key Laboratory of Scalable Computing and Systems,
Department of Computer Science and Engineering, Shanghai Jiao Tong University,
Shanghai 200240, China
{lizhaorui,gyuanning}@sjtu.edu.cn, {gao-xf,gchen}@cs.sjtu.edu.cn

Abstract. Tour routes designning is a non-trival step for the tourists who want to take an excursion journey in someplace which he or she is not familiar with. For most tourists this problem represents an excruciating challenge due to such unfamiliarity. Few existing works focus on using other tourists' experiences in the city to recommend a personalized route for the new comers. To take full advantage of tourists' historical routes in route recommendation, we propose LearningTour, a model recommending routes by learning how other tourists travel in the city before. Giving that the tourist's route is actually a special variance of time sequence, we treat such route as a special language and thus treat such recommendation process as a unique translation process. Therefore we use a sequence-to-sequence (seq2seq) model to proceed such learning and do the recommendation job. This model comprises a encoder and a decoder. The encoder encodes users' interest to the context vector and the decoder decodes the vector to the generated route. Finally, we implemented our model on several real datasets and demonstrate its effeciency.

Keywords: Travelling Salesman Problem · Points-of-Interest · Seq2Seq

1 Introduction

For the majority of tourists, it occurs to be troublesome to gather enough information and construct a reasonable route under realistic constraints. Some current studies [2,8] focus on data analysis on users' interest and the attributes of

This work was supported by the National Key R&D Program of China [2018YFB10 04703]; the National Natural Science Foundation of China [61872238, 61672353]; the Shanghai Science and Technology Fund [17510740200]; the Huawei Innovation Research Program [HO2018085286]; and the State Key Laboratory of Air Traffic Management System and Technology [SKLATM20180X].

© Springer Nature Switzerland AG 2019
G. Li et al. (Eds.): DASFAA 2019, LNCS 11448, pp. 453–456, 2019.
https://doi.org/10.1007/978-3-030-18590-9_65

Points-Of-Interest (POI) based on tourists' trajectory data. Then they proposed a Travelling Salesman Problem solver to provide a recommended route. Few of the existing works generate routes based on these trajectory data.

Among the existing works using trajectory data to do route recommendation, most of them either take the data as a source of users' interests [2,8] or recommend one single POI [1]. Only a few works [7] applied machine learning method in route recommendation. The common feature of these works is that they treat the route generation process as a pattern match process.

Different from the works mentioned above, we consider the route generation process as a kind of text generation process, which has been well studied [3]. The main reason is that tourists' routes and the texts are both, to some degree, a kind of time series. To address this trajectory-based recommendation system, we introduce a *LearningTour* model. This model is constructed based on the encoder-decoder (or called seq2seq model) [6], since the seq2seq model has been widely acclaimed in handling such text generation problems [5]. We adopt the design of the seq2seq model and transplant it to our recommendation process. The essential information is gathered from photographs taken in the POIs to further exploit the value of all data sources.

The remainder of this paper is structured as follows. In Sect. 2 we give a definition of the problem and discuss the framework of the LearningTour model. The experimental results are discussed in Sect. 3. The conclusion of this paper is stated in Sect. 4.

2   Recommendation Framework

2.1   Problem Definition

Definition 1. *A **POI Graph** is an undirected complete graph $G = (V, E)$, where V is the set of POIs in the city and $E = V \times V$. Each edge e_{ij} is labeled by a cost function $T(i, j)$. Each vertice v_i is labeled by a category C_{p_i}.*

Definition 2. *An **Interest Vector** is the vector representing the interest of the tourist in all categories of POIs. It is defined as $IV = \langle I_{C_1}, I_{C_2}, \cdots, I_{C_q} \rangle$, where $I_{C_j} (j = 1, 2, \cdots, q)$ is the User Interest for category C_j.*

The problem is transferred to the following one: Given an POI Graph $G = (V, E)$, a tourist's interest vector $\langle I_{C_1}, I_{C_2}, \cdots, I_{C_q} \rangle$, a start POI p_s and an end POI p_t, the goal is to generate a route $\langle p_s, p_{i_1}, \cdots, p_{i_m}, p_t \rangle$ for the tourist.

2.2   LearningTour Model

In the LearningTour model, an encoder-decoder model (also known as seq2seq model) is proposed to accomplish the route generation. Following the design of Liu [5], we use an LSTM to encode the target tourist's Interest Vector to a fixed-length context vector and then take this context vector as one of the inputs of an LSTM decoder to generate the route.

To be more specific, the final hidden state of the LSTM encoder is taken as the context vector and is set as the initial state of the decoder. In addition, the user's interest vector, the specified start POI p_s and end POI p_t is concatenated to a vector as the input of the encoder, i.e. $Input = \langle IV, p_s, p_t \rangle$. This vector is treated as a time sequence with $|IV| + 2$ timesteps.

For the decoder, we propose a dense layer after an LSTM with the same architecture of that in the encoder. The dense layer uses a softmax function to convert the output of the LSTM to the generated POIs. Moreover, one-hot encoding is used to encode all the POIs.

The generated route consists of m POIs $\langle y_1, y_2, y_3, \cdots, y_m \rangle$ with POI y_t at the time t. The route generation process is formulated as an inference over a probabilistic model. Therefore the goal of the whole inference process is to generate the sequence $\hat{y}_{1:m}$ that maximizes $P(y_{1:m}|IV)$. The whole recommendation process is formulated as the Eq. (1).

$$\hat{y}_{1:m} = \arg\max_{y_{1:m}} \prod_{t=1}^{m} P(y_t|y_{0:t-1}, p_s, p_t, IV) \tag{1}$$

3 Experiments

In our experiments, we use the dataset provided by Lim [4]. Random selection algorithm and GNU Linear Programming Kit (GLPK) are chosen as the baseline algorithms. These methods are evaluated based on two aspects: the score users get and the CPU running time the methods take. Three aspects, users' interests in POIs, POIs' popularity, and POIs' public reputation among tourists, determine the score users get.

Figure 1 shows that our method provides a better route for tourists than the two other baselines. In addition, the LearningTour model is one to two orders of magnitude faster than the GLPK method. It is shown that the model 'learns' the spatial structure of the POIs in the city from other users' experience, while each time the GLPK method provides recommendation as if facing a completely unfamiliar city and target user. Such unfamiliarity is shown to be not only a huge time overhead but also a lag behind in tourists' satisfaction.

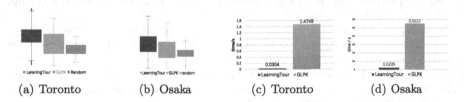

(a) Toronto   (b) Osaka   (c) Toronto   (d) Osaka

Fig. 1. (a) and (b) show the profits gathering by different methods on two cities. (c) and (d) show the average CPU time of different methods running on two cities.

Futhermore in both datasets we found there exists some POIs that have no records about the previous visited users. For the GLPK method, it takes a far longer time than usual when it needs to calculate through these POIs, while the LearningTour presents a steady running time facing the deciency of the data.

4   Conclusion

In this paper, we propose the LearningTour, a novel approach utilizing a encoder-decoder model for tour recommendation and taking other users' historical travel experience into account. This approach encodes users' interest to a context vector and decodes the vector to the recommended route. The result of our experiments based on real datasets shows that our approach can effectively and rapidly recommends an appealing route. Whereas we found our model can not recommend routes for a new city different from the city of the training data and also took a quite long time for the offline training. We also found that our model lacks the ability to recommend multiple-tour for one user. In the future, we will investigate the multi-cities training task and the similarities between cites to explant our model for other cities and try to ameliorate our model to support recommending routes during several days.

References

1. Baraglia, R., Frattari, C., Muntean, C.I., Nardini, F.M., Silvestri, F.: A trajectory-based recommender system for tourism. In: Huang, R., Ghorbani, A.A., Pasi, G., Yamaguchi, T., Yen, N.Y., Jin, B. (eds.) AMT 2012. LNCS, vol. 7669, pp. 196–205. Springer, Heidelberg (2012). https://doi.org/10.1007/978-3-642-35236-2_20
2. Dai, J., Yang, B., Guo, C., Ding, Z.: Personalized route recommendation using big trajectory data. In: IEEE International Conference on Data Engineering (ICDE), April 2015, pp. 543–554 (2015)
3. Goodfellow, I., et al.: Generative adversarial nets. In: Twenty-Eighth Conference on Neural Information Processing Systems (NIPS), pp. 2672–2680 (2014)
4. Lim, K.H., Chan, J., Leckie, C., Karunasekera, S.: Personalized tour recommendation based on user interests and points of interest visit durations. In: International Joint Conferences on Artificial Intelligence (IJCAI), vol. 15, pp. 1778–1784 (2015)
5. Liu, T., Wang, K., Sha, L., Chang, B., Sui, Z.: Table-to-text generation by structure-aware seq2seq learning. In: AAAI Conference on Artificial Intelligence (AAAI), pp. 4881–4888 (2018)
6. Sutskever, I., Vinyals, O., Le, Q.V.: Sequence to sequence learning with neural networks. In: Twenty-Eighth Conference on Neural Information Processing Systems (NIPS), pp. 3104–3112 (2014)
7. Wan, L., Hong, Y., Huang, Z., Peng, X., Li, R.: A hybrid ensemble learning method for tourist route recommendations based on geo-tagged socialnetworks. Int. J. Geogr. Inf. Sci. (IJGIS) 32(11), 2225–2246 (2018)
8. Zheng, Y., Zhang, L., Xie, X., Ma, W.-Y.: Mining interesting locations and travel sequences from GPS trajectories. In: International Conference on World Wide Web (WWW), pp. 791–800. ACM (2009)

TF-Miner: Topic-Specific Facet Mining by Label Propagation

Zhaotong Guo[1,2]($\boxtimes$), Bifan Wei[1,3], Jun Liu[1,2], and Bei Wu[1,2]

[1] Shaanxi Province Key Laboratory of Satellite and Terrestrial Network Tech. R&D,
Xi'an Jiaotong University, Xi'an 710049, Shaanxi, China
gzhtcrystal@stu.xjtu.edu.cn,{weibifan,liukeen}@xjtu.edu.cn,
xjtu_beiwu@163.com
[2] School of Electronic and Information Engineering, Xi'an Jiaotong University,
Xi'an 710049, Shaanxi, China
[3] School of Continuing Education, Xi'an Jiaotong University,
Xi'an 710049, Shaanxi, China

Abstract. Mining facets of topics is an essential task nowadays. Facet heterogeneity and long tail characteristic of information make facet mining tasks difficult. In this paper we propose a weakly supervised approach, called Topic-specific Facet (TF)-Miner, to mine TFs automatically by a Label Propagation algorithm (*LPA*). The process of propagation helps us mine complete facet sets. Experiments on several real-world datasets show that TF-Miner achieves better performance than the facet mining approaches which rely on the texts only.

Keywords: *Contents* section · Topic similarity · Label Propagation

1 Introduction

A topic refers to a term describing a specific concept which is widely used in a domain [2]. A facet is defined as one aspect of a topic [3]. For example, the facets of topic *Binary tree* include *definition*, *property*, *operation*, *storage* and so on.

Information related to a specific topic is usually distributed in different data sources, each of which generally just includes incomplete information [8]. Organizing the information of topics into different facets becomes an urgent issue. An important task of this issue is to find the facet sets for topics.

Some facet mining methods rely on structured data [2,5,9] or unstructured data [1,4,6]. These methods can hardly mine facets for the topics whose information is not very rich.

In this paper, we propose an automatic TF mining approach, TF-Miner.

The challenges are twofold: (1) **Facet heterogeneity**. The same facet for different topics have different meanings. (2) **Long tail characteristic**. Some topics have poor information. It is hard to mine their facets.

TF-Miner preprocesses *Contents* sections of Wikipedia article pages to get original facets to cope with the first challenge. Then, TF-Miner propagates according to topic similarities to cope with the second challenge.

© Springer Nature Switzerland AG 2019
G. Li et al. (Eds.): DASFAA 2019, LNCS 11448, pp. 457–460, 2019.
https://doi.org/10.1007/978-3-030-18590-9_66

2   Feature Analysis

To mine TFs, we first choose a data source, then analyze topic similarity.

Data Source. Wikipedia is generally recognized as the authoritative data source. The architecture of the Wikipedia article page is shown in Fig. 1. *Contents* section always includes lots of facets for the corresponding topic.

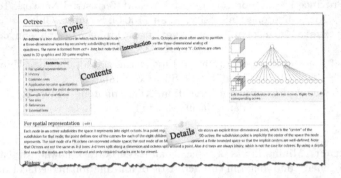

Fig. 1. The structure of Wikipedia article page

Facet Set Similarity Analysis. We quantify the facet set similarity (s_f) of topic t_x and t_y in Eq. 1. F_x is the facet set of topic t_x. According to Eq. 1, we calculate the s_f among the topics of domain *Data Structure*.

$$s_f = \frac{|F_x \cap F_y|}{|F_x \cup F_y|} \tag{1}$$

Wikipedia *Introduction* Section Similarity Analysis. We calculate the *Introduction* section similarities among the topics of domain *Data Structure* by method [7]. The results show that they are close to the facet set similarities. Thus, *Introduction* section similarity can be treated as topic similarity approximately.

3   TF-Miner

Given a topic set of a domain, the goal of our study is to mine a set of TFs F_i of each topic t_i. Based on Sect. 2, TFs are mined from *Contents* sections and completed by an *LPA* according to topic similarities. Refer to Fig. 2.

Firstly, we preprocess *Contents* sections to get origin facets. Preprocessing includes deleting the fixed clauses such as *"See also"* and extracting the headwords of clauses as origin facets. Inspired of *bag-of-words*, we use a 0–1 matrix F^0 to represent the origin facets for topics of a domain. Rows represent topics and columns represent facets. Secondly, we calculate *Introduction* section similarities among topics in the domain and represent them into a square matrix P. Finally, we propagate facets among topics within a domain. Refer to Algorithm 1.

For matrix F^c, we choose the top $h_i = 1.5 \times a_i$ facets for each topic as the facets in the final facet set. a_i is the number of original facets of topic t_i.

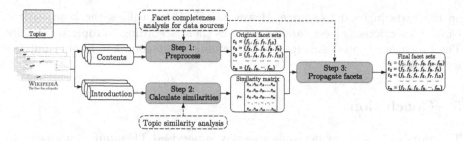

Fig. 2. Architecture of TF-Miner

Algorithm 1. Facet propagation process

   Input: P, F^0
   Output: F^c

1 $l = 0$;
2 **do**
3      $l = l + 1$;   //The 3-4 lines are the core of this algorithm
4      $F^l = P \cdot F^{l-1}$;   //Iterative process of the propagation
5   /* i represents topics, j represents candidate facets */
6   **for** $i = 1 : n$ **do**
7   **for** $j = 1 : m$ **do**
8   **if** $F_{ij}^0 \neq 0$ **then**
9                  $F_{ij}^l = F_{ij}^0$; //Avoid facets disappearing during propagation
10        $F^l[i] = F^l[i]/sum(F^l[i])$; //Normalize matrix F^l by row;
11 **while** $(F^l - F^{l-1})_{max} > 0.001$;
12 $F^c = F^l$;   //F^c is the convergent matrix

---

4 Experiments

We evaluated TF-Miner and compared it with baseline *QDMiner* [9] and *TF-Miner without propagation*. The result is shown in Table 1.

From Table 1 we find that TF-Miner outperforms the baselines in *precision* (p), *recall* (r) and *f1-score* ($f1$). The reason is that *QDMiner* strongly rely

Table 1. Performance of TF-Miner and baselines

Approach	Dataset (#topic)								
	Data Structure (170)			*Data Mining* (528)			*Computer Network* (351)		
	$p(\%)$	$r(\%)$	$f1(\%)$	$p(\%)$	$r(\%)$	$f1(\%)$	$p(\%)$	$r(\%)$	$f1(\%)$
QDMiner [9]	63.84	55.26	59.61	65.12	68.89	67.06	67.45	61.62	64.42
TF-Miner without propagation	82.59	53.64	65.10	81.64	61.53	70.17	75.05	50.02	60.03
TF-Miner	**86.34**	**85.88**	**86.11**	**84.06**	**83.43**	**83.74**	**80.48**	**76.91**	**78.65**

on the texts in its data source. However, TF-Miner utilizes the headwords in *Contents* sections as facets and propagates them according to topic similarities. The propagation process helps TF-Miner mine more complete facets compared with baselines.

5 Conclusion

The work in this paper develops a weakly supervised TF mining approach. We extract origin TFs from *Contents* sections. We use the *Introduction* section similarities of Wikipedia article pages of topics to represent the topic similarities. We add a facet propagation step after get original facet sets from data sources. The experimental results indicate that TF-Miner outperforms the baselines.

Acknowledgment. This work was supported by National Key R&D Program of China (2017YFB1401300, 2017YFB1401302), National Natural Science Foundation of China (61532015, 61532004, 61672419, and 61672418), Innovative Research Group of the National Natural Science Foundation of China (61721002), Innovation Research Team of Ministry of Education (IRT_17R86), Project of China Knowledge Centre for Engineering Science and Technology.

References

1. Benno, S., Tim, G., Dennis, H.: Search result presentation based on faceted clustering. In: Proceedings of the 21st ACM International Conference on Information and Knowledge Management, pp. 1940–1944. ACM (2012)
2. Bifan, W., Jun, L., Qinghua, Z., Wei, Z., Chenchen, W., Wu, B.: Df-Miner: Domain-specific facet mining by leveraging the hyperlink structure of wikipedia. Knowl.-Based Syst. **77**, 80–91 (2015)
3. Bifan, W., Jun, L., Qinghua, Z., Wei, Z., Xiaoyu, F., Boqin, F.: A survey of faceted search. J. Web Eng. **12**, 41–64 (2013)
4. David, C., Haggai, R., Naama, Z.: Enhancing cluster labeling using Wikipedia. In: Proceedings of the 32nd International ACM SIGIR Conference on Research and Development in Information Retrieval, pp. 139–146. ACM (2009)
5. Jeffrey, P., Stelios, P., Panayiotis, T.: Facet discovery for structured web search: a query-log mining approach. In: Proceedings of the 2011 ACM SIGMOD International Conference on Management of Data, pp. 169–180. ACM (2011)
6. Cutting, D.R., Karger, D.R., Pedersen, J.O., Tukey, J.W.: Scatter/gather: a cluster-based approach to browsing large document collections. In: Proceedings of the 15th Annual International ACM SIGIR Conference on Research and Development in Information Retrieval, pp. 318–329. ACM (1992)
7. Tom, K., Maarten, D.R.: Short text similarity with word embeddings. In: Proceedings of the 24th ACM International on Conference on Information and Knowledge Management, pp. 1411–1420. ACM (2015)
8. Liu, W., et al.: Faceted fusion of RDF data. Inf. Fus. **23**, 16–24 (2015)
9. Dou, Z., Jiang, Z., Hu, S., Wen, J.-R., Song, R.: Automatically mining facets for queries from their search results. IEEE Trans. Knowl. Data Eng. **28**(2), 385–397 (2016)

Fast Raft Replication for Transactional Database Systems over Unreliable Networks

Peng Cai[1,2(✉)], Jinwei Guo[1], Huan Zhou[1], Weining Qian[1], and Aoying Zhou[1]

[1] School of Data Science and Engineering, East China Normal University,
Shanghai, China
{pcai,wnqian,ayzhou}@dase.ecnu.edu.cn,
{guojinwei,zhouhuan}@stu.ecnu.edu.cn
[2] Guangxi Key Laboratory of Trusted Software,
Guilin University of Electronic Technology, Guilin, China

Abstract. Raft, a consensus algorithm, has been widely used in many open source database systems to enhance the availability and to guarantee the consistency. However, due to the constraint of coherent log entries, the transactional database systems adopting Raft replication do not perform well in the case of unreliable network environment. This is because with the relatively frequent occurrence of network failures, the serial log replication—which is guaranteed by the log coherency—can block the commit of transactions. In this paper, we propose the fast Raft replication (FRaft) protocol. FRaft adopts the *term coherency* property, which has a good tolerance for the unstable networks. Meanwhile, FRaft can be implemented easily by extending the basic Raft. Our experimental results show that our replication scheme has better throughput.

1 Introduction

Replication based on Paxos is widely used in many database systems for mission-critical applications, such as Google's Megastore [4] and Spanner [5]. However, the basic version of Paxos is well-known for its difficulty of understanding. In order to enhance the understandability and facilitate implementation, some multi-Paxos variants using *strong leadership* and *log coherency* features—which ensure that a log entry must be consistent with the leader's and there are no holes in the log—are proposed. Raft [6] is the most typical one of them and it has been used in many open source database systems [1–3].

In the Raft replication systems, the leader node synchronizes log entries to all followers. This follows the *strong leadership* property. Due to *log coherency*, logs are not allowed to have any hole on both leader and followers. In other words, log entries are acknowledged by a follower, committed by the leader and applied to all replicas in a serial order. Obviously, the serial order is suitable for the transactional database systems. This needs to guarantee the serializability for concurrent transactions. However, the synchronization of the serial order between the leader and followers increases the waiting time of a transaction in the commit phase, especially in the unreliable networks (e.g., package loss and increased network delay). There is no doubt that the increase of transactions' blocking time has a significant impact on the system performance.

In this work, we present the Fast Raft Replication (FRaft) protocol, which is designed for the transactional database systems over the unreliable networks. FRaft adopts the *term coherency* property, which allows holes to exist in a log without loss of correctness for transaction processing. Owing to *term coherency*, the log entry can be acknowledged by a follower out of the serial order. The out-of-order feature has a good tolerance for unstable wide-area networks. Therefore, FRaft can perform well in the unreliable network environment.

2 Fast Raft Replication Using Term Coherency

We have known that the *log coherency* has a strong constraint, which may have a significant impact on the performance of transactional database systems. Therefore, to accelerate the log replication, we have to relax the constraint of log coherency. Therefore, we present a new property, *term coherency*, which has some new rules as follows: (1) for the log in each replica server, the log entries are coherent except those in the last term; (2) for the log of each replica server, the log entries of the last term must contain the first one of the term.

As illustrated in Fig. 1, for each replica node, its local log contains consecutive entries except those in the last term. The log holes exist in the logs of Nodes 2 and 3, e.g., Node 2 misses the log entry with the index 9. However, all logs contain the log entry with the index 7, which is the first one of the current term. In this example, the `commitIndex` is 11 in the leader node. This is because all log entries whose indexes ≤ 11 are persisted in a majority of replicas.

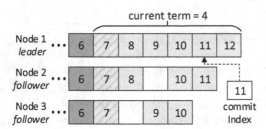

Fig. 1. An example of 3-way replication's logs supporting term coherency.

Parameters:	*term*	current term of the leader
	leaderId	server index of the leader
	prevTermLogId	log index of the last entry of the last term in the leader
	prevTerm	last term in the leader
	leaderCommit	*commitIndex* of the leader
Results:	*term*	current term
	logIndex	log index of the first one of missing entries

Fig. 2. An illustration of parameters and results of `FastAppendEntries`.

We now introduce the new RPC `FastAppendEntries`, which adopts the *term coherency* property. The parameters and results of `FastAppendEntries` are illustrated in Fig. 2, which are different from the original Raft. To keep track of the `commitIndex`, the leader in our protocol needs to know the responses of followers for each log entry in his term. Therefore, we replace the array `matchIndex[]`

with a two-dimensional array. In our protocol, we classify log sending threads into two categories: (1) there is one thread for each follower to replicate missing entries and to send heartbeats; (2) when the leader generates a new log entry for a write request from a client, it starts a new thread to send the log entry to followers by invoking `FastAppendEntries` asynchronously.

Recall from Fig. 2 that some parameters are the same as those in Raft. Our method treats these parameters according to the original `AppendEntries`. Next, we describe the processing of our new RPC in the leader and followers.

Processing in Followers: When a follower f receives a `FastAppendEntries` containing the entry e from the leader, it takes actions according to the parameters. We use log to denote the local log of the follower and let lst represent the index of the last log entry of the follower. There are four cases as follows:

Case 1: $e.term = log[lst].term$. If the local log entry $log[e.index]$ is empty, the follower will put the received entry into the specified position and flush it into local disk. Then, the follower locates the first hole in the last term and sets the result $logIndex$ to the index of the hole. Lastly, the follower returns the results including $logIndex$ and local $currentTerm$.

Case 2: $e.term > log[lst].term$ and $pervTerm = log[e.index - 1].term$. This shows that the follower receives the first log entry of a new term. If $log[e.index - 1].term < log[lst].term$, the follower will truncate the local log after $e.index - 1$ and append the received entry into the local log; otherwise, the follower will drop the log after $e.index - 1$ directly. Then, the follower sets the results in the same manner described in **Case 1** and returns.

Case 3: $e.term > log[lst].term$ and $prevTerm \neq log[e.index - 1].term$. Since the follower misses log entries in the term range $(log[e.index - 1], e.term)$ or it has invalid entries, it does not confirm the correctness of local log. Therefore, it needs to send back a failure message (i.e., $logIndex = -1$).

Case 4: $e.term < log[lst].term$. This means that the follower receives an expired message due to the unreliable networks. Therefore, the follower can ignore the request directly and does not response.

It should be noted that the variable lst is required to be updated when the follower receives a newer log entry in all above cases.

Processing in Leader: When the leader receives the response from the follower, it takes actions according to different cases as follows:

Case 1: $logIndex > start$. This indicates that the follower's log is complete before the current term of the leader. The leader sets the *next index* of the follower to $logIndex$. If $logIndex$ is not greater than the index of the last entry in the leader, the leader sends the specified entry to the follower.

Case 2: $0 < logIndex \leq start$. This response means that the follower is recovering its local log and returns the index of a missing entry. If $logIndex$ is greater than the value of corresponding entry in `nextIndex[]`, the leader will update the value to $logIndex$ and send the corresponding entry to the follower.

Case 3: $logIndex = -1$. This is a failure result. Note that the leader encounters this case only when it sends the first entry of a term t. Therefore, the leader needs to find the first log entry from previous term $(t-1)$ in the local log and sends it to the follower by `FastAppendEntries`.

3   Evaluation

We run the experiments on a cluster of 3 machines, each machine is equipped with a 2-socket Intel Xeon E5606, 96 GB RAM and 100 GB SSD. All machines are connected by a gigabit Ethernet switch. We use the 3 machines to deploy a 3-way replication system, i.e., one for the master node (leader) and two for slave nodes (followers). We compare our method FRAFT with other replication schemes RAFT and SYNC. RAFT is the implementation of the basic Raft, and SYNC is the traditional log replication method using *eager* mechanism.

Benchmark: We adopt YCSB to evaluate our implementation. By default, a transaction contains five operations, each one accesses a record uniformly. To generate enough log entries, we modify the workload to have a read/write ratio of 0/100. The size of each write operation is about 100 bytes.

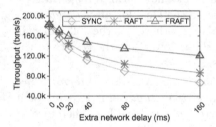

Fig. 3. Impact of network delay.

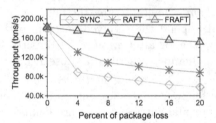

Fig. 4. Impact of package loss.

Results: We investigate the impact of network delay and package loss on different replication methods. With the growing of extra network delay, the throughput of SYNC declines fastest, which is illustrated in Fig. 3. The reason is that the master server can commit a transaction only when it confirms that the corresponding entry is persisted in all replicas. Due to the serial log replication, RAFT does not perform well when the network delay becomes large. Once the package loss encounters in the network, the performance of SYNC and RAFT drops rapidly, which is shown in Fig. 4. This is because when a package from the leader is lost, the leader must wait for the package timeout before sending a new package. Owing to the term coherency property, the leader can send the latest log entry regardless of whether it receives the response of the previous invocation of the function `FastAppendEntries`. Therefore, FRAFT performs much better than other methods in the unreliable network environment.

4   Conclusion

In this work, we introduced an efficient variant of Raft, which is called FRaft. FRaft relaxes the constraint of log coherency and allows holes to exist in log. Therefore, it is suitable for transactional database systems over the unreliable networks. Our experimental results demonstrate the effectiveness of our method.

Acknowledgments. This research is supported in part by National Key R&D Program of China (2018YFB1003402), National Science Foundation of China under grant number 61432006, and Guangxi Key Laboratory of Trusted Software (kx201602).

References

1. CockroachDB website. https://www.cockroachlabs.com/
2. etcd website. https://coreos.com/etcd/
3. TiDB website. https://github.com/pingcap/tidb
4. Baker, J., Bond, C., Corbett, J.C., Furman, J.J., et al.: Megastore: providing scalable, highly available storage for interactive services. In: CIDR, pp. 223–234 (2011)
5. Corbett, J.C., Dean, J., Epstein, M., Fikes, A., et al.: Spanner: Google's globally-distributed database. In: OSDI, pp. 261–264. USENIX Association (2012)
6. Ongaro, D., Ousterhout, J.K.: In search of an understandable consensus algorithm. In: ATC, pp. 305–319. USENIX Association (2014)

Parallelizing Big De Bruijn Graph Traversal for Genome Assembly on GPU Clusters

Shuang Qiu(✉), Zonghao Feng, and Qiong Luo

The Hong Kong University of Science and Technology,
Clear Water Bay, Hong Kong
{sqiuac,zfengah,luo}@cse.ust.hk

Abstract. De Bruijn graph traversal is a critical step in de novo assemblers. It uses the graph structure to analyze genome sequences and is both memory space intensive and time consuming. To improve the efficiency, we develop ParaGraph, which parallelizes De Bruijn graph traversal on a cluster of GPU-equipped computer nodes. With effective vertex partitioning and fine-grained parallel algorithms, ParaGraph utilizes all cores of each CPU and GPU, all CPUs and GPUs in a computer node, and all computer nodes of a cluster. Our results show that ParaGraph is able to traverse billion-node graphs within three minutes on a cluster of six GPU-equipped computer nodes. It is an order of magnitude faster than the state-of-the-art shared memory based assemblers, and more than five times faster than the current distributed assemblers.

1 Introduction

In genomic analysis pipelines, de novo assembly constructs genome sequences from short DNA fragments, without any reference sequence. Specifically, a De novo assembler first constructs a De Bruijn graph from short DNA fragments. Then it traverses the graph heuristically, searching local shortest paths. Thereafter, it outputs long DNA sequences (called contigs), representing the skeleton of the genome sequence. The performance issues in constructing big De Bruijn graphs are well addressed in recent work [2,7], but traversing big graphs efficiently with a limited number of machines remains an open problem.

Traversing a De Bruijn graph with hundreds of millions to billions of vertices takes hours on a single CPU core, and the memory consumption is tens to hundreds of gigabytes [4]. Parallelization is therefore commonly adopted in existing assemblers to speed up the traversal. In shared memory based parallel assemblers, the performance is limited by the memory size in a single machine [4] or the IO bandwidth for the data transfer [2]. In comparison, scalable distributed assemblers are able to handle big graphs [5,8]. However, the De Bruijn graph partitioning in these assemblers is based on a random strategy, ignoring connections among vertices. Consequently, the communication overhead is high in distributed assemblers. Additionally, all these assemblers are based on the CPU, whereas the GPUs, commonly equipped in the computer nodes, are not utilized.

© Springer Nature Switzerland AG 2019
G. Li et al. (Eds.): DASFAA 2019, LNCS 11448, pp. 466–470, 2019.
https://doi.org/10.1007/978-3-030-18590-9_68

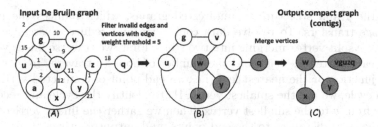

Fig. 1. De Bruijn graph traversal

The major difficulty that hinders the utilization of GPUs in De Bruijn graph traversal is the high divergence and read-write contention among GPU threads. To address this problem, we design algorithms in a vertex-centric manner, and split the vertex merging step into vertex traversing and result gathering steps. This way, graph traversal algorithms are converted into gather and scatter operations, and the costly write contention on many-core processors is resolved. Furthermore, we find that the identifiers of adjacent vertices share a common minimum substring with a high probability, and that this feature can be utilized to reduce the number of edge cuts between subgraphs. Therefore, we utilize a vertex partitioning algorithm [3,7] to reduce the communication overhead in ParaGraph.

2   ParaGraph

We design the workflow of ParaGraph as shown in Fig. 2. ParaGraph takes De Bruijn subgraphs as input, which can be generated from a graph construction tool [7]. In Step 1, we first filter invalid edges and their incident vertices, based on edge weights (as illustrated in Fig. 1(A) and (B)). Since a vertex in the graph is identified with two fixed-length strings, searching a vertex can be implemented as hash table lookup or binary search on sorted vertices. On multiple processors and computer nodes, our graph algorithms are designed with multiple iterations of neighbor updates for vertices. Therefore, we build a new index for vertices and update the index of neighbors of vertices. The efficiency in searching vertices is improved in graph traversal, and the index building overhead is offset by the performance gain. We finally split vertices into the set of linear vertices (vertices of at most two adjacent vertices) and the set of junctions (non-linear vertices).

Based on the results in Step 1, Step 2 traverses the De Bruijn graph and merges linear vertices (as illustrated in Fig. 1(B) and (C)). Specifically, we first

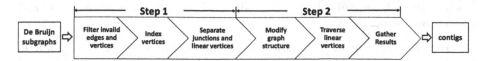

Fig. 2. Processing flow in ParaGraph

modify the graph structure to undirected graphs, which reduces the number of message transfers. To resolve data contention and reduce divergence across threads, we split vertex merging into two steps. First, we traverse linear vertices, and for each vertex v, we record the number of hops to traverse from v to the nearest junction or the nearest leaf (i.e., an end point of connected vertices). If v is in a cycle, we use the smallest vertex ID to identify the cycle and record the distance from v to the smallest vertex. Then we gather the linear vertices based on the recorded distances to the end points, and output contigs.

We design distributed graph processing algorithms, following the push based BSP (Bulk Synchronous Parallel) model [1], in which active vertices send messages to their neighbors, and the neighbors update the associated values based on the received messages. Specifically, we break down a graph algorithm as iterations of compute, communicate, and update operations, and define the compute operators and update operators on vertices. With this abstraction, traversing linear vertices takes $O(logD)$ number of iterations of compute, communicate and update operations, where D is the diameter of the graph.

In distributed graph algorithms, the graph partitioning method is critical in determining the number of messages across computers. We find that utilizing character distribution in vertex identifiers (two fixed-length strings) can reduce the number of edge cuts in partitioning the De Brujin graph. As such, we adopt the P-minimum-substring partitioning [3,7] and prove the following property with uniform and independent distribution assumptions on the start positions of P-minimum-substrings.

With the P-minimum-substring partitioning, the probability that a vertex v and its neighbor u are distributed to the same partition equals $\left(\frac{K-P-1}{K-P}\right)^2$, where K is the string length of the vertex and P is the P-minimum-substring length.

When $(K - P - 1)$ is close to $(K - P)$, the majority of adjacent vertices are partitioned into the same subgraph. For example, given $K = 27, P = 11$, this property indicates that more than 85 percent adjacent vertices are located in the same partition.

3   Performance Evaluation

We use two datasets in the evaluation. The dataset *Human Chr14* [4] contains 450 million valid vertices in the De Brujin graph. The dataset *7 Humans* [6], consisting of seven individual human genomes, contains 2.3 billion valid vertices. Experiments were conducted with one to six computer nodes. Each computer node contains two 2.3 GHz Intel Xeon E5-2670 12-core CPUs, and two Nvidia K80 GPUs. Each K80 consists of two GPUs, each with 12 GB memory. The main memory on each computer node is 128 GB. Each computer node uses Infiniband for network connection.

Overall Performance. We show the overall performance of ParaGraph in Fig. 3. On *Human Chr14*, the overall time of ParaGraph with all CPUs and GPUs is reduced to less than 1/3 of the time with only the CPUs. Moreover, the running time is reduced to 1/8 of the time with CPUs on a single machine, when ParaGraph runs on six computer nodes. Due to the limit of memory size on a single machine, we run ParaGraph on *7 Humans* on six computer nodes. The running time with both CPUs and GPUs is reduced to 1/3 of the time using only CPUs.

Comparison with Other Assemblers. We compare ParaGraph with the state-of-the-art parallel assemblers. All these CPU-based assemblers use all CPUs on each computer in experiments. Time measurement for each assembler begins at the time the input data is ready in memory and ends at the time the output results are generated in memory. As shown in Table 1, on *Human Chr14*, ParaGraph with CPUs and GPUs is 20 times faster than SOAPdenovo [4] and bcalm2 [2] on a single machine. It is about eight times faster than SWAP2 [5], and six times faster than PPA [8]. On *7 Humans*, only Bcalm2 and Para-Graph are able to run with the available amount of memory. ParaGraph on six computer nodes is 40 times faster than bcalm2 on a single computer.

Table 1. Running time (sec) comparison

Dataset	Human Chr14		7 Humans	
Software	Single[1]	Multi[2]	Single[1]	Multi[2]
SOAPdenovo	582	OM	OM	OM
bcalm2	485	OM	**2341**	OM
PPA	OM	59	OM	OM
SWAP2	209	68	OM	OM
ParaGraph-CPU	71	22	OM	144
ParaGraph-CPU-GPU	**24**	**9**	OM	**52**

1 Single computer node, 2 Six computer nodes
OM: Out of memory
SOAPdenovo: from tip removing to edge construction. bcalm2: graph compaction within and across buckets, with IO time excluded. SWAP2: graph simplification. PPA: listranking and contig merging, with load and dump time excluded.

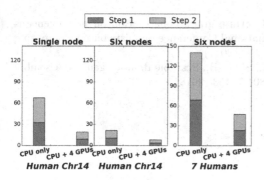

Fig. 3. Overall running time (sec) of Para-Graph

4   Conclusion

We propose ParaGraph to parallelize the De Bruijn graph traversal on GPU-equipped clusters. We implement multi-threaded algorithms on each GPU and CPU, use threads to manage message transfers and synchronizations among CPUs and GPUs in a computer node, and finally run concurrent processes on multiple computer nodes. To reduce the overhead in distributed graph traversal, we utilize the identifier distribution features in vertices, such that the majority of messages are within each processor. As a result, ParaGraph is efficient on multiple processors and multiple computer nodes. Source code of ParaGraph is available at https://github.com/ShuangQiuac/UNIPAR, integrated with our previous work ParaHash [7] to execute the entire workflow of De Bruijn graph construction and traversal.

References

1. Avery, C.: Giraph: large-scale graph processing infrastructure on Hadoop. In: Proceedings of the Hadoop Summit. Santa Clara, vol. 11, pp. 5–9 (2011)
2. Chikhi, R., Limasset, A., Medvedev, P.: Compacting de bruijn graphs from sequencing data quickly and in low memory. Bioinformatics **32**(12), i201–i208 (2016)
3. Li, Y., Kamousi, P., Han, F., Yang, S., Yan, X., Suri, S.: Memory efficient minimum substring partitioning. In: Proceedings of the VLDB Endowment, vol. 6, pp. 169–180. VLDB Endowment (2013)
4. Luo, R., et al.: Soapdenovo2: an empirically improved memory-efficient short-read de novo assembler. Gigascience **1**(1), 18 (2012)
5. Meng, J., Seo, S., Balaji, P., Wei, Y., Wang, B., Feng, S.: Swap-assembler 2: optimization of de novo genome assembler at extreme scale. In: 2016 45th International Conference on Parallel Processing (ICPP), pp. 195–204. IEEE (2016)
6. Minkin, I., Pham, S., Medvedev, P.: Twopaco: an efficient algorithm to build the compacted de bruijn graph from many complete genomes. Bioinformatics **33**(24), 4024–4032 (2016)
7. Qiu, S., Luo, Q.: Parallelizing big de bruijn graph construction on heterogeneous processors. In: 2017 IEEE 37th International Conference on Distributed Computing Systems (ICDCS), pp. 1431–1441. IEEE (2017)
8. Yan, D., Chen, H., Cheng, J., Cai, Z., Shao, B.: Scalable de novo genome assembly using pregel. arXiv preprint arXiv:1801.04453 (2018)

GScan: Exploiting Sequential Scans for Subgraph Matching

Zhiwei Zhang[1]($\boxtimes$), Hao Wei[2], Jianliang Xu[1], and Byron Choi[1]

[1] Hong Kong Baptist University, Hong Kong, China
{cszwzhang,xujl,bchoi}@comp.hkbu.edu.hk
[2] Amazon.com, Seattle, WA, USA
wehao@amazon.com

Abstract. Subgraph matching is to enumerate all the subgraphs of a graph that is isomorphic to the query graph. It is a critical component of many applications such as clustering coefficient computation and trend evolution. As the real-world graph grows explosively, we have massive graphs that are much larger than the memory size of the modern machines. Therefore, in this paper, we study the subgraph matching problem where the graph is stored on disk. Different from the existing approaches, we design a block-based approach, GScan, which investigates the schedule of the blocks transferred between the memory and the disk. To achieve high I/O efficiency, GScan only uses sequential I/O read operations. We conduct experimental studies to demonstrate the efficiency of our block-based approach.

Keywords: Graph matching · I/O efficient · Sequential access

1 Introduction

Given a query graph q and a data graph G, the subgraph matching query aims to find all the subgraphs of G that are isomorphic to q. Many applications are based on such queries, such as clustering coefficient computation [2,3]. Nowadays, the size of real graphs grows explosively and sometimes, they cannot entirely reside in the main memory. To deal with this, Kim et al. [1] propose the DualSim framework. It first computes a matching order for the nodes in the query graph. Then, the disk blocks are loaded according to the order. However, it can incur random disk accesses since the disk positions of the blocks to be loaded are generally not consistent with the generated matching order.

To solve the drawbacks of the existing approaches, we propose the GScan approach. We propose multiple blocks as a block combination and generate all the matching results in these blocks. A characteristics is that, only sequential I/O read cost will be employed. Also, we propose an ordering among the blocks, which aim to process as many subgraph matching candidates as possible during a single sequential scan.

G. Li et al. (Eds.): DASFAA 2019, LNCS 11448, pp. 471–475, 2019.
https://doi.org/10.1007/978-3-030-18590-9_69

2   Preliminaries

Graph Notation. Given an undirected labeled graph as $G(V, E)$, we use V and E to denote the set of nodes and edges of G, respectively. Every node u is associated with a set of labels, denoted as $\mathsf{label}(u)$. The subgraph isomorphism is defined below.

Definition 1 [Subgraph Isomorphism]. *Given a graph $q(V_q, E_q)$ and a graph $G(V, E)$, q is isomorphic to G if and only if there exist an injective mapping function f from V_q to V such that $\forall u \in V_q$, $\mathsf{label}(u) = \mathsf{label}(f(u))$ and $\forall(u, v) \in E_q$, $(f(u), f(v)) \in E(G)$.*

Each injective mapping function f is a matching (embedding) of q in G and can be represented as a set of node pairs (v_q, v_G) in which $v_q \in V_q$ is mapped to $v_G \in V(G)$. We maintain a graph G in the disk blocks $b_1, \cdots, b_g$. For a block b, its block number is denoted as $\mathsf{BN}(b)$. The block-graph of G, denoted as $G_B(V_B, E_B)$, is the undirected graph in which each node $u \in V_B$ represents a block $b_u \in \{b_1, b_2 \cdots, b_g\}$ and the edge $(u, v) \in E_B$ iff. there exists an edge $(u', v') \in G$ that $u' \in b_u$ and $v' \in b_v$. A *sequential* scan of the graph is to access the blocks in the increasing order of block numbers.

Problem Statement: Given an undirected graph G on disk and a query graph q with limited size of M memory, we study the I/O-efficient matching problem to enumerate all the subgraphs of G that are isomorphic to q with only sequential read I/O cost.

3   Our Approach

We propose the approach called GScan for the I/O-efficient subgraph matching problem. The matching procedure of GScan consists of two steps. The first is to generate all block combinations, such that the subgraph in these blocks is possible to contain the matchings of the query. The second step is to load the block into/out of the memory in an order and find the subgraph matchings in the memory. In this short paper, we focus on the second step, and study the block replacing algorithm, so that in a sequential scan, more block combinations can be checked. For ease of illustration, given a set of blocks S, the sequential access path of S, denoted as $\mathsf{SPath}(S)$, is the order to sequentially access all the blocks in S. For example, given $S = \{b_3, b_1\}$, $\mathsf{SPath}(S) = \{b_1, b_3\}$.

3.1   The Basic Approach

For each SPath sp, we should load in the blocks of sp to find the matchings. Since there can be many SPaths to be processed, we define the SPath -*sequential* relation between two SPaths. Given two SPaths sp and sp', sp' is SPath-sequential with sp (denoted as $sp' \rightarrow_S sp$), if for each block b appearing only in sp', $\mathsf{BN}(b)$ is larger than $\mathsf{BN}(b_i)$ for every block $b_i \in sp$. Otherwise, it is denoted as $sp' \nrightarrow_S sp$.

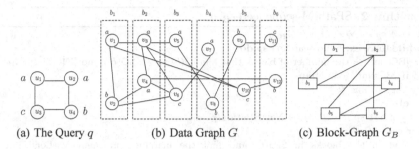

(a) The Query q (b) Data Graph G (c) Block-Graph G_B

Fig. 1. Graph examples

Algorithm 1. GScan-Basic(q, G, G_B)

1: $L \leftarrow \emptyset$, $sp \leftarrow$ NextSPath(G_B, q);
2: **while** sp is SPath-sequential to all SPaths in L **do**
3:     insert sp at the end of L, $sp \leftarrow$ NextSPath(G_B, q);
4: **for** each SPath$_i$ in L **do**
5:     load in the blocks in SPath$_i$ and find the matching in memory. Push out the block with the minimum block number if the memory is full;
6: go to line 2 if there is unchecked SPaths;

If $sp' \rightarrow_S sp$, they can be processed in a single sequential scan. Based on these, we design the algorithm for block transferring.

The basic algorithm, denoted as GScan-Basic, is shown in Algorithm 1. Initially, L is set as $\emptyset$ (line 1). Then all SPaths needed to be checked for subgraph isomorphism are generated by the procedure *NextSPath*. The *NextSPath* function finds the SPath, containing the candidates for subgraph matching. As many approaches have studied the candidate generation procedure, we do not show its details here. If sp is SPath-sequentially with all the SPaths in L, it is inserted at the end of L (lines 2–3). Otherwise, all the SPaths in L is processed in a sequential scan, and the block with the minimum block number will be loaded out when the memory is full (line 4–5). It needs to restart the sequential scan to process the unchecked SPaths if there is any (line 6).

Example. Consider the graph in Fig. 1. Assume we find two matching candidates of q generates the SPaths as $sp = $ "b_1, b_2, b_3" and $sp' = $"$b_2, b_3, b_4$". Then it has $sp' \rightarrow_S sp$. Thus, they can be processed in a sequential scan. Specifically, when b_4 needs to be loaded in, b_1 will be pushed out if the memory is full.

3.2 Advanced Block Scheduling

The advanced block scheduling algorithm addresses the problem with the memory size $M > |V_q|$. Firstly, we propose SPath-M-sequential relation. Given the SPath sequence $L = sp_1, sp_2 \cdots sp_q$, the SPaths in L are SPath-M-sequential if there exists a list of SPaths, $S_1, S_2 \cdots S_q$ such that $S_{i+1} \rightarrow_S S_i$ with $1 \leq i < q$ in which $|S_i| \leq M$, and sp_i is contained in S_i ($1 \leq i \leq q$).

Algorithm 2. SPath-M-sequence(G_B, q, sp)

1: $L \leftarrow \emptyset, sp \leftarrow NextSPath(G_B, q)$;
2: **while** there exists un-searched sp **do**
3:    |BCount| $\leftarrow$ the number of blocks that MinSO(b) $\leq$ MaxSO(b') and BN(b') < BN(b).
4:    **if** |BCount| > M **then**
5:       go to line 8;
6:    **else**
7:       insert sp into L, $sp \leftarrow$ NextSPath(G_B, q);
8: **for** each SPath$_i$ in L **do**
9:    load in the blocks in SPath$_i$ and find the matching in memory. Load out the block with the minimum MinSO value if the memory is full;
10: go to Line 2 if here is unchecked SPaths;

The idea behind for the SPath-M-sequential is that, we consider all blocks in the memory as a SPath and denote it as S_i. Assume all the blocks of sp_i exist in S_i. Thus, we only need to consider whether $S_{i+1} \rightarrow_S S_i$. Given the SPath sequence $sp_1, sp_2, \cdots sp_q$, assume that the block b first appears in sp_i. Thus, for S_j with $j < i$, if the maximum block number in S_j is larger than BN(b), S_j also contains b. On the other hand, if sp_i is the last SPath that contains the block b, b can be loaded out from the memory after sp_i has been processed. Therefore, for a block b, whether it should be kept in the memory is determined by the first and the last SPath containing it. We use MinSO(b) (MaxSO(b)) to denote the number of the first (last) SPath containing b in the SPath sequence. We have the following theorem.

Theorem 1 *Given the* SPaths *sequence* $L = $ "$sp_1, sp_2 \cdots sp_q$", *the sequence* L *is* SPath-M-*sequential if and only if for each block* b, *there exist less than* M *blocks* b_j *satisfying that* MaxSO(b_j) $\geq$ MinSO(b) *and* BN(b_j) < BN(b).

Based on Theorem 1, we design Algorithm 2. Initially L is set as $\emptyset$ and all SPaths are generated by the procedure *NextSPath* (line 1). Then all the blocks satisfying the conditions MinSO(b) $\leq$ MaxSO(b') and BN(b') < BN(b) are founded, and the number is BCount (lines 2–3). If BCount > M, then sp is not SPath-M-sequential with L (line 4). Otherwise, sp is inserted into L and the next candidate is generated (line 7); After finding all SPath-M-sequential SPaths, all the SPaths in L are processed in a sequential scan, and the block with the minimum MaxSO will be loaded out when the memory is full (line 8–9). The procedure carries on until all SPaths have been processed (line 6).

Example. Consider the graphs in Fig. 1 with SPaths as $f_1 = $ "b_1, b_2, b_3", $f_2 = $ "b_2, b_3, b_4", $f_3 = $ "b_2, b_5", $f_4 = $ "b_1, b_2, b_5, b_6". The memory $M = 4$. It can be found that BCount is 3. Thus, these SPaths can be processed in one sequential scan. Specifically, the blocks b_1, b_2, b_3, b_4 are loaded into the memory and f_1, f_2 can be processed. When processing f_3, b_5 is loaded in, and b_3, which has the minimum MaxSO value, is loaded out. After f_3 is checked, b_6 is loaded in to replace b_4, and process f_4.

4   Performance Evaluation

In this section, we present the performance of GScan-Basic and GScan-Advanced. All the algorithms are implemented using C++ and tested on a Mac Mini with 2.8 GHz Dual-Core Intel Core i5 CPU and 8 GB DDR3 RAM. We make a small adaption for the DaulSim approach in [1] as it is designed for unlabeled graphs, and denoted it as Dual-L. We test on the real massive datasets DBPedia, which has 4.56 million nodes and 30.97 million edges. The results on DBPedia are presented in Fig. 2. It can be found that, GScan-Advanced has the highest sequential I/O percentage and minimum time cost compared with Dual-L and GScan-Basic.

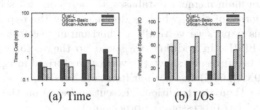

Fig. 2. Results of DBPedia

5   Conclusion

In this paper, we study the subgraph matching query with the graph stored on the disk. Different from the existing node-based approaches, we propose the GScan-Basic and GScan-Advanced approaches block scheduling algorithms for processing as many matching candidates as possible in every sequential scan. The experiments confirm the superiority of our GScan approach over the state-of-the-art approach.

Acknowledgement. This work was supported by the grants from NSFC 61602395, RGC 12201518, RGC 12232716, RGC 12258116, RGC 12200817, and RGC 12201615.

References

1. Kim, H., et al.: Dualsim: parallel subgraph enumeration in a massive graph on a single machine. In: Proceedings of SIGMOD (2016)
2. Wang, J., Cheng, J.: Truss decomposition in massive networks. Proc. VLDB Endow. **5**(9), 812 (2012)
3. Watts, D.J., Strogatz, S.: Collective dynamics of 'small-world' networks. Nature **393**, 440–442 (1998)

SIMD Accelerates the Probe Phase
of Star Joins in Main Memory Databases

Zhuhe Fang$^{(\boxtimes)}$, Zeyu He, Jiajia Chu, and Chuliang Weng

School of Data Science and Engineering,
East China Normal University, Shanghai, China
{zhfang,zyhe,chujiajia}@stu.ecnu.edu.cn, clweng@dase.ecnu.edu.cn

Abstract. In main memory databases, the joins on star schema tables cost the majority of time, which is dominated by the expensive probe phase. In this paper, we vertically or horizontally vectorize the probe phase using SIMD. In addition, we speed up the vectorized probe by prefetching. As our results show, the vertical vectorized integrated probe is up to 2.19X (2.63X) faster than its scalar version, as well as 3.24X (2.74X) faster than the traditional execution based on the right-deep-tree plans on CPU processors (co-processors).

Keywords: Star joins · Probe · SIMD · Prefetching

1 Introduction

In OLAP (On-Line Analytical Processing) applications, the relationship among tables is often modeled as a *star schema* [2], which is composed of a *fact table* and several *dimension tables*. Queries over such a schema obey a typical pattern: Filter predicates on the dimension tables select qualified tuples, then these tuples join with the fact table, followed by grouped aggregation and sorting operations. These queries are called *star join* queries [2]. A typical example of a star join query is listed below, where `lineorder` is the fact table, while others are dimension tables. It is modified from the queries of the Star Schema Benchmark (SSB) [4], which follows the design of the star schema.

SELECT count(*) FROM customer, supplier, part, date, lineorder WHERE lo_partkey = p_partkey AND lo_custkey = c_custkey AND lo_suppkey = s_suppkey AND lo_orderdate = d_datekey AND s_region = ? AND p_brand <?;

Traditionally, a star join query is translated into *a right-deep-tree plan (RP)* [2,7] to execute, where the left children are operators on dimension tables and the right child is the operations on the fact table. An example of right-deep-tree plans is shown in Fig. 1. Processing such a star join involves two phases. The first phase is called a *build* phase, during which the left dimension tables build hash tables. The second phase is called a *probe* phase, during which each tuple

© Springer Nature Switzerland AG 2019
G. Li et al. (Eds.): DASFAA 2019, LNCS 11448, pp. 476–480, 2019.
https://doi.org/10.1007/978-3-030-18590-9_70

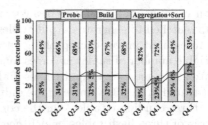

Fig. 1. A right-deep-tree plan

Fig. 2. The normalized execution time of SSB queries

from the fact table probes the hash tables. Finally, the joined tuples do grouped aggregation and sort operations in the whole query evaluation. In particular, the probe phase is extremely expensive. We verify this problem by evaluating star join queries of SSB (scale factor = 100) on Spark. Here we select the queries with at least two joins to execute. As illustrated in Fig. 2, the probe phase costs 53% to 82% of the overall execution time. Particularly, it can be up to almost 100% of the whole join time (including the build and the probe phases). This means the star join is dominated by the probe phase, while the build phase is negligible in most cases.

To reduce the response time of star joins, a part of previous work [2] uses indexing to optimize data access and prune out non-qualifying tuples from the fact table to do join. Another part of work directly avoids doing joins through precomputation [3], multidimensional cubes [5] or denormalization [8]. These previous work does not optimize the process of the expensive probe phase to fully use today's (co-)processors. On the other hand, modern (co-)processors provide higher data parallelism based on SIMD (Single Instruction Multiple Data). It is widely applied to optimize database operators, such as the sorting, scan, aggregation and join [6]. But it has never been utilized to accelerate the probe phase of star joins.

In this paper, we investigate how to efficiently apply SIMD to vectorize the probe phase of star joins. However, the traditional execution of the probe phase is hard to benefit from SIMD. Specifically, in the right-deep-tree plan, multiple probes are serially linked and executed in a pipelined manner. Each probe consumes tuples from its child and produces new ones to its parent. This way generates numerous useless intermediate results and causes function calling overhead, which cannot be alleviated by SIMD. Fortunately, this issue has been solved in previous work [1] by integrating a series of single probes into one operator called an *integrated probe (IP)* here, where a probing tuple has several join keys to look up several hash tables. So we convert the probe phase of star joins into the integrated probe and apply SIMD to accelerate the integrated probe.

2   Vectorizing the Integrated Probe

The integrated probe of a star join is more complex than a single probe of a two-table join. In the latter, each probing tuple only has one join key to compare

with variable number of keys in a corresponding hash bucket. This probing tuple
gets matched to generate output tuples if at least one comparison is true. But in
an integrated probe, each probing tuple has more than one join keys to match
variable number of tuples in corresponding hash buckets. The probing tuple gets
matched to generate output tuples if all of its keys get matched. For example, as
illustrated in Fig. 3, a single probe only has one key to match a hash bucket that
has three candidate tuples. On the contrary, Fig. 4 shows an integrated probe in
terms of three hash tables, which need to be looked up by three join keys of a
probing tuple. The qualified hash buckets have 2, 3, 1 tuples, respectively, where
corresponding keys are colored in tuples. Analogous to vectorizing the single
probe of a two-table join [6], the integrated probe can also be vectorized vertically
or horizontally. Here we assume a vector has four lanes to pack four keys.

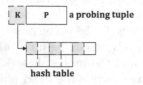

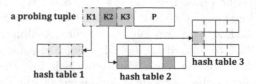

Fig. 3. One probing key in a single
probe

Fig. 4. Multiple probing keys in a
integrated probe

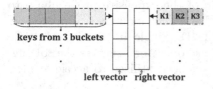

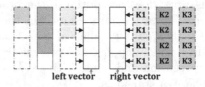

Fig. 5. Vertical vectorized integrated probe

Fig. 6. Horizontal vectorized integrated probe

First, we vectorize the integrated probe vertically, called a *Vertical Vectorized Integrated Probe (VVIP)*. It processes a different probing tuple per vector
lane. As illustrated in Fig. 5, each lane of the right vector is responsible for a
probing tuple, and each left lane serially loads keys from corresponding several
hash buckets, instead of only one in the single probe. If a probing join key gets
matched, a new probing join key and corresponding keys from a hash bucket
are loaded in the two lanes, respectively. If all join keys of a probing tuple are
matched, the payloads of the probing tuple and qualified left tuples generate an
output tuple, meaning this probing tuple is successfully done. Otherwise, if a
mismatch occurs, this probing tuple is aborted. On the done or aborted occa-
sion, another probing tuple will be arranged to process in the vector lane. By this
way, four lanes of a vector allow four probing tuples to be processed in parallel.

For example, the first lanes of the two vectors in Fig. 5 demonstrate the case in Fig. 4, where the three probing join keys need to serially compare 2, 3 and 1 keys from three corresponding hash buckets, respectively. If the three probing keys get matched, then an output tuple is generated; otherwise, this probing tuple is aborted. Later another probing tuple will be arranged to this right vector lane.

Second, we vectorize the integrated probe according to the horizontal vectorization for a single probe [6]. It is termed as a *Horizontal Vectorized Integrated Probe (HVIP)*. Rather than a vector lane, the whole vector are loaded multiple candidate join keys from a hash bucket to compare with a vector with identical join keys from a probing tuple. Figure 6 demonstrates how to horizontally vectorize the case in Fig. 4. Three probing join keys serially occupy the right vector, and candidate keys are accordingly loaded in the left vector. Note that the candidate keys from a hash bucket may exceed the vector size, so they need to be loaded into the left vector more times. If all probing join keys get matched, there will be generating an output. Otherwise, the process of the probing tuple will be aborted. Then a new probing tuple starts to probe hash tables horizontally.

3   Experiments

We conduct experiments on two hardware platforms: a single server with two Intel Xeon Silver 4110 CPUs (denoted as CPU) and a Xeon Phi co-processor 7120 (denoted as PHI). Each CPU has 8 cores and is equipped with 150 GB Memory. PHI has 61 cores and each core has 4 threads. It is equipped with 16 GB memory. We use ICC 17.0.2 to compile our code with −O3 optimization enabled. We focus on the probe phase of the star join queries and take the example query in Sect. 1 as a test query. By default, each tuple is composed of a 32-bit key and a 32-bit payload in dimension tables. Each tuple of the fact table includes specific probe keys and another 32-bit payload.

We compare the performance of different algorithms on CPU and PHI. These four algorithms are evaluated on SSB with scale factor 100 using the most threads but varying the selectivity of dimension tables. We set high (0.99), medium (0.5) and low (0) selectivity for dimension tables where the selectivity means how much percent of tuples is got rid of original tables [2]. On CPU, the raw data of dimension tables is larger than the capacity of the last level cache (LLC) of CPU, but it can reside in LLC for high selectivity. Besides, we also test the impact of prefetching for two vectorized algorithms. The prefetching is enabled by default. But in order to distinguish whether adopting the prefetching or not in some cases, each algorithm is appended with +p or -p.

As shown in Fig. 7, for two scalar code algorithms, IP is superior to the traditional RP by up to 1.48X and 1.33X on CPU and PHI, respectively. This improvement results from eliminating intermediate results in mismatch cases and function calling overhead in RP. However, IP also suffers heavy branch misses especially in the medium selectivity case. Furthermore, SIMD achieves considerable speedup in comparison to two kinds of scalar code algorithms. VVIP is faster than IP and RP by up to 2.19X and 3.24X on CPU, and 2.63X and

480   Z. Fang et al.

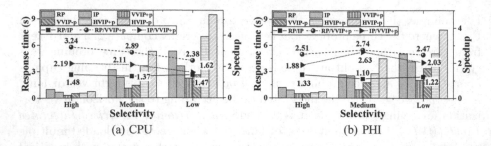

Fig. 7. The performance comparison

2.74X on PHI, respectively. This is attributed to not only the parallel processing but also eliminating branch misses using SIMD. Such two advantages perform differently depending on the data size and the selectivity of dimension tables. For example, the speedup is higher for vectorized code when setting medium selectivity. Because in such case, the scalar code algorithms, i.e., RP and IP, suffer from heavy branch misses. However, HVIP is rather slower than RP and IP for medium and low selectivity. This is attributed to its horizontal loop with expensive memory access. Thus the prefetching can achieve reasonable speedup in HVIP. Besides, the horizontal loop suffers from heavy branch misses through observing performance metrics. As a result, HVIP is seriously slower than VVIP. So we conclude that VVIP is the efficient way to accelerate the integrated probe.

References

1. Babu, S., Motwani, R., Munagala, K., Nishizawa, I., Widom, J.: Adaptive ordering of pipelined stream filters. In: SIGMOD, pp. 407–418 (2004)
2. Galindo-Legaria, C.A., et al.: Optimizing star join queries for data warehousing in Microsoft SQL server. In: ICDE, pp. 1190–1199 (2008)
3. Goldstein, J., Larson, P.: Optimizing queries using materialized views: a practical, scalable solution. In: SIGMOD, pp. 331–342 (2001)
4. O'Neil, P.E., O'Neil, E.J., Chen, X.: The star schema benchmark (SSB). Pat **200**, 50 (2007)
5. Padmanabhan, S., Bhattacharjee, B., Malkemus, T., Cranston, L., Huras, M.: Multi-dimensional clustering: a new data layout scheme in DB2. In: SIGMOD, pp. 637–641 (2003)
6. Polychroniou, O., Raghavan, A., Ross, K.A.: Rethinking SIMD vectorization for in-memory databases. In: SIGMOD, pp. 1493–1508 (2015)
7. Weininger, A.: Efficient execution of joins in a star schema. In: SIGMOD, pp. 542–545 (2002)
8. Zhang, Y., Zhou, X., Zhang, Y., Zhang, Y., Su, M., Wang, S.: Virtual denormalization via array index reference for main memory OLAP. TKDE **28**(4), 1061–1074 (2016)

A Deep Recommendation Model Incorporating Adaptive Knowledge-Based Representations

Chenlu Shen[1], Deqing Yang[1(✉)], and Yanghua Xiao[2,3]

[1] School of Data Science, Fudan University, Shanghai 200433, China
{clshen17,yangdeqing}@fudan.edu.cn
[2] School of Computer Science, Fudan University, Shanghai 200433, China
shawyh@fudan.edu.cn
[3] Shanghai Institute of Intelligent Electronics and Systems, Fudan University, Shanghai 200433, China

Abstract. Deep neural networks (DNNs) have been widely imported into collaborative filtering (CF) based recommender systems and yielded remarkable superiority, but most models perform weakly in the scenario of sparse user-item interactions. To address this problem, we propose a deep knowledge-based recommendation model in which item knowledge distilled from open knowledge graphs and user information are both incorporated to extract sufficient features. Moreover, our model compresses features by a convolutional neural network and adopts memory-enhanced attention mechanism to generate adaptive user representations based on latest interacted items rather than all historical records. Our extensive experiments conducted against a real-world dataset demonstrate our model's remarkable superiority over some state-of-the-art deep models.

1 Introduction

Despite of the superior performance of DNNs, some challenges are still not tackled well by previous deep recommendation models. The first one is *sparse observed user-item interactions*. In general, CF-based methods such as [2,4] suffer from few observed interactions, also known as the problem of *cold start* or *data sparsity*. The second one is *dynamic and diverse preferences of users*. In many deep recommendation models [2,3], user representations are generated based on all historical interacted items. However, user interests may shift as time elapses hence the latest records are more important than the early ones for preference inference. On the other hand, a user's representation is usually fixed which is not adaptive to the diversity of his/her preferences. To address above issues, we propose a novel deep recommendation model which not only imports external knowledge but also integrates some deep components as follows.

This work was supported by Chinese NSFC Project (No. U1636207, No. 61732004).

G. Li et al. (Eds.): DASFAA 2019, LNCS 11448, pp. 481–486, 2019.
https://doi.org/10.1007/978-3-030-18590-9_71

First, item representations are enriched by the knowledge distilled from open knowledge graphs (KGs) resulting in immunity from sparse observed user-item interactions, and high interpretability of recommendation results. In addition, user's *personal attribute embedding* is generated based on personal information, e.g., demographics or tags.

Second, our model also generates user's *historical preference embedding* to constitute comprehensive user representation, which is a temporal and adaptive vector generated through fusing latest interacted item embeddings by *memory-enhanced attention mechanism*.

Third, for predicting the probability that a given user likes a candidate item which is the key of achieving top-N recommendation, our model further utilizes a multi-layer perceptron (MLP) fed with element-wise product of the user embedding vector and the item embedding vector to generate a non-linearity output better than traditional inner products [2].

In summary, our contributions in this paper include:

1. We propose a deep knowledge-based recommendation model to overcome the sparsity of user-item interactions with external knowledge from KGs as auxiliary information.
2. We build a memory-enhanced attention mechanism which is helpful to generate user representations adaptive to their dynamic and diverse preferences.
3. Extensive experiments are conducted on Douban[1] movie dataset to justify that our model outperforms some state-of-the-art models.

2   Model Description

Without loss of generality, we explain the details of our model w.r.t. Douban movie recommendation in this section. Figure 1 depicts the framework of our model.

Knowledge-Aware Item/User Embedding. We first leverage knowledge as item features to discover latent relationships between items. A movie i's representation is projected into a knowledge-based tensor $\boldsymbol{P}_i \in \mathbb{R}^{f \times e}$, where f is feature number and e is embedding dimension. That is, each row of $\boldsymbol{P}_i$ relates to a feature embedding. Then, since each Douban user is specified by a set of tags, a user u's *personal attribute embedding* $\boldsymbol{p}_u^a$ is the average of u's tag embeddings. At the same time, we employ attention mechanism to generate u's *historical preference embedding* $\boldsymbol{p}_{uj}^h$ which is adaptive to different candidate movie j. Then, we get u's comprehensive representation w.r.t item j as follows,

$$\boldsymbol{p}_{uj} = \gamma \boldsymbol{p}_{uj}^h + (1 - \gamma)\boldsymbol{p}_u^a \tag{1}$$

where γ is a controlling parameter.

---

[1] https://movie.douban.com, a famous Chinese website of movie reviews.

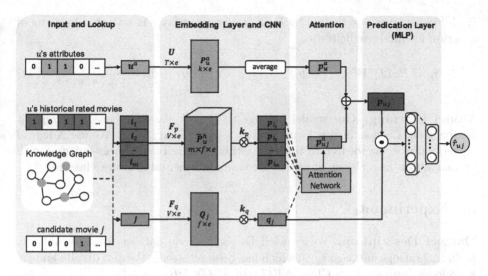

Fig. 1. The framework of our deep knowledge-based recommendation model.

Convolutional Neural Network for Embedding Compression. Given the shifted user preference, we only pay attention to the m latest interacted items to represent a user's preference. The union of m rated movies' representations is a feature cube (tensor) $\tilde{P}_u^h \in \mathbb{R}^{m \times f \times e}$. Then, we use CNN for compressing features and finally get a squeezed matrix $P_u^h = [p_{i_1}, p_{i_2}, ..., p_{i_m}] \in \mathbb{R}^{m \times e}$. Specifically, the convolutional operation is the inner product without activation.

Memory-Enhanced Attention Mechanism. Suppose p_i and q_j are the embedding of historical movie i and candidate j respectively, we adopt a two-layer fully connected neural network to calculate their similarity

$$sim(p_i, q_j) = g^\top ReLU\left(W(p_i \odot q_j) + b\right), i \in \mathcal{M}_u^+ \tag{2}$$

where W and b are the weight matrix and bias respectively, and g is a weight vector for output projection, $\odot$ denotes the element-wise product operator. As we emphasized before, we capture the historical preference of a user through a *memory component* that stores a fixed number of latest interacted items at time t as $\mathcal{M}_u^+ = \{i_1^t, i_2^t, ..., i_m^t\}$. Then, the attention weight α_{ij} and u's historical preference embedding are computed as follows

$$\alpha_{ij} = \frac{\exp\left(sim(p_i, q_j)\right)}{\sum_{i \in \mathcal{M}_u^+} \exp\left(sim(p_i, q_j)\right)}, \qquad p_{uj}^h = \sum_{i \in \mathcal{M}_u^+} \alpha_{ij} p_i \tag{3}$$

Prediction Layer. For final prediction, we adopt a MLP fed with user embeddings and item embeddings to enhance the capability of capturing non-linearity.

Each hidden layer is activated by *ReLU* and output layer is activated by sigmoid function for final prediciton $\hat{r}_{uj}$.

$$\boldsymbol{y}_l = ReLU(\boldsymbol{W}_l\boldsymbol{y}_{l-1} + \boldsymbol{b}_l), \qquad \hat{r}_{uj} = \sigma(\boldsymbol{h}^\top\boldsymbol{y}_L + b) = \frac{1}{1 + e^{\boldsymbol{h}^\top\boldsymbol{y}_L + b}} \qquad (4)$$

Model Learning. Our model belongs to binary classifier based on implicit feedback [3]. Each sample is formalized as a triplet $<u, j, r_{uj}>$. We use Adagrad [1] to optimize the objective function, which is the binary cross-entropy with regularization as a classic objective for training a neural network classifier.

3   Experiment

Dataset Description. We crawled Douban movie dataset from its website, including ratings and user tags, which has been released[2]. We also distilled movie knowledge from an open Chinese KG named *CN-DBpedia* [7].

Sample Collection and Evaluation Metrics. We conducted our experiments on different sparsity levels, according to a certain proportion s by sampling. Specifically, we evaluated model performance as s increases gradually.

Compared Models. *BPR* [5] is a MF-based model uses Bayesian Personalized Ranking loss as a ranking-ware objective function. A widely used deep model *NCF* [3] updates randomly initialized latent embeddings by observed interactions. *FISM* [4] is a representative factored item-based CF model that learns item similarity matrix as the product of two low dimensional latent factor matrices. *NAIS* [2] tailors attention mechanism to differentiate the weights of historical items in user representation extended from FISM. Then, our model is denoted as *KAC* (Knowledge+Attention+CNN) in the following texts. We further propose four variants to be compared which are named as *AC/KC/KAavg/KAcon*. Particularly, KAavg and KAcon replace CNN with average and concatenate.

Experiment Results. With the optimal settings of hyper parameters, we compared KAC with other models. Figure 2 displays top-1/3/5 movie recommendation performance of three evaluation metrics, i.e., HR (Hit Ratio), nDCG (Normalized Discounted Cumulative Gain), AP (Average Precision) and RR (Reciprocal Rank). The results show that KAC outperforms others in all scenarios, and the variants also prevail against the baselines. KAC's superiority justifies the significance of adaptive knowledge-based embeddings. Specifically, KAC's superiority over KA/AC/KAavg/KAcon shows the effects and necessities of corresponding components in our model.

---

[2] http://gdm.fudan.edu.cn/GDMWiki/Wiki.jsp?page=Network%20DataSet.

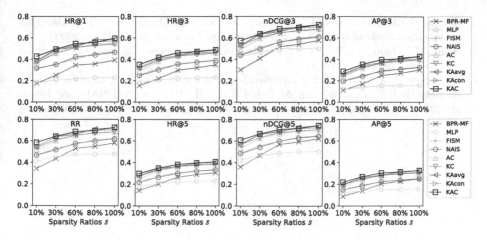

Fig. 2. Performance comparisons of recommendation with different sparsity ratios.

4   Related Work and Conclusion

As a pioneer work of deep recommendation model, NCF [3] fuses a generalized MF layer and an MLP together to learn user/item embeddings based on observed interactions. DKN [6] incorporates knowledge information of entities in news contexts from KGs. On the other hand, NAIS [2] pays adaptive attention to historical items for user representation according to different candidate items, which has been proven effective by empirical studies.

Comparatively, we propose a knowledge-based deep model in this paper, which is designed towards the recommendation of sparse user-item interactions. Our model not only strengthens user/item representation by importing knowledge and users' personal information to alleviate the sparsity problem, but also model employs an enhance attention mechanism with a memory component to better capture dynamic user preferences. We also adopt CNN to extract and compress user features in our model. The results of extensive evaluation not only justify our model's superiority but also confirm the significance of incorporating related DNN-based components.

References

1. Duchi, J., Hazan, E., Singer, Y.: Adaptive subgradient methods for online learning and stochastic optimization. J. Mach. Learn. Res. **12**(7), 257–269 (2011)
2. He, X., He, Z., Song, J., Liu, Z., Jiang, Y.G., Chua, T.S.: Nais: neural attentive item similarity model for recommendation. IEEE TKDE **30**, 2354 (2018)
3. He, X., Liao, L., Zhang, H., Nie, L., Hu, X., Chua, T.S.: Neural collaborative filtering. In: Proceedings of WWW (2017)
4. Kabbur, S., Ning, X., Karypis, G.: Fism: Factored item similarity models for top-n recommender systems. In: Proceedings of KDD (2013)

5. Rendle, S., Freudenthaler, C., Gantner, Z., Schmidt-Thieme, L.: BPR: Bayesian personalized ranking from implicit feedback. In: Proceedings of CUAI (2009)
6. Wang, H., Zhang, F., Xie, X., Guo, M.: DKN: deep knowledge-aware network for news recommendation (2018)
7. Xu, B., et al.: CN-DBpedia: a never-ending Chinese knowledge extraction system. In: Benferhat, S., Tabia, K., Ali, M. (eds.) IEA/AIE 2017. LNCS (LNAI), vol. 10351, pp. 428–438. Springer, Cham (2017). https://doi.org/10.1007/978-3-319-60045-1_44

BLOMA: Explain Collaborative Filtering via Boosted Local Rank-One Matrix Approximation

Chongming Gao[1], Shuai Yuan[1], Zhong Zhang[1], Hongzhi Yin[2], and Junming Shao[1(✉)]

[1] University of Electronic Science and Technology of China, Chengdu, China
junmshao@uestc.edu.cn
[2] The University of Queensland, Brisbane, Australia

Abstract. Matrix Approximation (MA) is a powerful technique in recommendation systems. There are two main problems in the prevalent MA framework. First, the latent factor is out of explanation and hampers the understanding of the reasons behind recommendations. Besides, traditional MA methods produce user/item factors globally, which fails to capture the idiosyncrasies of users/items. In this paper, we propose a model called Boosted Local rank-One Matrix Approximation (BLOMA). The core idea is to locally and sequentially approximate the residual matrix (which represents the unexplained part obtained from the previous stage) by rank-one sub-matrix factorization. The result factors are distinct and explainable by leveraging social networks and item attributes.

1 Introduction

Among all the recommendation methods, collaborative filtering (CF) methods based on Matrix Approximation (MA) are concise and effective. Usually, with the proliferation of additional content data such as social network and item attributes, researchers factorized the user-item rating matrix $R \in \mathbb{R}^{m \times n}$ into two low-rank matrices $U \in \mathbb{R}^{m \times r}$ and $V \in \mathbb{R}^{r \times n}$, so that $R \approx UV$, where r is the rank of latent space and is far smaller than the number of users/items m, n [2]. In this way, the preferences of users and properties of items are condensed in the low-rank latent vectors. By taking the inner product between the user and item vector, We can predict missing rating values. However, the latent factors do not possess intuitive meanings, making the recommendation system work like a blackbox. Understanding the reasons behind the model can make users accept the result. Consider the following example in recommendation system: when recommending the Chinese seafood noodle to a user, instead of plainly pointing out that "people also viewed", the system makes explanation like, "This recommendation is tailored to your tastes for Chinese cuisine (fitting 60%) and seafood (fitting 30%). Have a try?" To this aim, we propose a Boosted Local rank-One Matrix Approximation (BLOMA) model, Fig. 1 illustrates the basic logic, which has three major differences: (1) The rating matrix is factorized

© Springer Nature Switzerland AG 2019
G. Li et al. (Eds.): DASFAA 2019, LNCS 11448, pp. 487–490, 2019.
https://doi.org/10.1007/978-3-030-18590-9_72

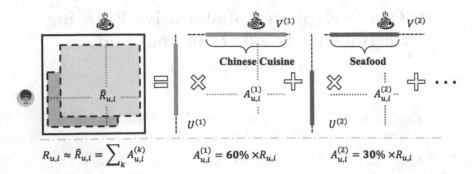

Fig. 1. Illustration of the explainable matrix approximation model. $U^{(1)}, U^{(2)}, V^{(1)}$ and $V^{(2)}$ are vectors, whose elements represented by black dash line are zeros.

locally on the part of the users and items, (2) the factorization is applied for several times sequentially on the residue matrix, and (3) each factorization is a rank-one decomposition. The three changes together make the topics extracted from the latent factors more distinct.

2   Method

The most classical recommendation models are based on Matrix Factorization (MF) techniques [3,6,8]. To improve the performance of the models, researchers proposed to integrate the heterogeneous knowledge [1,5,9,11,12]. However, it is often the case that users are not interested in all but a subset of items. Therefore, researchers proposed Local Matrix Approximation (LMA) methods to model a set of sub-matrices [4,9]. There are still two issues remained unresolved for LMA models: (1) How to select the subset of users and items to construct the sub-matrix? Shao et al. [7] adopted a co-clustering technique to simultaneously select users and items, and Yin et al. [10] proposed a bidirectional negative sampling method. In this work, we sample a user and a item, then use their neighbors on the social/item network and compute the arc-cosine distance in rating matrix to select a group of users/items to construct the sub-matrix. (2) How to assemble all models that are independently running on certain sub-matrices? Existing methods have to explicitly determine the number of local models, which is subjective. In this work, borrowing the idea from ensemble learning, we iteratively add a new local model to better approximate original rating matrix R in a forwarding stagewise manner. To be specific, in k-th stage, we are fitting the residue sub-matrix obtained from previous stage locally. The recommendation objective of k-th stage is as below:

$$\arg\min_{U^{(k)},V^{(k)}}\left\|\mathcal{I}\Big(R^{(k)}-g\big((\alpha U^{(k)}+(1-\alpha)SU^{(k)})\cdot(\beta V^{(k)}+(1-\beta)V^{(k)}T)\big)\Big)\right\|_F^2$$
$$+\frac{\lambda_1}{2}\|U^{(k)}\|_F^2+\frac{\lambda_2}{2}\|V^{(k)}\|_F^2,\tag{1}$$

where $R^{(k)} \in \mathbb{R}^{m \times n}$ is the selected residue rating matrix. $U^{(k)} \in \mathbb{R}^{m \times 1}, V^{(k)} \in \mathbb{R}^{1 \times n}$ are vectors representing latent factors of users/items. $S \in \mathbb{R}^{m \times m}$ and $T \in \mathbb{R}^{n \times n}$ are the normalized adjacent matrices of social network and item network with $\sum_u S(\cdot, u) = 1$ and $\sum_i T(i, \cdot) = 1$. α and β are the coefficients controlling the ratios of the user/item's own factor and its neighbors' factors. $\mathcal{I}(\cdot)$ is the indicator function. $g(x) = 1/(1 + e^{-x})$ is the logistic function. The residue matrix $R^{(k+1)}$ is computed in the forward stagewise manner as:

$$R^{(k+1)} = R^{(k)} - g\Big(\big(\alpha U^{(k)} + (1-\alpha)SU^{(k)}\big) \cdot \big(\beta V^{(k)} + (1-\beta)V^{(k)}T\big)\Big), \quad (2)$$

where $R^{(1)} = R$, which is the normalized rating matrix. Finally, we concatenate $U = [U^{(1)}, U^{(2)}, \cdots, U^{(K)}]$ by column and $V = [V^{(1)}; V^{(2)}; \cdots; V^{(K)}]$ by row. A local minimum of the objective function given by Eq. (1) can be found by alternately performing gradient descent on $U^{(k)}, V^{(k)}$.

3 Experiments

We conduct the experiments on Yelp[1], a well-known POI recommendation data set. For convenience, we extract a POI network from the given features: if the geographical distance between two POI is smaller than a given value $d = 50\,\mathrm{km}$, then they can be linked with a certain probability $p = 0.01$. We compare our BLOMA with the state-of-the-art models including RegSVD [6], NMF [3], LLORMA [4], SLOMA [9], RSTE [5] and TrustSVD [1]. To discover the meaning in the latent factors and explain the recommendation results, we compute the average Pointwise Mutual Information (PMI) of every pair of items with top-10 largest absolute values on each column of item factor matrix V.

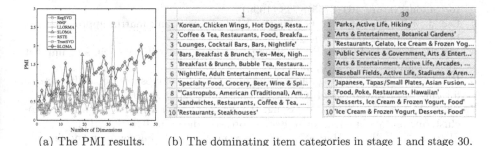

(a) The PMI results.   (b) The dominating item categories in stage 1 and stage 30.

Fig. 2. Evaluation of explainability on PMI metric and an exemplar illustration.

The result is shown in Fig. 2(a). We can see that BLOMA outperforms all other methods, and maintain the highest topic coherence with the dimension K increasing. This result reflects two conclusions. First, in the beginning, the user communities and items with similar semantic categories can be discovered by carefully selecting the sub-indices and using the rank-one decomposition.

---

[1] https://www.yelp.com/dataset.

And after iteratively subtracting a local approximation from the residue matrix, i.e., the unexplained component, the rating remained to be approximated is more distinct and easier to be explained. See Fig. 2(b) for an example. In the beginning, the subset of items is composed of items of common categories, e.g., Food and Bars. When BLOMA run about 30 stages, the subset of items begins to show the different categories in the check-in data, such as Arts and Parks.

4   Conclusion

In this paper, we present a novel ensemble approach of the local matrix approximation model. By exploring the heterogeneous information, the group of users and items sharing common characteristics are selected to form a sub-matrix. We apply a rank-one matrix factorization model to this sub-matrix in a forwarding stagewise manner. The experiments show good explainability in our model.

Acknowledgments. This work is supported by the National Natural Science Foundation of China (61403062, 61433014, 41601025), Science-Technology Foundation for Young Scientist of Sichuan Province (2016JQ0007), Fok Ying-Tong Education Foundation for Young Teachers in the Higher Education Institutions of China (161062) and National key research and development program (2016YFB0502300).

References

1. Guo, G., Zhang, J., Yorke-Smith, N.: A novel recommendation model regularized with user trust and item ratings. TKDE **28**, 1607–1620 (2016)
2. Koren, Y.: Factorization meets the neighborhood: a multifaceted collaborative filtering model. In: SIGKDD, pp. 426–434. ACM (2008)
3. Lee, D.D., Seung, H.S.: Learning the parts of objects by non-negative matrix factorization. Nature **401**(6755), 788 (1999)
4. Lee, J., Kim, S., Lebanon, G., Singer, Y.: Local low-rank matrix approximation. In: ICML, pp. 82–90 (2013)
5. Ma, H., King, I., Lyu, M.R.: Learning to recommend with explicit and implicit social relations. TIST **2**(3), 29 (2011)
6. Paterek, A.: Improving regularized singular value decomposition for collaborative filtering. In: Proceedings of KDD Cup and Workshop, vol. 2007, pp. 5–8 (2007)
7. Shao, J., Gao, C., Zeng, W., Song, J., Yang, Q.: Synchronization-inspired co-clustering and its application to gene expression data. In: ICDM (2017)
8. Shao, J., Yu, Z., Li, P., Han, W., Sorg, C., Yang, Q.: Exploring common and distinct structural connectivity patterns between schizophrenia and major depression via cluster-driven nonnegative matrix factorization. In: ICDM, pp. 1081–1086 (2017)
9. Yang, B., Lei, Y., Liu, J., Li, W.: Social collaborative filtering by trust. TPAMI **39**(8), 1633–1647 (2017)
10. Yin, H., Chen, H., Sun, X., Wang, H., Wang, Y., Nguyen, Q.V.H.: SPTF: a scalable probabilistic tensor factorization model for semantic-aware behavior prediction. In: ICDM, pp. 585–594. IEEE (2017)
11. Yin, H., Wang, W., Wang, H., Chen, L., Zhou, X.: Spatial-aware hierarchical collaborative deep learning for POI recommendation. TKDE **29**, 2537–2551 (2017)
12. Yin, H., Zou, L., Nguyen, Q.V.H., Huang, Z., Zhou, X.: Joint event-partner recommendation in event-based social networks. ICDE (2018)

Spatiotemporal-Aware Region Recommendation with Deep Metric Learning

Hengpeng Xu[1], Yao Zhang[1], Jinmao Wei[1]([✉]), Zhenglu Yang[1]([✉]), and Jun Wang[2]

[1] College of Computer Science, Nankai University, Tianjin, China
{xuhengpeng,yaozhang}@mail.nankai.edu.cn, {weijm,yangzl}@nankai.edu.cn
[2] College of Mathematics and Statistics Science, Ludong University, Yantai, Shandong, China
junwang@mail.nankai.edu.cn

Abstract. Personalized points of interests (POI) recommendation is an important basis for location-based services. A typical application scenario is to recommend a region with reliable POIs to a user when he/she travels to an unfamiliar area without any background knowledge. In this study, we explore spatiotemporal-aware region recommendation to manage this learning task. We propose a unified deep learning model that comprehensively incorporates dynamic personal and global user preferences across regions, along with spatiotemporal dependencies, into check-in region history. We model and fuse user preferences through a pyramidal ConvLSTM component, and capture the dynamic region attributes through a recurrent component. Two components are seamlessly assembled in a unified framework to yield next time region recommendation. Extensive experiments on real-word datasets demonstrate the effectiveness of the proposed model.

Keywords: Region recommendation · Spatio-temporal context · User preference · Deep learning

1 Introduction

When people travel to an unfamiliar area, their knowledge is limited to their areas that they are aware of but invalid for other areas. Therefore, recommending a set of nearby locations in an urban region is beneficial for travelers. In this study, we provide region recommendation to users, which contributes to mobility modeling, urban planning, and traffic management.

Considerable research effort has been exerted to adapt traditional spatial and temporal context learning to POI recommendation [2]. However, user mobility is sequential and complex to learn and transition patterns may not follow Markov chains assumed. Recently, several deep learning models have been combined with collaborative filtering models to learn the deep representation of items or users

© Springer Nature Switzerland AG 2019
G. Li et al. (Eds.): DASFAA 2019, LNCS 11448, pp. 491–494, 2019.
https://doi.org/10.1007/978-3-030-18590-9_73

in accordance with their associated side information [4]. These sequential models emphasize dependencies among sequence nodes; however, disregard critical spatial and temporal correlations. We argue that such correlations are essential for mitigating data sparsity, which occurs in the majority of deep learning models.

In this study, we utilize the ConvLSTM network to capture local relationships between POIs in one region and the dynamics of region patterns. Moreover, we adopt an RNN to identify the multiple factors that govern the transition regularities of human movements. Finally, to integrate the learned spatial-aware personal preferences among regions and semantic representation of POIs, we propose a framework to jointly perform matrix factorization and automatic extraction of the POIs' semantic representation, which are guided by user feedback information (e.g., check-in history) in a supervised manner. The contributions of our study are summarized as follows:

- We propose a novel deep learning region recommendation, namely, Deep-RegionRS, which is a principled and deep learning-based architecture to model the dynamic global preferences of users by considering the spatio-temporal dependencies across regions. To the best of our knowledge, this work is the first to utilize deep learning networks in region recommendation.
- We leverage the ConvLSTM model to encode spatial and temporal dependencies across regions, and employ expressive NNs to capture the high-level nonlinear interactions between personal preferences and global preferences of users. Furthermore, we combine two parts in an unified model for the next-time region recommendation.

2   The Proposed Model

To comprehensively utilize spatial and temporal dependence between user preferences and region characteristics, we partition a city into an $I * J$ grid map and then map user check-in historical data onto 2-D region check-in images.

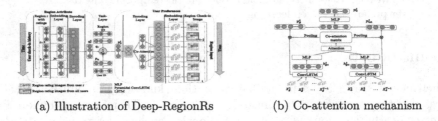

(a) Illustration of Deep-RegionRs   (b) Co-attention mechanism

Fig. 1. The main architecture of Deep-RegionRs and Co-attention mechanism.

2.1   Model Description

Our goal is to devise a framework to simultaneously extract the user preferences over regions, capture dynamic states of regions, quantify spatio-temporal mutual effects between regions, and recommend region. Hence, we introduce

Deep-RegionRs model, whose overall architecture is shown in Fig. 1a. Deep-RegionRs explores spatial and temporal dependencies over regions and utilizes ConvLSTM and LSTM to induce meaningful embeddings for users and regions.

Region Attribute. In our model, we use LSTM recurrent neural networks to capture temporal dependencies for regions. In using this method, we are able to incorporate past observations and predict future trajectories in an integrated manner. Region attributes change over time, and the attractiveness of a region depends on its popularity (i.e., users who have checked-in). Hence, each region can be represented as a sequence of ratings from users at previous times (i.e., $c_r^t \in R^N$). In inputting c_r^t into the encoding layer, we use transformation matrix W_e to project the information into an embedding space, and we obtain the latent vector y_i^t as the input to LSTM.

Pyramidal ConvLSTM Model. Intuitively, the user's preferences toward a nearby region may also be affected by another nearby region. Naturally, spatial related preference can be effectively handled by CNN given its powerful ability to hierarchically capture spatial-structural information. Hence, we need to design a CNN with many layers to capture the spatial dependency of any region. As shown in Fig. 1b, the three multilevel of feature maps that are connected with a few convolutions. Stacking a convolutional RNN layer will result in a relatively wide receptive field over nearby spatial regions and improve the upper layers, which then leads to a better representations and more accurate predictions. Naturally, we consider using ConvLSTM [6] as our recurrent neural network given its good performance and the few parameters required for optimization.

Co-attention mechanism. Considering the varying importance of mutual influence (i.e., global versus personal influences) to different users, we can not treat the strength of mutual influence on users equally. Thus, we propose to use the co-attention mechanism to process the hidden states of the two pyramidal ConvLSTM components whose inputs are the global users region check-in images and the personal -level user check-in images respectively, as is shown in Fig. 1b.

3 Experiments

3.1 Experimental Setup

We evaluate our model on public Foursquare check-in datasets collected from two big cities, namely, New York (NYC) and Tokyo (TKY) [7].

Baselines. We compare Deep-RegionRS with several representative methods (i.e., MF [3], NeuMF [1], ConvLSTM3 [6], RRN [5], and Deep-RegionRsNC) for the user's region recommendations.

Experimental Results. The performance of the compared approaches on two benchmarks under the metric of root mean squared error (RMSE) is illustrated in Table 1. It reveals that the Deep-RegionRS model remarkably outperforms RRN, NeuMF, ConvLSTM3, and the others on NYC (+0.051–0.089 improvement over the best compared method, i.e., RRN) and TKY (+0.058–0.113 improvement compared with the best candidate method RRN).

Table 1. Performance comparison on two datasets under the metric of RMSE

Method	TKY			NYC		
	d = 0.2 km	d = 0.5 km	d = 1 km	d = 0.2 km	d = 0.5 km	d = 1 km
MF	0.2310	0.2762	0.2307	0.3016	0.2901	0.3078
NeuMF	0.2015	0.2451	0.2150	0.2563	0.2670	0.2614
ConvLSTM3	0.1923	0.2421	0.1862	0.2452	0.2598	0.2473
RRN	0.1817	0.2397	0.1756	0.2307	0.2544	0.2451
Deep-RegionRSNC	0.1965	0.2427	0.1954	0.2375	0.2631	0.2754
Deep-RegionRS	**0.1725**	**0.2126**	**0.1654**	**0.2175**	**0.2325**	**0.2435**

4   Conclusions

In this paper, we leverage the pyramidal ConvLSTM model to encode the spatial and temporal dependencies over regions, and then employ the expressive NNs to capture the high-level nonlinear interactions between the global and the personal preferences of users. Furthermore, we employ a recurrent layer to capture the dynamic state changes of regions.

Acknowledgements. This work was supported in part by the National Natural Science Foundation of China under Grant No. 61772288, U1636116, 11431006, the Natural Science Foundation of Tianjin City under Grant No. 18JCZDJC30900, the Research Fund for International Young Scientists under Grant No. 61750110530, and the Ministry of education of Humanities and Social Science project under grant 16YJC790123.

References

1. He, X., Liao, L., Zhang, H., Nie, L., Hu, X., Chua, T.S.: Neural collaborative filtering. In: International World Wide Web Conference (WWW), pp. 173–182 (2017)
2. Liu, Q., Wu, S., Wang, D., Li, Z., Wang, L.: Context-aware sequential recommendation. In: IEEE International Conference on Data Mining (ICDM), pp. 1053–1058 (2016)
3. Mnih, A., Salakhutdinov, R.R.: Probabilistic matrix factorization. In: Conference on Neural Information Processing Systems (NIPS), pp. 1257–1264 (2008)
4. Wang, H., Wang, N., Yeung, D.Y.: Collaborative deep learning for recommender systems. In: ACM SIGKDD Conference on Knowledge Discovery and Data Mining (SIGKDD), pp. 1235–1244 (2015)
5. Wu, C.Y., Ahmed, A., Beutel, A., Smola, A.J., Jing, H.: Recurrent recommender networks. In: ACM International Conference on Web Search and Data Mining (WSDM), pp. 495–503 (2017)
6. Xingjian, S., Chen, Z., Wang, H., Yeung, D.Y., Wong, W.K., Woo, W.c.: Convolutional LSTM network: a machine learning approach for precipitation nowcasting. In: Conference on Neural Information Processing Systems (NIPS), pp. 802–810 (2015)
7. Yang, D., Zhang, D., Zheng, V.W., Yu, Z.: Modeling user activity preference by leveraging user spatial temporal characteristics in lbsns. IEEE Trans. Syst. Man Cybern. **45**(1), 129–142 (2015)

On the Impact of the Length of Subword Vectors on Word Embeddings

Xiangrui Cai, Yonghong Luo, Ying Zhang$^{(\boxtimes)}$, and Xiaojie Yuan

College of Computer Science, Nankai University, Tianjin, China
{caixiangrui,luoyonghong}@dbis.nankai.edu.cn,
{yingzhang,yuanxj}@nankai.edu.cn

Abstract. This paper hypothesizes that better word embeddings can be learned by representing words and subwords by different lengths of vectors. To investigate the impact of the length of subword vectors on word embeddings, this paper proposes a model based on the Subword Information Skip-gram model. The experiments on two datasets with respect to two tasks show that the proposed model outperforms 6 baselines, which confirms the aforementioned hypothesis. In addition, we also observe that, within a specific range, a higher dimensionality of subword vectors always improve the quality of word embeddings.

Keywords: Word embeddings · Subword information ·
Length of subword vectors

1 Introduction

A recent advancement for enriching word embeddings is to incorporate subword information. In particular, the Subword Information Skip-gram (SISG) model [1], which is built based on the Skip-gram model [4] and represents a word by the sum of its character n-grams vectors, has achieved the state-of-the-art performance in the downstream tasks.

Despite the effectiveness, the relationships between words and subwords have not been fully understood. For instance, *con* is a subword in *connect*, *conclude* and *contrary*, while it means "with" in *connect*, "thoroughly" in *conclude* and "opposite" in *contrary*. However, the subword *nect* also has its meaning in *connect*, which indicates a word is more expressive than a subword. Therefore, the relationships between words and subwords are not so straightforward. To the best of the present authors' knowledge, current methods embed words and subwords into the same latent space by default.

In this paper, we hypothesize that better word embeddings can be learned by representing words and subwords with different lengths of vectors. We build a model based on the SISG model [1]. The composition of subword vectors was realized by an affine transformation layer to represent words. Given a fixed length of word vectors, we vary the length of subword vectors to investigate the impact of the length. We conduct experiments on two evaluation tasks and two datasets. The results show that:

© Springer Nature Switzerland AG 2019
G. Li et al. (Eds.): DASFAA 2019, LNCS 11448, pp. 495–499, 2019.
https://doi.org/10.1007/978-3-030-18590-9_74

1. Representing words and subwords by the same length of vectors produces acceptable embedding results. However, within a specific range, a higher dimensionality of subword vectors always improve the quality of word embeddings.
2. The performance of the proposed model cannot improve when the length of subword vectors become too large, which could be explained by the ability of the model to handle the certain amount of subword information.
3. Affine transformation method achieves better performance than simply addition of subword vectors. Even a small length of subword vectors in the proposed model achieves competitive performance with the baselines.

2   Representation Model

The proposed model ATSG, named after Affine Transformation Skip-gram, is built based on the SISG model. Supposing there are m character n-grams in w_t, the model first projects them to the D_g-dimensional vector representations $z_{g_1}, z_{g_2}, \cdots, z_{g_m}$, where $z_{g_i} \in \mathbb{R}^{D_g}$. Then it takes affine transformation to convert the subword vectors to word vectors, i.e., $v_{w_t} \in \mathbb{R}^{D_w}$. Since we intend to investigate the impact of the length of subword representations on the quality of word embeddings, the lengths of word vectors and subword vectors are different in this work, i.e., $D_g \neq D_w$. Specifically, the transformation function is defined by:

$$v_{w_t} = \frac{1}{|\mathcal{G}_{w_t}|} \left(\sum_{g_i \in \mathcal{G}_{w_t}} (W \cdot z_{g_i} + b) \right) = W \cdot \left(\frac{1}{|\mathcal{G}_{w_t}|} \sum_{g_i \in \mathcal{G}_{w_t}} z_{g_i} \right) + b, \qquad (1)$$

where $W \in \mathbb{R}^{D_w \times D_g}$ and $b \in \mathbb{R}^{D_w}$ are the weights and bias respectively of the affine transformation. The objective of the proposed model is to maximize the average log-likelihood of the contexts given the central words $p(w_{t+j}|w_t)$. Given an input central word w_t and a context word w_{t+j}, the pairwise loss function is approximated by the negative sampling:

$$\mathcal{L}(w_t, w_{t+j}) = -\log \sigma(v'_{w_{t+j}}{}^\top v_{w_t}) - \sum_{i=1}^{k} \mathbb{E}_{w_i \sim P_n(w)} [\log \sigma(-v'_{w_i}{}^\top v_{w_t})], \qquad (2)$$

where v_w and v'_w are the input and output vectors respectively, w_i refers to the negative sampled context word. The word w's input vector v_w is composed of subword representations by Eq. (1), and its output vector v'_w is given directly without subword transformation, which is the same way as that in the Skip-gram model.

Parameters in the proposed model are learned by the Stochastic Gradient Descent (SGD) algorithm, where the gradients of the loss function with respect to the parameters are updated through backpropagation.

3   Experiments

The experiments are conducted on Enwik9[1] and Wesbury[2] datasets with regard to the human similarity judgement and the analogy tasks.

We compare against 6 neural embedding models, i.e., CBOW and Skip-gram (SG) [3,4], SICBOW and SISG [1], mCBOW (mCBOW) [5] and csmRNN [2], with regard to the human similarity judgement and the analogy tasks. We use "ATSG_D_g" to denote the ATSG model with different lengths of the character n-gram vectors. We take the recommended settings to train the baselines and similar settings for ATSG, except that the starting learning rate of ATSG is 0.01. Since mCBOW and csmRNN have not published the source codes, we only compare with the results presented in their papers. The dimensionality of word embeddings is set to be 100 for Enwik9 and 50 for Wesbury.

3.1   Human Similarity Judgement

This task is to compute the Spearman's Rank Correlation Coefficient (SRCC) of the human labeled similarity and the similarity of word embeddings. We benchmark the models on five evaluation sets, i.e., MC, RG, WS353, SCWS and RW.

The results of the human similarity judgement task on the two datasets are shown in Tables 1 and 2 respectively. As we can observe, the proposed model ATSG performs better than the baselines on the two datasets. ATSG_120 and ATSG_70 perform the best for Enwik9 and Wesbury respectively, where the lengths of character n-gram vectors are larger than those of word vectors. We guess that a larger length of character n-gram vectors is more capable of subword uncertainty and as a result benefits word embeddings. We also observe that the quality of word embeddings does not improve as D_g becomes too large.

Table 1. SRCC results on Enwik9 (%).

Model	MC	RG	WS353	SCWS	RW
CBOW	70.84	67.00	**67.67**	63.67	32.71
SG	73.75	64.57	66.70	63.96	34.11
SICBOW	41.35	41.10	42.51	54.93	30.71
SISG	63.29	57.61	60.12	63.59	42.52
ATSG_40	67.68	68.75	62.81	62.75	40.56
ATSG_60	71.08	69.70	65.88	63.30	42.43
ATSG_80	72.91	70.95	66.55	63.61	43.22
ATSG_100	72.50	70.26	66.32	64.28	43.51
ATSG_120	**75.09**	**73.76**	67.05	64.07	**43.96**
ATSG_140	72.28	72.07	66.81	**67.19**	43.30
ATSG_160	69.52	72.06	65.66	64.22	43.39

Table 2. SRCC results on Wesbury (%).

Model	MC	RG	WS353	SCWS	RW
CBOW	74.49	74.21	64.04	62.54	35.84
SG	76.58	76.62	63.00	61.63	35.84
SICBOW	75.24	61.47	48.67	60.00	37.64
SISG	77.31	74.60	60.73	61.12	**40.75**
csmRNN	71.72	65.45	64.58	43.65	22.31
mCBOW	**81.62**	67.41	65.19	53.40	32.13
ATSG_20	69.28	70.85	57.33	60.26	35.49
ATSG_30	72.48	72.06	60.36	61.76	37.40
ATSG_40	76.49	73.53	63.88	62.17	39.22
ATSG_50	77.29	74.26	64.45	62.82	39.41
ATSG_60	78.07	75.20	65.23	62.74	39.49
ATSG_70	78.49	**77.03**	**65.74**	**63.05**	39.58
ATSG_80	78.07	75.28	64.88	62.90	39.46

---

[1] http://mattmahoney.net/dc/enwik9.zip.

[2] http://www.psych.ualberta.ca/{~}westburylab/downloads/westburylab.wikicorp.down load.html.

Furthermore, when both words and character n-grams are represented by the same length of vectors, i.e., $D_w = D_g$, the ATSG model performs better than SISG.

3.2 Analogy

The word analogy task was originally introduced by Mikolov et al. [3]. Given the first three words that are represented by v_a, v_b and v_c, this task is to answer the question by finding the word that maximizes the cosine similarity to the vector $(v_b - v_a + v_c)$. It can be formulated by $\arg\max_{d' \in V} \frac{(v_b - v_a + v_c) \cdot v_{d'}}{\|(v_b - v_a + v_c)\| \cdot \|v_{d'}\|}$. We report the analogy results on both Enwik9 and Wesbury datasets in Tables 3 and 4.

As shown in Table 3, the best model is ATSG_120, which achieves a good margin compared with the baselines. Even when D_g is 60, the ATSG model is still competitive to the baselines. Similarly, in Table 4, the ATSG_60 model achieves the best performance on the Wesbury dataset. When D_g is smaller than 40, the ATSG models become worse than the baseline models. However, we also observe that setting a very large D_g can not improve the quality of word embeddings, such as ATSG_160 for Enwik9.

Table 3. Analogy results on Enwik9 (%).   **Table 4.** Analogy results on Wesbury (%).

Model	Total	Semantic	Syntactic
CBOW	31.82	25.44	37.14
SG	28.64	22.02	34.12
SICBOW	20.84	10.76	29.22
SISG	32.53	19.08	**43.70**
mCBOW	32.49	21.77	41.40
ATSG_40	27.19	21.56	31.87
ATSG_60	32.69	26.00	38.25
ATSG_80	34.55	26.38	41.33
ATSG_100	34.59	26.53	41.28
ATSG_120	**35.61**	**26.90**	<u>42.85</u>
ATSG_140	<u>34.89</u>	26.17	42.13
ATSG_160	35.21	<u>26.77</u>	42.22

Model	Total	Semantic	Syntactic
CBOW	37.41	34.01	40.23
SG	31.82	25.44	37.14
SICBOW	35.55	29.68	40.42
SISG	38.35	33.11	42.71
ATSG_20	18.95	17.57	20.10
ATSG_30	29.42	27.13	31.32
ATSG_40	35.15	31.25	38.38
ATSG_50	38.95	33.50	<u>43.34</u>
ATSG_60	**39.41**	**34.51**	**43.48**
ATSG_70	38.98	<u>34.39</u>	42.81
ATSG_80	<u>39.00</u>	33.94	43.21

4 Conclusion

This paper proposes the problem of representing words and subwords by different lengths of vectors. The proposed ATSG model shows its superiority on human similarity judgement and analogy tasks comparing to the baselines. The results also suggest that within a suitable range, a higher dimensionality of subword vectors can always improve the quality of word embeddings.

Acknowledgements. We thank the reviewers for their valuable comments. This research is supported by National Natural Science Foundation of China (No. U1836109 and No. 61772289).

References

1. Bojanowski, P., Grave, E., Joulin, A., Mikolov, T.: Enriching word vectors with subword information. In: ACL, pp. 135–146 (2017)
2. Luong, T., Socher, R., Manning, C.D.: Better word representations with recursive neural networks for morphology. In: CoNLL, pp. 104–113 (2013)
3. Mikolov, T., Chen, K., Corrado, G., Dean, J.: Efficient estimation of word representations in vector space (2013)
4. Mikolov, T., Sutskever, I., Chen, K., Corrado, G.S., Dean, J.: Distributed representations of words and phrases and their compositionality. In: NIPS, pp. 3111–3119 (2013)
5. Qiu, S., Cui, Q., Bian, J., Gao, B., Liu, T.Y.: Co-learning of word representations and morpheme representations. In: COLING, pp. 141–150 (2014)

Using Dilated Residual Network to Model Distantly Supervised Relation Extraction

Lei Zhan[1], Yan Yang[1(✉)], Pinpin Zhu[2], Liang He[1,3], and Zhou Yu[4]

[1] Department of Computer Science and Technology, East China Normal University,
Shanghai, China
`topzlei@foxmail.com`, {`yanyang,lhe`}`@cs.ecnu.edu.cn`
[2] Xiaoi Robot Technology Co., Ltd., Shanghai, China
`pp@xiaoi.com`
[3] Shanghai Key Laboratory of Multidimensional Information Processing,
Shanghai, China
[4] Computer Science Department, University of California, Oakland, USA
`joyu@ucdavis.edu`

Abstract. Distantly supervised relation extraction has been widely used to find relational facts in the text. However, distant supervision inevitably brings in noise that can lead to a bad relation contextual representation. In this paper, we propose a deep dilated residual network (DRN) model to address the noise of in distantly supervised relation extraction. Specifically, we design a module which employs dilated convolution in cascade to capture multi-scale context features by adopting multiple dilation rates. By combining them with residual learning, the model is more powerful than traditional CNN model. Our model significantly improves the performance for distantly supervised relation extraction on the large NYT-Freebase dataset compared to various baselines.

Keywords: Distant supervision · Relation extraction · Knowledge graph

1 Introduction

Relation extraction (RE) is the task of predicting the relations between two entities in a sentence [3], it is a fundamental problem for enriching knowledge bases (KBs). One of the main challenges in relational extraction task is to generate lebeled training data because it is time-consuming and labor-intensive. Distant supervision strategy is an effective method for relation extraction due to its ability to automatically label training data. Despite its effectiveness, it is plagued by the wrong label problem.

Recently, the neural network based approaches have achieved good performance without any human interventions [4,5]. Among all the neural network approaches for distantly supervised relation extraction, one of the well-known architectures is the Convolutional Neural Network (CNN). Zeng et al. [5] proposed a convolutional neural network based model on this task and achieved the

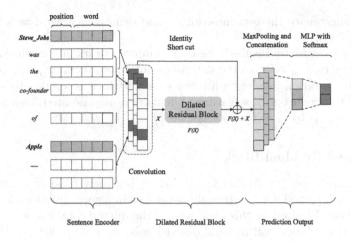

Fig. 1. The architecture of DRN model.

state-of-the-art performance. Following their success, Zeng et al. [4] proposed a piece-wise max-pooling strategy to improve the CNNs. For the distant supervision noisy problem, there are two important issues that need to be addressed. The first is the representation of each sentence, and the second is the selection of the sentence. To the best of our knowledge, most existing approaches focused on the wrong label problem and ignored the well representation of the valuable sentences with entities. The traditional CNNs cannot capture enough context relation features of each sentence which keep long dependency between two entities. Moreover, some shallow CNN models do not keep out the noisy data problem and select a better sentence.

To solve these two distant supervision problem. We propose a deep dilated residual network (DRN). DRN integrates a rich and multi-level relation features for enhancing the distantly supervised relation extraction model. Specifically, we revisit the application of dilated convolution, which allows us to effectively enlarge the field of views of filters to represent long dependency between two entities. Then we combine it with residual learning to form a deep network which can obtain better sentence representations automatically. Finally, experiments on the NYT-Freebase dataset [1] show that our proposed DRN model is effective to boost the performance of this task.

2 Our Proposed Model

2.1 Sentence Encoder

As shown in Fig. 1, given a set of sentences and two corresponding entities e_1 and e_2, our model aims to find the most possible relation for these two entities. Let x_i be the i-th word in the sentence. Each word will access two embedding look-up tables to get the word embedding WF_i and the position embedding PF_i.

Then we concatenate the two embeddings and denote each word as a vector of $v_i = [WF_i; PF_i]$.

To get the contextual embedding of a sentence, we use a convolutional neural layer. Let $v_{(i:i+j)}$ refer to the concatenation of words v_i, $v_{(i+1)},...,v_{(i+j)}$. A convolution operation involves a filter $w \in \mathbb{R}^{h \times d}$ which is applied to a window of h words to produce a new feature. A feature c_i is generated from a window of word $v_{(i:i+h-1)}$ by $c_i = f(w \cdot x_{i:i+h-1} + b)$.

2.2 Dilated Residual Block

Dilated Convolution (or Atrous Convolution) is a new type of convolutional neural network [2] and it was originally developed in algorithm atrous for wavelet decomposition. Inspired by this idea, we use the dilated convolution in our task to capture rich relation features and discard noisy data. Specifically, We use the dilated convolution of different rates to obtain features at different levels for better classification.

Here are three convolutional filter $w_1, w_2, w_3 \in \mathbb{R}^{h \times 1}$ and D_r means dilated convolution with dilation rate r. For the first dilated convolutional layer, c' can be calculated by using $c' = D_{r=1}(w_1 \cdot c + b_1)$, where $D_{r=1}$ denotes that the dilation rate is 1. For the second dilated convolutional layer, c'' is given as $c'' = D_{r=2}(w_2 \cdot c' + b_2)$, where $D_{r=2}$ denotes that the dilation rate is 2. For the third dilated convolutional layer, c''' is given as $c''' = D_{r=5}(w_3 \cdot c'' + b_3)$, $D_{r=2}$ denotes that the dilation rate is 5. Here b_1, b_2, b_3 are bias terms. For the residual learning operation: $c = 0.3 \times c + 0.7 \times c'''$.

2.3 Prediction Output

We then apply a max-pooling operation over the feature and take the maximum value $\widehat{c} = max(c)$. Next we take all features into one high level extracted feature $z = [\widehat{c}_1, \widehat{c}_2, ..., \widehat{c}_m]$ (note that here we have m filters). Then, these features are passed to a fully connected softmax layer whose output is the probability distribution over all possible relations. In the test procedure, the learned weight vectors are scaled by p such that $\widehat{w} = pw$ and used to score unseen instances.

3 Results and Analysis

We compare the proposed model with baselines. our DRN model achieves the best performance which shows that the proposed structure is superior.

The distantly supervised relation extraction needs to fully consider the local part of sentence information and the global semantic information for accurate classification. The intuition of DRN helps this task in some aspect. First, a deeper neural network can obtain richer features like hidden semantic representations and each layer of dilated convolution network can acquire different levels of features like hidden lexical or syntactic representations. Residual learning helps to bypass the syntactic to connect lexical and semantic space directly.

Second, dilated convolution allows us to enlarge the field of view of filters to incorporate larger context and it can work in multi-rate processing. We set the dilated convolution of the small rate to focus on the features around the entities and set the big rate dilated convolution to obtain the features between two entities. These multi-scales features contribute to the effect of the classifier. Finally, DRN tackles the gradient vanishing problem by using residual network which is an inevitable problem in very deep neural network. We use deep networks to automatically learn the importance of different sentences rather than building attentions. It decreases the effect of noise in distant supervision data. In addition to the above two representative baselines, we compare our model with the recent progress in distantly supervised relation extraction.

The results from recent work on distantly supervised relation extraction are summarized in Table 1. We observe that our DRN model outperforms all models that do not select training instances artificially. Even without piecewise max-pooling and instance-based attention, our model performs better than PCNN+ATT. We also achieve better performance than ResCNN-9 on a deeper CNN model. It proves the advantage of the dilated residual block in deep network.

Table 1. P@N for relation extraction with different models.

P@N(%)	100	200	300	Mean
CNN-B (Zeng et al. 2014)	41.0	40.0	41.0	40.7
CNN+ATT (Lin et al. 2016)	76.2	68.6	59.8	68.2
PCNN (Zeng et al. 2015)	72.3	69.7	64.1	68.7
PCNN+ATT (Lin et al. 2016)	76.2	73.1	67.4	72.2
ResCNN-9 (Huang et al. 2017)	79.0	69.0	61.0	69.7
ResCNN-13 (Huang et al. 2017)	76.0	65.0	60.3	67.1
DRN-13 (proposed)	83.0	73.5	67.3	74.6

4   Conclusion and Future Work

In this paper, we introduce a model combined residual learning and dilated convolution to extract lexical and sentence level features for distantly supervised relation extraction. We demonstrate that deeper layers and larger field of view are useful for noisy problems and show the impressive result of dilated convolutions on relation extraction task. The performances of our model are significantly improved. In future, we will find more applications of our model.

Acknowledgement. This research is funded by the Science and Technology Commission of Shanghai Municipality (No. 18511105502), and Xiaoi Research.

References

1. Riedel, S., Yao, L., McCallum, A.: Modeling relations and their mentions without labeled text. In: Balcázar, J.L., Bonchi, F., Gionis, A., Sebag, M. (eds.) ECML PKDD 2010. LNCS (LNAI), vol. 6323, pp. 148–163. Springer, Heidelberg (2010). https://doi.org/10.1007/978-3-642-15939-8_10
2. Yu, F., Koltun, V.: Multi-scale context aggregation by dilated convolutions. CoRR abs/1511.07122 (2015). http://arxiv.org/abs/1511.07122
3. Zelenko, D., Aone, C., Richardella, A.: Kernel methods for relation extraction. J. Mach. Learn. Res. **3**, 1083–1106 (2003). http://dl.acm.org/citation.cfm?id=944919.944964
4. Zeng, D., Liu, K., Chen, Y., Zhao, J.: Distant supervision for relation extraction via piecewise convolutional neural networks. In: Proceedings of the 2015 Conference on Empirical Methods in Natural Language Processing, pp. 1753–1762. Association for Computational Linguistics (2015). https://doi.org/10.18653/v1/D15-1203
5. Zeng, D., Liu, K., Lai, S., Zhou, G., Zhao, J.: Relation classification via convolutional deep neural network. In: Proceedings of COLING 2014, the 25th International Conference on Computational Linguistics: Technical Papers, pp. 2335–2344. Dublin City University and Association for Computational Linguistics (2014)

Modeling More Globally: A Hierarchical Attention Network via Multi-Task Learning for Aspect-Based Sentiment Analysis

Xiangying Ran[1,2], Yuanyuan Pan[1,2], Wei Sun[1,2], and Chongjun Wang[1,2(✉)]

[1] National Key Laboratory for Novel Software Technology, Nanjing University, Nanjing, China
lebronran@gmail.com, yypan112@gmail.com, weisun_@outlook.com,
chjwang@nju.edu.cn
[2] Department of Computer Science and Technology, Nanjing University, Nanjing, China

Abstract. Aspect-based sentiment analysis (ABSA) is a fine-grained sentiment analysis problem, which has attracted much attention in recent years. Previous methods mainly devote to employing attention mechanism to model the relationship between aspects and context words. However, these methods tend to ignore the overall semantics of sentence and dependency among the aspect terms. In this paper, we propose a Hierarchical Attention Network (HAN) to solve the aforementioned issues simultaneously. Experimental results on standard SemEval 2014 datasets demonstrate the effectiveness of the proposed model.

Keywords: Aspect-based · Sentiment analysis · Multi-task learning

1 Introduction

Aspect-based sentiment analysis (ABSA) is a fine-grained sentiment analysis problem, which has attracted much attention in recent years [3,5]. Given a pair of sentence and aspect, ABSA aims at predicting the sentiment polarity (i.e., *positive*, *neutral*, or *negative*) expressed towards the given aspect. Recent works [1,7] try to apply attention mechanism to enforce the model to pay more attention to the related part of the sentence about the given aspect, and finally get a weighted contextual representation for sentiment prediction.

This paper is supported by the National Key Research and Development Program of China (Grant No. 2016YFB1001102), the National Natural Science Foundation of China (Grant Nos. 61876080, 61502227), the Fundamental Research Funds for the Central Universities No.020214380040, the Collaborative Innovation Center of Novel Software Technology and Industrialization at Nanjing University.

G. Li et al. (Eds.): DASFAA 2019, LNCS 11448, pp. 505–509, 2019.
https://doi.org/10.1007/978-3-030-18590-9_76

We propose a novel model named Hierarchical Attention Network (HAN) to enhance previous attention-based methods. HAN firstly attempts to incorporate the overall semantics of sentence into the existing attention mechanism via a multi-task learning way, and then HAN employs another attention layer to learn to fuse the aspect terms appearing in the same sentence to model the potential dependency. Finally, the sentence representations of the two layers are combined via a novel gate mechanism to obtain the final aspect-specific sentence representation.

2   Proposed Model

Given the sentence $\mathcal{S} = \{w_1, w_2, \ldots, w_n\}$ and the aspect term $\mathcal{A} = \{w_{k+1}, w_{k+2}, \ldots, w_{k+m}\}, 0 \leq k \leq n - m$ explicitly appearing in the given sentence, ABSA aims at predicting the sentiment polarity of the aspect term. Here we associate each word w_i with a continuous word embedding x_i.

2.1   Semantic Attention-Based LSTM

In this section, we introduce the proposed **Semantic Attention-based LSTM** (**SAtt-LSTM**) firstly. Here we incorporate the overall sentence semantics into the existing attention mechanism via a multi-task learning way.

Firstly, we employ LSTM to capture sequential information in the sentence and produce a list of hidden vectors: $\mathbf{H} = \{h_1, h_2, \ldots, h_n\} \in \mathbb{R}^{d_h \times n}$. Next, we use the last hidden state h_n of LSTM as the semantic of the sentence and incorporate it into attention mechanism to get the final aspect-specific sentence representation r^s as follow:

$$r^s = \sum_{i=1}^{n} \alpha_i h_i \tag{1}$$

And α_i is obtained by:

$$\mathbf{a} = \frac{1}{m} \sum_{i=1}^{m} x_{k+i} \tag{2}$$

$$\gamma_i = W_m tanh\left(W_h \left[h_i; h_n; \mathbf{a}\right] + b_h\right) \tag{3}$$

$$\alpha_i = softmax\left(\gamma_i\right) \tag{4}$$

Especially, to guide **SAtt-LSTM** to learn more reasonable sentence semantics for sentiment prediction, we introduce a novel task: **Polarity Number Prediction (PNP)**, the main aim of the task is identifying the number of unique sentiment polarities contained in the given sentence based on the sentence semantics. The PNP task is trained with sentiment prediction jointly to improve the performance.

2.2   Attention-Based Aspect Fusion

Assuming there are K aspects in the sentence, the aspect-specific sentence representations of all aspects are $\mathbf{R} = \{r_1^s, r_2^s, \ldots, r_K^s, r_{K+1}^s\}$, r_{K+1}^s denotes the aspect-specific sentence representation of dummy aspect, which is randomly initialized and fine tuned during training. Without loss of generality, we take the i-th aspect as an example to describe aspect fusion here. Firstly, we employ Bi-LSTM to produce a list of hidden vectors[1]: $\mathbf{H}^a = \{h_1^a, h_2^a, \ldots, h_{K+1}^a\} \in \mathbb{R}^{2d_a \times (K+1)}$:

Next, attention mechanism is applied to model the dependency among aspects. The final *peer-aware* aspect-specific sentence representation r_i^a can be calculated as follows:

$$r_i^a = \sum_{j \in \{1,2,\ldots,K+1\}\setminus i} \beta_j h_j^a \tag{5}$$

And β_j is obtained by:

$$\nu_j = W_b tanh \left(W_a \left[h_j^a; h_i^a \right] + b_a \right) \tag{6}$$

$$\beta_j = \frac{\exp(\nu_j)}{\sum_{q \in \{1,2,\ldots,K+1\}\setminus i} \exp(\nu_q)} \tag{7}$$

And then, we use a gate mechanism to combine them : $r^{final} = g \odot r^s + (1 - g) \odot r^a$, and $g \in \mathbb{R}^{2d_a}$ is calculated as: $g = \sigma \left(W_g \left[r^s; r^a \right] + b_g \right)$

At last, we use r^{final} as the final sentence representation to sentiment prediction and train it together with the task of PNP. We use the coefficient λ to tradeoff the loss function of these two tasks to improve the performance of ABSA task.

3   Experiments

3.1   Experiment Settings

We conduct experiments on two standard datasets: Restaurant and Laptop, which are from SemEval 2014 [5], containing reviews of restaurant and laptop domains respectively. For these two datasets, following the setting of Tang et al. [6], we remove the training examples with the "conflict" label.

We use pre-trained GloVe vectors [4], the dimension is 300. To ease overfitting, we apply dropout with dropout rate 0.5. The λ value we adopt is 0.1 for all datasets. The hidden dimensions d_h and d_a are 300 and 150. Adam [2] is adopted as the optimizer here with learning rate 0.001 and batch size 25.

---

[1] We run two LSTM forwards and backwards respectively and concatenate the two hidden vectors in i-th time step to produce h_i^a.

3.2 Result Comparisons and Analysis

Result Analysis. Experimental results can be seen in Table 1. The main evaluation metrics are Accuracy and Macro-averaged F1-score. As shown in Table 1, our HAN model consistently outperforms all compared methods on both two datasets, which verifies the efficacy of our model. On the other hand, we can see from Table 1 that our SAtt-LSTM model performs better than most of baselines except RAM[2], which verifies the effectiveness of our proposed semantic attention-based LSTM. Besides, HAN performs better than SAtt-LSTM. And the improvement of HAN relative to SAtt-LSTM is the attention-based aspect fusion, which can also prove the advantage of considering the dependency among the aspects appearing in the same sentence by attention mechanism.

Table 1. Experimental results (%). The results with symbol ♮ are retrieved from the corresponding papers. N/A means *not available*. The best results are in bold.

Model	Laptop		Restaurant	
	Accuracy	Macro-F1	Accuracy	Macro-F1
SVM [5]	70.49♮	N/A	80.16♮	N/A
GCNN [8]	72.84	59.70	79.31	65.55
TD-LSTM [6]	72.10	64.03	78.66	67.84
ATAE-LSTM [7]	72.22	66.98	78.12	64.14
RAM [1]	74.29	70.26	79.64	68.43
SAtt-LSTM	74.14	68.57	79.64	67.99
HAN	75.71	71.25	80.18	69.22

4 Conclusion

In this paper, we propose a Hierarchical Attention Network (HAN) for aspect-based sentiment analysis, which utilizes the overall sentence representation and fusion among aspects to enhance the vanilla attention mechanism. The proposed model can be viewed as a general algorithm framework to improve the performance of existing attention-based methods further. Experimental results on SemEval 2014 Dataset demonstrate the effectiveness of the proposed model.

---

[2] It is worth noting that RAM utilizes additional position information.

References

1. Chen, P., Sun, Z., Bing, L., Yang, W.: Recurrent attention network on memory for aspect sentiment analysis. In: Proceedings of the 2017 Conference on Empirical Methods in Natural Language Processing, pp. 452–461 (2017)
2. Kingma, D.P., Ba, J.: Adam: a method for stochastic optimization. arXiv preprint arXiv:1412.6980 (2014)
3. Liu, B.: Sentiment analysis and opinion mining. Synth. Lect. Hum. Lang. Technol. 5(1), 1–167 (2012)
4. Pennington, J., Socher, R., Manning, C.D.: Glove: global vectors for word representation. In: Proceedings of the 2014 Conference on Empirical Methods in Natural Language Processing, pp. 1532–1543 (2014)
5. Pontiki, M., Galanis, D., Pavlopoulos, J., Papageorgiou, H., Androutsopoulos, I., Manandhar, S.: SemEval-2014 task 4: aspect based sentiment analysis. In: Proceedings of the 8th International Workshop on Semantic Evaluation, pp. 27–35 (2014)
6. Tang, D., Qin, B., Feng, X., Liu, T.: Effective LSTMs for target-dependent sentiment classification. In: Proceedings of COLING 2016, the 26th International Conference on Computational Linguistics: Technical Papers, pp. 3298–3307 (2016)
7. Wang, Y., Huang, M., Zhao, L., et al.: Attention-based LSTM for aspect-level sentiment classification. In: Proceedings of the 2016 Conference on Empirical Methods in Natural Language Processing, pp. 606–615 (2016)
8. Xue, W., Li, T.: Aspect based sentiment analysis with gated convolutional networks. In: Proceedings of the 56th Annual Meeting of the Association for Computational Linguistics, pp. 2514–2523 (2018)

A Scalable Sparse Matrix-Based Join for SPARQL Query Processing

Xiaowang Zhang[1,2(✉)], Mingyue Zhang[1,2], Peng Peng[3], Jiaming Song[1,2], Zhiyong Feng[1,2], and Lei Zou[4]

[1] College of Intelligence and Computing, Tianjin University, Tianjin, China
xiaowangzhang@tju.edu.cn
[2] Tianjin Key Laboratory of Cognitive Computing and Application, Tianjin, China
[3] College of Computer Science and Electronic Engineering, Hunan University,
Changsha, China
[4] Institute of Computer Science and Technology, Peking University, Beijing, China

Abstract. In this paper, we present gSMat, a SPARQL query engine for RDF datasets. It employs join optimization and data sparsity. We bifurcate gSMat into three submodules e.g. Firstly, SM Storage (Sparse Matrix-based Storage) which lifts the storage efficiency, by storing valid edges, introduces a predicate-based hash index on the storage and generate a statistic file for optimization. Secondly, Query Planner which holds Query Parser and Query Optimizer. The Query Parser module parses a SPARQL query and transformed it into a query graph and the latter generates the optimal query plan based on statistical input from SM Storage. Thirdly, Query Executor module executes query in an efficient manner. Lastly, gSMat evaluated by comparing with some well-known approaches like gStore and RDF3X on very large datasets (over 500 million triples). gSMat is proved as significantly efficient and scalable.

1 Introduction

Over decades, Resource Description Framework (RDF) has become very common to model information on Web in the form of a triple: (*subject, predicate, object*) with its official query language SPARQL. The existing query engines for SPARQL can be categorized into two major kinds in basis of storage strategies e.g. relation-based approach and graph-based approach [1]. The existing stores take an RDF triple as a tuple in a ternary relation and the latter stores RDF data as a directed-labeled graph (e.g., gStore). They improve performance for query processing but limited improvement for large-scale datasets of real words such as Watdiv DBpedia and YAGO with millions of triples. It is due to little consideration of an important feature called *sparsity* of those practical RDF data. The sparsity of RDF data means that the neighbors of each vertex in an RDF graph take a quite small proportion of the whole vertices. The sparsity of RDF data means that the neighbors of each vertex in an RDF graph take a quite small proportion of the whole vertices. In fact, the sparsity of RDF data exists everywhere. For an instance, There are over 99.41% nodes with at most

© Springer Nature Switzerland AG 2019
G. Li et al. (Eds.): DASFAA 2019, LNCS 11448, pp. 510–514, 2019.
https://doi.org/10.1007/978-3-030-18590-9_77

43° (sum of out-degrees and in-degrees) in DBpedia (42,966,066 nodes in total). SPARQL is built on basic graph pattern (BGP) and SPARQL algebra operators. The join for concatenating variables is the core operation of SPARQL query evaluation, as BGP is the join of triple patterns. In an RDF graph, the product of their adjacent matrices can compute the join (concatenation) of two vertices. For instance, BitMat [2] employs matrices to compute join of SPARQL over RDF data. In proposing gSMat framework, our contributions for this demonstration are as followed.

- We proposed SM Storage (Sparse Matrix-based Storage) a predicate-based hash index storage, which lifts the storage efficiency.
- We have improve SPARQL join query evaluation over large-scale RDF data by employing sparsity.
- We transformed the problem of SPARQL query processing to the multiplication operations of multiple sparse matrices, and each triple pattern corresponds to a sparse matrix in the SPARQL query. A sparse matrix is a kind of matrix which consists of large numbers of zero values with very few dispersed non-zero ones. Comparing to general matrices, sparse matrices characterize the sparsity of RDF data well.

Finally, we compared our approach with existing methodologies. It significantly improve BGP query evaluation as compared to others, especially over practical RDF data with 500 million triples.

2   Overview of gSMat

Sparse Matrix-Based Storage: This module maintains a collection of Sparse Matrix-based (SM-based) tables indexed by predicates and the statistics of predicates from RDF dataset. The SM-based storage provides a compact and efficient method to store RDF physically and supports high-performance algorithm for query processing (see Fig. 1). We develop a data dictionary by encoding a string of raw data into positive integers so that we can construct two bijective mappings h_p and h_{so}. The goal of this encoding is to reduce space usage and disk I/O. Secondly, we store RDF Cube (a logical model of SM-based storage) as sets of matrices via a index of predicates. Finally, a series of sparse matrices are generated from the RDF cube indexed by predicates.

Query Planner: This module has two submodules, *Query Parser* and *Query Optimizer*. The query parser parses a SPARQL query as a query graph and the latter generates the optimal query plan on the bases of SM-based storage which owns a sparse matrix. The cost of query is directly proportional to the no of matrices (the intermediate results). The Simple approach is adopted, if no of matrices are greater, the cost of the query will be greater. The query planner passed the optimal plan to query execution module by calculating the query cost.

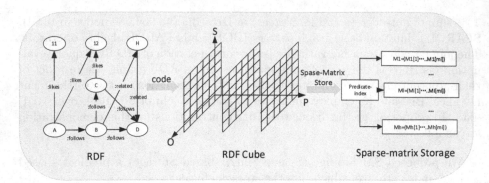

Fig. 1. The framework of SM-Storage

Query Executor: It is well known that joins of relations and matrix multiplication is essentially the same problem. The entire reason is that both are essentially reachability in two steps in a graph. The SPARQL query is transformed into the multiplications of a series of sparse matrices. Therefore, table-based join operations are converted to sparse matrix-based join (SM-based join) that approximates multiplications for sparse matrices. After obtaining an optimized query, a series of sparse matrices can be obtained based on the predicates of each triple pattern in the query (assuming the predicate is not a variable, which is known). A SPARQL query is a continuous SM-based join operation. For a whole SPARQL query, we need to loop a join operation in a query plan. Matrix-based multiplication depends on our sparse matrix multiplication algorithm. Due to the length of the article, it is not described here. Experiments show that the sparse matrix proportionality can effectively reduce the intermediate results of the query and increase the query efficiency.

3   Experiments

We performed experiments on a machine with following specifications, Intel(R) Xeon(R) E5-2603 v4 1.7 GHz amd processor, Linux ubuntu 14.04 as OS and main memory is 72 GB. The experiments are conducted on three leading RDF benchmarks e.g watdiv (https://dsg.uwaterloo.ca/watdiv/), YAGO (http://yago-knowledge.org/) and DBpedia (http://dbpedia.org/) with million of triples. We have compared gSMat with well known and efficient query engines like gStore [3], RDF-3X [4]. For Watdiv bench mark we employ four types of queries snowflake(S), linear(L), star(S), and complex(C) The results shows in Tables 1 and 2 that gSMat proved to be efficient than RDF-3X and gStore. In comparison gstore remains unsuccessful to execute watdiv with 500 million triples. As for DBpedia we compared for the same four type of queries as shown in Fig. 2(b), but for YAGO we compared new designed seven queries, as it does not provide benchmark queries, As an experimental result, gSMat is more efficient in processing scable RDF data as shown in Fig. 2(a).

Table 1. Query runtime for Watdiv 100, 300, and 500

	Watdiv 100				Watdiv 300			
	C	F	L	S	C	F	L	S
gSMat	3378.1	3784.1	1932.5	2503.3	18106.2	9185.3	5033.3	5350.3
RDF-3X	11980.3	8405.6	16252.1	3820.6	39571.1	26231.3	56879.2	16066.5
gStore	15447.0	29204.8	20128.0	10808.7	111490.8	1047971.5	311390.4	2996687.5

Table 2. Query runtime for Watdiv 500 million triples

	Watdiv 500			
	C	F	L	S
gSMat	35509.1	16071.4	9719.6	9072.4
RDF-3X	66513.7	46906.9	92685.6	27456.7

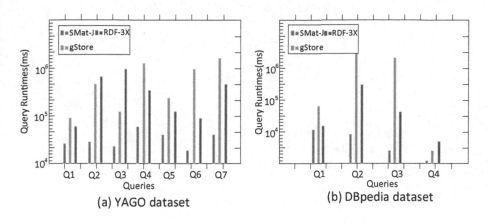

Fig. 2. Average query runtime for YAGO and DBpedia (ms)

4   Conclusion

By visualizing above experimental results and discussion, it is concluded that gSMat has outclassed the RDF3X and gStore, especially, for large scale RDF datasets. In the future we will try gSMat framework on GPU, which will helpful in fasting query execution [5] and try engage distributed parallelism.

Acknowledgement. This work is supported by the National Natural Science Foundation of China (NSFC) (61672377), National Key Research and Development Program of China (2016YFB1000603), and Key Technology Research and Development Program of Tianjin (16YFZCGX00210).

References

1. Abdelaziz, I., Harbi, R., Khayyat, Z.: A Survey and experimental comparison of distributed SPARQL engines for very large RDF data. PVLDB **10**(13), 2049–2060 (2017)
2. Atre, M., Chaoji, V., Zaki, M.J., Hendler, J.A.: Matrix "Bit" loaded: a scalable lightweight join query processor for RDF data. In: Proceedings of WWW, pp. 41–50 (2010)
3. Zou, L., Ozsu, M.T., Chen, L., Shen, X., Huang, R., Zhao, D.: gStore: a graph-based SPARQL query engine. VLDB J. **23**(4), 565–590 (2014)
4. Neumann, T., Weikum, G.: The RDF-3X engine for scalable management of RDF data. VLDB J. **19**(1), 91–113 (2010)
5. Song, J., Zhang, X., Peng, P., Feng, Z., Zou, L.: MapSQ: a MapReduce-based framework for SPARQL queries on GPU. In: Proof of WWW Compassion, pp. 81–82 (2018)

Change Point Detection for Streaming High-Dimensional Time Series

Masoomeh Zameni[1]($\boxtimes$), Zahra Ghafoori[1], Amin Sadri[2], Christopher Leckie[1], and Kotagiri Ramamohanarao[1]

[1] University of Melbourne, Melbourne, Australia
{masoomeh.zameni,zahra.ghafoori,caleckie,kotagiri}@unimelb.edu.au
[2] RMIT University, Melbourne, Australia
amin.sadri@rmit.edu.au

Abstract. An important task in analysing high-dimensional time series data generated from sensors in the *Internet of Things* (IoT) platform is to detect changes in the statistical properties of the time series. Accurate, efficient and near real-time detection of change points in such data is challenging due to the streaming nature of it and the presence of irrelevant time series dimensions. In this paper, we propose an unsupervised Information Gain and permutation test based change point detection method that does not require a user-defined threshold on change point scores and can accurately identify changes in a sequential setting only using a fixed short memory. Experimental results show that our efficient method improves the accuracy of change point detection compared to two benchmark methods.

Keywords: Change detection · Information Gain · Permutation test

1 Introduction

Change point detection is detecting time points when the statistical properties of a time series, such as the mean or variance, changes. This problem actively studied for a variety of applications such as human activity segmentation (Fig. 1). Existing change point detection methods are batch [1] or sequential [2]. Batch change point detection methods require the whole time series data to be stored for detecting change points. Sequential change point detection methods, however, detect change points as data points arrive. In real-world applications, data is usually streaming which necessitates sequential methods. Most change point detection approaches estimate the probability distribution of the data points in the windows before and after the time point t and then assess the divergence between these two distributions using a user-defined threshold. However, estimating the distribution and setting the user-defined threshold are highly domain dependant. Moreover, in high-dimensional time series only some dimensions may be responsible for a change, which degrades the performance of probability distribution-based techniques. In this paper, we propose a *Sequential Information*

© Springer Nature Switzerland AG 2019
G. Li et al. (Eds.): DASFAA 2019, LNCS 11448, pp. 515–519, 2019.
https://doi.org/10.1007/978-3-030-18590-9_78

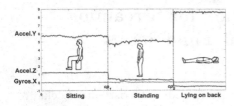

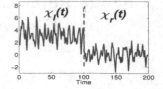

Fig. 1. Change point detection for activity segmentation.

Fig. 2. A sample change point at time point t.

Gain Permutation-based (SequentialIGPerm) method that sequentially detects change points in high-dimensional time-series. Our distribution-free method uses a permutation test to validate candidate change points.

$X = \langle x(1), \cdots, x(t), \cdots \rangle$ denotes an *m-dimensional streaming time series*, where $x(t) \in \Re^m$ arrives at time t from m sensors. Let $\chi_l(t) := \{x(t - N + i)\}_{i=1}^{N}$ and $\chi_r(t) := \{x(t + i)\}_{i=1}^{N}$ be the windows that include N data points before and after time t, respectively (Fig. 2) that are generated according to some unknown probability density functions pdf_1 and pdf_2. Time t is a change point if $pdf_1 \neq pdf_2$. A change point detection method divides X into k non-overlapping *segments*. Each change point cp_j denotes the point between two consecutive segments. SequentialIGPerm aims to sequentially detect the change points at the arrival of a new sequence of the data.

2   Background

The *Information Gain* (IG) metric proposed in [3] used for calculating the change point scores in a batch setting. The increase in IG, the expected reduction in the entropy, for segmenting the high-dimensional time series X into k segments (using $k - 1$ change points) is defined using the cost function $\mathcal{L}_k$ as $\mathcal{L}_k(X, T) = H(X) - \sum_{j=1}^{k} \frac{|\mathbf{s}_j|}{|X|} H(\mathbf{s}_j)$, where $T = \{cp_1, cp_2, \cdots, cp_{k-1}\}$ is the set of detected change points, $\mathbf{s}_j$ is the jth segment, $|.|$ is the length operator, $H(X)$ is the entropy of the whole time series as a segment, and $H(\mathbf{s}_j)$ is the entropy of the segment $\mathbf{s}_j$ measured as $H(\mathbf{s}_j) = -\sum_{i=1}^{m} p_{ji} \log p_{ji}$, where m is the number of time series dimensions, and p_{ji} ($p_{ji} \leq 1$, $\sum_{i=1}^{m} p_{ji} = 1$) is the mass ratio of time series dimension c_i in segment $\mathbf{s}_j$ (the area under time series dimension c_i in segment $\mathbf{s}_j$ divided by the sum of the areas under all dimensions in segment $\mathbf{s}_j$) defined as $p_{ji} = \frac{\sum_{q \in \mathbf{s}_j} q(c_i)}{\sum_{d=1}^{m} \sum_{q \in \mathbf{s}_j} q(c_d)}$, where $q(c_i)$ denotes the value of the qth point of the ith time series dimension within the segment $\mathbf{s}_j$. The greatest reduction in the entropy of the signals is achieved when the time series is divided into the most coherent segments (the lowest variance segments) [3].

3   Proposed Method

For each data point $x(t)$ in a new data sequence, SequentialIGPerm computes its change point score (IG) using the data points in the two sub-windows $\chi_l(t)$

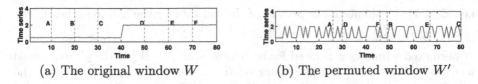

(a) The original window W (b) The permuted window W'

Fig. 3. Example of permutation test when a significant change point exists at $t = 41$. $\mathcal{L}(W, t) = 0.57$, whereas $\mathcal{L}(W', t)$ is reduced to 0.02 ($\mathcal{L}(W, t) \gg \mathcal{L}(W', t)$).

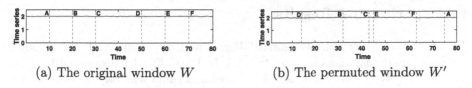

(a) The original window W (b) The permuted window W'

Fig. 4. Example of permutation test when no change point exists at $t = 41$. $\mathcal{L}(W, t) = 0.0087$, whereas $\mathcal{L}(W', t)$ is 0.0086 ($\mathcal{L}(W, t) \approx \mathcal{L}(W', t)$).

and $\chi_r(t)$. This indicates how much the IG of segmenting $W = \{\chi_l(t), \chi_r(t)\}$ is increased when $x(t)$ is considered as a change point. As adding each change point always increases IG (see [3] for the proof), we propose a method that tests if the increase of IG is likely to be due to statistical change in the data. We use a permutation test that generates a new window $W' = \{\chi'_l(t), \chi'_r(t)\} = \{x(\Pi(i)); x(i) \in W, i = \{1, .., |W|\}\}$, such that $\Pi(i = t) = t$ and $\Pi(i \neq t) \in \{1, 2, \cdots, |W|\} - \{t\}$ is the permutation function. W' is generated by randomly permuting data points around $x(t)$ in W. The data distribution is the same for both W and W' but the order of data points in W and W' is different. Then, we compute the increase in IG, i.e., $\mathcal{L}_1(W, t)$ and $\mathcal{L}_1(W', t)$, when W and W' are segmented into two segments considering time $t = cp_j$ as the change point.

Theorem 1. *Define the alternative hypothesis $H_1 : \mathcal{L}(W, t) \gg \mathcal{L}(W', t)$ against the null hypothesis $H_0 : \mathcal{L}(W, t) \approx \mathcal{L}(W', t)$. If a change point occurs at time t, H_0 would be rejected.*

Proof. Consider two scenarios: (a) no change point occurs at time t, and (b) a change point occurs at time t. We prove that H_0 is rejected in scenario (b) but is not rejected in scenario (a). In scenario (a) $\chi_l(t) \in pdf_1$ and $\chi_r(t) \in pdf_2$ and $pdf_1 \approx pdf_2 \Rightarrow \mathcal{L}(W, t) \approx \epsilon$. In contrast, $\chi'_l(t) \in \{pdf_1, pdf_2\}$ and $\chi'_r(t) \in \{pdf_1, pdf_2\}$ and $pdf_1 \approx pdf_2 \Rightarrow E(\mathcal{L}(W', t)) \approx \epsilon$, where E is the expected value and ϵ represents a small value. Thus, $\mathcal{L}(W, t) \approx E(\mathcal{L}(W', t))$, H_0 is not rejected and t is rejected as a change point. In scenario (b), $\chi_l(t) \in pdf_1$ and $\chi_r(t) \in pdf_2$ and $pdf_1 \neq pdf_2 \Rightarrow \mathcal{L}(W, t) \gg \epsilon$. In contrast, $\chi'_l(t) \in \{pdf_1, pdf_2\}$ and $\chi'_r(t) \in \{pdf_1, pdf_2\}$ and $pdf_1 \approx pdf_2 \Rightarrow E(\mathcal{L}(W', t)) \approx \epsilon$. Thus, $\mathcal{L}(W, t) \gg E(\mathcal{L}(W', t))$, H_0 is rejected and t is accepted as a change point (Figs. 3, 4).

To test H_0, we generate a number of randomly permuted windows W'. Then, we check how often $(\mathcal{L}(W, t) - \mathcal{L}(W', t)) > \epsilon$. This probability of an increase in IG, i.e., $\mathcal{L}(W, t)$, by chance is the p-value of the data under the null hypothesis. The p-value is computed as $p = \frac{\sum_{g=1}^{N_p} I(\mathcal{L}(W'_g, t) - \mathcal{L}(W, t) < \alpha)}{N_p}$, where N_p is the number of permuted windows generated for each time stamp t, W'_g is the gth permuted window and $\alpha \in [0, 1]$ is a parameter set in an unsupervised manner using the training data. If p is low, e.g., below 0.05, then $\mathcal{L}(W, t)$ is significant change point is confirmed at time t (Fig. 5).

4   Experimental Results

We compare our method, SequentialIGPerm, with two sequential change point detection methods, namely ecp [4], and RuLSIF [2]. In SequentialIGPerm, $N = 20$, i.e., $|W| = 40$. The reason for choosing a small window size is to reduce the detection delay. For each W, $k_p = 20$ randomly permuted windows W' are generated to calculate the p-value. We observed a subtle change in the accuracy when we increased k_p from 20 to 2000. We set $k_p = 20$ in our experiments. In the p-value calculation, α is set to 0.3, 0.1, and 0.03, respectively for the synthetic, DSA, and RFID datasets based on our training data in an unsupervised manner. We assume that training data X_{train} including k change points is available to set our only parameter α. The value of k is estimated using the method proposed in [3]. The value of α is set based on the average of the minimum $\mathcal{L}$ for these k true change points and the maximum $\mathcal{L}$ for the non-change points in X_{train}. The time points with a p-value lower than 0.05 are considered as change points for the test

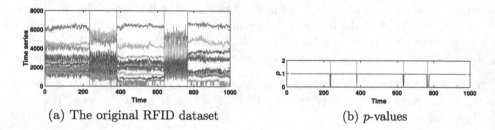

(a) The original RFID dataset   (b) p-values

Fig. 5. p-values after applying the permutation test on the RFID dataset.

Table 1. AUC and run-time results

Data	Type	Length	m	#CP	AUC			Run-time (sec)		
					SIGP[a]	ecp [4]	RuLSIF [2]	SIGP	ecp	RuLSIF
GME25 [5]	Synthetic	26000	25	25	1	0.99	0.9	0.02	0.01	0.1
GME100 [5]	Synthetic	101000	100	100	1	0.99	0.9	0.09	0.01	0.13
Syn1000 [3]	Synthetic	10000	1000	200	1	0.55	0.88	1.3	0.01	1.79
DSA [6]	Real	14000	45	111	**0.96**	0.71	0.88	0.04	0.01	0.11
RFID [7]	Real	6812	12	44	**0.91**	0.69	0.85	0.01	0.01	0.1

[a]SequentialIGPerm.

data in all our experiments. The distance margin $\delta = 20$ is set for all datasets. The parameters in ecp are set as recommended in [4]. The results for RuLSIF are reported based on the best results obtained by setting parameter values recommended in [2]. Table 1 reports the *Area Under the ROC Curve* (AUC) and run-time of the methods. The results show that SequentialIGPerm detects the true change points more accurately with than the benchmarks with comparable run-time, which confirms the accuracy of our method is not degraded by irrelevant dimensions in high-dimensional time series. The time complexity of our method, ecp, and RuLSIF are $O(nmk_p)$, $O(n|W|^2)$, and $O(nm^2|W|^2 L)$, respectively, where L is the number of retrospective subsequences that are considered for change point detection at each time t.

5   Conclusion

We proposed a novel distribution and user-defined threshold-free approach to detect change points in streaming high-dimensional time series data using an information gain metric and a permutation test. Our experimental results showed the advantages of our method compared to the examined benchmark techniques.

References

1. Vasko, K.T., et al.: Estimating the number of segments in time series data using permutation tests. In: ICDM, pp. 466–473 (2002)
2. Liu, S., et al.: Change-point detection in time-series data by relative density-ratio estimation. Neural Netw. **43**, 72–83 (2013)
3. Sadri, A., et al.: Information gain-based metric for recognizing transitions inhuman activities. In: PMC, vol. 38, pp. 92–109 (2017)
4. Matteson, D.S., et al.: A nonparametric approach for multiple change point analysis of multivariate data. J. Am. Stat. Assoc. **109**(505), 334–345 (2014)
5. Rajasegarar, S., et al.: Measures for clustering and anomaly detection in sets of higher dimensional ellipsoids. In: IJCNN, pp. 1–8 (2012)
6. Daily and sports activities data set (2014). https://archive.ics.uci.edu/ml/datasets/Daily+and+Sports+Activities
7. Yao, L., et al.: Unobtrusive posture recognition via online learning of multi-dimensional RFID received signal strength. In: ICPADS, pp. 116–123 (2015)

Demos

Distributed Query Engine for Multiple-Query Optimization over Data Stream

Junye Yang[1], Yong Zhang[1(✉)], Jin Wang[2], and Chunxiao Xing[1]

[1] RIIT, TNList, Department of Computer Science and Technology,
Tsinghua University, Beijing, China
yjy17@mails.tsinghua.edu.cn, {zhangyong05,xingcx}@tsinghua.edu.cn
[2] Computer Science Department, University of California, Los Angeles, USA
jinwang@cs.ucla.edu

Abstract. Query processing over data stream has attracted much attention in real-time applications. While many efforts have been paid for query processing of data streams in distributed environment, no previous study focused on multiple-query optimization. To address this problem, we propose EsperDist, a distributed query engine for multiple-query optimization over data stream. EsperDist can significant reduce the overhead of network transmission and memory usage by reusing operators in the query plan. Moreover, EsperDist also makes best effort to minimize the query cost so as to avoid resource bottle neck in a single machine. In this demo, we will present the architecture and work-flow of EsperDist using datasets collected from real world applications. We also propose a user-friendly to monitor query results and interact with the system in real time.

1 Introduction

Query processing over data stream has become an integral part of many real-time applications, from network traffic monitoring to financial data analysis [1] and sensor data querying [4]. A considerable amount of efforts has been paid for query processing of data streams in distributed environment, such as balancing workload, approximate query processing and quality of service. Recently many distributed stream processing systems have emerged to provide high efficiency combined with high throughput at low latency, such as Apache Storm, S4 and Samza. However, the problem of multiple-query optimization over stream data in distributed environment has never been tackled before. This is a meaningful and important problem since placement of the operators in a query plan to particular machines (nodes) in a distributed system, is an important design decision impacting the query processing performance [3]. In many real world scenarios, different queries on a same data stream could share a large portion of common computations. Take the stock market analysis as an example. Suppose there are two users want to retrieve the high-volume trades of the most active

ⓒ Springer Nature Switzerland AG 2019
G. Li et al. (Eds.): DASFAA 2019, LNCS 11448, pp. 523–527, 2019.
https://doi.org/10.1007/978-3-030-18590-9_79

stocks and then join with the stock steam to get other information. However, they have different understanding of the "high-volume". In other words, one thinks 1,000 is a high volume but another needs 5,000. In this case, we can save a great deal of computing resources by reusing the operators. Moreover, optimizing multiple queries in a distributed query plan will also help achieve the goal of load balancing, which is a primary problem in modern distributed stream processing engines.

In this demonstration, we propose EsperDist, an efficient distributed query engine for multiple-query optimization over data stream. We first propose a framework for query plan generation: Given a query on the data stream, we first generate the corresponding central query plan. We then decide the relations between operators so as to make an effective schedule [2]. Next we generate the distributed query plan by considering the reuse of operators in each node for different queries. Unlike previous studies that only reuse the operators that are directly related to each other [5], we also consider the indirected correlated operators. Finally, we devise a cost model to select the distributed query plan for execution according to the usage of different types of resources such as CPU, bandwidth and memory.

Based on above idea, we implement the distributed query engine based on the centralized stream processing system Esper. We deploy the Esper query engine on each node and execute the tasks allocated to them separately. The query is described with EPL, which is a structured query language for data stream supported by Esper. We further provide a user-friendly query interface built on top of EsperDist. The interface displays query results in near real-time as a visualized view, and help users interact with the system.

The rest of the paper is organized as following. We introduce the overall architecture of EsperDist and the deployment of system in Sect. 2, followed by system demonstration with a real world application in Sect. 3.

2   System Design

Figure 1 shows the overall architecture and deployment of our EsperDist system. **Worker** has three modules: 1. Publish-Subscribe Processor, 2. Scheduler and 3. Monitor. Publish-Subscribe Processor decides the publish and subscribe operations of Workers, which serves as an interface for communication between Esper engines on different nodes. The function of Scheduler is to schedule the communication between publishers and subscribers globally. The Monitor collects necessary information, e.g., the CPU/memory usage, the statistical information of different attributes in real time so as to help estimate the cost of plans and make the choice. **Coordinator** is the most important component in the system. Its task is to generate the global query plans as well as the execution plan. It has two main modules: query processor and controller. The query processor aims at generating the query plans and the controller manipulates the execution of the query plan, including submitting the operators to corresponding nodes and modifying the operators in each node after new plan arrives.

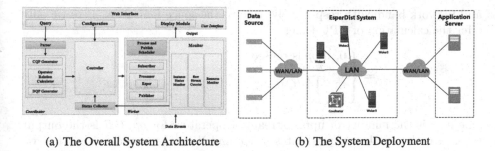

(a) The Overall System Architecture   (b) The System Deployment

Fig. 1. System overview

Next we introduce the query optimization techniques of EsperDist. There are mainly three steps: centralized query plan generation, distributed query plan generation and cost-based plan evaluation.

Centralized Query Plan Generation. Given a query Q on the coordinator, we first transform Q into syntax tree and obtain the centralized query plans, denoted as W, without considering the distribution of operators. For each plan, if there is m-way join, there would be $m!$ possible plans even just considering left deep trees. To reduce search space, we propose two pruning strategies: Firstly, we only keep the plans whose query tree is a balanced tree. Secondly, we decide the execution order of operators by their "selectivity" so as to minimize the network transmission.

Distributed Query Plan Generation. As we have generated the set W, we need to further allocate the operators to each node. Given two operators c and d, c is *equivalent* to d (denoted as $c \equiv d$) iff they have the same predicates and their upper stream operators are equivalent. If the upper stream operators of d is included in those of c and the predicates of d are a subset of c's predicates, we say that d is *implicated* with c (denoted as $c \succ d$). For each centralized query plan $P_i^C \in W$ and each operator $op \in P_i^C$, we retrieve the set of operators that are equivalent to op, denoted as $E(op)$ and the set of operators that are implicated with op, denoted as $I(op)$. For operators in $E(op)$, we can directly reuse it. For operators in $I(op)$, we indirectly reuse it by adding secondary filter between op and the operator in $I(op)$ so that we can use the output of op as the intermediate results.

Cost-Based Plan Evaluation. We propose a cost model to select a best plan for execution from V. The foundation of cost model is estimating the *Output Rate* of each operators. The output rate of an operator op can be estimated according to that of its upper stream operators, denoted as $op.ups$ and its equivalent operators $E(op)$. We consider cost from three aspects: the CPU time, the memory usage and network bandwidth. This can be done based on the observation that the cost of an operator can be calculated by referring to its upper stream operators. Then we can estimate the incremental cost of CPU time, memory usage and network bandwidth respectively. Due to the space limitation, here we only offer the calculation of CPU time:

$$\Delta c_{op} = \begin{cases} 0, & \text{if } op \text{ is reused} \\ m \cdot \prod_{i=0}^{N-1} (OR_{op.ups[i]} \cdot W_{op.ups[i]}), & \text{otherwise} \end{cases}$$

where N is the number of upper stream of operators for op, OR is the output rate, m is the number of predicates of op, and W is the window size. After we obtain the total cost of each distributed query plan, we can select the one with minimum cost.

3   Case Study and Demonstration

Figure 2(a) shows the main application GUI. We present a real world application for demonstration. Suppose a investor wants to analyze high-volume trading events made in stock markets. Suppose we want to keep track of the volume larger than 5,000 on the most active stocks with a positive change in the past 1 h, we can write an EPL query as following:

```
1    SELECT s.symbol, s.volume, s.price, t.volume
2    FROM Stock(s.change > 0).win:time(1 h) as s,
3         Trade(t.volume > 5000).win:time(1 h) as t
4    WHERE s.ID = t.stockID
5    ORDER BY s.volume
6    OUTPUT every 60 seconds;
```

And here we have already registered a query before, which is:

```
1    SELECT t.time, s.symbol, s.change
2    FROM Stock(s.change > 0).win:time(1 h) as s,
3         Trade(t.volume > 1000).win:time(1 h) as t
4    WHERE s.ID = t.stockID
5    ORDER BY t.time
6    OUTPUT every 60 seconds;
```

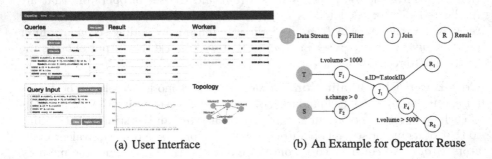

(a) User Interface   (b) An Example for Operator Reuse

Fig. 2. GUI and a running example

As is shown in Fig. 2(b), by comparing the new query with the global query plan, i.e., the previous query, we find that the new join operator is Implicated with J_1 because of $t.volume > 5000 \succ t.volume > 1000$, $s.change > 0 \equiv s.change > 0$ and $s.ID = t.stockID \equiv s.ID = t.stockID$. It means that we can reuse the join operator and add a filter operator following it to get the final result. Thus we can save a large portion of common computation. Note that we need to adjust the output of J_1, since the selected properties of the two queries are different.

Acknowledgement. This work was supported by NSFC (91646202), National Key R&D Program of China (SQ2018YFB140235), and the 1000-Talent program.

References

1. Ao, X., Luo, P., Wang, J., Zhuang, F., He, Q.: Mining precise-positioning episode rules from event sequences. IEEE Trans. Knowl. Data Eng. **30**(3), 530–543 (2018)
2. Babcock, B., Babu, S., Datar, M., Motwani, R.: Chain: operator scheduling for memory minimization in data stream systems. In: SIGMOD, pp. 253–264 (2003)
3. Cherniack, M., et al.: Scalable distributed stream processing. In: CIDR (2003)
4. Gu, J., Wang, J., Zaniolo, C.: Ranking support for matched patterns over complex event streams: the CEPR system. In: ICDE, pp. 1354–1357 (2016)
5. Repantis, T., Gu, X., Kalogeraki, V.: Synergy: sharing-aware component composition for distributed stream processing systems. In: van Steen, M., Henning, M. (eds.) Middleware 2006. LNCS, vol. 4290, pp. 322–341. Springer, Heidelberg (2006). https://doi.org/10.1007/11925071_17

Adding Value by Combining Business and Sensor Data: An Industry 4.0 Use Case

Guenter Hesse[(✉)] [iD], Christoph Matthies[iD], Werner Sinzig,
and Matthias Uflacker

Hasso Plattner Institute, University of Potsdam,
August-Bebel-Str. 88, 14482 Potsdam, Germany
{guenter.hesse,christoph.matthies,
werner.sinzig,matthias.uflacker}@hpi.de

Abstract. Industry 4.0 and the Internet of Things are recent developments that have lead to the creation of new kinds of manufacturing data. Linking this new kind of sensor data to traditional business information is crucial for enterprises to take advantage of the data's full potential. In this paper, we present a demo which allows experiencing this data integration, both vertically between technical and business contexts and horizontally along the value chain. The tool simulates a manufacturing company, continuously producing both business and sensor data, and supports issuing ad-hoc queries that answer specific questions related to the business. In order to adapt to different environments, users can configure sensor characteristics to their needs.

Keywords: Industry 4.0 · Internet of Things · Data integration

1 Introduction

The developments in the areas of Internet of Things (IoT) and sensor technologies drive advances in modern manufacturing settings. Industrial manufacturing enterprises recognize this technological progress and are using the new Industry 4.0 capabilities to generate added value. For example, on a daily basis, a single sensor located on a General Electric gas turbine engine can produce 500 GB of data [2]. Injection molding machines, as an example of a common manufacturing device, can even generate multiple terabytes of sensor data per day [4].

However, these new possibilities also pose unique challenges, e.g., regarding data integration, as the characteristics of IoT and business data differ [3]. Linking these two kinds of data holds the key for unlocking the full potential that lies within the collected data treasure. Contrary to horizontal integration, which describes the integration of business data along the value chain, vertical integration refers to the connection between business and sensor data. Whereas in horizontal integrations only homogenous business data needs to be merged, vertical integration requires integration of a variety of data characteristics.

© Springer Nature Switzerland AG 2019
G. Li et al. (Eds.): DASFAA 2019, LNCS 11448, pp. 528–532, 2019.
https://doi.org/10.1007/978-3-030-18590-9_80

In the presented demo that is available online[1], we tackle the challenges of understanding the complex data relations in an Industry 4.0 setting. We present an approach for simulating different types and amounts of sensors in the context of an industrial manufacturing company, which also produces business data as part of its regular activities. Furthermore, we enable issuing ad-hoc queries on the collected data in an easy-to-use fashion by employing SQL. This flexibility allows analyzing and combining all kinds of available data. This allows for horizontal as well as vertical integration in this synthetic and configurable scenario.

2   Developed Demo System

The system is developed using the Play framework [1] and Scala as the programming language. The demo is realized as a single page application (SPA) which allows controlling data generation for both, business data and sensor data [5]. As the application is preconfigured with the default settings of a fictional engine producing factory employing IoT sensors, it can immediately be run and explored. However, the number and characteristics of sensors that are sending data can be adapted using on-screen controls. Two live-updating line charts visualize the data ingestion rate for both kinds of data. This data is inserted into a columnar in-memory database to enable real-time query execution.

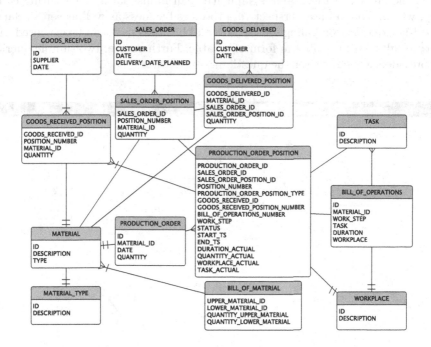

Fig. 1. Entity relationship diagram in crow's foot notation for the business data

---

[1] https://github.com/Gnni/DemoDataIntegration.

The used data model visualized in Fig. 1 is inspired by the schemas of real Enterprise Resource Planning (ERP) systems [6]. Particularly, the idea of having a head and an item table for, e.g., sales orders, is adapted in order to be as close to real-world scenarios as possible.

IoT data is stored in another table with the columns *ID, WORKPLACE_ID, SENSOR_ID, DATE,* as well as columns related to specific sensor measurements, namely *TEMPERATURE_VALUE, TEMPERATURE_UNIT, NOISE_VALUE, NOISE_UNIT, VIBRATION_VALUE,* and *VIBRATION_UNIT.* As there are only three kinds of sensors, the last columns are specific to these. When, e.g., a new temperature value is sent and inserted, the columns storing information about the other two sensor types stay empty for that row.

Horizontal integration is achieved using IDs whereas the process of vertical integration makes use of a time-based approach. Particularly, a link between sensor data and ERP data can be created as *PRODUCTION_ORDER_POSITION* stores the information when a product entered or left a certain workplace. Since sensor data also comes with a timestamp, a connection between measurements and workplaces, and thus, between IoT data and products can be established.

3   Features and Demo Scenario

The demo shown in Fig. 2 allows simulating an industrial manufacturing company, which, from a data perspective, produces business as well as sensor data.

Ad-hoc queries combining sensor data and business data can be executed. All query results are presented in form of a table. Furthermore, two sample queries are provided, which answer the questions:

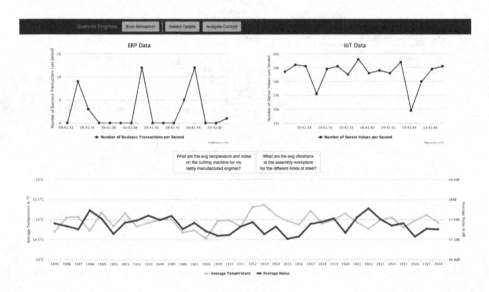

Fig. 2. Screenshot a selected part of the demo application

- What are the average temperature and noise on the workplace cutting machine for my recently manufactured products?
- What are the average vibrations at the assembly workplace dependent on the supplier?

At the top of Fig. 2, there are three buttons. The one in the upper-left allows starting and stopping data generation. The two diagrams on the top visualize the input rate of business and IoT data. The sensor characteristics are defined in a JSON file. Clicking on the button in the middle opens the sensor config area which allows, e.g., adding certain kinds of sensors to workplaces that produce data at a definable input rate. This area is not shown in Fig. 2.

Below the upper two diagrams, there are two more buttons triggering the execution of one of the two mentioned predefined queries. The lower diagram shows the result of the first query, i.e., average temperature and noise values for the lastly manufactured products at the cutting machine. Not part of Fig. 2 is the result table which presents the raw data belonging to issues queries as well as the query formulation area, where the predefined queries can be adapted or any ad-hoc query can be inserted and executed.

4   Conclusion and Future Work

The presented tool allows experiencing horizontal and vertical integration in scenarios with a real-world character. IoT data can be configured and influences on, e.g., performance can be analyzed. Next to predefined queries that answer valuable questions, any ad-hoc query on the collected data can be executed. Results of given queries are visualized in a diagram. To the best of our knowledge, the presented demo application is the first of its kind, i.e., a program providing an explorable Industry 4.0 environment with focus on scenarios close to real-world systems and use cases. Experiments with regard to data integration strategies, data volumes, and resulting impact analysis on query performance can be done easily. As the code is provided, adaptions and further developments are possible.

References

1. Play - The High Velocity Web Framework For Java and Scala. https://www. playframework.com. Accessed 06 Nov 2018
2. Davenport, T.H., Dyché, J.: Big Data in Big Companies, May 2013. http://docs. media.bitpipe.com/io_10x/io_102267/item_725049/Big-Data-in-Big-Companies. pdf. Accessed 28 Feb 2017
3. Hesse, G., Reissaus, B., Matthies, C., Lorenz, M., Kraus, M., Uflacker, M.: Senska - towards an enterprise streaming benchmark. In: Performance Evaluation and Benchmarking for the Analytics Era - TPC Technology Conference, pp. 25–40 (2017)
4. Huber, M.F., Voigt, M., Ngomo, A.N.: Big data architecture for the semantic analysis of complex events in manufacturing. In: 46. Jahrestagung der Gesellschaft für Informatik, Informatik, pp. 353–360 (2016)

5. Mikowski, M., Powell, J.: Single Page Web Applications: JavaScript End-to-End, 1st edn. Manning Publications Co., Greenwich (2013)
6. Plattner, H.: A common database approach for OLTP and OLAP using an in-memory column database. In: ACM SIGMOD International Conference on Management of Data, pp. 1–2 (2009)

AgriKG: An Agricultural Knowledge Graph and Its Applications

Yuanzhe Chen[1], Jun Kuang[1], Dawei Cheng[3,4], Jianbin Zheng[1],
Ming Gao[1,2(✉)], and Aoying Zhou[1]

[1] School of Data Science and Engineering, East China Normal University,
Shanghai, China
{yzchen,jkuang}@stu.ecnu.edu.cn, {jbzheng,mgao,ayzhou}@dase.ecnu.edu.cn
[2] Key Laboratory of Advanced Theory and Application in Statistics
and Data Science - MOE, Shanghai, China
[3] Shanghai Jiao Tong University, Shanghai, China
dawei.cheng@sjtu.edu.cn
[4] Keydriver Inc, Shanghai, China

Abstract. Recently, with the development of information and intelligent technology, agricultural production and management have been significantly boosted. But it still faces considerable challenges on how to effectively integrate large amounts of fragmented information for downstream applications. To this end, in this paper, we propose an agricultural knowledge graph, namely AgriKG, to automatically integrate the massive agricultural data from internet. By applying the NLP and deep learning techniques, AgriKG can automatically recognize agricultural entities from unstructured text, and link them to form a knowledge graph. Moreover, we illustrate typical scenarios of our AgriKG and validate it by real-world applications, such as **agricultural entity retrieval**, and **agricultural question answering**, etc.

1 Introduction

Agriculture is the industry that accompanied the evolution of humanity, and fulfilled faithfully its core mission of the food supply. With decreasing workforce in the rural areas, advancing in the artificial intelligence, and developing the IoT technologies, it is desired to improve the efficiency and productivity of the agricultural industry. An agricultural knowledge graph repository will work as the foundation to achieve these goals.

Knowledge graph, which can be general-purpose and domain-specific, is a backbone of many applications, such as search engine, online question answering, and knowledge inference, etc. As a result, there are various knowledge graphs, including Wikidata[1], DBpedia[2], etc., for accessing to structured knowledge. Although there are some general knowledge graphs which contain some entities

---

[1] https://www.wikidata.org.
[2] https://wiki.dbpedia.org.

© Springer Nature Switzerland AG 2019
G. Li et al. (Eds.): DASFAA 2019, LNCS 11448, pp. 533–537, 2019.
https://doi.org/10.1007/978-3-030-18590-9_81

and relations about the agriculture, there is not a domain-specific knowledge graph for agricultural applications.

With the development of Web and IoT techniques, a wealth of fragmented data is crawled from Internet, generated by sensors or collected by agricultural drones. It is helpful and valuable to extract the agricultural knowledge from the fragmented data. Based on the agricultural knowledge, farmers will be able to take more informed and rapid decisions, make decisions to maximize return on crops, and be provided the advice and recommendations on the specific farm problems. Therefore, in this paper, we demonstrate an Agricultural Knowledge Graph in Chinese, namely AgriKG, which can be applied to support some agricultural applications, and further improve the efficiency and productivity of the agricultural industry.

The goal of this Demo system can be summarized as follows:

- Automated knowledge growth: AgriKG is able to identify the agricultural entities and relations from raw text, and incrementally adds the incoming knowledge triples into the knowledge base.
- Agricultural entity retrieval: AgriKG provides the entity retrieval in different fashions. Users are allowed to retrieve the agricultural entities via submitting a keyword search or image retrieval.
- Agricultural question answering: to enable AI-driven agriculture, AgriKG is able to address the questions via applying the subgraph matching.

2   System Overview and Key Techniques

As illustrated in Fig. 1, AgriKG consists of five key components: (i) crawlers collect the raw text and semi-structured data from Web; (ii) NLP module is a key component which provides a set of tools for the raw text understanding; (iii) entity recognition identifies the agricultural entities from the raw text; (iv) relation extraction aims at finding the attributes of entities and extracting relations from the raw text; (v) the applications of AgriKG include agricultural entity retrieval and question answering, etc.

Crawler. To construct an agricultural knowledge graph, AgriKG crawls the taxonomy from Wikidata, collects the attributes and images about entities from Hudong Baike, and acquires the massive agricultural raw text from some agricultural Web sites, such as China Agriculture, Xinnong, China National Seed Association, etc.

NLP Module. Since massive agricultural information appears in the raw text, the NLP module is applied to extract information, understand the raw text. It consists of a set of tools, such as text representation [1], word segmentation, and POS tagging [2], etc.

Entity Recognition. All entities in AgriKG is grouped into 16 predefined categories, including *animal, plant, chemical, climate, agricultural products, disease, nutrients, agricultural implements, agricultural terminology,* etc. Given a piece of

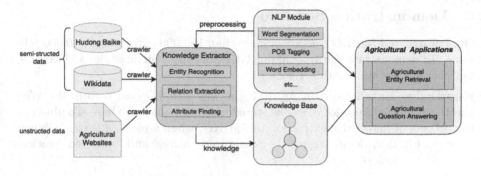

Fig. 1. The framework of AgriKG

text, we enumerate all spans, which are considered as the candidates of entities, after word segmentation and POS tagging. If a span is an entry in Hudong Baike, it is considered as an entity, and further classified into one of the 16 categories. In addition, to collect the ground-truth data for training, an auxiliary tool is developed to help the entity annotation.

Relation Extraction and Attribute Finding. One part of relations, such as *instance of, has part, subclass of, parent taxon, material used, natural product of taxon*, etc., in AgriKG extracts from Wikidata. The other part of relations, including *suitable planting, growing climate*, etc., extracts via using the remote-supervised approach [3] to train a neural relation extractor [4]. All entities and relations are stored in Neo4j, and the remaining data is stored in MongoDB.

Agricultural Applications. To achieve the precision farming, we develop two smart agricultural applications: agricultural entity retrieval and question answering for agricultural knowledge.

To support smart farming applications, such as weed monitoring and pest controlling, users can retrieve agricultural entities via submitting the traditional keyword search or image retrieval. For a keyword search, AgriKG returns the exactly matched entity. For an image retrieval request, AgriKG recognizes the most similar entities via using ImageMatch API and Elasticsearch for image similarity search[3].

AgriKG also provides question answering, which consists of three key components: entity linkage, user intention understanding and answer ranking. A question request will trigger AgriKG to recognize the entities mentioned in the question [5]. Furthermore, the user intention is modelled as a multi-constraint question graph. It will be constructed based on the detected entities after the question annotation [6]. By doing so, question answering is transferred into a subgraph matching problem. Finally, after the ranking scores of candidates are calculated by a Siamese convolution neural network (CNN) [6], the answer will be subgraphs of the knowledge graph with the largest ranking scores.

---

[3] https://image-match.readthedocs.io.

3   Demonstration Scenario

Our constructed AgriKG consists of more than 150,000 entities and 340,000 relations. To demonstrate the system, our GUI not only visualizes the architecture, but also lets the users interact with it.

Knowledge Extraction. In AgriKG, the raw text is crawled from the Web, and the extracted knowledge will be stored into the knowledge base. To illustrate the process of knowledge extraction in AgriKG, when a piece of text in Chinese is given, Fig. 2(a) demonstrates the recognized entities and extracted relations from the input text.

Entity Retrieval. In AgriKG, we can retrieve entities from the knowledge graph in two manners. In the traditional manner, it is a keyword search, which returns the exactly matched entity to us. In the other manner, it is an image retrieval. We can require an image retrieval in AgriKG when we have some photos of plants or pests. When we upload a picture of *agave*, Fig. 2(b) illustrates the result of the image retrieval. AgriKG will tells us exactly what the species is. With this functionality, we can identify unknown species whenever and wherever, and access to corresponding agricultural knowledge, such as planting strategy, pest controlling, etc.

Question Answering. Users are allowed to ask some simple questions (only involving single relation) or multi-constraint questions (involving multiple relations). AgriKG transfers a question into a multi-constraint query graph, and returns the most similar subgraphs via subgraph matching. Figure 2(c) demonstrates the answer of question *"what plants are suitable for growing in Chongming County"*. Therefore, AgriKG enables us to obtain the answers of agriculture-related questions in real time.

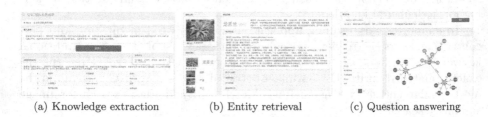

| (a) Knowledge extraction | (b) Entity retrieval | (c) Question answering |

Fig. 2. System demonstration

4   Conclusion

To overcome the challenges on how to effectively integrate large amount of information for agricultural applications, in this paper, we propose a knowledge-based system, namely AgriKG, to automatically integrate the massive agricultural information into a knowledge graph, and to provide some services, such as agricultural entity retrieval, agricultural question answering, and so on.

Acknowledgments. This work has been supported by the National Key Research and Development Program of China under grant 2016YFB1000905, and the National Natural Science Foundation of China under Grant No. U1811264, 61877018, 61672234, and 61502236. It has been also supported by the Shanghai Agriculture Applied Technology Development Program, China (Grant No. T20170303).

References

1. Bojanowski, P., Grave, E., Joulin, A., Mikolov, T.: Enriching word vectors with subword information. arXiv preprint arXiv:1607.04606 (2016)
2. Li, Z., Sun, M.: Punctuation as implicit annotations for Chinese word segmentation. Comput. Linguist. **35**(4), 505–512 (2009)
3. Mintz, M., Bills, S., Snow, R., Jurafsky, D.: Distant supervision for relation extraction without labeled data. In: ACL, pp. 1003–1011 (2009)
4. Lin, Y., Shen, S., Liu, Z., Luan, H., Sun, M.: Neural relation extraction with selective attention over instances. In: ACL, vol. 1, pp. 2124–2133 (2016)
5. Yang, Y., Chang, M.: S-MART: novel tree-based structured learning algorithms applied to tweet entity linking. In: ACL 2015, pp. 504–513 (2015)
6. Bao, J., Duan, N., Yan, Z., Zhou, M., Zhao, T.: Constraint-based question answering with knowledge graph. In: COLING 2016, pp. 2503–2514 (2016)

KGVis: An Interactive Visual Query Language for Knowledge Graphs

Xin Wang[1,2(✉)], Qiang Fu[1,2], Jianqiang Mei[3], Jianxin Li[4], and Yajun Yang[1,2]

[1] College of Intelligence and Computing, Tianjin University, Tianjin, China
{wangx,tomqcust,yjyang}@tju.edu.cn
[2] Tianjin Key Laboratory of Cognitive Computing and Application, Tianjin, China
[3] Tianjin University of Technology and Education, Tianjin, China
meijianqiang@tute.edu.cn
[4] School of Information Technology, Deakin University, Melbourne, Australia
jianxin.li@deakin.edu.au

Abstract. With the rise of artificial intelligence, knowledge graphs have been widely recognized as a cornerstone of AI. In recent years, more and more domains have been publishing knowledge graphs in different scales. However, it is difficult for end-users to query and understand those knowledge graphs consisting of hundreds of millions of nodes and edges. To improve the availability, accessibility, and usability of knowledge graphs, we have developed an interactive visual query language, called KGVis, which can guide end-users to gradually transform query patterns into query results. Furthermore, KGVis has realized the novel capability of flexible bidirectional transformations between query patterns and query results, which can significantly assist end-users to query large-scale knowledge graphs that they are not familiar with. In this paper, we present the syntax and semantics of KGVis, discuss our design rationale behind this interactive visual query language, and demonstrate various use cases of KGVis.

Keywords: Knowledge graphs · Visual query language · Interactive ·
Bidirectional transformation

1 Introduction

With the application of artificial intelligence, AI has become a powerful tool to realize practical requirements in various fields. Recently, knowledge graphs, as a cornerstone of AI, have been widely published on the Web. Currently, a large number of query languages for knowledge graphs are designed for professional users, i.e., SPARQL and Cypher, while many interactive visual graph languages only bind to one specific query language, i.e., VISAGE [2] or QueryVOWL [1]. In order to allow knowledge graphs in various fields to be accessed and used by all end-users, we have developed an interactive visual query language, called KGVis, which can break the bound between query patterns and query results.

© Springer Nature Switzerland AG 2019
G. Li et al. (Eds.): DASFAA 2019, LNCS 11448, pp. 538–541, 2019.
https://doi.org/10.1007/978-3-030-18590-9_82

Furthermore, KGVis can gradually guide end-users to construct the query pattern and transform it into the query result. In this paper, we introduce the syntax and semantics of KGVis, discuss our design rationale behind the scene, and demonstrate effectiveness and user-unfriendliness of KGVis.

Although, in this paper, SPARQL is used as the underlying query language for knowledge graphs, KGVis is independent of a specific low-level graph query language. By analyzing and summarizing the definition of graph patterns, we define a complete set of visual notations for KGVis, as shown in Table 1, and translate them into SPARQL. In KGVis, a circle and an adjacent directed edge are used to construct triple pattern in a query pattern, while double circles and rectangles are used to represent operators and parameters in the query pattern. We use dashed and solid lines to indicate the state of variables and constants which are assigned to variables, respectively. If the number of the matched subjects, predicates, or objects are more than one, we represent them by a solid circle labeled with List and the corresponding number ③ in Fig. 2.

Table 1. The visual syntax of KGVis and their corresponding semantics in SPARQL

Syntax	SPARQL Semantics	Syntax	SPARQL Semantics
s/o	A variable subject or object {?s p o} { s p ?o}	s/o	A constant subject or object { s ?p ?o} {?s ?p o}
predicate	A variable predicate { s ?p o}	predicate	A constant predicate {?s p ?o}
e	Operators: optional, union, limit, filter, order by	edit text	Parameters of operators: ⩾, ⩽, ascending, descending, etc.

2   System Architecture of KGVis

For end-users without background knowledge of query languages for knowledge graphs, it is difficult to retrieve the information they need from the knowledge graph. With the help of KGVis, end-users can easily construct the query pattern combined with interactive operations. Figure 1 depicts the system architecture behind the interactive operations of end-users.

First, the end-user creates a subject ① and inputs Epicurus according to the demand. Second, the SPARQL code ② that is equivalent to the previous interaction is generated, and an HTTP request ③ is sent to the corresponding SPARQL endpoint to query the DBpedia knowledge graph. Finally, the query result ④ is shown through the interface layout. By interactively constructing the query pattern with the positive feedback from the returned fragment of the query result, users can construct the final query pattern (result) that meets their demands.

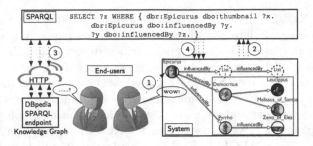

Fig. 1. The system architecture of the KGVis SPARQL implemented

3 Demonstration

The interactive system shown in this paper has been published on the Github[1].

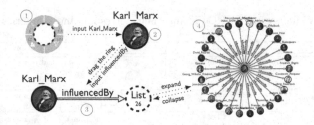

Fig. 2. The construction process of the complete query pattern

Use Case 1: The Interactive Visual Query on Knowledge Graph.
KGVis supports different SPARQL endpoints, in this paper, we use the DBpedia
SPARQL endpoint as the underlying knowledge graph. First, the end-user creates a variable subject ① whose surrounding tabs (i.e., i. order by, ii. optional,
iii. filter, iv. union) are the operators. After inputting Karl_Marx, a thumbnail of
Karl_Marx will return to the user. Then user can drag the border of the subject
② to create a new predicate and an object. In accordance with the demand,
the user can input the predicate or the object to query the result. In this use
case, the user inputs influencedBy as a predicate ③. When the user expands the
results of the third step, it is transformed into the query result ④.

Use Case 2: The Flexible Bidirectional Transformation of KGVis.
KGVis also supports end-users to flexibly bidirectionally transform between
query patterns and query results. As shown in the red rectangle in the Fig. 1,
after end-users get the results that who influencedBy Epicurus, end-users can
expand the results by transforming the query pattern to query results. Likewise,
end-users can also transform backward to restore the original query pattern.

---

[1] https://andywx.github.io/kgvis/KGVis/index.html.

Use Case 3: The Deep Exploration of KGVis. The further query based on the previous query, which we called "Deep Exploration", is also supported in KGVis. Figure 3 illustrates an expanded layout for deep exploration by selecting Aristotle ⑤ as the subject for the next query. As shown in the query result ⑥ ⑦, end-users would be gradually guided to deeply explore the knowledge graph to discover the expected answer even if they do not know how to write this complex query at the beginning.

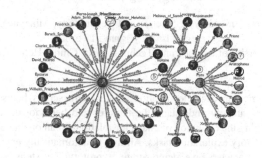

Fig. 3. The deep exploration in KGVis

4   Conclusion

We have developed an interactive visual query language, called KGVis, to help end-users inquire information from knowledge graphs even if they do not have the background knowledge of query languages for knowledge graphs. KGVis provides visual notations to guide end-users to interactively construct a simple query pattern until the final query result graph is gradually obtained. Another advantage of KGVis is the flexible bidirectional transformation between query patterns and query results, which has become a useful tool for end-users to get the insight of the large-scale knowledge graphs.

Acknowledgments. This work is supported by the National Natural Science Foundation of China (61572353 and 61402323) and the Natural Science Foundation of Tianjin (17JCYBJC15400).

References

1. Haag, F., Lohmann, S., Siek, S., Ertl, T.: QueryVOWL: a visual query notation for linked data. In: Gandon, F., Guéret, C., Villata, S., Breslin, J., Faron-Zucker, C., Zimmermann, A. (eds.) ESWC 2015. LNCS, vol. 9341, pp. 387–402. Springer, Cham (2015). https://doi.org/10.1007/978-3-319-25639-9_51
2. Pienta, R., Tamersoy, A., Endert, A., Navathe, S., Tong, H., Chau, D.H.: Visage: interactive visual graph querying. In: Proceedings of the International Working Conference on Advanced Visual Interfaces, pp. 272–279. ACM (2016)

OperaMiner: Extracting Character Relations from Opera Scripts Using Deep Neural Networks

Xujian Zhao[1]([✉]), Xinnan Dai[1], Peiquan Jin[2], Hui Zhang[1], Chunming Yang[1], and Bo Li[1]

[1] Southwest University of Science and Technology, Mianyang, China
jasonzhaoxj@gmail.com
[2] University of Science and Technology of China, Hefei, China

Abstract. Retrieving character relations from opera scripts helps performers and audience accurately understand the features and behavior of roles. Meanwhile, discovering the evolution of character relations in an opera benefits many opera-oriented story exploration tasks. Aiming to automatically extract relations among opera characters, we demonstrate a prototype system named OperaMiner, which extracts relations for opera characters based on a hybrid deep neural network. The major features of OperaMiner are: (1) It provides a uniform reasoning framework for character relations considering language structure information as well as explicit and implicit expressions in opera scripts. (2) It explores the deep features in opera scripts, including character embeddings features, word embeddings features, and the linguistic features in artistic texts. (3) It presents a hybrid learning architecture enhancing CNN and Bi-LSTM with a CRF layer for character relation extraction. After a brief introduction to the architecture and key technologies of OperaMiner, we present a case study to demonstrate the main features of OperaMiner, including the generation of the character relation graph, the demonstration of major roles, and the evolution sequence of character relations.

Keywords: Relation extraction · Opera scripts · Character relation · Deep learning

1 Introduction

Character relation, often contained in a large number of opera scripts, is considered as a direct expression of the interrelation between characters in opera. For the performers such as opera singers in opera performing, singing style, facial expression, and voice emotion are all closely associated with the relationships among the roles they impersonate in an opera. For example, a gentle and sweet performing style is more suitable for singers who need to play lover roles in an opera. In addition, the evolution of role relations among opera characters normally reflects the development of story; therefore a good grasp of character relations is also useful for audience to understand the opera's storyline clearly.

© Springer Nature Switzerland AG 2019
G. Li et al. (Eds.): DASFAA 2019, LNCS 11448, pp. 542–546, 2019.
https://doi.org/10.1007/978-3-030-18590-9_83

In this paper, we present a new system called OperaMiner for character relation extraction from opera scripts in Chinese. Although there are some previous works focusing on relation extraction [1], this study differs from existing works in several aspects, which are summarized as follows:

(1) It provides a uniform reasoning framework for character relations considering language structure information as well as explicit and implicit expressions in opera scripts. Basically, character relation usually appears in a few character-included expressions, which are typically classified into explicit and implicit expressions. However, explicit and implicit expressions have various depicting styles in scripts. Specifically, explicit expressions are able to be transformed into a pre-defined pattern, but implicit expressions cannot be handled in the same way.

(2) It explores the deep features in opera scripts, including character embeddings features, word embeddings features, and the linguistic features in artistic texts. In OperaMiner, the three types of features are integrated into one hybrid embedding for character relation learning.

(3) It presents a hybrid learning architecture enhancing CNN and Bi-LSTM with a CRF layer for character relation extraction. The CRF layer is able to add some constraints to the final label tagged by Bi-LSTM to ensure that the predicted result is semantically legal.

2 Architecture and Key Technologies of OperaMiner

Figure 1 shows the architecture of OperaMiner. There are four layers in the system. The details of the layers are presented as follows.

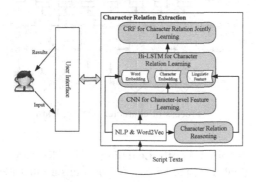

Fig. 1. Architecture of OperaMiner

Character Relation Reasoning. Owing to the significant difference in depicting styles between explicit and implicit expressions, we have to deal with a challenging issue for character relation extraction from opera scripts: how to model the character relationship with a uniform framework? Practically, we use various linguistic parsing solutions to reason the character entities and the relationship among them. Firstly, in order to infer two character entities associated with each other in explicit or implicit expressions, we utilize the solution of Semantic Dependency Parsing (SDP) to determine all the word pairs related to each other semantically, and then model their relationship through revealing the syntactic sentence structure by Dependency Parsing (DP).

CNN for Character-Level Feature Learning. The previous study [2] has shown that CNN is an effective approach to extract character-level deep features and encode them into neural representations. In OperaMiner, the first neural layer is defined as a CNN-based learning framework which is used to extract character-level features of a given word in opera scripts. The CNN approach in OperaMiner is similar to that in Chiu and Nichols' work [3], except that we only use the character embeddings by Word2Vec as the inputs to CNN, and the character type features are not included.

Bi-LSTM for Character Relation Learning. In this layer, firstly the character-level embedding vector as the output of CNN is appended into a mixed structure, consisting of the word-level embedding vector by Word2Vec and the linguistic feature vector by reasoning model, to establish a hybrid embedding vector before feeding into the Bi-LSTM network. Then, we employ a bi-directional LSTM to solve sequence labeling for character relation learning. As shown in Fig. 2, a forward LSTM computes a representation $\vec{h}$ of the sequence from left to right for each word, and another backward LSTM computes a representation $\overleftarrow{h}$ of the same sequence reversely. Then, considering the representation of a word $h_i = [\vec{h}_i; \overleftarrow{h}_i]$ as defined in [4], the Bi-LSTM is carried out to get the confidence scores for tags.

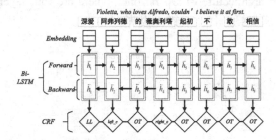

Fig. 2. Process of character relation learning

CRF for Character Relation Jointly Learning. The CRF layer is able to add some constraints to the final label tagged by Bi-LSTM to ensure that the predicted result is legal in semantics. In OperaMiner, sequence labeling for sentences is carried out using five types of tagging denotations: *left_c* (left character entity), *right_c* (right character entity), *LL* (Lover Relation), *FM* (Family Relation) and *OT* (Other). Based on the tagging rules, we model label sequence jointly using a conditional random field (CRF) instead of decoding each label independently. Generally, the output vectors of Bi-LSTM are fed to the CRF layer to jointly decode the best label sequence.

3 Demonstration

Figure 3 shows the user interface of OperaMiner. Users are first required to train a deep learning model through selecting a training file. Then, users can select a file path that stores the test data, and the system will automatically extract the relations among all characters in one opera play. Currently, Family and Lover relationships are considered in our system. As shown in the user interface, the extracting results are shown in a list as well as in a relation graph. In the zone of Character Relation Analysis, we can find the major characters by large circles in the relation graph. In addition, the result of relation evolution shows the chapter of opera's climax, which has more characters as well as relationships among characters. The Character Relation Search part in OperaMiner supports querying for a specific character. We provide a video at http://staff. ustc.edu.cn/~jpq/OperaMiner.avi for the detailed demonstration.

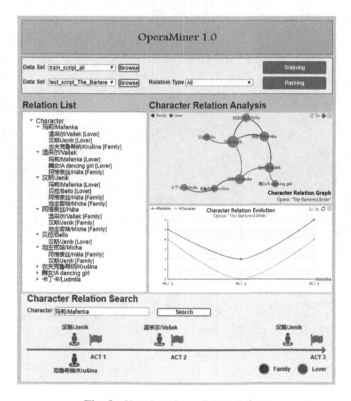

Fig. 3. User interface of OperaMiner

Acknowledgements. This work is partially supported by the National Key Research and Development Program of China (No. 2018YFB0704400 and No. 2018YFB0704404), the Humanities and Social Sciences Foundation of the Ministry of Education (No. 17YJCZH260) and the National Science Foundation of China (No. 61672479).

References

1. Kadry, A., Dietz, L.: Open relation extraction for support passage retrieval: merit and open issues. In: SIGIR, pp. 1149–1152 (2017)
2. Ma, X., Hovy, E.: End-to-end sequence labeling via bi-directional LSTM-CNNs-CRF. In: ACL, pp. 1064–1074 (2016)
3. Chiu, J., Nichols, E.: Named entity recognition with bidirectional LSTM-CNNs. In: ACL, pp. 357–370 (2015)
4. Luo, L., Yang, Z., Yang, P., et al.: An attention-based BiLSTM-CRF approach to document-level chemical named entity recognition. Bioinformatics **34**(8), 1381–1388 (2017)

GparMiner: A System to Mine Graph Pattern Association Rules

Xin Wang[1(✉)], Yang Xu[1], Ruocheng Zhao[2], Junjie Lin[3], and Huayi Zhan[4]

[1] Southwest Jiaotong University, Chengdu, China
xinwang@swjtu.edu.cn, xuyang@my.swjtu.edu.cn
[2] Birkbeck, University of London, London, UK
rzhao01@mail.bbk.ac.uk
[3] Xinjiang University, Ürümqi, China
jjlin50382@gmail.com
[4] Northwestern University, Evanston, USA
huayi.zhan@u.northwestern.edu

Abstract. With the rapid development, social network analysis has been receiving significant attention. One popular direction in the filed is to mine graph-pattern association rules (GPARs). In the demo, we present GparMiner, a system for mining GPARs, on big and distributed social networks. The system has following characteristics: (1) it supports parallel mining computation, to handle sheer size of real-life social networks; (2) it provides graphical interface to help users monitor the mining progress and have a better understanding of GPARs.

1 Introduction

Association rules have been studied for discovering regularities between items in relational data [2]. In recent years, graph-pattern association rules are introduced in [4–6], and used for social media marketing, community structure analysis, social recommendation, link prediction, and so on.

It is, however, nontrivial to efficiently mine GPARs on big social networks G, since (1) GPARs mining involves subgraph isomorphism, which is an NP-complete problem, and computational costly; (2) real-life social networks are typically large, *e.g.,* Facebook has 2.2 billion active users [1], it is prohibitively expensive to mine GPARs on such big graphs; and (3) due to safety, privacy protection and physical limitations, social networks are often distributively stored, which makes mining even more challenging.

To tackle the issues, we present GparMiner, a prototype system to mine GPARs. It implements the techniques that we presented in [6], and possesses following features.

Distributed Frequent Pattern Mining. GparMiner applies a novel technique to mine frequent patterns in parallel, and constructs a *DFS code graph* for rule generation.

© Springer Nature Switzerland AG 2019
G. Li et al. (Eds.): DASFAA 2019, LNCS 11448, pp. 547–552, 2019.
https://doi.org/10.1007/978-3-030-18590-9_84

Graph-Pattern Association Rules Generation. With *DFS code graph* $\mathcal{G}_c$, (1) GparMiner is able to generate GPARs via an efficient algorithm whose computational complexity depends on the size of $\mathcal{G}_c$, independent of large size of G; and (2) GparMiner can even produce "representative" GPARs, without paying extra cost.

To the best of our knowledge, GparMiner is among the first effort to mine graph-pattern association rules on large and distributed social networks. The prototype of GparMiner was deployed and tested by one of our industrial collaborators, and shows its performance in GPARs mining.

Demo Overview. We demonstrate functionality of GparMiner as follows. (1) We show how frequent patterns are mined, in parallel. (2) We illustrate how GPARs are generated.

Below, we first present the foundation (Sect. 2) and the functional components (Sect. 3) of GparMiner. We then propose a detailed demonstration plan (Sect. 4).

2   Graph Pattern Association Rules

In this section, we review fundamental concepts first, followed by GPARs.

Preliminary Concepts. We next review fundamental concepts and notations.

DFS Code & DFS Code Graph. The definitions of *DFS code* and *DFS code graph* are introduced in [7]. To make the paper self-contained, we cite them as follows (rephrased).

Given a graph G with edge set E, its *DFS tree* T_G can be built by performing a depth first search in G from a specified node. Given T_G, a *DFS code* α of G is an edge sequence $(e_0, e_1, \ldots, e_m)$, that is constructed based on the binary relation $\prec_{E,T}$, such that $e_i \prec_{E,T} e_{i+1}$ for $i \in [0, |E| - 1]$. We refer readers to [7] for more details about binary relation $\prec_{E,T}$. A graph G can have multiple *DFS code* when different starting node and growing edges are used. While one can sort them by *DFS lexicographic order*, then the minimum one can be chosen as the canonical label of G. Given code α, we refer its corresponding pattern as $Q(\alpha)$.

A *DFS code tree* T_c is a directed tree with a single root, where (a) the root is a virtual node, (b) each non-root node, denoted as v, corresponds to a *DFS code*, (c) for a node v with *DFS code* $(e_0, e_1, \ldots, e_k)$, its child must have a valid *DFS code* in the form of $(e_0, e_1, \ldots, e_k, e')$, and (d) the order of the *DFS code* of v'_α siblings satisfies the *DFS* lexicographic order. Similarly, a rooted *DFS code graph* G_c is a directed graph with a single root as virtual node, and non-root node corresponding to a *DFS code*.

Graph Fragmentation. A fragmentation F of a graph $G = (V, E, f_A)$ is $(F_1, \ldots, F_n)$, where each fragment F_i is specified by $(V_i \cup F_i.O, E_i, L_i)$ such that (1) $(V_1, \ldots, V_n)$ is a partition of V; (2) $F_i.O$ is the set of nodes v' such that there exists an edge $e = (v, v')$ in E, $v \in V_i$ and node v' is in another fragment;

and (3) $(V_i \cup F_i.O, E_i, L_i)$ is a subgraph of G induced by $V_i \cup F_i.O$. We assume *w.l.o.g.* that each F_i is stored at site S_i for $i \in [1.n]$.

Graph-Pattern Association Rules. A *graph-pattern association rule R* is defined as $Q_l \Rightarrow Q_r$, where Q_l and Q_r (1) are both patterns, and (2) share nodes but have no edge in common. We refer to Q_l and Q_r as the *antecedent* and *consequent* of R, respectively.

The rule states that in a graph G, if there is an isomorphism mapping h_l from Q_l to a subgraph G_1 of G, then there likely exists another mapping h_r from Q_r to another subgraph G_2 of G, such that for each $u \in V_l \cap V_r$, if it is mapped by h_l to v in G_1, then it is also mapped by h_r to the same v in G_2. The support of a pattern Q in a graph G, denoted by $supp(Q, G)$, indicates the appearance frequency of Q in G. Considering *anti-monotonic* property, we adopt minimum image [3] as support metric.

GPARs Mining. Along the same line as traditional strategy applied by association rule mining, GparMiner mines GPARs as following. (1) It mines frequent patterns Q_R with support above threshold from graph G. (2) It generates a set of GPARs satisfying confidence constraint with frequent patterns Q_R.

Remark. On distributed graphs, GparMiner applies asynchronous message passing to mine frequent patterns, in parallel. As optimization strategies, GparMiner (1) prunes edges that don't satisfy support threshold; (2) determines whether a candidate pattern is redundant via verifying whether its *DFS code* is minimum; (3) reduces excessive rules via "representative" GPARs that are generated from a set of *maximal frequent patterns*.

3   The System Overview

The architecture of GparMiner, shown in Fig. 1, consists of the following components: (1) A *Graphical User Interface* (GUI), which provides a graphical interface to help users manage graph data, set parameters, monitor mining process, and understand results. (2) A *Coordinator* that communicates with GUI to receive

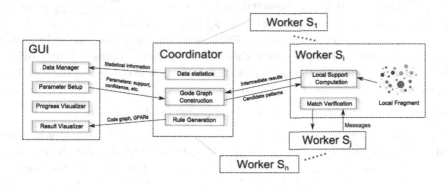

Fig. 1. Architecture of GparMiner

and response requests, coordinates with *works*, and generates GPARs. (3) Multiple *workers*, which communicates with each other in an asynchronous mode to mine frequent patterns.

Below we briefly present key components of GparMiner. Interested readers can refer to [6] for more details.

Graphical User Interface. The GUI provides users an interface to interact with the system, *e.g.*, graph data overview, parameter setup, and results browsing. Specifically, it (1) provides a panel to show statistical information about specified graph data; (2) is equipped with an input panel, with which users can set (a) support and confidence thresholds, (b) lower and upper bound of size of maximal frequent patterns, and (c) rounds of iterations; (3) carries a panel to show mining progress; and (4) visualizes *DFS code graph* and GPARs such that users can browse the results with more intuition.

Coordinator. The *coordinator* interacts with GUI and *workers* as following. (1) It responses to users' requests, with results *e.g.*, statistical information, visualized results, or actions, *e.g.*, startup of mining computation. (2) It maintains a *DFS code graph* to keep track of frequent patterns. The construction of *DFS code graph* follows an iterative manner, and in each round iteration, the *coordinator* assembles intermediate results from each *worker*, eliminates infrequent patterns, and generates and broadcasts valid candidates to all *workers*. (3) It generates GPARs with (maximal) frequent patterns.

Candidate Pattern Generation. Upon receiving intermediate results, *i.e.*, candidate patterns along with their supports, from *workers*, the *coordinator* calculates global support for each pattern, and eliminates patterns whose support is below threshold. For each valid pattern $Q(\alpha)$, the *coordinator* (1) uses a node v_α that represents $Q(\alpha)$ to expand *DFS code graph* $\mathcal{G}_c$; (2) generates new patterns via forward or backward extension on the rightmost path of $Q(\alpha)$.

Rule Generation. After a *code graph* $\mathcal{G}_c = (V_c, E_c)$ is constructed, the *coordinator* repeatedly traverses $\mathcal{G}_c$ from each node v_α via reverse breadth first search, and produces GPARs as following. Specifically, for each ancestor $v_{\alpha''}$ of v_α encountered during the traversal, it generates another pattern Q_r by excluding edges of $Q(\alpha'')$ from Q_α. If Q_r is connected and the confidence $\frac{\mathsf{supp}(Q(\alpha),G)}{\mathsf{supp}(Q(\alpha''),G)}$ is no less than the threshold, a new GPAR $Q(\alpha'') \Rightarrow Q_r$ is generated.

As there may generate excessive GPARs, which results in difficulty in understanding and application, one may choose to generate "representative" GPARs with *maximal frequent patterns*, that corresponding to leaf nodes in $\mathcal{G}_c$, such that the rules can be dramatically reduced.

Workers. Each *worker* conducts local computation, in parallel.

Local Computation. Upon receiving a code α_c, each *worker* S_i computes matches and image of pattern $Q(\alpha_c)$ as following. (1) S_i first checks whether the new edge in $Q(\alpha_c)$ is a forward edge or backward edge, and generates candidate matches $G_{Q(\alpha_c)}$. If $G_{Q(\alpha_c)}$ is a valid match, S_i then updates auxiliary structures, and propagates truth to neighbor sites; otherwise, S_i broadcasts unknown to neighbor sites, to request verification. Upon receiving truth from neighbor sites, S_i updates auxiliary structures, and further propagates truth value to neighbor sites. Upon receiving unknown, S_i determines whether the candidate is a valid match or not, and propagates truth or unknown to neighbor sites for further propagation.

Fig. 2. Visual interfaces of GparMiner

4   Demonstration Overview

The demonstration is to show the following: (1) the use of GUI to set up parameters for GPARs mining, and browse various results; (2) the efficiency of distributed frequent pattern mining; and (3) the effectiveness of reducing rules with "representative" GPARs. The back-end of the system is implemented in Java and deployed on a machine with 2.9 GHz CPU, 8 GB Memory.

Interacting with the GUI. We invite users to use the GUI, from parameter setup to intuitive illustration of various results. Figure 2 shows three panels on parameter setup, *DFS code graph*, and typical GPAR, respectively.

Efficiency of Parallel Computation. We will show the efficiency of parallel computation supported by GparMiner. As will be seen, when the number of sites increases from 4 to 8, the time used for frequent pattern mining can be reduced by 42%, on average.

Effectiveness of Rule Reduction. We aim to show the effectiveness of reducing rules via generating "representative" GPARs. As will be seen, "representative" GPARs only accounts for 1.4% of the entire rule set, on average.

5   Summary

This demonstration aims to show the key idea and performance of the system GparMiner. The GparMiner is able to efficiently mine GPARs from large, distributed social networks. Moreover, GparMiner allows us to generate a small part of GPARs as "representative" rules, for easy interpretation. We contend that GparMiner can serve as a promising mining tool on social networks.

Acknowledgement. Xin Wang is supported in part by the NSFC 71490722,71472185, and Fundamental Research Funds for the Central Universities, China.

References

1. Facebook statistics; second quater (2018). https://www.statista.com/topics/751/facebook/
2. Agrawal, R., Imieliński, T., Swami, A.: Mining association rules between sets of items in large databases. SIGMOD Rec. **22**(2), 207–216 (1993)
3. Elseidy, M., Abdelhamid, E., Skiadopoulos, S., Kalnis, P.: GRAMI: frequent subgraph and pattern mining in a single large graph. PVLDB **7**(7), 517–528 (2014)
4. Fan, W., Wang, X., Wu, Y., Xu, J.: Association rules with graph patterns. PVLDB **8**, 1502–1513 (2015)
5. Namaki, M.H., Wu, Y., Song, Q., Lin, P., Ge, T.: Discovering temporal graph association rules. In: CIKM (2017)
6. Wang, X., Xu, Y.: Mining graph pattern association rules. In: Hartmann, S., Ma, H., Hameurlain, A., Pernul, G., Wagner, R.R. (eds.) DEXA 2018. LNCS, vol. 11030, pp. 223–235. Springer, Cham (2018). https://doi.org/10.1007/978-3-319-98812-2_19
7. Yan, X., Han, J.: gSpan: graph-based substructure pattern mining. In: ICDM (2002)

A Data Publishing System Based on Privacy Preservation

Zhihui Wang[1,2(✉)], Yun Zhu[1,2], and Xuchen Zhou[1,2]

[1] School of Computer Science, Fudan University, Shanghai, China
zhhwang@fudan.edu.cn
[2] Shanghai Key Laboratory of Data Science, Fudan University, Shanghai, China

Abstract. For data openness and sharing, we need to publish data and protect sensitive data at the same time. This paper provides the users with a system to realize privacy-preserving data publishing, which is implemented based on differential privacy. It has the following characteristics: (1) the raw data are first imported into a database and then are used to generate synthetic data for publishing; (2) a user can choose different privacy preservation levels for the synthetic data; (3) a subset of the attributes can been chosen to be synthesized while keeping the others untouched.

1 Introduction

In recent years, privacy-preserving data publishing has become more and more important. In practice, a data owner often wants to provide data but without revealing any private information. To address this problem, we introduce a data publishing system which supports differential privacy and thus provides powerful capability of privacy preservation. Even for an attacker with strong background knowledge and reasoning ability, the system can still prevent the privacy leakage. The system can be used for data publishing in various fields, such as health care, financial investment and so on. Since the information in these fields has potential private information, it is necessary to be published with privacy preservation.

In this demonstration, we present a data publishing system based on privacy preservation. The system has a wide range of applications, and also works well for high-dimensional data sets. It generates synthetic data based on the joint distribution of raw data to meet the requirement of different privacy, and then publishes the synthetic data according to the selection of the subset of attributes, the number of data records and the privacy preservation level.

2 System Design

Figure 1 is the main architecture of the system, which has three main function modules: *data import*, *data synthesis*, and *data publishing*. With the module of *data import*, a user can import raw data sets or data files into the MongoDB [1] database. For the module of *data synthesis*, a user can select privacy preservation level, generate

© Springer Nature Switzerland AG 2019
G. Li et al. (Eds.): DASFAA 2019, LNCS 11448, pp. 553–556, 2019.
https://doi.org/10.1007/978-3-030-18590-9_85

synthetic data files, and restore the synthetic data files back into the database for the future data publishing. For the module of *data publishing*, a user can choose the synthetic data to be released with various subsets of attributes, privacy preservation levels, and number of data records.

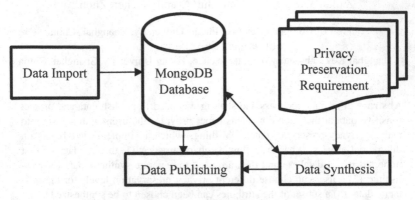

Fig. 1. The architecture of system

2.1 Differential Privacy

Differential privacy [5] has attracted many attentions since it was proposed. It aims at preserving the data privacy even when an attacker has the strong background knowledge. The basic knowledge of differential privacy is introduced in [3]. Although differential privacy provides a powerful tool for theoretic analysis of privacy preservation, it is rarely seen a practical data publishing system for preserving differential privacy.

2.2 Implementing Differential Privacy

Differential privacy has been used in many aspects [6–8]. For the purpose of data publishing, we implement differential privacy based on data synthesis. Specifically speaking, the system is implemented with the construction of Bayesian network in order to deal with the correlation between attributes and to address the problem of publishing high-dimensional datasets [2]. Furthermore, we improve the original approach at three aspects: firstly, it allows the user to choose only a subset of attributes of the raw data to be synthesized; secondly, it allows the data to be synthesized with different degrees of privacy preservation; thirdly, it allows the data to be synthesized for different number of records.

3   Demonstration Overview

In the demonstration, Adults dataset [4] is mainly used, which has 15 attributes and 45222 data records. The data are first imported into a MongoDB database, and are used to generate synthetic data and then publish to other users according to the privacy preservation requirement. Figure 2 is the data synthesis interface of the system. A user can choose the data sets to be synthesized and also the degree of privacy preservation. Generally speaking, the smaller the value of privacy preservation level is, the more the privacy of raw data is preserved. The data synthesis interface also shows the sample graph of the joint distribution of raw data and that of synthetic data for the chosen value of privacy preservation level.

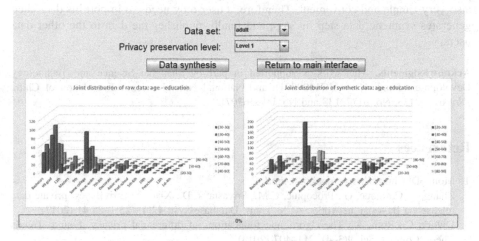

Fig. 2. Data synthesis interface

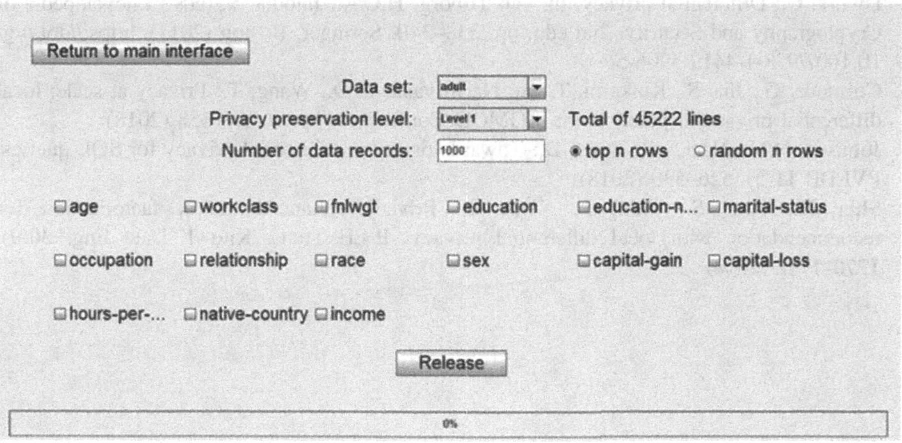

Fig. 3. Data publishing interface

Figure 3 is the data publishing interface of the system. After a dataset is chosen via the interface, all the attributes in the dataset will be shown too. The user can choose the privacy preservation level, the number of data records and the subset of attributes. Then, a mixture of raw data and synthetic data is released for publishing to other data users.

4 Summary

In this demonstration, we present a data publishing system which is based on differential privacy. Since the differential privacy can resist the attacks with the strong background knowledge, the system has good potentialities to realize privacy-preserving data publishing. For the users who worry about the privacy disclosure due to data publishing, the system offers a good solution. Besides, the operations of the system are also very simple and convenient. Therefore, a user only needs to import his data, then generates synthetic data step by step, and finally publishes the data to the other data users.

Acknowledgments. This work is supported in part by Shanghai Science and Technology Development Fund (No. 16JC1400801), and National Natural Science Foundation of China (No. 61572135, No. 61772138 and No. U1636207).

References

1. MongoDB. https://www.mongodb.com/
2. Zhang, J., Cormode, G., Procopiuc, C.M., Srivastava, D., Xiao, X.: PrivBayes: private data release via Bayesian networks. ACM Trans. Database Syst. **42**(4), 25:1–25:41 (2017)
3. Dwork, C., Roth, A.: The algorithmic foundations of differential privacy. Found. Trends Theor. Comput. Sci. **9**(3–4), 211–407 (2014)
4. Dua, D., Taniskidou, E.K.: UCI Machine Learning Repository. University of California, School of Information and Computer Science, Irvine, CA (2017). http://archive.ics.uci.edu/ml
5. Dwork, C.: Differential privacy. In: van Tilborg, H.C.A., Jajodia, S. (eds.) Encyclopedia of Cryptography and Security, 2nd edn, pp. 338–340. Springer, Boston (2011). https://doi.org/10.1007/978-1-4419-5906-5
6. Cormode, G., Jha, S., Kulkarni, T., Li, N., Srivastava, D., Wang, T.: Privacy at scale: local differential privacy in practice. In: SIGMOD Conference, pp. 1655–1658 (2018)
7. Johnson, N.M., Near, J.P., Song, D.: Towards practical differential privacy for SQL queries. PVLDB **11**(5), 526–539 (2018)
8. Shin, H., Kim, S., Shin, J., Xiao, X.: Privacy enhanced matrix factorization for recommendation with local differential privacy. IEEE Trans. Knowl. Data Eng. **30**(9), 1770–1782 (2018)

Privacy as a Service: Publishing Data and Models

Ashish Dandekar[1(✉)], Debabrota Basu[1], Thomas Kister[1], Geong Sen Poh[2], Jia Xu[2], and Stéphane Bressan[1]

[1] School of Computing, National University of Singapore, Singapore, Singapore
{ashishdandekar,debabrota.basu}@u.nus.edu, {dcsktpf,steph}@nus.edu.sg
[2] NUS-Singtel Cyber Security R & D Lab, Singapore, Singapore
{geongsen.poh,jia.xu}@singtel.com

Abstract. The main obstacle to the development of sustainable and productive ecosystems leveraging data is the unavailability of robust, reliable and convenient privacy management tools and services. We propose to demonstrate our Privacy-as-a-Service system and Liánchéng, the Cloud system that hosts it. We consider not only the publication of data but also that of models created by parametric and non-parametric statistical machine learning algorithms. We illustrate the construction and execution of privacy preserving workflows using real-world datasets.

1 Introduction

A Wired's online article titled "The Privacy Revolt: The Growing Demand For Privacy-as-a-Service" is asking every company the privacy question: "What are you doing to provide Privacy-as-a-Service?"[1]. Indeed, the main obstacle to the development of sustainable and productive ecosystems leveraging data, including data market places, recommendation systems and crowd sourcing systems, is the unavailability of robust, reliable and convenient privacy management tools and services. This entails developing privacy risk assessment and privacy preservation algorithms, and integrating them into a service architecture.

We demonstrate our *Privacy-as-a-Service* (PaaS) system and Liánchéng, our *Workflow-as-a-Service* (WaaS) cloud that hosts it. Liánchéng is a data sharing cloud system that provides a graphical workflow language. We extend it by incorporating privacy risk assessment and privacy preservation operators. We refer to this extension as a *Privacy-as-a-Service* model. Privacy-as-a-Service provides operators to publish not only anonymised data but also models created by statistical machine learning with differential privacy guarantees. We illustrate the construction and execution of privacy preserving workflows in these Workflow-as-a-Service and Privacy-as-a-Service models and systems for publishing data and statistical machine learning models using a census dataset and a medical dataset.

---

[1] https://www.wired.com/insights/2015/03/privacy-revolt-growing-demand-privacy-service/.

G. Li et al. (Eds.): DASFAA 2019, LNCS 11448, pp. 557–561, 2019.
https://doi.org/10.1007/978-3-030-18590-9_86

2   Liánchéng: Workflow as a Service

Liánchéng is a private data sharing Cloud service. The processing of data in Liánchéng is programmable by means of a Workflow-as-a-Service model. Liánchéng is deployed on a hardware infrastructure consisting of 128 commodity servers.

Data Sharing. This aspect is reminiscent of services such as Dropbox[2]. Each Liánchéng user gets a private account on which she can upload, download, organise and manage her data. Liánchéng provides both a Web interface and a desktop computer synchronisation agent. The internal sharing mechanism (user-to-user) relies on access control lists on directories. Liánchéng also provides additional publishing mechanisms, such as public access through URLs, for files.

Workflow as a Service. Liánchéng provides a Workflow-as-a-Service model and interface that offers a compromise between the traditional Infrastructure-as-a-Service (IaaS) and Platform-as-a-Service (PaaS) models. While, on the one hand, IaaS is fully customisable, it requires computing skills and efforts that constitute unnecessary obstacles. On the other hand, PaaS often limits users to the options proposed and has insufficient programmability. Liánchéng realises the compromise by offering an interactive GUI-based workflow language and domain specific operators. A Liánchéng workflow is a directed acyclic graph whose vertices represent operators and whose edges represent data flow. An operator can have an arbitrary number of parameters and has at least one input or output interface. The interaction and visualisation are reminiscent of that of Yahoo Pipes[3] and other graphical workflow design software.

Privacy as a Service. We extend Liánchéng with disclosure risk assessment and privacy preservation operators. We refer to such a cloud system functionality as *Privacy-as-a-Service*. We provide statistical disclosure risk assessment techniques such as uniqueness, overlap [8], and more advanced techniques [7,15] as well as privacy preservation mechanisms such as k-anonymity [14] and differential privacy [5]. In particular, we propose a set of machine learning algorithms offering differential privacy guarantees by means of the functional mechanism [16] and of functional perturbation [6].

Figure 1 shows a screenshot of the composition window of a Liánchéng workflow that applies a differentially private linear regression onto a dataset and uses the parameters of the model to generate a synthetic version of the provided dataset that preserves the targeted utility.

3   Private Data and Differentially Private Models

We consider both the publication of data and the publication of models created by the analysis of data by statistical machine learning algorithms. In both cases, there are risks of breach of privacy [2,12].

---

[2] https://www.dropbox.com.
[3] http://radar.oreilly.com/2007/02/pipes-and-filters-for-the-inte.html.

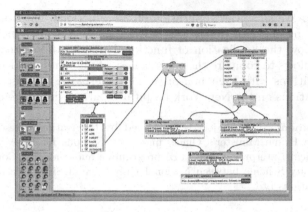

Fig. 1. A Liánchéng workflow generating a differential privacy compliant dataset.

In order to preserve privacy while publishing data, we use traditional anonymisation techniques such as k-anonymity [14], l-diversity [10] and t-closeness [9]. Alternatively, we generate fake but realistic datasets by using machine learning models that are trained on private datasets. We use machine learning algorithms for Linear regression, Decision tree, Random forest, Neural network to generate fully as well as partially synthetic datasets [4]. We use statistical disclosure risk assessment techniques to assess the risk of disclosure of the synthetic datasets.

In order to preserve privacy while publishing parametric models, we publish the parameters of statistical machine learning algorithms perturbed with the functional mechanism [16] with a differential privacy guarantee [3]. For non-parametric models, as they require to release the training dataset along with the parameters to compute the output [11], we release a non-parametric model as a service wherein the training data and model parameters reside at the server and users send their queries to get the answers. We use functional perturbation [6] to provide differential privacy guarantees for non-parametric models that use kernels [13] such as Kernel density estimation, Kernel SVM and Gaussian process regression.

4   Demo Scenario

We show experiments on the 2000 US census dataset [1] that consists of 1% sample of the original census data. We select 212,605 records, corresponding to heads of the households, and 6 attributes, namely, *Age, Gender, Race, Marital Status, Education, Income*. We start by uploading the data into Liánchéng. We initiate the workflow with a *filtering* operator for data cleaning. We further extend the workflow by adding different operators. For instance, we use Linear regression operator to fit a regression model on a selection of attributes. We show the use of a trained model to synthetically generate a sensitive attribute such as *Income* in the dataset. We show the application of the functional mechanism operator to release the model with differential privacy guarantees. For non-parametric models, we show the application of functional perturbation operator. We use different workflows to compare the effectiveness of differentially private machine learning algorithms with their non-private counterparts. We also show similar privacy evaluations on the New York hospital inpatient discharge dataset[4].

Acknowledgement. This project is supported by the National Research Foundation, Singapore Prime Minister's Office, under its Campus for Research Excellence and Technological Enterprise (CREATE) programme and under its Corporate Laboratory@University Scheme between National University of Singapore and Singapore Telecommunications Ltd.

References

1. Minnesota population center. Integrated public use microdata series - international: Version 5.0 (2009). https://international.ipums.org
2. Regulation (EU) 2016/679 general data protection regulation (text with EEA relevance). Official J. Eur. Union **L**(119), 1–88 (2016). https://eur-lex.europa.eu/eli/reg/2016/679/oj
3. Dandekar, A., Basu, D., Bressan, S.: Differential privacy for regularised linear regression. In: Hartmann, S., Ma, H., Hameurlain, A., Pernul, G., Wagner, R.R. (eds.) DEXA 2018. LNCS, vol. 11030, pp. 483–491. Springer, Cham (2018). https://doi.org/10.1007/978-3-319-98812-2_44
4. Dandekar, A., Zen, R.A.M., Bressan, S.: A comparative study of synthetic dataset generation techniques. In: Hartmann, S., Ma, H., Hameurlain, A., Pernul, G., Wagner, R.R. (eds.) DEXA 2018. LNCS, vol. 11030, pp. 387–395. Springer, Cham (2018). https://doi.org/10.1007/978-3-319-98812-2_35
5. Dwork, C., Roth, A., et al.: The algorithmic foundations of differential privacy. Found. Trends® Theor. Comput. Sci. **9**(3–4), 211–407 (2014)
6. Hall, R., Rinaldo, A., Wasserman, L.: Differential privacy for functions and functional data. J. Mach. Learn. Res. **14**(Feb), 703–727 (2013)
7. Heyrani-Nobari, G., Boucelma, O., Bressan, S.: Privacy and anonymization as a service: PASS. In: Kitagawa, H., Ishikawa, Y., Li, Q., Watanabe, C. (eds.) DASFAA 2010. LNCS, vol. 5982, pp. 392–395. Springer, Heidelberg (2010). https://doi.org/10.1007/978-3-642-12098-5_33
8. Hundepool, A., et al.: Statistical Disclosure Control. Wiley, Chichester (2012)
9. Li, N., Li, T., Venkatasubramanian, S.: t-closeness: privacy beyond k-anonymity and l-diversity. In: IEEE 23rd International Conference on Data Engineering, 2007. ICDE 2007, pp. 106–115. IEEE (2007)
10. Machanavajjhala, A., Kifer, D., Gehrke, J., Venkitasubramaniam, M.: L-diversity: privacy beyond k-anonymity. ACM Trans. Knowl. Discov. Data (TKDD) **1**(1), 3 (2007)
11. Murphy, K.P.: Machine Learning: A Probabilistic Perspective. The MIT Press, Cambridge (2012)

---

[4] https://health.data.ny.gov/Health/Hospital-Inpatient-Discharges-SPARCS-De-Identified/u4ud-w55t.

12. Shokri, R., Stronati, M., Song, C., Shmatikov, V.: Membership inference attacks against machine learning models. In: 2017 IEEE Symposium on Security and Privacy (SP), pp. 3–18. IEEE (2017)
13. Smola, A.J., Schölkopf, B.: Learning with Kernels, vol. 4. Citeseer (1998)
14. Sweeney, L.: k-anonymity: a model for protecting privacy. Int. J. Uncertainty Fuzziness Knowl. Based Syst. **10**(05), 557–570 (2002)
15. Zare-Mirakabad, M.-R., Jantan, A., Bressan, S.: Privacy risk diagnosis: mining *l*-diversity. In: Chen, L., Liu, C., Liu, Q., Deng, K. (eds.) DASFAA 2009. LNCS, vol. 5667, pp. 216–230. Springer, Heidelberg (2009). https://doi.org/10.1007/978-3-642-04205-8_19
16. Zhang, J., Zhang, Z., Xiao, X., Yang, Y., Winslett, M.: Functional mechanism: regression analysis under differential privacy. Proc. VLDB Endow. **5**(11), 1364–1375 (2012)

Dynamic Bus Route Adjustment Based on Hot Bus Stop Pair Extraction

Jiaye Liu, Jiali Mao$^{(\boxtimes)}$, YunTao Du, Lishen Zhao, and Zhao Zhang

School of Data Science and Engineering, East China Normal University,
Shanghai, China
{51184501030,10153903105,5118510035}@stu.ecnu.edu.cn
{jlmao,zhzhang}@dase.ecnu.edu.cn

Abstract. The crowdedness of buses caused by limited public transportation capacity has already severely influenced the convenience and comfort of inhabitant trip. Existing measures that reducing dispatching interval and replenishing more buses can soothe this case while aggravate traffic jam. To address the issue of inconvenient public transit characterized by packed buses, we propose a data-driven route adjustment framework, called Dynamic Bus Line Adjustment System, to recommend new operating route for the existing bus line by building direct route between extracted hot bus stop pair. *DBLAS* mainly involves extracting hot bus stop pair based on passenger volume estimation, and planning optimal route between hot bus stop pair using taxi traces. Finally, we develop a demo system to demonstrate the effectiveness of *DBLAS*.

Keywords: Bus route · Passenger volume · Hot bus stop pair

1 Introduction

During peak hours, a majority of bus lines are crowded due of limited capacity of existing urban public transport system. The intuitive solutions aim to shorten the dispatching interval between consecutive buses [3] and supplement with buses. Although it reduces residents' awaiting time for a bus, traveling time to the destination would probably become longer. This is due to that the condition of urban traffic block would become more aggravated because of the impacts of above measures. To improve residents' traveling experience, it necessitates an effective way to shorten traveling time based on the premise that the pressure of urban traffic remains unchanged.

There have been several researches on bus route adjustment over the years, but all of them attempt to build new operating route to cover more regions of the city [2] or enable ease of inhabitant trip at midnight [1]. Obviously that employing fixed operation scheme of bus routes cannot address our aforementioned issue, instead, designing various operation schemes for a same bus line is more advisable.

© Springer Nature Switzerland AG 2019
G. Li et al. (Eds.): DASFAA 2019, LNCS 11448, pp. 562–566, 2019.
https://doi.org/10.1007/978-3-030-18590-9_87

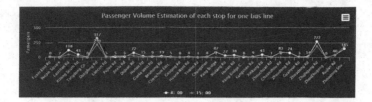

Fig. 1. Illustration of passenger volume estimation

Observed that during morning rush hour most of residents generally take a trip by bus from the uptown and through the downtown to the workplace in the other side of city. For instance, through estimating passenger volume for each stop in a bus line using swiping card data at 8 a.m, we can see that passenger volumes of two bus stops are much greater than the others for a bus line. Additionally, such two bus stops do not locate on the downtown and distant from each other, as shown in Fig. 1. It indicates that a large percentage of residents who live at one end of city have to travel by bus through most congested roads to their destinations (e.g., workplace, public service agency, etc.) which locate on the other end of city. Identifying hot bus stop pair that is away from downtown, and building a direct route between them enables designing additional operating route for any bus line. In view of that, our route adjustment mechanism attend to extract hot bus stop pair through estimating passenger volume of each bus stop on any bus line, and obtain direct route between the extracted stops by planning less congested route using historical taxi traces.

2   Overview

We propose a two-phase framework which consists of hot bus stop pair extraction and route planning between hot bus stops, as illustrated in Fig. 2. At the first phase, we seek for the popular bus stops away from downtown by estimating passenger volume for each bus stop on one bus line, and then extract hot bus stop pair from them according to the specified threshold. At the second phase, we derive the direct route between hot bus stop pair through mining optimal route from the historical taxi traces. Noted that our route planning task is premised on avoiding the worst of rush-hour congestion. Therefore, it is imperative to take traffic congestion situation into account when planning route.

Data Analysis. Three different data sets from Shanghai, China are used. (i) *Transit Smart Card Data* records the data of swiping the bus card; (ii) *Bus Route Data* collects the information of different bus routes; (iii) *Taxi Trajectory Data* contains the trajectories of taxis. The details of three data sets are shown in Fig. 3. *Transit Smart Card Data* and *Bus Route Data* are used through data fusion to extract hot bus stop pair, and *Taxi Trajectory Data* is applied to derive the optimal route between hot bus stop pair.

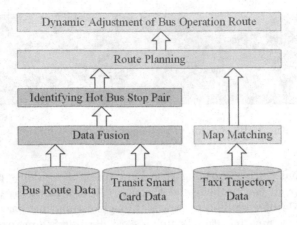

Fig. 2. DBLAS system

Identification of Hot Bus Stop Pair. A bus route refers to an ordered set of bus stops, which is denoted as $P = \{p_1, p_2, ..., p_a, p_{a+1}, ..., p_{a+b}, ..., p_m\}$. Generally, bus stop setting strategy of a bus line adopts starting with uptown through downtown to uptown. Thus, bus stops located in downtown can be represented by $PD = \{p_a, p_{a+1}, ..., p_{a+b}\}$, $PD \subset P$. The core task of hot bus stop identification is to estimate passenger volume for each bus stop on a line. But it is difficult to accurately calculate passenger flow of each bus stop in case of only the total sum of all the bus stops is available. Given a time period (e.g., from 7:00 am to 10:00 am), we heuristically estimate passenger flow based on the ratio of the bus stoppage time of one bus stop to the total sum of bus stoppage time of all the bus stops, the detail process is illustrated in Algorithm 1. Let $T(p_i)$ denote the sum of stoppage time of all the buses that stop at one bus stop p_i, $i \in \{1, 2, ..., m\}$, $CSum$ denote the amount of transit smart card records within a specified time period. Through data fusion, the total sum of stoppage time of all the bus stops can be calculated by $T_{Sum} = \sum_{i=1}^{m} T(p_i)$ (lines 2–4), and hence passenger flow of bus stop p_i can be estimated by $StopVolume_{p_i} = \frac{T(p_i)}{T_{Sum}} \times CSum$ (Lines 6–7). Given a volume threshold τ, a bus stop $p_i \in P$ is defined as a hot bus stop, iff $p_i \notin PD$ and $StopVolume_{p_i} \geq \tau$. Accordingly, a hot bus stop pair is composed of two hot bus stops that are distant from each other.

Transit Smart Card Data		Bus Route Data		Taxi Trajectory Data	
Shanghai	3/2018	Shanghai	3/2018	Shanghai	4/2018
Size	9.95 GB	Size	57GB	Size	231.9GB
# of Records	8billion	# of Route	1265		
Format		Format		Format	
CardID	Date&Time	RouteID	BusID	TaxiID	Date&Time
RouteID		StationID	StartTime&EndTime	Speed	Longitude&Latitude

Fig. 3. Data sets

Algorithm 1: Passenger Volume of Each Bus Stops Calculation

Input: $CSum$: the total number of transit smart card records;
$BusSet$: bus route data;
Output: $StopVolume$: a list of $StopVolume_{p_i}$;
1 $StopVolume \leftarrow \oslash$;
2 $T_{Sum} = \sum_{i=1}^{m} T(p_i)$;
3 **foreach** $busstop$ p_i in $BusSet$ **do**
4     $StopVolume.add(\frac{T(p_i)}{T_{Sum}} \times CSum)$;

5 **return** $StopVolume$;

Route Planning. We propose a route planning method incorporated with road congestion situation. The speed of each road segment within a given time period, denoted as $S_j (j \in \{1, 2, ..., n\})$, can be obtained by average speed calculation of its matched historical taxi trajectory data after map matching. Each optimal route extracted between hot bus stop pair using algorithm [4] consists of a list of road segments, the length (denoted as $length(line_k)$) of which is obtained by the sum of all the traversed road segments' lengths. Combined with the derived speed of road segment, the traveling cost of a route can be calculated by the sum of the ratio of each traversed road segment's length len_j to S_j, i.e., $length(line_k) = \sum_{j=1}^{n} \frac{len_j}{S_j}$. The direct route between hot bus stop pair is derived by choosing the route with shortest traveling cost.

3 Demonstration

Via the system interface, different existing bus line can be selected for adjustment. Our demo system takes Shanghai No. 451 bus line as an example.

Identification of Hot Bus Stop Pair. A hot bus pair, as shown in Fig. 4(a), <Zhoudong Road Zhujiagang Road→ Dongfang Lushan Road> is firstly derived by running $DBLAS$ system, here the value of τ is empirically set to 200. We can see that there is a large software park near *Dongfang Lushan Road*, it means that a majority of residents who live near Zhoudong Road travel by No.451 bus line to there.

Route Planning. Then we plan the path between the two extracted hot bus stops using the method in [4], and obtain K (K is set to 4) paths with quite short lengths, as shown in Fig. 4(b). Incorporated with road congested situation, we select a route with minimum traveling cost (depicted in blue line) as final result, who is near the original bus route (depicted in red line), as illustrated in Fig. 4(c). Last but not least, we suggest thatnew derived route and original one should be used alternately within different time periods.

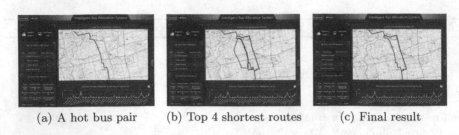

(a) A hot bus pair   (b) Top 4 shortest routes   (c) Final result

Fig. 4. DBLAS system demonstration (Color figure online)

Acknowledgements. The authors are very grateful to the editors and reviewers for their valuable comments and suggestions. This work is supported by NSFC (No. 61702423, U1501252).

References

1. Chen, C., Zhang, D., Li, N., Zhou, Z.: B-planner: planning bidirectional night bus routes using large-scale taxi GPS traces. Trans. Intell. Transp. Syst. **15**(4), 1451–1465 (2014)
2. Chuah, S., Wu, H., Lu, Y., Yu, L., Bressan, S.: Bus routes design and optimization via taxi data analytics. In: CIKM, pp. 2417–2420 (2016)
3. Huang, Q., Jia, B., Jiang, R., Qiang, S.: Simulation-based optimization in a bidirectional A/B skip-stop bus service. IEEE Access **5**, 15478–15489 (2017)
4. Luxen, D., Vetter, C.: Real-time routing with openstreetmap data. In: SIGSPATIAL, GIS 2011, pp. 513–516 (2011)

DHDSearch: A Framework for Batch Time Series Searching on MapReduce

Zhongsheng Li[1], Qiuhong Li[2(✉)], Wei Wang[2], Yang Wang[2], and Yimin Liu[3]

[1] JiangNan Institute of Computing Technology, Wuxi, China
lizhsh@yean.net
[2] School of Computer Science, Fudan University, Shanghai, China
{qhli09,weiwang1,081024004}@fudan.edu.cn
[3] Third Affiliated Hospital of Second Military Medical University, Shanghai, China
liuyiminzsh@aliyun.com

Abstract. We present DHDSearch, a framework for distributed batch time series searching on MapReduce. DHDSearch is based on a two-layer DHDTree. The upper DHDTree serves as a route tree to distribute the time series. While the lower DHDTrees serve the batch searching in parallel. Compared with traditional time series searching methods, DHDSearch has better scalability and efficiency.

1 Introduction

Many studies on similarity search over time-series databases have been conducted in the past decade. In this demo, we build a distributed index to conduct similarity time series searching. Euclidean distance is used. We know that the property of balance is important for tree index. Unfortunately, many popular time series index, such as iSAX [1] and DSTree [3], are not balanced. It means the number of records in nodes may differ dramatically, which may cause data skewness. In addition, for batch time series searching on massive datasets, it is critical for a parallel searching strategy. DHDTree [2] is proposed to cluster time series data by utilizing a dimension hierarchical decomposition tree to organize the data. Notice that DHDTree can reorganize the time series collections to achieve the best separation effect. We utilize it for time series searching by proposing a distributed searching method based on it. As Hadoop becomes more and more popular and has good scalability and fault tolerance, we select Hadoop to implement parallel DHDTree. Some distributed tree index on MapReduce assume that the computing nodes share a route tree which is the upper part of the whole tree. Each computing node constructs the subtree based on the route tree respectively. However, we need to solve the following problems for implementing a distributed searching framework for batch time series searching on MapReduce.

The work is supported by National Natural Science Foundation of China (U1509213), Shanghai Software and Integrated Circuit Industry Development Project (170512), National Key Research and Development Program (Grant Nos. 2016YFE0100300, 2016YFB1000700).

G. Li et al. (Eds.): DASFAA 2019, LNCS 11448, pp. 567–570, 2019.
https://doi.org/10.1007/978-3-030-18590-9_88

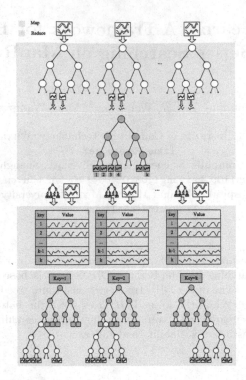

Fig. 1. Two-layer DHDTree on MapReduce

1. How to alleviate the effect of dimension curse when employing data partitioning for time series data?
2. How to design the route tree to guarantee both the balance and data locality?
3. How to guarantee the construction of subtree can take place in main memory of computer nodes?

2   DHDSearch: A Batch Time Series Searching on MapReduce

2.1   Two-Layer DHD Tree on MapReduce

We present a two-layer DHDTree on MapReduce. On the top layer, we build a DHDTree as a global route tree shared by all subtrees using a sampling method; on the lower layer, we build the local DHDTrees in computing nodes on the records. The key challenge is how to balance the data load among clusters. Fortunately, DHDTree can solve this problem by its balance property. Two parameters should be specified, one is the sampling rate, the other is the height of the routing DHDTree.

2.2   Constructing Two-Layer DHDTree on MapReduce

We use a MapReduce job to construct the upper DHDTree.

Constructing Upper DHDTree. For a given set TS of time series, the construction of the upper DHDTree needs three steps as follows.

1. Deciding the Height of Upper-DHD tree. We can estimate the height of the route tree according to the number of reducer. For example, we have a data set with the size of 100 G. Each reducer can only construct DHDTree index for 200 M data. We need at least 500 reducers to fulfil the whole construction, which means we need a route tree with at least 500 leaf nodes. Because the route tree is a balanced binary tree, so the height is at least 9 to satisfy $2^{n-2} \geq 500$.
2. Initialization. For massive data sets, it is impossible to load all time series into main memory to construct the route tree. We use two strategies to reduce the info which needs be loaded into main memory. First, we do sampling. Second, we perform dimension reduction by adopting DHD representations for time series. The sampling and dimension reduction can be done on the map phase.
3. Insertion. We use only one reducer to construct the upper DHDTree in main memory of the reducer. We insert the records one by one. Here a record represents an DHD representation for a time series.

Constructing Lower DHDTrees

1. In map phase, the data are processed in parallel. By default, the number of *map* tasks is decided by the number of data blocks. By inserting the item into the Upper-DHD tree, the lower DHDTree is selected.
2. In reduce phase, the time series are collected for a specified lower DHDTrees. The lower DHDTrees is constructed and is stored as a file in HDFS.
3. Root nodes of the lower DHDTrees are collected. The synopsis in the root nodes are used for accurate searching.

2.3   Batch Time Search Searching Utilizing Two-Layer DHDTree

We consider both approximate searching and accurate searching. In map phase, the queries are processed in different mappers. The upper DHDTree is shared by all the mappers. By querying the upper DHDTree, the nearest leaf node is found, thus the lower DHDTree is specified for each query. For approximate search, the queries for the same lower DHDTree are collected and processed in the same reducer. For accurate searching, the synopsis of the root nodes are used for pruning.

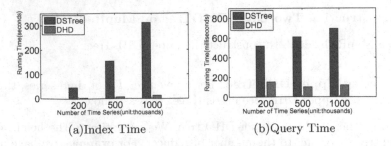

(a)Index Time   (b)Query Time

Fig. 2. Index building time and query time (TAO)

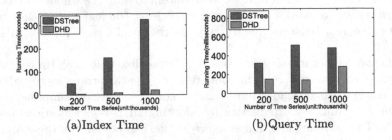

(a)Index Time   (b)Query Time

Fig. 3. Index building time and query time (UCI)

3 Demonstration and Evaluation

We evaluation DHDSearch framework with a cluster of 4 computing nodes, each with 4 GB memory. We use TAO and UCI datasets. TAO data set is from the Tropical Atmosphere Ocean project tao, which includes 12218 streams, each one of length 962. We use a sliding window of step 20 to generate 268796 subsequences with length 512. UCI data set is from the UCI archive. A similar distributed DSTree index is used as the baseline. The implementation way is almost the same as the two-layer DHDTree. Here we evaluate approximate searching. The results are illustrated in Figs. 2 and 3.

References

1. Camerra, A., Palpanas, T., Shieh, J., Keogh, E.J.: isax 2.0: Indexing and mining one billion time series. In: ICDM, pp. 58–67 (2010)
2. Li, Q., et al.: Clustering time series utilizing a dimension hierarchical decomposition approach. In: Candan, S., Chen, L., Pedersen, T.B., Chang, L., Hua, W. (eds.) DASFAA 2017. LNCS, vol. 10177, pp. 247–261. Springer, Cham (2017). https://doi.org/10.1007/978-3-319-55753-3_16
3. Wang, Y., Wang, P., Pei, J., Wang, W., Huang, S.: A data-adaptive and dynamic segmentation index for whole matching on time series. PVLDB **6**(10), 793–804 (2013)

Bus Stop Refinement Based on Hot Spot Extraction

Yilian Xin, Jiali Mao$^{(\boxtimes)}$, Simin Yu, Minxi Li, and Cheqing Jin

School of Data Science and Engineering, East China Normal University,
Shanghai, China
{51185100031,10153901225,minxli}@stu.ecnu.edu.cn
{jlmao,cqjin}@dase.ecnu.edu.cn

Abstract. During rush hour, numerous residents travel to their destinations by a multi-mode transfer way (e.g. bus & taxi) due to lack of direct buses, which sharply increases trip expense and even heavy traffic. The root of such inconvenience in bus service is that obsolete and incorrect bus stop information cannot satisfy residents' time-dependent travel demand. In this work, we put forward a framework, called *BSRF*, to optimize the existing bus route using the mined bus stops from trajectory data of taxis' short-haul order, including identifying candidate bus stop based on hot drop-off point and matching new bus stop with the existing bus line. We build a demo system to showcase the effectiveness of *BSRF*, which can offer reliable suggestion on bus stop setting for public transport companies.

Keywords: Bus stop refinement · Distance reachable ·
Direction matchable

1 Introduction

Public transportation service plays a significant role in alleviating congestion, boosting traffic efficiency and improving traveling experience. However, due to that bus stop adjustment lags behind the rapid development of urban construction, bus stop information has become obsolete and obviously cannot satisfy time-dependent traveling requirements of residents. As a result, more and more citizens travel to their destinations by transferring between various traffic modes (e.g. a combination of bus and taxi) because of lack of direct buses. This not only sharply increases trip expense as well as inconvenience, but also possibly aggravates traffic jam in a way. It necessitates an effective bus stop setting method based on the actual traveling request of residents.

Recently, there have been a few researches committed to the issue of mining bus stops from bus trajectory data or taxi trace data, but most of which aim at building a new bus route by finding out all the popular locations [1–3]. According to the observation, during morning rush hour, a majority of residents go out by using bus first and then hailing taxi in order not to be late. Based on the above

© Springer Nature Switzerland AG 2019
G. Li et al. (Eds.): DASFAA 2019, LNCS 11448, pp. 571–575, 2019.
https://doi.org/10.1007/978-3-030-18590-9_89

observation, distinct from the prior works, we attempt to update the existing bus route with new extracted bus stop using trajectory data of taxis' short-haul order. More specifically, this paper makes the following contributions:

- We propose a framework for optimizing bus route with new extracted bus stops, called as *BSRF*, which mainly consists of identifying candidate bus stop, matching new bus stop with the existing bus line.
- We design a novel bus stop detection mechanism, the task of which is to find out hot drop-off points through clustering drop-off locations from taxis' trajectories, and then identify new bus stop by searching its matched bus line.
- We build a demo system to demonstrate the effectiveness of *BSRF*.

2   Major Modules and System Architecture

Data Analysis. Two distinct data sets from Shanghai, China are used. (i) *Taxi Trajectory Data* collects 30 day's Taxi Trajectories of over 30,000 taxis, and each record contains car ID, business status, longitude, latitude, timestamp, direction and speed. (ii) *Bus Route Data* records 1,265 Shanghai bus line station data, which contains the information like bus ID, bus line name, station name, longitude, latitude, current location, timestamp and station relationship, etc.

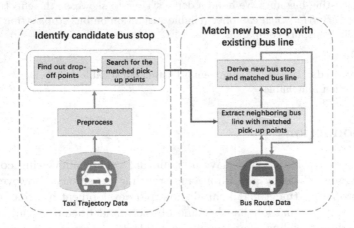

Fig. 1. The framework of BSRF system

As shown in Fig. 1, we propose a two-phase framework, called *BSRF*. The candidate bus stops are identified using *Taxi Trajectory Data* and then the existing bus line can be adjusted through matching it with new bus stop based upon *Bus Route Data*.

Identifying Candidate Bus Stop. Firstly, taxi trajectories are preprocessed with several requirements: (i) classify them by the car ID and the date, (ii) filter the data with the business status $\neq operating$ and the passenger status is $\emptyset$, and

(iii) extract trajectories of taxi's short-haul order. After map-matching, we apply *DBSCAN* algorithm to cluster drop-off locations to find out drop-off point, and identify drop-off point with higher occurrence frequency larger than the specified frequency threshold α as candidate bus stop.

Matching Candidate Bus Stop and Existing Bus Line. Based on the identified candidate bus stops, we confirm a new bus stop and the bus line that it belongs to by implementing a matching strategy. More specifically, we search the corresponding pick-up points P_p for extracted hot drop-off points P_d. This is based on the observation that residents usually travel using bus first then taxi in morning rush hour due to lacking of direct buses. As a result, when the corresponding pick-up point of hot drop-off point is near a certain bus stop on existing bus line, such hot drop-off point can be regarded as a new bus stop. Meanwhile, their neighboring bus stops ($Stop_p$, $Stop_n$) on existing bus lines are sought out.

Next we consider whether the relationship between new bus stop and the prior or next stop of a same bus line satisfies *distance reachable* and *direction matchable*. If the road network distance between candidate bus stop and the (prior & next) bus stop of the existing bus line is not larger than average distance between bus stops on that line, the candidate bus stop and bus line are viewed as *distance reachable*. In addition, for a matched bus line that new bus stop belongs to, we shall ensure moving direction between the prior bus stop and new one keep the same with that between new one and the next one. Let Dir_p denote the moving direction calculated by the positions of the prior bus stop and new one, and Dir_n denote the moving direction calculated by the positions of new bus stop and the next bus stop, *included angle* between two consecutive directions, denoted as Dir can be obtained by $Dir = \arctan(\frac{|Dir_p - Dir_n|}{1 + Dir_p \times Dir_n})$. If Dir satisfies the specified angle threshold θ, the candidate bus stop and bus line are *direction matchable*.

In the end, we attend to adjust the existing bus line with new bus stop. Based on *Taxi Trajectory Data*, we extract the optimal path between the prior bus stop and the new one, as well as that between the new bus stop and the next one respectively using open source API [4].

3 Demonstration

Based on *BSRF* framework, we build a demo system online to demonstrate the utility of the *BSRF*. All codes, written in Java and Python in the backend, and HTML, CSS and JavaScript in the frontend. As shown in Fig. 2, our demo system consists of *a notice board* and *a real-time visualization platform*. The users could acquire the background knowledge about total number of taxi and bus, bus stop ranking by short-haul order drop-off occurrence and nearby possible bus lines to be adjusted in the notice board.

On the visualization platform, we show the complete process of *BSRF*.

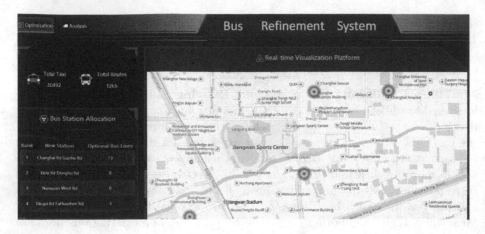

Fig. 2. *BSRF*- the demo system

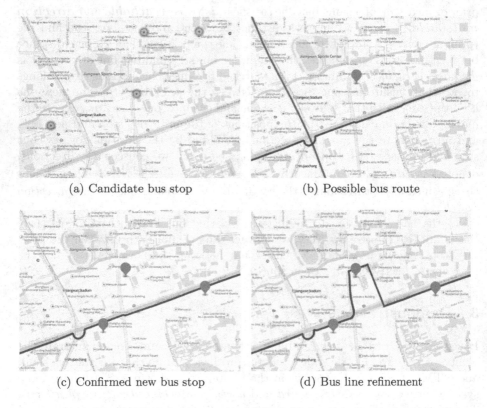

(a) Candidate bus stop

(b) Possible bus route

(c) Confirmed new bus stop

(d) Bus line refinement

Fig. 3. The whole process of *BSRF* (Color figure online)

Identifying Candidate Bus Stop. After preprocessing, the drop-off points are derived by clustering the drop-off locations using trajectories of taxi's short-haul order, they are identified as bus stop candidates according to the specified frequency threshold α (here α is set to 25), as shown in Fig. 3(a).

Matching Candidate Bus Stop and Existed Bus Line. Next, we find the pick-up points of hot drop-off points and then the neighboring bus line of such pick-up points, to derive the possible bus lines (depicted in different colors), as illustrated in Fig. 3(b). Further, we confirm the bus stop to be added and its corresponding bus line that satisfies *distance reachable* and *direction matchable*, as shown in Fig. 3(c). Finally, we adjust the operating route of the existed bus line with new identified bus stops, as depicted in Fig. 3(d).

Acknowledgements. The authors are very grateful to the editors and reviewers for their valuable comments and suggestions. This work is supported by NSFC (No. 61702423, U1501252).

References

1. Chen, C., Zhang, D., Li, N., Zhou, Z.: B-planner: planning bidirectional night bus routes using large-scale taxi GPS traces. Trans. Intell. Transp. Syst. **15**(4), 1451–1465 (2014)
2. Chuah, S., Wu, H., Lu, Y., Yu, L., Bressan, S.: Bus routes design and optimization via taxi data analytics. In: CIKM, pp. 2417–2420 (2016)
3. Garg, N., Ramadurai, G., Ranu, S.: Mining bus stops from raw GPS data of bus trajectories. In: COMSNETS, pp. 583–588 (2018)
4. Luxen, D., Vetter, C.: Real-time routing with openstreetmap data. In: SIGSPA-TIAL, pp. 513–516 (2011)

Adaptive Transaction Scheduling for Highly Contended Workloads

Jixin Wang[1], Jinwei Guo[1], Huan Zhou[1], Peng Cai[1,2(✉)], and Weining Qian[1]

[1] School of Data Science and Engineering, East China Normal University,
Shanghai, China
{wangjixin,guojinwei,zhouhuan}@stu.ecnu.edu.cn
{pcai,wnqian}@dase.ecnu.edu.cn
[2] Guangxi Key Laboratory of Trusted Software,
Guilin University of Electronic Technology, Guilin, China

Abstract. Traditional transaction scheduling mechanism—which is a key component in database systems—slows down the performance of concurrency control greatly in such environments for highly contended workloads. Obviously, to address this issue, there are two effective methods: (1) avoiding concurrent transactions that access the same high-contention tuple at the same time; (2) accelerating the execution of these high-contention transactions. In this demonstration, we present a new transaction scheduling mechanism, which aims to achieve the above goals. An adaptive group of *first-class* queues is introduced, where each queue is allocated to a specified worker thread and takes charge of transactions accessing specified high-contention tuples. We implement a system prototype and demonstrate that our transaction scheduling mechanism can effectively reduce the abort ratio of high-contention transactions and improve the system throughput dramatically.

1 Introduction

Nowadays, with the rapid growth of data traffic in many applications, database servers need to face numerous online transactions at the same time. Fortunately, the servers equipped with thousands of CPU cores have the ability to handle a huge number of requests. However, existing databases cannot scale up to thousands of cores, especially when the workload is highly contended [3]. This is because that traditional transaction scheduling mechanism randomly chooses a worker thread to execute a new transaction. In the highly contended workload, the probability that two concurrent transactions in different worker threads access the same tuple is very high. This leads to a result that one of both conflicted transactions may be aborted or blocked, which has a negative impact on system performance. To address this issue, there are two effective methods as follows: (1) Avoid concurrent transactions that access the same high-contention tuple. (2) Accelerate the execution of the high-conflict transactions.

Zhang et al. provided an intelligent scheduling to put a new arrived transaction into the queue with the highest likelihood of conflict [4], which aims to

G. Li et al. (Eds.): DASFAA 2019, LNCS 11448, pp. 576–580, 2019.
https://doi.org/10.1007/978-3-030-18590-9_90

achieve the first goal. However, this method does not accelerate the execution of any transaction. MySQL enterprise edition utilizes the high-priority queue to accelerate the execution of a likely-conflicted transaction [1]. But it is not designed for the first point. Although H-Store [2] is regarded as a solution to achieve both goals, its distributed feature leads to its struggle for load-balance.

In this work, we propose a new transaction scheduling mechanism, which aims to achieve the above two goals at the same time. First, we utilize the *first-class* queues, which are responsible for handling transactions that access the specified high-contention tuples. As a result, we can prevent the concurrent transactions from accessing the same high-contention tuple at a time point. Second, we can increment the number of *first-class* queues. Since each *first-class* queue is allocated to a specified worker thread, the computing resources can be added for the high-contention transactions. Consequently, the execution time of a high-contention transaction is decreased effectively, which can reduce the probability of conflict. In the next sections, we will introduce our system architecture and the key techniques and provide a demo for our system.

2   System Architecture and Key Techniques

The overall architecture we designed is shown in Fig. 1, which consists of two main components: statistics module and transaction scheduling module. The statistics module is responsible for collecting the historical information of transactional operations, which is used to acquire: (1) which tuple is high-contention; (2) which two high-contention tuples in a transaction unit are accessed frequently. We provide the details of this module in Sect. 2.1.

In the scheduling module, there is a transaction scheduler, which is responsible for forwarding a transaction to a queue. In our system, these queues are classified into two categories: one *public* queue and a variable number of *first-class* queues. We will introduce the adaptive strategy of varying the number of *first-class* queues in Sect. 2.2. It should be noted that each *first-class* queue is assigned to a specified worker thread, and it is in charge of a set of high-contention tuples. The intersection of tuple sets of any two different *first-class* queues is empty. We will describe how to distribute the high-contention tuples to different *fist-class* queues in Sect. 2.3. If a new transaction wants to access a high-contention tuple, it is forwarded to the *first-class* queue which takes charge of this tuple. Otherwise, the transaction—which does not contain any high-contention tuples—would be forwarded to the *public* queue. If a worker thread is not assigned to a first-class queue, it will get and execute a transaction from the *public* queue. Otherwise, it can only handle the transactions in its own *first-class* queue.

As shown in the left picture of Fig. 1, there are two new transactions T_1 and T_2. Since T_2 wants to access the high-contention tuples c and e, it is forwarded to the *first-class* queue f_1, which is in charge of these tuples. On the other hand, the transaction T_1—which does not access any high-contention tuples—is forwarded to the *public* queue.

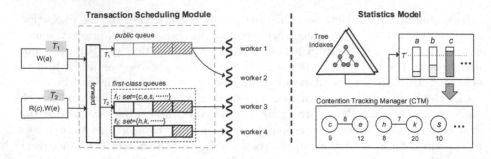

Fig. 1. Illustration of our system architecture.

2.1 Statistics Information

In our system, each tuple has an implicit field *temperature* (abbr. *temp*), which is used to track the access number of this tuple in the past period of time. If a tuple is accessed by a transaction, we increment its *temp*. On the other hand, if the tuple is not accessed by any transaction over a specified period of time, we decrement its *temp*. A tuple, whose *temp* exceeds a pre-configured threshold $\mathcal{T}$, is added to the contention tracking manager (CTM). Note that CTM also tracks the number of the *correlation* of any two tuples. If two tuples in CTM are accessed by a transaction, we increment the *correlation* of these two tuples. As illustrated in the right picture of Fig. 1, the *correlation* of tuples c and e is 8, which illustrates that c and e are accessed simultaneously by eight transactions. When a transaction accesses a tuple in CTM and finds its *temperature* is less than $\mathcal{T}$, it removes the tuple from CTM.

2.2 Adaptive Adjustment of the Number of First-Class Queues

The number of *first-class* queues is determined adaptively by two aspects: (1) the total number of transactions accessing the high-contention tuples; (2) the degree of emergency for handling the high-contention transactions. Therefore, we introduce an equation to acquire the *first-class* queues' number N_f: $N_f = (\lambda \cdot n_f \cdot (N - N_f))/n_p$, where n_p and n_f are the number of transactions entering the *public* queue and the *first-class* queues respectively, N is the total number of the worker threads, and λ is the emergency factor that is decided by the abort ratio of transactions. Note that with the increase of abort ratio, λ, which is always ≥ 1, becomes larger in order to allocate more resources for the high-contention transactions. In our system, once the workload is changed, the number of *first-class* queues are varied to adapt to the new data access model.

2.3 Element Allocation for the First-Class Queues

Recall that each *first-class* queue has its own set of high-contention tuples. In other words, a transaction only enters into the *first-class* queue whose set contains the high-contention tuples accessed by the transaction, and it is only executed by the specified worker thread. The number of *first-class* queues has been acquired in the above section. Therefore, we need a reasonable approach to allocate the high contention tuples to the different *first-class* queues, which leads to the result that (1) each worker thread executes transactions in the load-balanced situation; (2) a transaction enters into the *first-class* queue whose set contains as many of high-contention tuples accessed by the transaction. Consequently, we first order the tuples in CTM according to their *temps*, and then allocate the tuple and its correlative tuples to a *first-class* queue in a round-robin manner.

3   Demonstration

We have implemented a system prototype to demonstrate the effectiveness of our proposed method. We can choose the worker number, scheduling type (e.g., ADAPTIVE is our proposed method), the workload type (e.g., YCSB) and the contention of the corresponding workload (e.g., θ in YCSB). We can dynamically change the workload while the system is running.

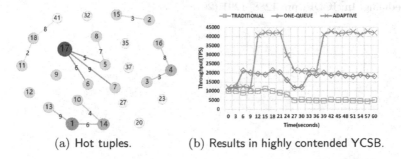

(a) Hot tuples.   (b) Results in highly contended YCSB.

Fig. 2. Experimental demonstration. The right picture shows the throughput of different transaction scheduling methods under the highly contended YCSB workloads.

Hot tuples and associations at some point in the system running process are shown in Fig. 2(a). Experimental results are shown in Fig. 2(b). Note that TRADITIONAL is used to denote the database that does not adopt any optimization method and ONE-QUEUE is used to represent the database that adopts a single queue for the highly contended workload. For the first 12 s, the system uses our method to configure the queues. From 12 s to 20 s, owing to our scheduling strategy, our method ADAPTIVE outperformes the other schemes. When the system runns for 20 s, we change the workload. Then our system detectes workload changes and reconfigures the queues, and in 40 s we achieve our previous high performance, which conforms to our expectation. Consequently, our method is suitable for the highly contended workload in the modern hardware equipped with multiple CPU cores.

4   Conclusion

Traditional transaction scheduling mechanism does not perform well for the highly contended workload in the modern server equipped with multi-sockets. In this demonstration, we present a new adaptive mechanism for transaction scheduling and show that this method has a good performance.

Acknowledgments. This research is supported in part by National Key R&D Program of China (2018YFB1003402), National Science Foundation of China under grant number 61432006, and Guangxi Key Laboratory of Trusted Software (kx201602).

References

1. MySQL Enterprise Thread Pool. https://dev.mysql.com/doc/refman/8.0/en/thread-pool.html
2. Kallman, R., Kimura, H., Natkins, J., et al.: H-store: a high-performance, distributed main memory transaction processing system. PVLDB **1**(2), 1496–1499 (2008)
3. Wang, T., Kimura, H.: Mostly-optimistic concurrency control for highly contended dynamic workloads on a thousand cores. PVLDB **10**(2), 49–60 (2016)
4. Zhang, T., Tomasic, A., Sheng, Y., Pavlo, A.: Performance of OLTP via intelligent scheduling. In: ICDE, pp. 1288–1291 (2018)

IMOptimizer: An Online Interactive Parameter Optimization System Based on Big Data

Zhiyu Liang, Hongzhi Wang$^{(\boxtimes)}$, Jianzhong Li, and Hong Gao

Harbin Institute of Technology, Harbin, Heilongjiang, China
olympic2008dreams@163.com,
{wangzh,lijzh,honggao}@hit.edu.cn

Abstract. Intelligent manufactory is a typical application of big data analysis. Flexible production line is an essential fundamental of intelligent manufactory. Producing different types of similar products alternately in one line with fixed stations but varying parameters is a typical kind of flexibility. In this case, the quality of products is directly determined by the parameter setting. However, the relation between parameters and product quality are too complicated to model. Consequently, current solution is bound to tune the parameters manually, which highly relies on expertise and is very costly. Inspired by recommender systems, we develop IMOptimizer, a novel online interactive processing parameter setting system. IMOptimizer holds the features of *Configurable, Interactive, High Efficiency* and *Friendly UI*. To the best of our knowledge, our system is the first big-data-driven generic platform focusing on online process optimization. In this demonstration, we will present our prototype.

1 Introduction

Intelligent manufacturing is one of the most promising applications of big data analysis [1]. In real production line, manufacturing processes of a product are divided into a series of individual sections, and each section contains a sequence of steps with different targets [2]. Thus, each step requires specific equipments, conditions and workers, involving various kinds of parameters. At the end of each sections, products are detected by some metrics to evaluate their quality and determine whether they are good or waste products. The quality is closely related to the corresponding processing parameters.

A significant fundamental of intelligent manufactory is flexible production line. While flexibility has many manifestations, a case is that the production line has the ability to adapt various types of products rapidly just by fine-tuning several parameters online [3, 4]. However, in common circumstance, a production line consists of hundreds of processing steps, therefore the processing parameters are high dimensional. Additionally, interdependency exists in manufacturing processes widely among different steps, which leads to complicated interconnections of the processing parameters. Thus, it is difficult to model the relation between the parameters and the quality of products for parameter setting. In consequence, when producing different types of

G. Li et al. (Eds.): DASFAA 2019, LNCS 11448, pp. 581–584, 2019.
https://doi.org/10.1007/978-3-030-18590-9_91

products in a flexible production line, existing solution is tuning parameters manually sections by sections, which highly depends on the expertise and becomes expensive.

To provide logical reference for processing parameters setting and reduce the dependency on manual experiences, we propose a similarity-based online interactive processing parameter setting system, which is, to our best knowledge, the first big-data-driven interactive configurable platform focusing on the online processing parameter optimization. Our system has the following benefits:

- *Configurable.* Our system can be easily configured to adapt different kinds of production lines by setting the number and type of processing parameters and metrics of product quality.
- *Interactive.* To ensure the stability and robustness of the production line, our system just recommends a list of feasible parameter setting projects for reference, while the final decision is made by related staff. The manufactory process is shown visually to users, and they could easily choose a recommended project or determine a user-defined parameter setting. As a result, the time and cost spent on acquiring expertise is substantially reduced.
- *High Efficiency.* Since the flexible production line is required to adapt different products timely, our system adopts efficient search algorithm for similarity match.
- *Friendly UI.* Considering that most staffs in manufactory domains are lack of experience on complicated computer operation, we design a friendly user interface in our system to show the running state and parameter setting projects. Production line configuration and decision making are also simple.

The remaining parts of this paper are organized as followings. Section 2 discusses the system architecture. Section 3 describes the search and recommendation method of our system. In Sect. 4, the demo scenarios are introduced.

2   System Architecture

In this section, we discuss the architecture of our system.

The architecture is shown in Fig. 1. To support various types of production line, the system is designed to be configurable with parameters and quality metrics in the *Interaction* module. The data of each series of steps are loaded from data acquisition module in production line timely, and the features are extracted from the standard data of quality metrics by *Data Loading* module. While all quality metrics are acceptable, current piece of data is stored to *Data Storage* module, and the system prepares for the next piece of data. Otherwise, *Search&Recommendation* module sets up to inform users of the quality problems, search for possible combinations of processing parameters from *Data Storage* module and recommend a list of the optimal combinations to users.

Visualization module shows the processing parameters and product quality timely, and displays the projects of parameter setting recommended by the system. Users could select one of the projects, set user-defined parameters, or tune some parameters based on the recommended projects to generate the final combination of parameters by this module. The final project is executed with the *Parameter Setting* module.

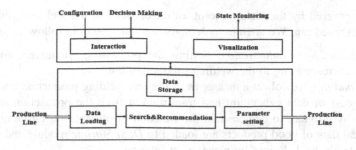

Fig. 1. System architecture

3   Search&Recommendation Method

In this section, we describe the search and recommendation method adopted to generate the projects of parameter setting.

Because the similar types of products in one production line are usually manufactured by similar processing parameters, inspired by recommender system [6], we developed a similarity-based processing parameter search and recommendation technology in our system. First, for a new waste product, the system searches for the records corresponding to the products whose types are the top-k most similar with the type of the waste product from *Data Storage* module. Next, the system compares the parameters of the waste product with the selected records, and resorts the records in the descending order by the number of unchanged parameters. This step aims to recommend the projects which tune the minimum amounts of parameters to improve the robustness of the production line. Finally, the parameters within the resorted records are recommended to users in sequence. The similarity of types and manufacturing parameters of products guarantees the feasibility of the recommendation results to some extent.

The types of products are represented by feature vectors withdrawn from the quality metrics which contain rich information because the diverse types of measured data (e.g. numerical values, images, and waveforms). To guarantee the efficiency in big data processing, we developed an improved LSH algorithm based on [5] in our system to search for similar types of products. The algorithm adopts the randomized index structure for acceleration issues. The metric of similarity is angular distance.

4   Demonstration

To demonstrate the features of our system, we choose one section of a real production line which does welding work on special-shaped connectors. This section contains 7 steps. The first 6 steps conduct welding operation, and the last step detects the quality of the welding seam. This section contains dozens of processing parameters including welding time, temperature, volume of soldering tin and position of the manipulator and 5 metrics of welding quality i.e. the shape of the welding joint, plumpness of the welding joint, area covered by the welding joint, offset distance of the welding joint and the length of exposed pin. We attempt to demonstrate our system in following steps.

1. *Production Line Configuration*: Setting the processing parameters and quality metrics corresponding to the welding production line.
2. *Data Loading*: We collect a dataset from the real welding production and simulate the process of data generation and transmission from the production line to our system by program to demonstrate the function of our system. Large amounts of historical data of good products are loaded to *Data Storage* module and simulated new data are loaded piece by piece at set intervals.
3. *Parameter Setting Recommendation Result Review*: Each time the data of waste product are loaded into our system, the *Search&Recommendation* module runs and the parameter settings of the most similar types of products are recommended to users. Figure 2 shows the result displayed on the interface where the number of projects k is set to be 5.

Layer/Pass	Preheat temp.	Air flow	current	Volts	Speed	X1_ActualPosit ion	Y1_ActualPosit ion	Z1_ActualPositi on	X1_ActualA: leration
2	127	253235	0.85	1.31	424954	198	158	119	-31.3
1	125	250532	0.88	1.29	432936	198	158	119	-12.5
1	125	250862	0.79	1.31	434688	198	158	119	-18.8
2	122	247847	0.82	1.33	431411	198	158	119	-25.2
1	124	252136	0.80	1.30	432939	198	158	119	0

Fig. 2. Recommended projects of parameter setting

Acknowledgments. This paper was partially supported by NSFC grant U1509216, U1866602, The National Key Research and Development Program of China 2016YFB1000703, NSFC grant 61472099,61602129.

References

1. O'Donovan, P., Leahy, K., Bruton, K., et al.: Big data in manufacturing: a systematic mapping study. J. Big Data **2**(1), 20 (2015)
2. Production line. https://en.wikipedia.org/wiki/Production_line
3. Browne, J., Dubois, D., Rathmill, K., et al.: Classification of flexible manufacturing systems. FMS Mag. **2**(2), 114–117 (1984)
4. ElMaraghy, H.A.: Flexible and reconfigurable manufacturing systems paradigms. Int. J. Flex. Manuf. Syst. **17**(4), 261–276 (2005)
5. Andoni, A., Indyk, P., Laarhoven, T., et al.: Practical and optimal LSH for angular distance. In: Advances in Neural Information Processing Systems, pp. 1225–1233 (2015)
6. Ricci, F., Rokach, L., Shapira, B. (eds.): Recommender Systems Handbook. Springer, Boston (2015). https://doi.org/10.1007/978-1-4899-7637-6

Tutorials

Cohesive Subgraphs with Hierarchical Decomposition on Big Graphs

Wenjie Zhang[1,2]([envelope]), Fan Zhang[1,2], Ying Zhang[3], and Lu Qin[3]

[1] Guangzhou University, Guangzhou, China
[2] University of New South Wales, Sydney, Australia
wenjie.zhang@unsw.edu.au
[3] CAI and School of Software,
University of Technology Sydney, Ultimo, Australia

1 Introduction

Graphs have been widely used to represent the relationships of entities in a large spectrum of applications such as social and information networks, collaboration networks, and biology. Graph decomposition is one of the fundamental problems. Recent works on the hierarchical graph decomposition rely on the cohesive subgraph models to (1) analyze the community structures on large scale networks with different granularity; and (2) measure the importance/significance of each individual vertex. The hierarchical decomposition of graph has a wide range of applications such as social contagion, community detection, event detection, anomalies detection, network analysis, network visualization, internet topology, influence study, protein function prediction, and user engagement in social networks.

2 Tutorial Outline

This tutorial is tailored for DASFAA attendees who are expected to be aware of the broad area of database and data management but may or may not be actively working in graph and social networks.

2.1 Introduction, Modeling, Applications and Challenges (25 min)

The aim of this part is to provide necessary background to the audience. It will consist of an introduction to the research field to highlight the popularity, opportunities and challenges in cohesive subgraph models and hierarchical decomposition of graphs. Specifically, we will first introduce the cohesive subgraph computation and general concepts of hierarchical graph decomposition (e.g., modular decomposition [1]). Then, we will discuss popular models and arising opportunities in this research area using several real-world examples, including degree based decomposition (k-core) [2], tie-strength based decomposition (k-truss) [3], density based decomposition [4], edge-connectivity based decomposition (k-ECC) [5], and the variants.

G. Li et al. (Eds.): DASFAA 2019, LNCS 11448, pp. 587–589, 2019.
https://doi.org/10.1007/978-3-030-18590-9

2.2   Computation Under Different Settings (40 min)

In this part, we focus on the decomposition algorithms under different models and computing environments.

k-core and Its Variants. We will first introduce the linear algorithms proposed for k-core computations and decomposition. The k-core computation under different computing environments will be introduced, including I/O efficient algorithms, semi-external memory model [6], parallel computation [7], distributed algorithms in Spark, Hadoop and Pregel like systems [8]. Then the computation of the variants of the k-core problem will be introduced [9, 10].

k-truss and Its Variants. We will introduce the k-truss algorithm [3] as well as the enhanced version [11]. Then we will introduce I/O efficient k-truss computing algorithm based on external memory model, followed by the parallel computing algorithm. We will also introduce the computation algorithms for several variants of k-truss.

Density Based Decomposition. We will introduce the computation of maximum density subgraph and *locally-dense* subgraph [4]. Then we will show how to decompose a graph based on locally dense subgraphs, including exact [12] and approximate [12] algorithms.

Connectivity Based Decomposition. We will first introduce the state-of-the-art k-edge connected component (k-ECC) computation algorithm [13], then we present the I/O efficient k-edge connectivity based hierarchical graph decomposition algorithms [5] on external memory model.

2.3   Future Research Directions (15 min)

We will discuss several important future research directions in hierarchical graph decomposition including new models, managing uncertain graph data, distributed and approximate solutions, etc.

3   Previous Tutorial Presentation and Attendance

The presenters will give a tutorial "Hierarchical Decomposition of Big Graphs" in ICDE 2019 in which the attendance will be database and data management researchers. In this tutorial we will introduce more fundamental concepts and computing techniques regarding cohesive subgraph models.

4   Presenters

Wenjie Zhang (wenjie.zhang@unsw.edu.au) is an Associate Professor in School of Computer Science and Engineering, the University of New South Wales, Australia. Her research interests include graph data analysis and spatial data processing.

Fan Zhang (fan.zhang3@unsw.edu.au) is a research associate in the School of Computer Science and Engineering, University of New South Wales. He received PhD in Computer Science from University of Technology Sydney in 2017.

Ying Zhang (Ying.Zhang@uts.edu.au) is an Australian ARC Future Fellow (2017–2021) and Associate Professor at the University of Technology, Sydney (UTS). He has been the head of the database group at the Centre for Artificial Intelligence (CAI) since 2014. His research focuses on efficient query processing and analytic on large scale data.

Lu Qin (Lu.Qin@uts.edu.au) is now a senior lecturer in the Centre for Artificial Intelligence (CAI), University of Technology, Sydney. His research interests include big graph processing, I/O efficient algorithms on massive graphs, and keyword search in relational database.

References

1. Habib, M., Paul, C.: A survey of the algorithmic aspects of modular decomposition. Comput. Sci. Rev. **4**(1), 41–59 (2010)
2. Malliaros, F.D., Papadopoulos, A.N., Vazirgiannis, M.: Core decomposition in graphs: concepts, algorithms and applications. In: EDBT, pp. 720–721 (2016)
3. Cohen, J.: Trusses: Cohesive subgraphs for social network analysis. National Security Agency Technical Report, p. 16 (2008)
4. Qin, L., Li, R., Chang, L., Zhang, C.: Locally densest subgraph discovery. In: SIGKDD, pp. 965–974 (2015)
5. Yuan, L., Qin, L., Lin, X., Chang, L., Zhang, W.: I/O efficient ECC graph decomposition via graph reduction. PVLDB **9**(7), 516–527 (2016)
6. Wen, D., Qin, L., Zhang, Y., Lin, X., Yu, J.X.: I/O efficient core graph decomposition at web scale. In: ICDE, pp. 133–144 (2016)
7. Wang, N., Yu, D., Jin, H., Qian, C., Xie, X., Hua, Q.: Parallel algorithm for core maintenance in dynamic graphs. In: ICDCS, pp. 2366–2371 (2017)
8. Thomo, A., Liu, F.: Computation of k-core decomposition on giraph. CoRR, vol. abs/1705.03603 (2017)
9. Zhang, F., Zhang, W., Zhang, Y., Qin, L., Lin, X.: OLAK: an efficient algorithm to prevent unraveling in social networks. PVLDB **10**(6), 649–660 (2017)
10. Zhang, F., Zhang, Y., Qin, L., Zhang, W., Lin, X.: Finding critical users for social network engagement: the collapsed k-core problem. In: AAAI, pp. 245–251 (2017)
11. Wang, J., Cheng, J.: Truss decomposition in massive networks. PVLDB **5**(9), 812–823 (2012)
12. Danisch, M., Chan, T.H., Sozio, M.: Large scale density-friendly graph decomposition via convex programming. In: WWW, pp. 233–242 (2017)
13. Chang, L., Yu, J.X., Qin, L., Lin, X., Liu, C., Liang, W.: Efficiently computing k-edge connected components via graph decomposition. In: ACM SIGMOD, pp. 205–216 (2013)

Tracking User Behaviours: Laboratory-Based and In-The-Wild User Studies

Gianluca Demartini$^{(\boxtimes)}$and Shazia Sadiq

School of ITEE, University of Queensland,
GP South Building, Staff House Road, St Lucia, QLD 4072, Australia
g.demartini@uq.edu.au, shazia@itee.uq.edu.au

1   Motivation for the Tutorial

Database systems are today being used for the most disparate applications. While most of the research done in the field focus on systems, more and more attention is being paid to users. Understanding how the developed systems are being used and how their use could be made more effective and efficient requires to collect and analyze usage data. Such user evaluations follows well-defined protocols, data collections tools, and analysis methods. In this tutorial, we will introduce the area of user behaviours and present methods to log and analyze behavioural data from users. This tutorial has not been presented elsewhere in the past.

2   Brief Outline of the Tutorial

The tutorial will be 90 min long and will consist of two main parts. The first part cover tools and methods to collect and analyze user behaviour data at scale by means of logging tools that can be embedded into web browsers and thus scale-out to large number of potential users. The second part of the tutorial will cover additional methods which are relevant for user studies conducted in a laboratory setting such as eye-tracking technology.

2.1   Introduction - User Studies Motivation and Examples (10 min)

We will start motivating the need for user studies in the field of database and data analytics. We will then introduce the basic elements of a user studies with a focus on the data to be collected during the user/system interaction.

2.2   Part I - Tracking In-Browser Behavioural Features (40 min)

In this part of the tutorial we will present recent research that have analyzed user interaction data. The specific focus will be on data collected within behavioural logs by tools (e.g., [1]) including events like, for example, user clicks, page views, mouse over, and keyboard actions. Example studies using this type of data include [2, 4] where large-scale log data has been used to understand crowd worker or search engine user behaviours. Other than academic research, we will also discuss how industry researchers use behavioural log data at scale to improve web-scale systems [3].

© Springer Nature Switzerland AG 2019
G. Li et al. (Eds.): DASFAA 2019, LNCS 11448, pp. 590–591, 2019.
https://doi.org/10.1007/978-3-030-18590-9

2.3 Part II - Running In-Lab User Behaviour Studies (40 min)

The increasing availability of neuro-physiological measurement devices such as eye tracking and EEG, at lower costs, and with a decreasing level of intrusiveness is presenting unprecedented opportunities in a number of areas that require study of human behaviour and interaction with technology artefacts. This includes the development of adaptive software that responds to the emotional and cognitive state of the user (e.g., online query refinement), but also provides the basis for a more in-depth understanding of the challenges that occur when interacting with, or developing technology artefacts (e.g., conceptual models, data visualizations). In this part we will outline the main methods for studying user behaviour in lab settings, with a special focus on collecting and understanding user behaviour through neuro-physiological devices.

References

1. Dang, B., Hutson, M., Lease, M.: Mmmturkey: a crowdsourcing framework for deploying tasks and recording worker behavior on amazon mechanical turk. In: 4th AAAI Conference on Human Computation and Crowdsourcing (HCOMP) (2016)
2. Kutlu, M., McDonnell, T., Barkallah, Y., Elsayed, T., Lease, M.: Crowd vs. expert: what can relevance judgment rationales teach us about assessor disagreement? In: SIGIR, pp. 805–814. ACM, New York (2018)
3. White, R.W.: Interactions with search systems. Cambridge University Press (2016)
4. Zhuang, M., Demartini, G., Toms, E.G.: Understanding engagement through search behaviour. In: CIKM, pp. 1957–1966. ACM, New York (2017)

Mining Knowledge Graphs for Vision Tasks

Xiaojun Chang[1(✉)], Fengda Zhu[2], Xiaoran Bi[1], Weili Guan[3], Zongyuan Ge[4], and Minnan Luo[5]

[1] Faculty of Information Technology, Monash University, Clayton, Australia
cxj273@gmail.com
[2] Centre for Artificial Intelligence, University of Technology Sydney, Ultimo, Australia
[3] Hewlett Packard Enterprise, Singapore, Singapore
[4] Monash e-Research Centre, Melbourne, Australia
[5] Department of Computer Science and Technology, Xi'an Jiaotong University, Xi'an, China

Abstract. Semantic technologies, such as knowledge graph, have been of great interest to the community of different areas. Recent advances in knowledge acquisition, alignment, and utilization have resulted in a bunch of new approaches for knowledge graph learning in computer vision tasks. This tutorial focuses on the end-to-end utilization of knowledge graph in computer vision tasks.

Keywords: Knowledge graph · Computer vision · Zero-shot learning · Visual reasoning

1 Introduction

Knowledge graph (KG) is one of the most well-known technique from the Semantic Web community, and has been widely used in web search, text classification, entity linking etc. Recently, KG has been of interest to the vision community [3, 4, 5, 7]. Researchers in the field of computer vision have applied KG to various applications, including zero-shot learning [5], image classification [3], relationship detection [6], etc.

2 Tutorial Outline

This tutorial is tailored for DASFAA attendees who are expected to be aware of the broad area of database and data management but may or may not be actively working in knowledge graph learning.

2.1 Introduction, Modeling and Challenges (25 min)

The aim of this part is to provide general background to the audience. We will give a brief introduction of knowledge graph learning, the modeling of knowledge graph learning, and the main challenges of KG. This will give the audience a big picture of what is KG and how KG works.

© Springer Nature Switzerland AG 2019
G. Li et al. (Eds.): DASFAA 2019, LNCS 11448, pp. 592–594, 2019.
https://doi.org/10.1007/978-3-030-18590-9

2.2   Knowledge Graph for Vision Tasks

In this part, we focus on discussing some important vision tasks using knowledge graph learning.

Zero-Shot Learning. We will introduce how to incorporate structured knowledge graphs into the zero-shot learning process [2, 5]. We will also discuss how to leverage different relations defined in the knowledge graphs [2].

Image Classification. We will discuss how to use knowledge graph as extra information to improve image classification [3]. We will also discuss the possibility of integrating Neil [1] with the procedure for a robust system.

Relationship Detection. We will introduce how to distill linguistic knowledge into a deep neural network for visual relationship detection [6].

2.3   Future Research Directions (15 min)

We will discuss several important future research directions in knowledge graph learning, including new models, applications to new vision tasks, etc.

3   Previous Tutorial Presentation and Attendance

The presenters presented a Tutorial "Towards Efficient Text Retrieval with Knowledge Graph Learning" for the course "Large-Scale Multimedia Analysis" at Carnegie Mellon University (CMU), in which the attendance were graduate students from Language Techniques Institute of CMU. In this tutorial, we will focus more on vision tasks.

4   Presenters

Xiaojun Chang (xiaojun.chang@monash.edu) is a lecturer in the Faculty of Information Technology, Monash University, Australia.

Fengda Zhu (zhufengda@yahoo.com) is a research assistant in the Centre for Artificial Intelligence (CAI), University of Technology, Sydney.

Xiaoran Bi (xbii0003@monash.edu) is a research associate in the Faculty of Information Technology, Monash University, Australia.

Weili Guan (honeyguan@gmail.com) is a research associate in Hewlett Packard enterprise Singapore.

Minnan Luo (minnluo@xjtu.edu.cn) is an Associate Professor in Department of Computer Science and Technology, Xi'an Jiaotong University, China.

Zongyuan Ge (zongyuan.ge@monash.edu) is currently a Senior Research Fellow at Monash and also serve as a Deep Learning Specialist at NVIDIA East Asia Research Centre.

References

1. Chen, X., Shrivastava, A., Gupta, A.: NEIL: extracting visual knowledge from web data. In: ICCV (2013)
2. Lee, C., Fang, W., Yeh, C., Wang, Y.F.: Multi-label zero-shot learning with structured knowledge graphs. In: CVPR (2018)
3. Marino, K., Salakhutdinov, R., Gupta, A.: The more you know: using knowledge graphs for image classification. In: CVPR (2017)
4. Sadeghi, F., Divvala, S.K., Farhadi, A.: Viske: visual knowledge extraction and question answering by visual verification of relation phrases. In: CVPR (2015)
5. Wang, X., Ye, Y., Gupta, A.: Zero-shot recognition via semantic embeddings and knowledge graphs. In: CVPR (2018)
6. Yu, R., Li, A., Morariu, V.I., Davis, L.S.: Visual relationship detection with internal and external linguistic knowledge distillation. In: ICCV (2017)
7. Zhu, X., Anguelov, D., Ramanan, D.: Capturing long-tail distributions of object subcategories. In: CVPR (2014)

Enterprise Knowledge Graph from Specific Business Task to Enterprise Knowledge Management

Rong Duan[1](✉) and Yanghua Xiao[2]

[1] Huawei Technology, Shenzhen, China
rong.duan@huawei.com
[2] Fudan University, Shanghai, China
shawyh@fudan.edu.cn

Abstract. Benefit from digitization, many industries consider digital transformation as the top priority, especially for the traditional industries. One of the major challenges is how to manage the explosion data, and extract knowledge from it. Besides volume increase, multiple data format, diverse information storage systems, different quality types, developing knowledge base, and especially the interactive knowledge consumption are the new challenges that modern **E**nterprise **K**nowledge Management system is facing. The predefined ontology based **EKM** [Dietz08, Sure02] could not fulfill these requirements. Data driven **K**nowledge **G**raph, after coined by google, is rapidly adapted by different societies, and also in **EKM** field. [Masuch14]. Some **EKG**s have been constructed [Sabou18] to solve specific problems. Due to the complication of enterprise knowledge, there are different types of **EKG**, and need different technologies to construct [Pan17, Pechsiri10, Paulheim17]. In this tutorial, we will introduce three types of **EKG**, and illustrate the difference among open domain, specific domain, and enterprise knowledge graph. We will focus on the steps, challenges, techniques and future research in constructing each type of **EKG**s.

Rong Duan

Dr. Rong Duan currently is Chief Data Scientist of Corporate Data Management Department, Huawei Technologies Co., Ltd. She is guiding a group to construct multiple enterprise knowledge graphs in supporting varies business applications. Before joined Huawei, Rong was principal inventive scientist at AT&T Labs, big data research and adjunct professor at Stevens Institute of Technology. Has been working in AT&T Labs for more than 20 years, Rong has extensive experience on statistical learning, data mining, predictive modeling and data analysis for business data. Her research interests include data mining, statistical learning theory and methods, Spatial-temporal Risk management, Data Integration and Data Quality Assessment. Rong was former Chair for the Data Mining Section of INFORMS, and Data Mining cluster co-chair for INFORMS International Beijing. Rong also served as a program co-chair for the 1st and the 2nd International Symposium on System Informatics and Engineering.

© Springer Nature Switzerland AG 2019
G. Li et al. (Eds.): DASFAA 2019, LNCS 11448, pp. 595–596, 2019.
https://doi.org/10.1007/978-3-030-18590-9

596   R. Duan and Y. Xiao

Yanghua Xiao

Dr. Yanghua Xiao is a full professor of the School of Computer Science at Fudan University. His research interest includes big data management and mining, graph database and knowledge graph. He was a visiting professor of Human Genome Sequencing Center at Baylor College Medicine, and visiting researcher of Microsoft Research Asia and Alibaba. He won 10+ research awards granted by governments or industries, including CCF Natural Science Award (second level), ACM(CCF) Shanghai distinguished young scientists and Alibaba Research Fellowship Award. Recently, he has published 100+ papers in international leading journals and top conferences, including TKDE, SIGMOD, VLDB, ICDE, IJCAI, AAAI. He is the PI or Co-PI of 30+ projects supported by 10+ national and local funding agency and big companies including Microsoft, IBM, HUAWEI, China Telecom, China Mobile, Baidu, XiaoI Robot etc. He is the director of Knowledge Works Research Laboratory at Fudan University. His team at Fudan published a lot of Chinese knowledge graphs, which serve industries with 1 billions of API calls. He is the chief scientist or senior advisors of many top Chinese big data companies or AI companies. He has ever given more than 10 keynote speeches or tutorials in international leading conference including SIGMOD2012, IDEAL2017, ADMA2018, PIC2017, CCKS2018 etc.

References

[Dietz08] Dietz, J.L., Hoogervorst, J.A.: Enterprise ontology in enterprise engineering. In: Proceedings of the 2008 ACM Symposium on Applied Computing, pp. 572–579. ACM, March 2008

[Pan17] Pan, J.Z., Vetere, G., Gomez-Perez, J.M., Wu, H. (eds.): Exploiting linked data and knowledge graphs in large organisations, p. 281. Springer, Heidelberg (2017). https://doi.org/10.1007/978-3-319-45654-6

[Masuch14] Masuch, L.: Enterprise knowledge graph-one graph to connect them all. In: Proceedings of the Unlimited Human Potential M-Prize, M-Prize Proceedings. MPrize (2014)

[Sure02] Sure, Y., Staab, S., Studer, R.: Methodology for development and employment of ontology based knowledge management applications. ACM Sigmod Rec. **31**(4), 18–23 (2002)

[Sabou18] Sabou, M., et al.: Exploring enterprise knowledge graphs: a use case in software engineering. In: Gangemi, A., et al. (eds.) ESWC 2018. LNCS, vol. 10843, pp. 560–575. Springer, Cham (2018). https://doi.org/10.1007/978-3-319-93417-4_36

[Pechsiri10] Pechsiri, C., Piriyakul, R.: Explanation knowledge graph construction through causality extraction from texts. J. Comput. Sci. Technol. **25**(5), 1055–1070 (2010)

[Pujara13] Pujara, J., Miao, H., Getoor, L., Cohen, W.: Knowledge graph identification. In: Alani, H., et al. (eds.) ISWC 2013. LNCS, vol. 8218, pp. 542–557. Springer, Heidelberg (2013). https://doi.org/10.1007/978-3-642-41335-3_34

[Paulheim17] Paulheim, H.: Knowledge graph refinement: a survey of approaches and evaluation methods. Semant. Web **8**(3), 489–508 (2017)

Knowledge Graph Data Management

Xin Wang[1,2](✉)[iD]

[1] College of Artificial Intelligence and Computing,
Tianjin University, Tianjin, China
wangx@tju.edu.cn
[2] Tianjin Key Laboratory of Cognitive Computing and Application,
Tianjin, China

Abstract. Knowledge graph technology is a cornerstone of artificial intelligence. In recent years, an increasing number of large-scale knowledge graphs have been publicly available. In this tutorial, we will comprehensively introduce the state-of-the-art research and development on theories and technologies of knowledge graph data management, consisting of knowledge graph data models, query languages, storage schemes, query processing, reasoning, and related frameworks and systems for knowledge graph data management.

Keywords: Knowledge graphs · Data management · Data models

1 Motivation of the Tutorial

With the rise of artificial intelligence, knowledge graphs have been widely considered as a cornerstone of AI. In recent years, an increasing number of large-scale knowledge graphs have been constructed and published, by both academic and industrial communities, such as DBpedia [1], YAGO [2], Wikidata [3], Google Knowledge Graph, Microsoft Satori, Facebook Entity Graph, and others. In fact, a knowledge graph is essentially a large network of entities, their properties, semantic relationships between entities, and ontologies the entities conform to. Such kind of graph-based knowledge data has been posing a great challenge to the traditional data management theories and technologies. On the other hand, the database community has been putting a lot of effort into graph databases for nearly two decades. However, there are still considerable gaps between the new requirements of knowledge graph data management and the current state of techniques in databases. This tutorial is to offer the audience a comprehensive introduction to the state-of-the-art research on knowledge graph data management, which includes knowledge graph data models, query languages, storage schemes, query processing, and reasoning. We will also describe the latest development trends of various database management systems for knowledge graphs.

Supported by the National Natural Science Foundation of China (61572353) and the Natural Science Foundation of Tianjin (17JCYBJC15400).

G. Li et al. (Eds.): DASFAA 2019, LNCS 11448, pp. 597–599, 2019.
https://doi.org/10.1007/978-3-030-18590-9

2   Outline of the Tutorial

The contents of this tutorial can be summarized as follows:

- Data models of knowledge graphs, which includes the RDF graph model and the property graph model.
- Query languages for knowledge graphs, which includes SPARQL for RDF graphs, and Cypher, Gremlin, PGQL, and G-CORE for property graphs.
- Storage schemes for knowledge graphs, which includes relational-based approaches and native graph approaches.
- Query processing over knowledge graphs, which includes subgraph pattern matching queries, navigational queries, and analytical queries.
- Reasoning over knowledge graphs, which includes reasoning support of knowledge graph storage and query processing.
- Knowledge graph data management systems, which includes RDF triple stores, native graph databases, distributed graph processing frameworks, and benchmarking tools on these systems.
- Future research directions of knowledge graph data management.

3   Biography of the Presenter

This tutorial is presented by Dr. Xin Wang, who is an associate professor and vice dean at School of Artificial Intelligence, College of Intelligence and Computing, Tianjin University. He received his Ph.D. degree from Nankai University, China in 2009. He worked as a visiting scholar at The University of Western Australia from May 2018 to Jul. 2018, a visiting scholar at Griffith University from Oct. 2015 to Oct. 2016. He is a senior member of China Computer Federation (CCF), a member of CCF Technical Committee on Databases, a member of CCF Technical Committee on Information Systems. His research interests include knowledge graph data management, large-scale graph databases, and big data distributed processing. He has been the primary investigator of two research projects funded by the National Natural Science Foundation of China (NSFC). He has published more than 70 research papers in various international conferences and journals, including IEEE TPDS, ICDE, WWW, CIKM, ISWC, ER, DASFAA, WISE, WebDB, etc. He served as a co-chair of the 4th International Workshop on Big Data Quality Management (BDQM 2019), a co-chair of the 1st International Workshop on Knowledge Graph Management and Analysis (KGMA 2018), a publicity co-chair of DASFAA 2018, a workshop co-chair of JIST 2016, and PC members of WWW 2019, WISE 2018, APWeb-WAIM 2018, DASFAA 2019/2018/2017, ADMA 2018, JIST 2018/2017/2016, and WAIM 2016.

The initial version of this tutorial has been given on a seminar at The University of Western Australia in Jun. 2018 when the presenter was visiting there.

References

1. DBpedia. https://wiki.dbpedia.org. Accessed 6 Jan 2019
2. YAGO. https://www.mpi-inf.mpg.de/departments/databases-and-information-systems/research/yago-naga/yago/. Accessed 6 Jan 2019
3. Wikidata. https://www.wikidata.org. Accessed 6 Jan 2019

Deep Learning for Healthcare Data Processing

Weitong Chen[1]($\boxtimes$) and Guodong Long[2]

[1] The University of Queensland, Brisbane, Australia
w.chen9@uq.edu.au
[2] University of Technology Sydney, Ultimo, Australia
guodong.long@uts.edu.au

Abstract. Deep learning techniques have revolutionized many fields including computer vision, natural language processing, speech recognition, and is being fundamentally changed healthcare industries. Vary types of data have been emerging in modern healthcare research, including electronic health records (EHR), Clinical imaging, and Continuing monitoring data, which are noise, complex, high-dimensional, multi-modality, poorly annotated and generally unstructured. Healthcare applications pose many significantly different challenges to existing deep learning models. For instance, interpretation of prediction, missing value, and privacy preservation. In this tutorial, we will discuss the challenges and solutions to the problems in healthcare applications, as well as data sets and demos.

Keywords: Deep learning · Healthcare research
Healthcare applications

This Tutorial will present the following topics with details:

- An introduction to healthcare data and applications
- Overview of deep learning techniques on healthcare data processing
- Summering the challenges on deep learning research
- Discuss the challenges for applying deep learning for healthcare applications

1 Motivation

Deep learning models have revolutionized many fields including computer vision [1], natural language processing [6], robotics applications [5], and is being fundamentally changed healthcare industries. The accumulation of healthcare data, over 18.8 million EHR from over 5.4 million people available in Australia[1], has attracted much attention from the machine learning [2, 3] and data mining communities [1, 4]. However, various types of data have been emerging in modern healthcare research, including EHRs, imaging, and monitoring data, which are noise, complex, high-dimensional, heterogeneous, poorly annotated and generally

---

[1] https://myhealthrecord.gov.au.

G. Li et al. (Eds.): DASFAA 2019, LNCS 11448, pp. 600–601, 2019.
https://doi.org/10.1007/978-3-030-18590-9

unstructured. Healthcare research poses many significantly different challenges to existing deep learning models. Despite the promising results achieved by using deep learning techniques, there several unsolved challenges facing the healthcare research that is worth discussing. For instance, Data Volume, Data Quality, and Interpret-ability.

2   About the Presenters

Mr. Weitong Chen is now a PhD candidate in the University of Queensland, Australia. He received the BS and the MS degrees from the Griffith University, Australia, and the University of Queensland, Australia, in 2011 and 2013, respectively. His current main research interests include Medical Data Analytic, Deep Learning, Data Mining, Pattern Recognition, and Social Computing. He has published near 20 peer-reviewed papers in prestigious journals and top international conferences including IEEE ICDE, AAAI, WWW, WWWJ, ECML, CIKM, and SIAM SDM. He has been actively engaged in professional services by serving as conference organizers, conference PC members, and reviewer of journals such as ADMA, WWW, MobiQuitous, JCST, KAIS, MobiSPC, WWWJ, etc.

Dr. Long obtained his PhD degree in computer science from UTS in 2014. Before joined UTS in 2010, he has more than six years industry R&D working experience. He currently leading multiple collaboration projects with Australian Government Department of Health. He is currently leading a research group to conduct application-driven research on machine learning and data mining. Particularly, his research interests focus on several application domains, such as NLP, Healthcare, Smart Home, Education and Social Media. He is a senior lecturer at the Centre for Artificial Intelligence (CAI), Faculty of Engineering and IT, UTS. His research focuses on data mining, machine learning, and natural language processing. He has more than 40 research papers published on top-tier journals and conferences, including IEEE Trans on PAMI/CYB/KDE, ICLR, AAAI, IJCAI, and ICDM.

References

1. Akhtar, N., Mian, A.: Threat of adversarial attacks on deep learning in computer vision: A survey. arXiv (2018)
2. Che, Z., Purushotham, S., Cho, K.: Recurrent neural networks for multivariate time series with missing values. arXiv (2016)
3. Chen, W., Wang, S., Long, G., Yao, L., Sheng, Q.Z., Li, X.: Dynamic illness severity prediction via multi-task rnns for intensive care unit. In: 2018 ICDM. IEEE (2018)
4. Chen, W., et al.: EEG-based motion intention recognition via multi-task RNNS. In: SDM, SIAM (2018)
5. Sünderhauf, N., et al.: The limits and potentials of deep learning for robotics. IJRR (2018)
6. Young, T., Hazarika, D., Poria, S., Cambria, E.: Recent trends in deep learning based natural language processing. CIM (2018)

Author Index